Nanocrystalline Materials: Their Synthesis-Structure-Property Relationships and Applications

Elsevier Internet Homepage

http://www.elsevier.com

Consult the Elsevier homepage for full catalogue information on all books, journals and electronic products and services.

Elsevier Titles of Related Interest

Handbook of Nanostructured Materials and Nanotechnology, H.S. Nalwa
ISBN: 0125137605, 2000

Nano and Microstructural Design of Advanced Materials, M. Meyers, M. Sarikaya, R. Ritchie
ISBN: 0080443737, 2003

Nanomaterials, S. Mitura
ISBN: 0444503455, 2000

Related Journals

The following journals related to nanocrystalline materials can all be found at http://www.sciencedirect.com

Nano Today
Materials Today
Materials Science and Engineering: A
Materials Science and Engineering: B
Acta Materialia
Scripta Materialia
Materials Letters
Superlattices and Microstructures
Materials Chemistry and Physics

To Contact the Publisher

Elsevier welcomes enquiries concerning publishing proposals: books, journal special issues, conference proceedings, etc. All formats and media can be considered. Should you have a publishing proposal you wish to discuss, please contact, without obligation, the Publisher responsible for Elsevier's materials science books publishing programme:

Amanda Weaver
Publisher, Materials Science
Elsevier Limited
The Boulevard, Langford Lane
Kidlington, Oxford
OX5 1GB, UK

Tel.: +44 1865 84 3634
Fax: +44 1865 84 3920
E-mail: a.weaver@elsevier.com

General enquiries including placing orders, should be directed to Elsevier's Regional Sales Offices – please access the Elsevier internet homepage for full contact details.

Nanocrystalline Materials: Their Synthesis-Structure-Property Relationships and Applications

S.C. Tjong
Department of Physics and Materials Science,
City University of Hong Kong,
Kowloon, Hong Kong

ELSEVIER

Amsterdam - Boston - Heidelberg - London - New York - Oxford
Paris - San Diego - San Francisco - Singapore - Sydney - Tokyo

ELSEVIER B.V.
Radarweg 29
P.O. Box 211, 1000
AE Amsterdam
The Netherlands

ELSEVIER Inc.
525 B Street, Suite 1900
San Diego, CA 92101-4495
USA

ELSEVIER Ltd
The Boulevard, Langford Lane
Kidlington, Oxford OX5 1GB
UK

ELSEVIER Ltd
84 Theobald's Road
London WC1X 8RR
UK

First edition 2006

ISBN-13: 978-0-08-044697-4
ISBN-10: 0-08-044697-3

Printed in Great Britain.

06 07 08 09 10 10 9 8 7 6 5 4 3 2 1

CONTENTS

Preface

In recent years, nanocrystalline materials have attracted great attention from chemists, physicists, and materials scientists due to their unique and special chemical, physical, and mechanical properties. Nanocrystalline materials with new functionalities show great promise for use in industrial applications. For example, one-dimensional nanostructures (tubes, wires, rods, belts, etc) represent an interesting and important class of materials with potential applications from nanoscale devices to reinforcing fillers in novel polymer composites. Prospects on nanomaterials mainly depend on the success in fabrication processes. Nanomaterials can be synthesized via vapor, liquid, and solid state processing routes. The most general approaches for fabricating one-dimensional nanostructures have adopted the vapor-phase evaporation and/or vapor–liquid–solid growth mechanism, and controlled solution growth at elevated temperatures. The development of these nanostructures for industrial applications is still in an embryonic stage. The properties of nanomaterials are now recognized to be dependent on their chemical composition, phase, structure, size, and shape. Substantial progress has been made in the synthesis and processing of nanomaterials in the past decade. However, there are several issues that need to be addressed to develop these materials further. These include synthesis of high purity, well-aligned or self-assembled building blocks for device and structural applications, achieving microstructural stability of nanograins for maintaining the mechanical strength, improving mechanical ductility of nanomaterials, etc. Among these, exploration of novel methods for the large-scale synthesis of low-cost self-assembled nanostructures is a challenging research topic. Accordingly, there has emerged a demand to study their synthesis–structure–property relationships in order to understand the fundamental concepts underlying the observed physical and mechanical properties. The book describes the fundamental theories and concepts that illustrate the complexity of the problem in developing novel nanocrystalline materials. It reviews the most up-to-date progress in the synthesis, microstructural characterization, physical, and mechanical behavior, and application of nanomaterials. The book serves as a research and reference source for graduate students, chemists, physicists, materials scientists, and engineers working in this exciting subject.

The chapters are contributed by leading scientists in their respective subject field. Qian and coworkers gave a comprehensive review on recent advances in the solution synthesis, characterization and application of various types of semiconducting 1D nanostructures prepared via solvothermal, hydrothermal, sonochemical and γ-irradiation-assisted methods. Zeng reviewed his recent work on the solution synthesis and architectural design of various shapes of inorganic nanomaterials via template-free approaches including Ostwald ripening and Kirkendall effect for solid evacuation in forming hollow nanostructures. Sugimura presented the solution and vapor phase synthesis routes for preparing molecular self-assembled monolayers (SAMs) on Si. The organic molecules spontaneously assembled and well-organized onto silicon surfaces due to interactions between adsorbed molecules and substrate. Micro- and nanopatterning as well as the manipulation of molecules and nanochemical reactions based on atomic force microscopy were described. This self-assembly approach allows the fabrication of ultra thin films using molecules as building blocks and permits the design of new functional electronic materials or nanodevices at the supramolecular level. Wang and Fung reported the vapor phase synthesis and growth mechanism of Si and ZnSe nanowires prepared from laser ablation and molecular-beam epitaxy. They explained the growth of nanowires in terms of vapor–solid–liquid (VLS) or oxide-assisted growth mechanism. Si nanowires were reported to exhibit enhanced photothermal effect that may find applications in smart ignition systems, self-destructing systems for electronic devices, nanosensors, etc. The structures of such nanomaterials were characterized by high-resolution transmission electron microscopy (HRTEM). The formation of passivated metallic nanoparticles (e.g. Fe, Cr, Co, and Ni) was also discussed. The orientation relationship between the oxide and substrate of some nanoparticles obeyed either Bain or Nishiyama–Wassermann relationship. Park summarized her recent work on the synthesis of GaN and ZnO nanowires and their heterostructures as well as SiC, Si_3N_4, and BN nanostructures by means of the CVD method. HRTEM, electron energy-loss spectroscopy (EELS), X-ray photoelectron spectroscopy (XPS), X-ray absorption near edge structure (XANES) and other physical measurements were used to characterize the structure-property relationship of the nanomaterials. Wu demonstrated the low-temperature growth of well-aligned β-Ga_2O_3-, ZnO- and doped ZnO-based nanostructures as well as their hybrids on various substrates using the metal organic chemical vapor deposition (MOCVD) method. The structure, photoluminescence and magnetic properties of the well-aligned nanorods/nanowires were determined. Tang reported the use of hyperbranched polymers as precursors for forming magnetic nanoceramics. This represents a novel synthesis route for nanoceramics with low coercivity or near-zero magnetic loss, which may find applications in various electromagnetic systems. Tjong discussed the anomalous mechanical behavior of bulk nanocrystalline metals in which deviation from normal Hall-Petch effect occurs. Refinement of the grain sizes to nanometer regime led to specific changes in the mechanisms of deformation owing to reduced availability of mobile dislocations. Okuyama and coworkers addressed the effects of nanoparticle sizes on the optical, electrical conductivity and luminescent properties of the nanoparticle-polymer composites. Finally, Tjong summarized the use of low-cost clay silicates as nanofiller reinforcements for polymers to achieve high strength and high flame retardancy at low loading levels. The incorporation of nanoceramics with an extremely large surface area and high aspect ratio into polymers improves their mechanical and physical performances significantly.

S. C. Tjong, CEng CSci FIMMM
City University of Hong Kong

CHAPTER 1

Solution Route to Semiconducting Nanomaterials

Changlong Jiang[a], Jianbo Liang[a], and Yitai Qian[a,b]*
[a]*Hefei National Laboratory for Physical Sciences at Microscale*
University of Science and Technology of China
Hefei, Anhui 230026, P.R. China

[b]*Department of Chemistry*
University of Shandong
Jinan, Shandong 250100 P.R. China

1.1 Introduction

Semiconducting nanomaterials are attractive because their physical properties are different from those of the bulk due to the quantum-size effect [1,2]. Also, they provide opportunities to study the effect of spatial confinement and problems related to surfaces or interfaces, which is important for chemistry. Recently, one-dimensional (1D) nanomaterials such as nanowires, nanobelts, nanorods, and nanotubes have become the focus of intensive research owing to their potential applications in electronic, optoelectronic, electrochemical, electromechanical, and other fields [3–7].

Various strategies have been employed to synthesize semiconducting nanomaterials such as vapor–liquid–solid (VLS) growth, solid–solution–solid (SSS) synthesis, solid–liquid–solid (SLS) process, electron beam lithography, and template-mediated growth [8–12]. Compared to physical methods, chemical methods have been more versatile and effective in the synthesis of nanomaterials. Among them, solution-based process is a low cost, facile, and promising approach for fabrication of semiconducting nanocrystallines.

In this chapter, we discuss recent advances in the solution synthesis, characterization, and applications of semiconducting nanomaterials. Section 1.2 explicitly reviews

*Corresponding author

solvothermal route to semiconducting nanocrystals. In the following section, we introduce the mixed solvothermal preparation of the semiconductors. In Section 1.4, we evaluate the hydrothermal synthesis method. The final section illustrates other solution-based chemical preparation of semiconducting nanocrystallines, including sonochemical method, irradiation synthesis, refluxing, and room temperature route.

1.2 Semiconducting Nanomaterials Synthesized by Solvothermal Method

Solvothermal synthesis is similar to the hydrothermal method except that organic solvents are used instead of water. This method can effectively prevent the products from oxidizing, especially useful in the synthesis of a variety of non-oxides. We have demonstrated the solvothermal synthetic method in the preparation of a rich field of nanocrystalline semiconductors including II–VI and III–V binary compounds. The reaction is processed in a sealed container. Solvents can be brought to temperatures well above their boiling points by the increase in autogenous pressures resulting from heating. By designing reactions, selecting solvents, controlling the reaction conditions, a series of semiconducting nanocrystals has been prepared.

In this section, we introduce the synthesis of nanocrystals in organic solvents, including multiamine and benzene. Then we illustrate iodine-transportation method as an effective strategy in syntheses of peculiar compounds such as cubic GaN.

1.2.1 Synthesis of II–VI compounds in ethylenediamine

The solvent ethylenediamine (en) has been proven to be effective in the formation of semiconductor 1D nanostructures. Taking CdS as an example, the formation process of CdS nanorods prepared by the reaction of thiourea and cadmium nitrate in ethylenediamine was investigated. The en molecules played the role of a director for the nanorods growth and replacing it by other organic solvents failed to produce the CdS nanorods. An accordion-like folding process was proposed to account for the formation of CdS nanorods [13]. The nanorods formation was thought to proceed in several steps. First, cadmium nitrate and thiourea reacted to produce the lamellar CdS with many folds on it. After that, the folds on the lamella agglomerated together. Then these folds broke into needle-like fragments. Finally, these needles grew into well-crystallized nanorods. It was believed that the en molecules in the solution was not relevant to this transformation process.

A monodentate ligand, n-butylamine, was found to be a shape controller for the formation of CdS and MSe (M = Zn, Cd, or Pb) nanorods [14]. These works further prove that one anchor atom in a ligand is necessary and sufficient for the formation of nanorods, even though more anchor atoms may be present in a ligand. Nanorods can be synthesized by choosing an appropriate mono-dentate ligand as solvent. The possible of the surface-passivating mine molecular conformation is depicted in Fig. 1.1

By choosing a designed chemical reaction route, quantum-confined CdS nanowires can be obtained. For example, CdS nanowires with diameters around 4 nm and lengths

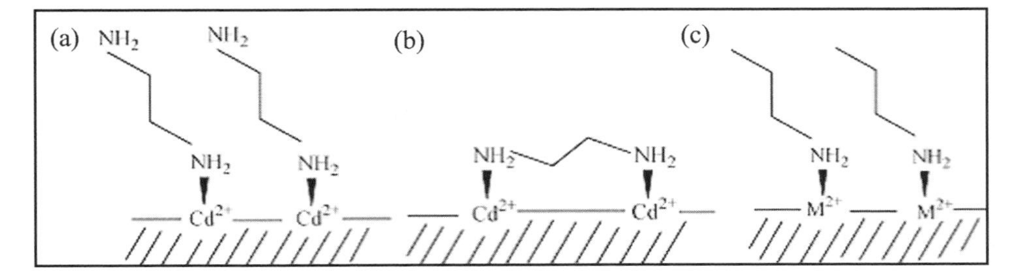

Fig. 1.1. Schematic illustration of the possible molecular conformation of the surface-passivating multiamine.

ranging from 150 to 250 nm were grown from cadmium bis(diethyldithiocarbamate) $[Cd(DDTC)_2]_2$ by removal of the four thione groups with ethylenediamine (en) at 117°C for 2 min [15]. The obtained nanowires obviously show size confinement effects at room temperature. Under photoluminescent excitation at 370 nm, the nanowires emit blue light at 440 and 460 nm with a 55 nm blue shift relative to bulk CdS. By employing a simple polymer-controlled method, long CdS nanowires with high aspect ratios were fabricated in bidentate ligand ethylenediamine under solvothermal condition (see Fig. 1.2) [16]. Hence the cadmium ions were well distributed in the polymer matrix and the solvent may guide the crystal-growth direction.

Using spherical CdS nanocrystals as starting materials, rod-, twinrod-, and tetrapod-shaped CdS nanocrystals have been successfully prepared through a highly oriented solvothermal recrystallization technique [17]. CdS nanorods with the longest and highest quality were obtained in 18 ml of ethylenediamine at 150°C using 0.2 g of spherical CdS nanocrystals; twinrod CdS nanocrystals can be obtained from starting solution consisting spherical CdS nanocrystals and sulfur powder (1:6), the recrystallization temperature was 180°C, and water cooling was used to lower the temperature of the resulting solution from 180°C to room temperature within 2 h; tetrapod CdS nanocrystals can be harvested by keeping 0.3 g of CdS nanocrystals in ethylenediamine at 180°C for 5 h and then

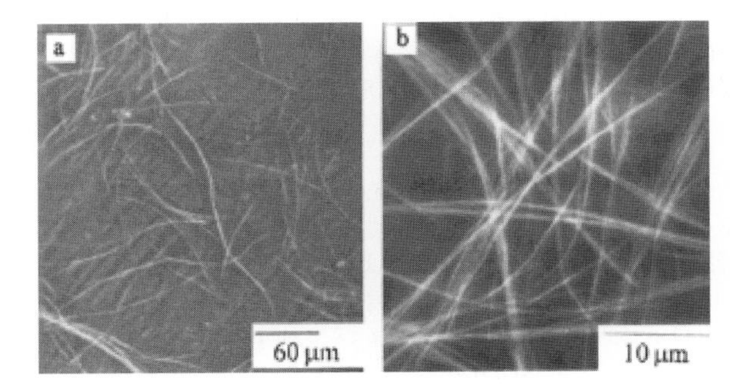

Fig. 1.2. (a) Low and (b) high magnification SEM images of the CdS nanowires synthesized via polymer-controlled solvothermal route.

cooling the resulting solution to room temperature at a rate of 10°C/h. By choosing ligand solvents, nanocrystalline CdE can be synthesized by the reaction of CdC_2O_4 with elemental E (E = S, Se, or Te) in polyamines such as en, diethylenetriamine, and triethylenetetramine. CdS nanorods and nanoparticles were synthesized at 180°C in en and pyridine, respectively [18]. The solvothermal reaction of sulfur, selenium, or tellurium with cadmium metal powder in different organic solvents in the temperature range 120–200°C was investigated systematically to prepare cadmium chalcogenides, CdE (E = S, Se, Te) [19].

Ethylenediamine have also been used as a solvent to synthesize other II–VI compounds. For example, ZnS ultrathin nanosheets have been prepared by Yu *et al.* in ethylenediamine solution under solvothermal conditions [20]. Experiments showed that a lamellar intermediate $ZnS(en)_2$ as first formed in the reaction [21–23]. On evaporation at a high temperature, this lamellar intermediate gradually decomposed to ZnS nanoflakes with thickness of only 1.1 nm, which was composed of only three cell units perpendicular to the planar surface. The bandgap of the ultrathin nanoflakes is 5.08 eV, which is 1.28 eV blue shift when compared with the bulk ZnS. If the reaction proceeds in an ethanol solution, only ZnS nanoparticles can be obtained.

If diethylenetriamine (DETA) was selected as a solvent, ZnS-DETA intermediate can also be formed under solvothermal conditions [24]. Further evaluating the reaction temperature, flexible ZnS nanobelts can be harvested. The reaction temperature can be greatly reduced to 180°C. The products can also show quantum-size effects (see Fig. 1.3). The above results show the kinds of bidentate ligands that are essential for the formation of the ZnS 2D nanostructures with quantum-size effects.

The 3D assembles of NiS, composed of nanoneedles with diameters in the range of 50 nm, have been successfully obtained by using $Ni(Ac)_2{\cdot}6H_2O$ and dithiazone as reagents and ethylenediamine as a solvent in the temperature range of 220°C. The reaction temperature can greatly affect the morphologies of the products. If the reaction proceeds at higher temperature, NiS nanobelts can be obtained in large scale [25].

Generally, the coordinating agents proved to be effective in controling the size and shape of a series of binary compounds. The main effect can be embodied in the following two

Fig. 1.3. Morphology of the ZnS nanobelts and their corresponding optical properties.

aspects: first, it can coordinate with metal ions and thus dissolve the raw materials before the aging process; second, the molecules can selectively absorb on the surface, resulting in the anisotropic growth behavior of products. Time-dependent investigation including XRD, TEM, and IR technologies have proved the above two roles during the reaction process.

1.2.2 Benzene-thermal reaction to III–V semiconductors

Benzene is the most commonly used organic solvent in the preparation of III–V binary compounds. Compared with water, it can prevent the hydroxylation of the starting materials; pure product can be obtained under the benzene-thermal conditions. Moreover, due to the lower boiling point of benzene when compared with water, the pressure in the sealed container is much higher than that in the pure water system. Thus, some metastable phase material that formed under rigorous condition can be obtained under the benzene-thermal conditions. Here, we take the III–V binary compounds such as GaP, GaN etc. as examples that introduce the development of benzene-thermal reaction in nanomaterial synthesis.

GaP is an indirect-gap semiconductor and has many important uses in microelectronic devices. The most important way to prepare GaP is the metalloorganic route. Dougall and co-workers have reported that GaP semiconductor clusters can be synthesized in zeolite by using the reaction [26]:

$$(Me_3)Ga + PH_3 \rightarrow 3CH_4 + GaP.$$

At refluxing temperature, GaP nanocrystallines were obtained from $GaCl_3$ and $P(SiMe_3)_3$ or $(Na/K)_3P$ and GaX_3 in toluene or other solvents respectively [27,28]. Another important method for preparing GaP is the solid-state metathesis, in which GaX_3 ($X = F, Cl, I$) and Na_3P are used as the raw materials [29,30]; this process requires an extra washing step to remove starting materials and by-products.

Xie *et al.* developed the synthesis of GaP nanorods and nanospheres through a mild benzene-thermal route at 240 and 300°C [31]. Na, P, and $GaCl_3$ were selected as the starting materials. The products were mainly nanorods when the reaction was carried out at 240°C. If the reaction temperature was elevated to 300°C, only nanospheres were obtained. The SLS mechanism is most likely to function under the present conditions because of the very low melting point of metal gallium (302.65 K) and its coexistence with the final products. If the reaction were prolonged to 24 h, a new Ga–P phase was formed under the current benzene-thermal conditions. This result indicates that GaP is in fact metastable in the solvothermal process. The obtained GaP nanocrystals show great catalytic activities in the thermic conversion of benzene into 6-phenylfulvene [32]. With GaP nanocrystals being used in a close reaction system, 6-phenylfulvene is successfully obtained via a high yield thermic conversion from benzene, which provides the possibility of applying nanocrystals to mediate organic reactions (see Fig. 1.4). Similarly, iron group phosphide nanocrystals can be prepared under the solvothermal condition using benzene as solvent [33–35].

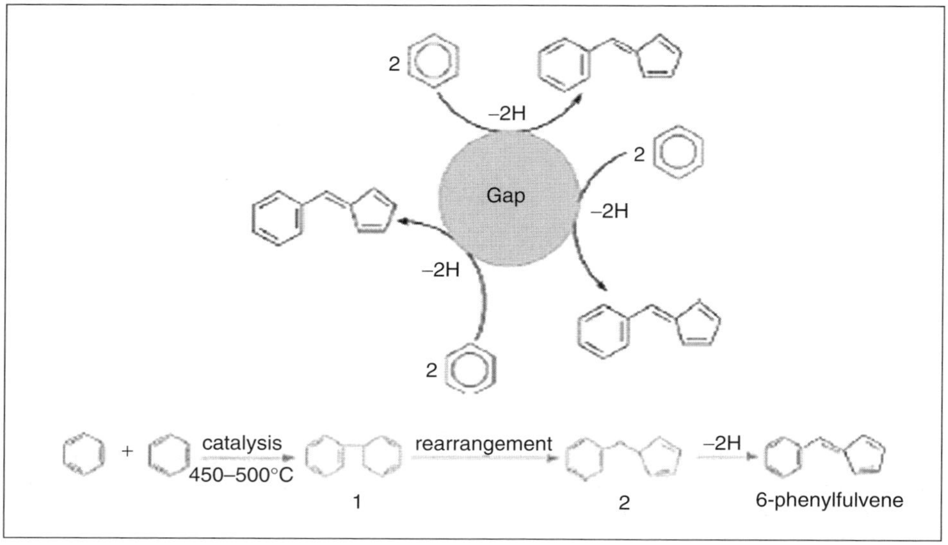

Fig. 1.4. Scheme of the atom economic protocol for the thermic rearrangement and dimerization of benzene induced by GaP nanocrystals at 450–500°C.

We have used the solvothermal method to prepare nanocrystalline InP [36]. Since 1,2-dimethoxyethane (DME) dissolves $InCl_3$, it was selected as the solvent. The autoclave was kept at 180°C for 12 h. The liquid–solid reaction in an autoclave was as follows:

$$InCl_3 + Na_3P \xrightarrow{180^\circ C} InP + 3NaCl.$$

The TEM results showed that the grains of InP were ~12 nm in diameter and were approximately spherical in shape. The nature of the solvent affects the reaction process and product quality. DME is an open chain polyether and it can act as a bidentate ligand that could facilitate the formation of chelate complexes of group III halides. It can also break the dimeric structure of $InCl_3$ and form ionic coordination complexes by expanding the coordination sphere of the metal center [37]. It is suggested that the dissociation of $InCl_3$ and the formation of ionic complexes may prevent growth of InP crystallites and limit the size of the InP crystallites to 12 nm. When benzene was selected as the solvent in the solvothermal process, the InP grain size could reach 40 nm since benzene has no coordination effect and acts only as a solvent. The as-prepared InP nanocrystallites exhibited a quantum size effect in the UV–vis and photoluminescence (PL) spectra. The Bohr excitation radius of InP was calculated as ~28 nm [38].

InP and GaP nanowires have been synthesized from an Ullmann-like reaction of indium and gallium with triphenyl phosphine at about 350–400°C for 8 h [39].

$$In(or\ Ga) + PPh_3 \rightarrow InP(or\ GaP) + biphenyl\ (bPh).$$

This equation is analogous to the traditional Ullmann reaction:

$$ArI + Cu \rightarrow Ar\text{–}Ar + CuI,$$

which has been used to prepare many symmetrical or unsymmetrical biaryls through the coupling of aryl halides with copper or a noncopper catalyst such as Pd [40–44]. In our current work, the Ullmann-like reaction was used to enhance the preparation of 1D inorganic InP and GaP semiconductor nanocrystals.

The benzene-thermal synthetic route was also applied to synthesize GaN nanocrystals [45]. $GaCl_3$ reacted with Li_3N in benzene in an autoclave at 280°C for 12 h. Benzene was selected as the solvent because it is stable and dissolves $GaCl_3$. The as-prepared GaN nanocrystallites exist mainly as hexagonal structures. A small amount of rocksalt-structure GaN coexist. In previous work, rocksalt-structure GaN was only observed at 37 GPa with high-pressure energy-dispersed X-ray diffraction using diamond-anvil techniques, and this rocksalt structure disappeared as the pressure was released [46]. The cell constant (»4.100) of the rocksalt-structure GaN was confirmed by HRTEM. This is the first time that rocksalt-structure GaN was observed at ambient pressure and temperature. When the reaction temperature was increased to 300°C, HRTEM results showed the third phase, zincblende-structure GaN, which was previously formed during a film preparation process [47]. The photoluminescence spectrum of the as-prepared GaN showed an emission peak at 370 nm, which is in agreement with that of bulk GaN, and no quantum-size effect was observed. This is because the as-prepared GaN is too large ($\sim$30 nm) while the calculated Bohr excitation radius for GaN is only 11 nm.

In general, the benzene-thermal reaction proved to be useful in the synthesis of nanocrystalline III–V compounds such as GaN, GaP, InN, InP, FeP etc. There are two general advantages when the reactions proceed in the organic solvents: first, it can prevent the products from oxide formation, which is inevitable when the reactions proceed in water; second, because of the autogenous pressure in the sealed container, we can obtain crystalline products even when the reaction proceeds at a relatively low temperature.

1.2.3 Iodine transfer route to pure metastable rocksalt GaN

As mentioned above, we have obtained metastable rocksalt GaN under benzene-thermal reaction. The products are a mixture of hexagonal and cubic phase GaN. How can we obtain pure metastable rocksalt GaN under the benzene-thermal reaction? Xie and coworkers have designed an iodine transfer route to achieve this goal [48]. The reaction is as follows:

$$GaI_3 + 3NaNH_2 + 3I_2 \xrightarrow{\text{benzene, 483 K}} GaN + 3NaI_3 + 2NH_3$$

In this reaction, GaI_3 and $NaNH_2$ are maintained at 210°C and under 3 MPa pressure. Here I_2 acts as the transporting agent.

The XRD patterns of the products showed that main products were indeed pure cubic phase GaN. The designed approach to rocksalt GaN is essentially based on the solvothermal metathesis reaction between gallium halides (GaX_3) and alkali metal nitrides (M_3N), in which the following efficient strategies are put forward to favor the formation of metastable GaN: (1) carrying out the reaction under milder conditions for a longer time will be helpful to the formation of metastable phase and (2) the nitride source with stronger ionicity will favor the formation of metal nitride with stronger iconicity. (3) Na_3N has a stronger ionicity than Li_3N, which will be helpful to the formation of ionic GaN with rocksalt structure. In addition, the release of NH_3 increases the pressure in the system rapidly, which may also be helpful to the formation of rocksalt GaN. The active additive I_2 was added to the metathesis reaction to act as a transporting agent. Here, iodine helps the NaI product to dissolve in benzene in the form of NaI_3 based on the above reaction, which can ensure the completion of the metathesis reaction of GaI_3 and Na_3N.

1.3 Development of Mixed Solvothermal Method

In addition to ethylenediamine and benzene, we have also developed the solvothermal synthesis in the mixed solvents. For example, various metal sulfides (PbS, Cu_2S, and Ag_2S) and their assemblies into nanocrystal superlattices (NCSs) have been synthesized by utilizing pure dodecanethiol as a solvent [49] (see Fig. 1.5).

Ultrathin nanowires of hexagonal-phase Cu_2S are obtained by thermal decomposition of CuS_2CNEt_2 in a mixed surfactant solvent of dodecanethiol and oleic acid at 160°C. Cu_2S nanowires can be synthesized in a controlled manner with a diameter as thin as 1.7 nm and length up to tens of micrometers; they are usually aligned in the form of bundles with a thickness of hundreds of nanometers. UV–vis spectroscopy measurement reveals that the resultant ultrathin nanowires show a strong quantum-size effect [50] (Fig. 1.6).

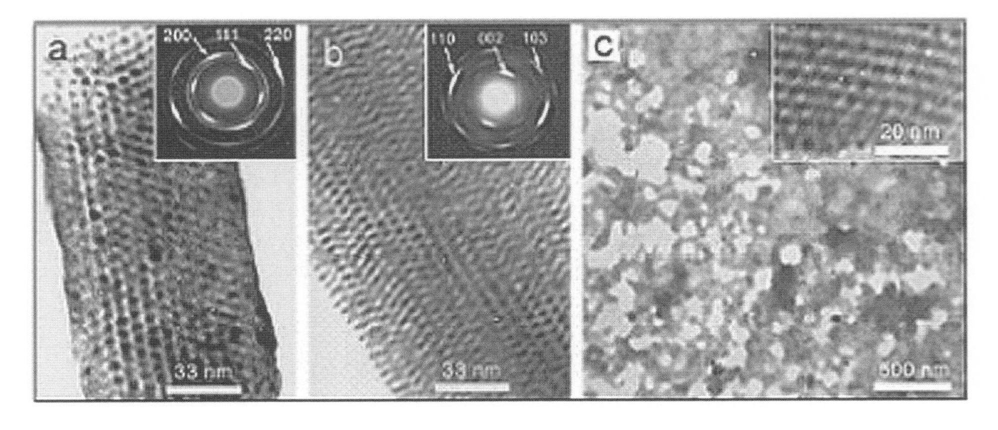

Fig. 1.5. TEM images of (a) PbS, (b) Cu_2S, and (c) Ag_2S nanocrystal superlattices (prepared with 0.4 mmol of TAA).

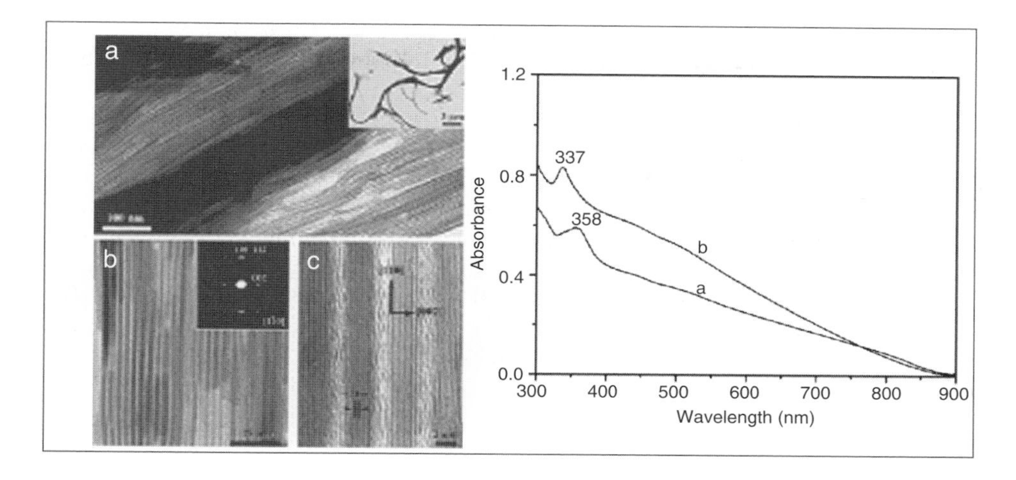

Fig. 1.6. TEM images and the absorptive spectrum of the Cu_2S nanowires.

PbSe nanowires with an average size of 8 × 350 nm were synthesized by the reaction of $Pb(NO_3)_2$ and Se powder in the mixture of arachidic acid (AA, n-$C_{19}H_{39}COOH$) and octadecylamine (ODA) [51]. Ultralong Bi_2S_3 nanoribbons have been synthesized by a solvothermal process, using a mixture of aqueous NaOH solution and glycerol as the solvent. These nanoribbons are 50–300 nm wide, 20–80 nm thick, and up to several millimeters long (see Fig. 1.7). The initial formation of the precursor polycrystalline $NaBiS_2$ phase is crucial to the formation of the Bi_2S_3 nanoribbons. A solid–solution–solid transformation growth process was observed during the aging process [52].

Rod-shaped PbS nanocrystals with diameters of 30–150 nm and lengths of up to several micrometers were prepared by a biphasic solvothermal interface reaction between

Fig. 1.7. Ultralong Bi_2S_3 nanoribbons synthesized in aqueous NaOH solution and glycerol-mixed solvent.

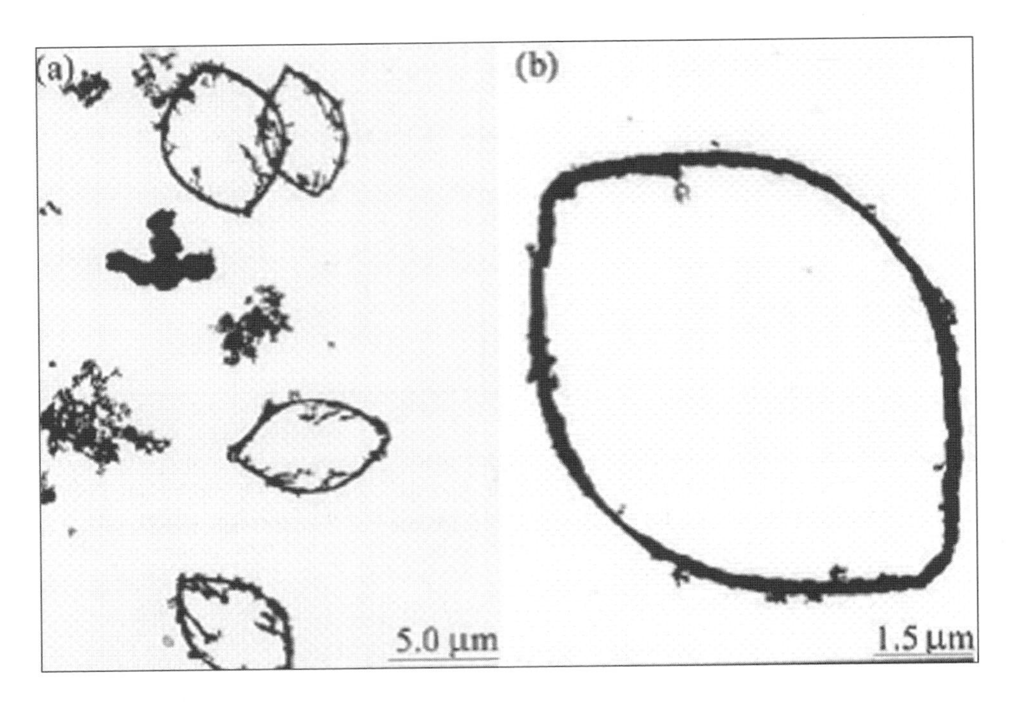

Fig. 1.8. TEM image of the closed PbS nanowires obtained from the poly[N-(2-aminoethyl)acrylamide] in ethylenediamine/H_2O (3:1, v/v) solvent.

140 and 160°C [53]. Closed PbS nanowires with regular geometric morphologies (ellipse and parallelogram shape) were synthesized in the presence of poly[N-(2-aminoethyl)acrylamide] in ethylenediamine/H_2O (3:1, v/v) mixed solvent between 110 and 150°C (see Fig. 1.8) [54].

Similarly, $CdIn_2S_4$ nanorods were prepared in en/ethanol mixed solvents [55]. The ethanol acts as the transportation reagent for $InCl_3$, while the ethylenediamine essentially serves as the nucleophile for the formation of the inorganic $[Cd_2S_2]$ core by scission of the thione groups of cadmium bis(diethyldithiocarbanate) $[Cd(DDTC)_2]_2$. These newly formed cores assemble into one-dimensional $[CdS]n$ clusters and act as intermediate templates for the growth of CdS nanorods. The solvent ratio between en/ethanol show great effects on the size of the $CdIn_2S_4$ nanorods.

As a combination of the hydrothermal and solvothermal methods, synthesis of nano-materials in mixed solvents showed attractive effects in control over the shape of these semiconductors. The above-mentioned examples show, in the mixed solvents, that the shape, size, and phase of various classes of semiconductors can be controlled. The ratio of the mixed solvents, temperature, and the concentration of the raw materials are among those parameters that can affect the quality of the products. As a newly developed method, the scope and details of this method need to be further investigated.

1.4 Hydrothermal Route to Semiconducting Nanomaterials

In the hydrothermal process, water at elevated temperatures plays an essential role in the precursor material transformation because the vapor pressure is much higher and the state of water at elevated temperatures is different from that at room temperature. The solubility and reactivity of the reactants also change at high pressure and high temperatures, and high pressure is favorable for crystallization of products. During the synthesis of nanocrystals, parameters such as water pressure, temperature, reaction time, and the respective precursor-product system can be tuned to maintain a high simultaneous nucleation rate and good size distribution.

1.4.1 Metal chalcogenide nanomaterials

Novel three-dimensional (3D) PbS dendritic nanostructures have been prepared by Kuang *et al.* through a surfactant-assisted hydrothermal progress, the individual PbS dendrites have 3D structures with one trunk (long axis) and four branches (short axes); the nanorods in each branch are parallel to each other and in the same plane, and are perpendicular to the trunk. Both the surfactant CTAB and thiourea might play important roles in the formation of 3D PbS dendritic nanostructures [56]. Recently, Zhang *et al.* have synthesized one-dimensional (1D) PbTe nanostructure by a hydrothermal reaction between lead foil and tellurium powder, the formation of 1D PbTe nanostructures can be explained by an in situ hydrothermal rolling-up mechanism whereby PbTe is formed hydrothermally and deposited on the lead substrate, the lead underneath the PbTe layer is then selectively etched by a CTAB solution, thus allowing the PbTe to roll up into 1D structures, and the synthesis process was also applied to prepare other 1D tellurides [57].

A template-free hydrothermal method was applied to prepare uniform Bi_2S_3 and Bi_2Se_3 nanoribbons (see Figs. 1.9a and b) using hydrochloric acid, $BiCl_3$, and S (or Se) powders as the starting materials. The key step was only adjusting the amount of hydrochloric acid in the crystalline growth, which served to suppress secondary nucleation and slow down growth rate [58]. Assisted by polymer reagent PVA, Sb_2Se_3 nanoribbons with diameters in the range of 25–100 nm and lengths of tens of micrometers were obtained in aqueous solutions (Figs. 1.9c and d) [59].

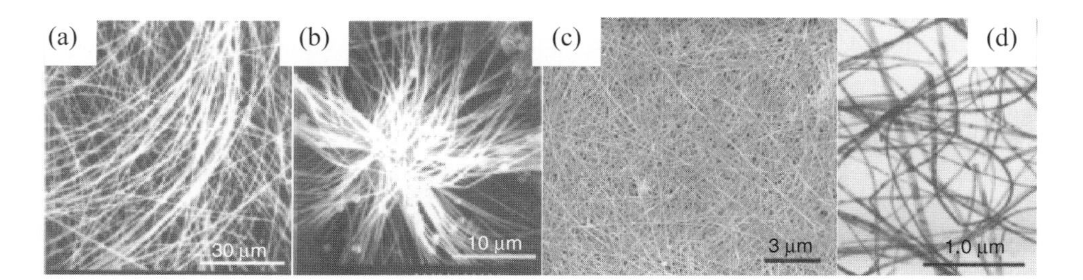

Fig. 1.9. SEM patterns of the Bi_2S_3 (a) and Bi_2Se_3 (b) nanoribbons, FE-SEM (c) and TEM (d) images of the Sb_2Se_3 nanoribbons.

Using acrylamide and SDBS as surfactants, CuS nanocones and nanobelts were synthesized in aqueous solution under hydrothermal conditions [60]. Lu *et al.* demonstrated one-step synthesis and assembly of copper sulfide nanoparticles to nanowires, nanotubes, and nanovesicles by an organic amine-assisted hydrothermal process [61]. CuS flakes and Cu_2S nanodisks were formed in hydrothermal microemulsions [62], dedocanethiol has the key control over the stoichiometries of copper sulfide and the resulting morphologies. Xie *et al.* developed an in situ template-controlled route to CuS nanorods via transition metal liquid crystals [63]. IB-VIA nonetoichiometric nanocrystallines of tellurides were generated by a hydrothermal reaction of metal salts with tellurium in the presence of $N_2H_4 \cdot H_2O$ [64], some interesting nonstoichiometric phases such as $Cu_{2.86}Te_2$, Cu_7Te_5, $Cu_{2-x}Te$, and Ag_7Te_4 were observed in the products at lower temperature.

Applied with an elemental-direct reaction, Li *et al.* reported a low temperature hydrothermal synthesis of ZnSe and CdSe nanocrystallines [65]. ZnSe hollow nanospheres have been hydrothermally prepared from the precursor $ZnCl_2(N_2H_4)_2$ and Na_2SeO_3 by using $N_2H_4 \cdot H_2O$ as the reducing agent [66], the hollow nanospheres might be formed by a soft template of gas bubbles of N_2 generated during the reaction. CdSe nanorods and dendritic fractals were also synthesized through a controlled solution-phase hydrothermal method by selectively choosing appropriate complexing agents to adjust the dynamics of the reaction progress [67].

A simple elemental-direction reaction route has been developed to prepare α-MnS nanorods under hydrothermal conditions using manganese and sulfur powder as starting materials in the temperature range of 240–260°C [68], the α-MnS nanorods might be formed via an in situ sulfiding $Mn(OH)_2$ nanorods generated by the reaction of manganese and water. Chen *et al.* reported a solution synthesis of transition metal dichalcogenides at low temperature, including pyrite FeS_2, CoS_2, NiS_2, $NiSe_2$, and layered MoS_2, and $MoSe_2$ [69]. $FeTe_2$ nanocrystals [70] and nanorods [71] had been synthesized by a solution progress at low temperature.

1.4.2 III–V Nanomaterials

Due to their direct bandgaps, III–V group semiconductors have been widely used as active materials in fabricating optoelectronic devices, such as light-emitting diodes (LEDs), in the visible and ultraviolet regions [72]. Synthesis of these semiconductors in aqueous solution proved to be convenient, simple, and green-chemical progress. Recently, our group developed a facile chemical process to prepare III–V semiconductors nanocrystals under hydrothermal condition. Spherical InAs nanocrystals of 30–50 nm have been hydrothermally prepared by the reaction between $InCl_3 \cdot 4H_2O$, As_2O_3, and Zn powder in HCl aqueous solution [73]. Aided by surfactant potassium stearate, InP nanocrystals were also synthesized in aqueous ammonia via a hydrothermal progress [74,75].

GaP and InP nanowires were first prepared in aqueous solution; cetyltrimethylammonium cations (CTA^+) can be combined with anionic inorganic species via a co-condensation mechanism to form lamellar inorganic-surfactant intercalated mesostructures, which serves as both microreactors and reactants for the growth of the nanowires, as shown in Fig. 1.10; This process could also be employed to synthesize other inorganic nanowires

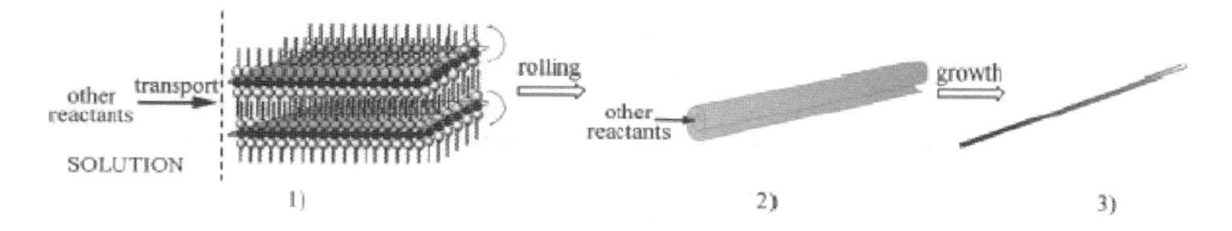

Fig. 1.10. Schematic formation illustration for the nanowire growth from lamellar inorganic-surfactant mesostructures.

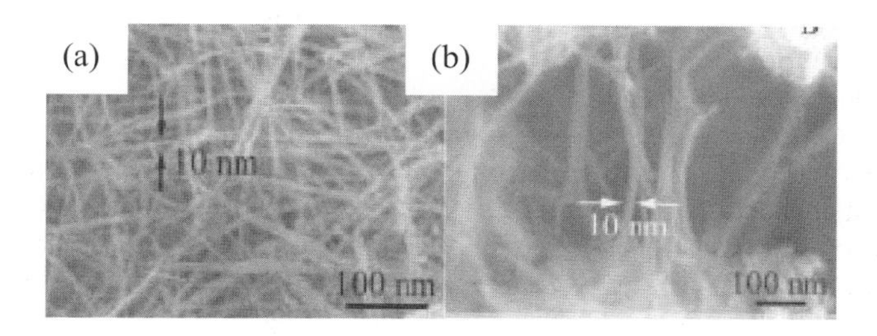

Fig. 1.11. SEM images of the GaP (a) and InP (b) nanowires.

including γ-MnO_2, ZnO, SnS_2, and ZnS nanowires in aqueous solution [76]. SEM images in Figs. 1.11a and b display the GaP and InP nanowires prepared under mild conditions [77].

1.4.3 Other semiconductors

Elemental nanomaterials Se is an important group VI B semiconductors, which has found applications in rectifiers, solar cells, photographic exposure meters, and xerography. Many efforts were undertaken to synthesize Se nanostructures. In the presence of a surfactant, trigonal Se nanowires were prepared by a hydrothermal reaction between $(NH_4)_2S_2O_3$ and $Na_2Se_2O_3$ [78]. Xia *et al.* demonstrated a soft, solution-phase approach to Se nanowires with uniform diameters as small as 10 nm and controlled lengths up to several hundred micrometers [79]. Photoconductivity measurements on individual nanowires (32 nm in diameter) indicated a ~145 times increase in photocurrent when the incident light intensity was increased from 0 to 3 $\mu W\ \mu m^{-2}$. Single crystalline Se nanotubes were reported by a micelle-mediated solution progress [80]. Se nanobelts and nanowires were also obtained by Ma *et al.* under suitable conditions.

Thin Te nanotubes and nanobelts were formed in disproportionation process of Na_2TeO_3 in aqueous ammonia [81], the TEM images shown in Figs. 1.12a and b are the as-prepared Te nanobelts, a nanobelt bending in a circle (Fig. 1.12c) displayed its bending property. Many nanotubes, with open ends and a common cylindrical shape (see Fig. 1.12d), were likely to have been created from thin Te nanobelts. The formation mechanism of Te nanotubes is shown schematically in Fig. 1.13. Liu *et al.* synthesized trigonal Te nanotubes

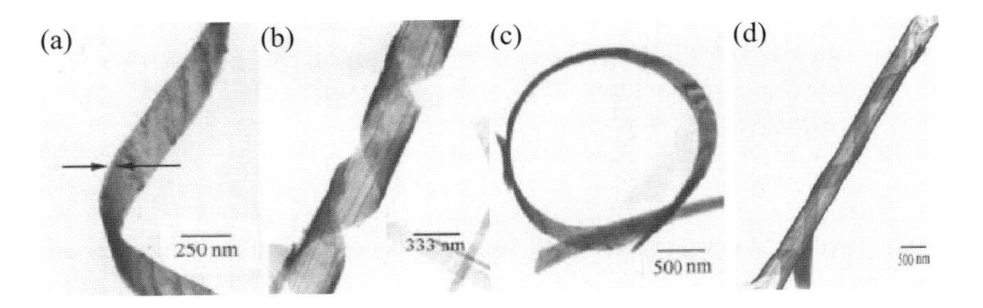

Fig. 1.12. TEM patterns of the Te nanobelts (a–c) and nanotubes (d).

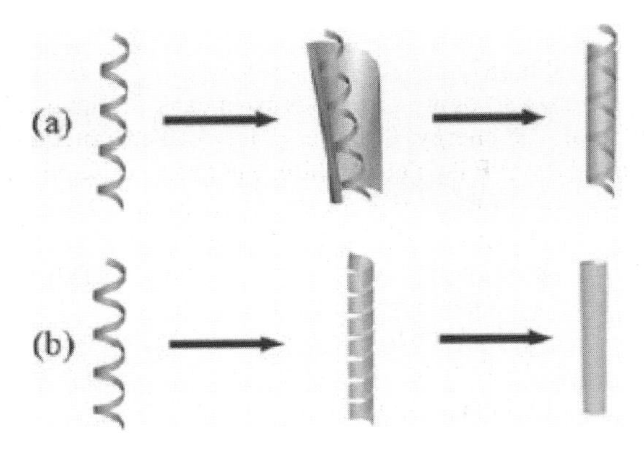

Fig. 1.13. Schematic of the proposed template-roll-growth mechanism (a) and template-twist-join-growth mechanism (b) for Te nanotubes from the nanobelts.

and nanowires by running the hydrothermal reaction in different media [82], namely in alkaline and acidic solutions respectively. Adopted by a biomolecule-assisted reduction process, Te nanowires were obtained by Lu *et al.* under hydrothermal condition [83], the presence of alginic acid is important for the formation of the nanowires. Single crystalline nanowires and nanorods of trigonal Te had been prepared by Rao Xia *et al.* via a simple solution route. The procedure involves the disproportionation of NaHTe, obtained by the reduction of Te with $NaBH_4$, and employed solvothermal and hydrothermal conditions. Te nanobelts and nano junctions were fabricated by a self-seeding progress [84].

Oxide nanomaterials ZnO is a wide bandgap (3.4 eV) and a large excition binding energy (60 meV) semiconductor, which has found applications in light-emitting diodes, room-temperature nanolasers, gas sensors, piezoelectric devices, and solar cells [85]. Well-separated ZnO nanotip arrays were grown vertically on ZnO films using a soft chemical route under simple and mild conditions [86]. The field emission of the ZnO nanotip arrays shows a turn-on field of about 10.8 V/μm at a current density of 0.1 μA/cm and emission current density up to about 1 mA/cm^2 at a bias field of 19.5 V/μm. Vayssieres *et al.* presented the growth of arrayed nanorods and nanowires of

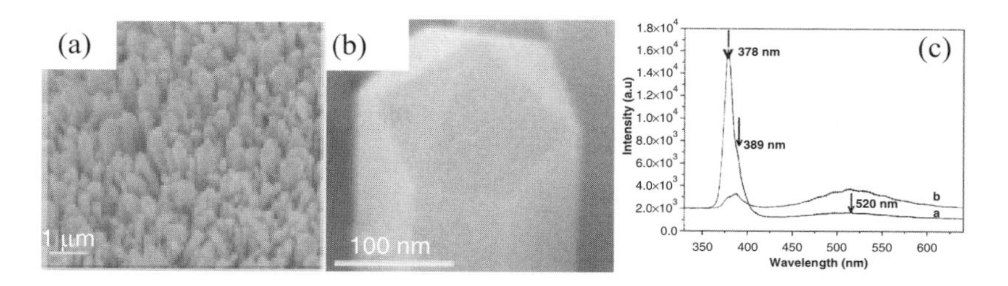

Fig. 1.14. FESEM images (a, b) of the ZnO nanorod arrays. PL spectrum (c)-curve 'a' from the ZnO nanorod arrays and curve 'b' from ZnO nanoparticles.

ZnO from aqueous solutions [87], the synthesis strategy of which consists of monitoring of the nucleation, growth, and aging processes by means of chemical and electrostatic control of the interfacial free energy. Using a variety of substrates including Si wafers, (polyethylene terephthalate) (PET), and sapphire, ZnO nanorods and nanotubes were fabricated in aqueous solutions [88]. A H_2O_2-assisted hydrothermal method was designed to synthesize ZnO nanorod arrays with high optical property (see Fig. 1.14); H_2O_2, at least, takes a double role of promoting the 1D arrays structures and enhancing its exciton emission effectively [89]; ZnO nanorods in the diameter regime of 50 nm were also prepared by Lie *et al.* through a hydrothermal progress [90]. Highly oriented ZnO nanorod and nanotube arrays could be formed on ZnO-film-coated substrates of any kind (glass, silicon, PET, etc.) via a low temperature liquid-phase method [91].

MnO_2 exists in many polymorphic forms (such as α, β, γ, and δ), which are different in that the basic unit [MnO_6] octahedra are linked in different ways [92]. Controlled synthesis of MnO_2 nanomaterials attracted various research groups in the past decades. Li *et al.* have probed the formation of MnO_2 nanocrystals with different phases under mild solution conditions: α-MnO_2 nanorods were obtained by the hydrothermal reaction between $MnSO_4$ and $KMnO_4$ [93]; under hydrothermal condition α- and β-MnO_2 single crystalline nanowires could be formed by simply adjusting the reaction agents [94], the XRD patterns shown in Figs. 1.15a, b, and c represent the TEM images of the α- and β-MnO_2 nanowires; α-, β-, γ-, and δ-MnO_2 nanorods/nanowires with different

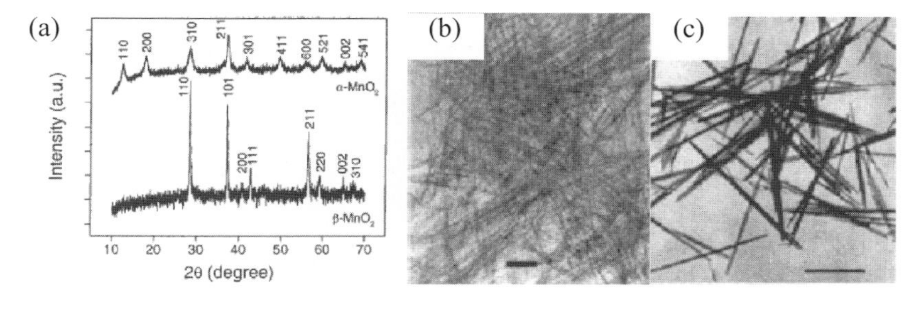

Fig. 1.15. XRD pattern (a) and TEM images of the α- (b) and β-MnO_2 (c) nanowires.

aspect ratios were also fabricated through a hydrothermal process based on the redox reactions of MnO_4^- and/or Mn_2^+, the results demonstrated that δ-MnO_2, which has a layer structure, serves as an important intermediate to other forms of MnO_2, and is believed to be responsible for the initial formation of MnO_2 one-dimensional nanostructures [95]. Xie *et al.* adopted a coordination-polymer-precursor process and developed the synthesis of well-aligned γ-MnO_2 nanowires from aqueous solution [96].

Controlled growth of oxide nanocrystallines from aqueous solutions proved to be an effective, simple, and green-chemical progress compared to other synthetic routes. $In(OH)_3$ nanocubes were fabricated by a hydrothermal reaction between In powder and H_2O_2 in aqueous solution, and their sizes can be moderated by varying the temperature. By calcining the $In(OH)_3$ nanocubes in air at 400°C In_2O_3 nanocubes were also obtained [97], the PL emission of the In_2O_3 nanocubes, originating from oxygen vacancies, lies at the blue-green region peaked at 450 nm. By an $Sn/H_2O_2/H_2SO_4$ hydrothermal oxidation, 3 nm SnO_2 nanocrystallines were prepared from aqueous solution [98]. Guo *et al.* designed a surfactant-assisted hydrothermal route to synthesize SnO_2 nanorods [99], where the surfactant might play a template-directed role.

1.5 Chemical Solution Routes to Semiconducting Nanomaterials

1.5.1 Irradiation method

The γ-irradiation method is a very useful technique to synthesize nanoscale materials. The main advantages of this method include two aspects. One is that the synthesis can be carried out at room temperature under atmospheric pressure; the reaction conditions are not rigid. Another one is its lower production cost.

Various nanocrystals could be obtained by this process including arsenic [100] and metal sulfides [101]. Semiconductor ZnS nanowires were prepared by this method in a liquid crystal template, the final product consists of ordered nanowires with a diameter of 5 nm (see Fig. 1.16a) [102]. Xie *et al.* reported the synthesis of CdSe/polymer nanocables under γ-irradiation [103], one can see from Fig. 1.16b that the 6 nm CdSe nanowires are wrapped in a 80 nm PVAc tubule, Fig. 1.16c shows the HRTEM image of the CdSe nanowire.

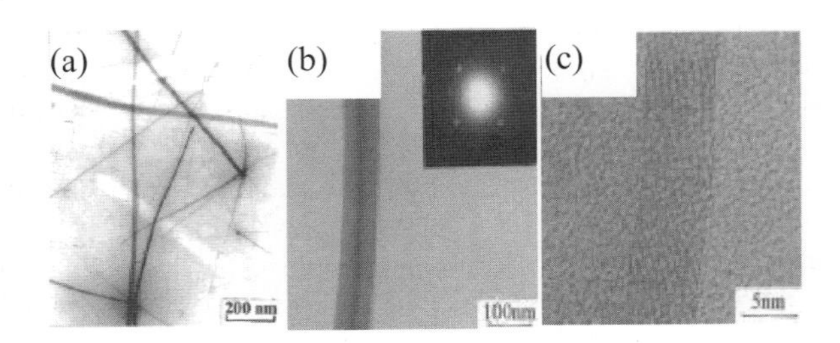

Fig. 1.16. TEM image of the ZnS nanowires, TEM (b) and HRTEM (c) patterns of the CdSe/polymer nanocables.

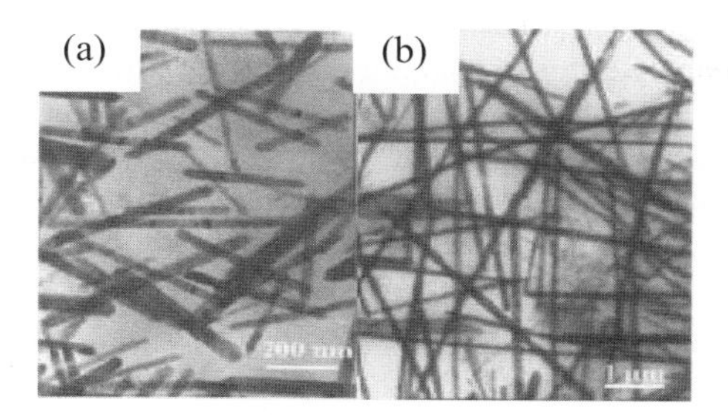

Fig. 1.17. Te nanorods (a) and nanowires (b) synthesized by microwave-assisted heating in ionic liquids.

Microwave irradiation is another way to produce inorganic compounds since 1986 [104]. Compared to conventional methods, microwave synthesis has the advantages of very short reaction time and product particles with narrow-size distribution [105].

Metal sulfide nanocrystallines were synthesized by the reaction between metal salts and thiourea in ethylene glycol (EG) under microwave irradiation [106]. Li *et al.* have prepared polymer-stabilized tetragonal ZrO_2 nanopowders with an average size of 2 nm by microwave heating in an aqueous solution [107]. Through the microwave-assisted ionic liquid method Bi_2S_3 and Sb_2S_3 nanorods were synthesized, the results show that the ionic liquid played an important role in the morphology control of the samples [108]. By employing TM_{01}-type microwave irradiation as a template, the CdS nanotubes were prepared at room temperature [109]. Zhu *et al.* presented the formation of Te nanorods and nanowires (see Fig. 1.17a and b) using microwave-assisted heating in ionic liquids; both the ionic liquid and the microwave heating are found to be vital to the shape control of Te nanocrystals [110].

1.5.2 Sonochemical method

Sonochemistry arises from acoustic cavitation phenomenon, that is the formation, growth, and implosive collapses of babbles in a liquid medium. The extremely high temperatures, pressures, and cooling rates attained during acoustic cavitation lead to many unique properties in the irradiation solution [111]. The remarkable advantages of this method include a rapid reaction rate, the controllable reaction condition, and the ability to form nanoparticles with uniform shapes, narrow-size distribution, and high purities [112].

This process was successfully applied to prepare metal chalcogenide nanocrystallines [113,114] and ZrO_2 nanopowders [115]. Xie *et al.* presented a safe sonochemical route to synthesize iron, cobalt, and nickel monoarsenides [116], the ultrasonic irradiation and the solvent ethanol are both important in the formation of the sample. Wang *et al.* have prepared Bi_2S_3 [117] and Sb_2S_3 [118] nanorods using sonochemical process in solution, and the diameters and lengths of the nanorods could be controlled by selecting different complexing agents. One-dimensional PbSe nanorods were synthesized by combining ultrasonic and γ-ray irradiation techniques [119]. Hexagonal CdS nanotubes were prepared at room temperature under ultrasonic irradiation in aqueous solution [120]. Trigonal Se nanowires were fabricated by Xia *et al.* via sonochemical process [121]; the sonication was used as the driving force for both nucleation and dispersion. Single- and multiple-heterojunction Bi_2Se_3 nanowires with diameters of 10–40 nm were synthesized in aqueous solution by an electrically assisted sonochemical method at room temperature [122].

1.5.3 Refluxing process

Over the recent years, the refluxing process was emerging as an effective, convenient, less energy-demanding and timesaving synthetic technique for inorganic nanomaterials, including metals, alloys, metal oxides, and sulfides. This progress was frequently assisted by a polyol-mediated method, especially ethylene glycol is widely used to prepare metal nanoparticles due to its strong reducing powder, relatively high boiling point (~197°C) [123], and coordination ability with transition metal ions [124].

Uniform urchin-like patterns of Bi_2S_3 nanorods had been prepared by refluxing bismuth salt solution and thiourea in ethylene glycol at 197°C for 30 min [125]. Cao *et al.* reported the synthesis of ZnSe nanorods through a surfactant-assisted refluxing method in solution [126]. CdS nanotubes and nanowires were fabricated via an in situ micelle-template-interface reaction route by refluxing the mixture and adjusting the concentration of the surfactant [127]. Sb_2E_3 (E = S, Se) nanowires were fabricated by a rapid ethylenediamine-aided refluxing route under mild conditions [128]. A solution-phase approach was developed by Xia *et al.* to synthesize uniform trigonal Se nanowires converted from amorphous Se colloidal particles, which were obtained by refluxing progress in aqueous solution [129]. Refluxing the metal salts in ethylene glycol Xia *et al.* prepared glycolate precursor nanowires, by calcining the precursor semiconducting metal oxide (including TiO_2, SnO_2, In_2O_3, and PbO) nanowires could be transformed from corresponding glycolate precursor nanowires [130].

1.5.4 Room temperature synthesis

Recent development in the synthesis stratagem of nanomaterials is seeking the simple, facile, novel, and mild process. Chemical approaches may provide a more promising route to nanostructures than conventional methods in terms of cost, throughput, and potential for large-scale production [131]. Room temperature synthesis avoids exterior energy demands and only drives by interior reaction force; this method is widely used to prepare inorganic nanomaterials in solution.

Various metal chalcogenide nanocrystals have been synthesized at room temperature under ambient pressure [132–135]: Nanocrystalline selenides were synthesized at room temperature starting from metal chlorides, KBH_4, and Se powder in ethylenediamine [136]. Using CdS nanorods as the template, Ag_2S nanorods were fabricated via a solution-phase ion-exchange route at room temperature [137]. Templated against trigonal Se nanowires, Xia *et al.* presented the formation of Ag_2Se nanowires at room

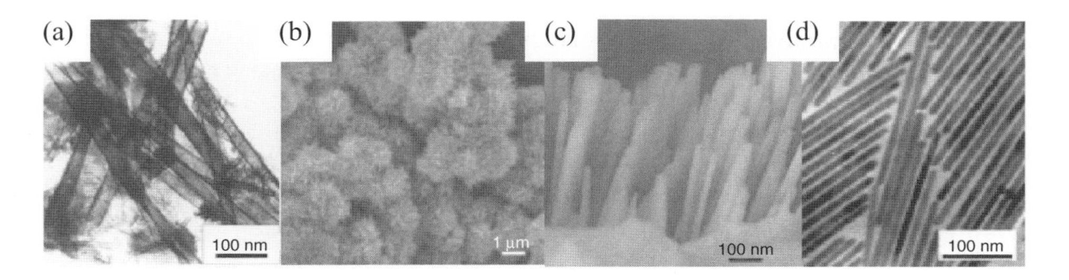

Fig. 1.18. TEM image of the $Cu_{2-x}Se$ nanotubes (a) FESEM images of ZnO urchin-like assemblies (b) and nanorod arrays (c), TEM pattern of the Te nanorods (d).

temperature [138–139], which demonstrated an effective route for the conversion of one nanostructured material to another. Direct reaction of Se nanorods in aqueous $AgNO_3$ gave Se/Ag_2Se core/shell nanocables, which furnished Ag_2Se nanotubes when the Se cores were removed [140]. $Cu_{2-x}Se$ nanotubes were prepared in a room temperature redox reaction in ethylenediamine solution [141], as shown in Fig. 1.18a. Mixed Se powder, $PbCl_2$, and KBH_4 in ethylenediamine solution gave the PbSe nanowires at room temperature [142]. A solution surface-erosion route was employed by Xie *et al.* to produce one-dimensional ZnO nanostructures at room temperature [143], ZnO urchin-like assemblies (Fig. 1.18b) and nanorod arrays (see Fig. 1.18c) could be selectively obtained with different manipulations. A surfactant-assisted approach was designed to synthesize monodisperse Te nanorods (Fig. 1.18d) [144], which were grown from a colloidal dispersion of amorphous Te and trigonal Te nanoparticles at room temperature.

Acknowledgment

This work was supported by the National Natural Science Funds and the 973 Projects of China.

References

[1] C.B. Marry, C.R. Kagan and M.G. Bawendi, Science 270 (1995) 1335.

[2] M.L. Steigerwald and L.E. Brus, Acc. Chem. Res. 23 (1990) 183.

[3] M.G. Bawendi, M.L. Steigerwald and L.E. Brus, Annual Rev. phys. Chem. 41 (1990) 477.

[4] M.G. Bawendi, W.L. Wilson and L. Rothberg, Phys. Rev. Lett. 65 (1990) 1623.

[5] S. Mann, Nature 332 (1988) 119.

[6] P.V. Braum, P. Osenar and S.I. Stupp, Nature 380 (1996) 325.

[7] C.B. Marry, D.J. Norris and M.G. Bawendi, J. Am. Chem. Soc. 115 (1993) 8706.

[8] A.M. Morales and C.M. Lieber, Science 279 (1998) 208.

[9] B. Gates, Y. Yin and Y. Xia, J. Am. Chem. Soc. 122 (2000) 12582.

[10] T.J. Trentler, K.M. Hickman, S.C. Geol, A.M. Viano, P.C. Gibbons and W.E. Buhro, Science 270 (1995) 1791.

[11] E. Leobandung, L.J. Guo, Y. Wang and S.Y. Chou, Appl. Phys. Lett. 67 (1995) 938.

[12] W.Q. Han, S.S. Fan, Q.Q. Li and Y.D. Hu, Science 277 (1997) 1287.

[13] J. Yang, J.H. Zeng, S.H. Yu, L. Yang, G.E. Zhou and Y.T. Qian, Chem. Mater. 12 (2000) 3259.

[14] J. Yang, C. Xue, S.H. Yu, J.H. Zeng and Y.T. Qian, Angew. Chem. Int. Ed. 41 (2002) 4697.

[15] P. Yan, Y. Xie, Y.T. Qian and X.M. Liu, Chem. Commun. 14 (1999) 1293.

[16] J.H. Zhan, X.G. Yang, D.W. Wang, S.D. Li, Y. Xie, Y.N. Xia and Y.T. Qian, Adv. Mater. 12 (2000) 1348.

[17] M. Chen, Y. Xie, J. Lu, Y.J. Xiong, S.Y. Zhang, Y.T. Qian and X.M. Liu, J. Mater. Chem. 12 (2002) 748.

[18] S.H. Yu, Y.S. Wu, J. Yang, Z.H. Han, Y. Xie, Y.T. Qian and X.M. Liu, Chem. Mater. 10 (1998) 2309.

[19] Y.D. Li, H.W. Liao, Y. Ding, Y. Fan, Y. Zhang and Y.T. Qian, Inorg. Chem. 38 (1999) 1382.

[20] S.H. Yu and M. Yoshimura, Adv. Mater. 12 (2002) 296.

[21] X.Y. Huang, J. Li and H.X. Fu, J. Am. Chem. Soc. 122 (2000) 8789.

[22] X.Y. Huang, H.R. Heulings IV, V. Le and J. Li, Chem. Mater. 12 (2000) 3754.

[23] H.R. Heulings IV, X.Y. Huang, J. Li, T. Yuen and C.L. Lin, Nano Lett. 1 (2001) 521.

[24] W.T. Yao, S.H. Yu and L. Pan, Small 1 (2005) 320.

[25] W.Q. Zhang, L.Q. Xu, K.B. Tang, F.Q. Li and Y.T. Qian, Eur. J. Inorg. Chem. 15 (2005) 653.

[26] J.E.M. Dougall, H. Eckert, G.D. Stucky, N. Herron, Y. Wang, K. Moller and T. Bein, J. Am. Chem. Soc. 111 (1989) 8006.

[27] S.R. Aubuchon, A.T. Mcphail, R.L. Wells, J.A. Giambra and J.K. Bowser, Chem. Mater. 6 (1994) 82.

[28] S.S. Kher and R.L. Wels, Mater. Res. Soc. Symp. Proc. 351(1994) 293.

[29] R.E. Treece, G.S. Macala, L. Rao, D. Franke, H. Eckert and R.B. Kaner, Inorg. Chem. 32 (1993) 2745.

[30] R.E. Treece, G.S. Macala and B.R. Kaner, Chem. Mater. 4 (1992) 9.

[31] S.M. Gao, J. Lu, Y. Zhao, N. Chen and Y. Xie, Eur. J. Inorg. Chem. 9 (2003) 1822.

[32] S.M. Gao, J. Lu, Y. Zhao, N. Chen and Y. Xie Chem. Commun. 23 (2002) 2880.

[33] J.Q. Hu, B. Deng and K.B. Tang, Solid State Commun. 121(2002) 493.

[34] J.Q. Hu, B. Deng and K.B. Tang, Solid State Sci. 3 (2001) 275.

[35] J.Q. Hu, B. Deng and W.X. Zhang, Int. J. Inorg. Mater. 3 (2001) 639.

[36] Y. Xie, W.Z. Wang, Y.T. Qian and X.M. Liu, Chin. Sci. Bull. 41 (1996) 1964.

[37] G. Wilkinson, Comprehensive Coordination Chemistry, Pergamon: Oxford, Vol. 3, 1987.

[38] B.K. Ridley, Quantum Process in Semiconductors, Clarendon: Oxford, 1982.

[39] Q. Yang, K.B. Tang, Q.R. Li, H. Yin, C.R. Wang and Y.T. Qian, Nanotechnol. 15 (2004) 918.

[40] M. Sainsbury, Tetrahedron 36 (1980) 3327.

[41] A. Cairncross and W.A. Sheppard, J. Am. Chem. Soc. 90 (1968) 2186.

[42] A. Klapars, J.C. Antilla, X.H. Huang and S.L. Buchwald, J. Am. Chem. Soc. 123 (2001) 77.

[43] N.A. Dhas, C.P. Raj and A. Gedanken, Chem. Mater. 10 (1998) 1446.

[44] M.C. Harris and S.L. Buchwald, J. Org. Chem. 65 (2000) 5327.

[45] Y. Xie, Y.T. Qian, W.Z. Wang, S.Y. Zhang and Y. Zhang, Science 272 (1996) 1926.

[46] H. Xia, Q. Xia and A.L. Ruoff, Phys. Rev. B 47 (1993) 12925.

[47] N. Kuwano, Y. Nagatomo and K. Kobayshi, Jpn. J. Appl. Phys. 33 (1994) 18.

[48] F. Xu, Y. Xie, X. Zhang, S.Y. Zhang and L. Shi, New J. Chem. 27 (2003) 565.

[49] Z.P. Liu, J.B. Liang, D. Xu, J. Lu and Y.T. Qian, Chem. Commun. 24 (2004) 2724.

[50] Z.P. Liu, D. Xu, J.B. Liang, J.M. Shen, S.Y. Zhang and Y.T. Qian, J. Phys. Chem. B 109 (2005) 10699.

[51] Y.F. Liu, J.B. Cao, J.H. Zeng, C. Li, Y.T. Qian and S.Y. Zhang, Eur. J. Inorg. Chem. 2 (2003) 644.

[52] Z.P. Liu, S. Peng, Q. Xie, Z.K. Hu, Y. Yang, S.Y. Zhang and Y.T. Qian, Adv. Mater. 15 (2003) 936.

[53] M.S. Mo, M.W. Shao and H.M. Hu, J. Cryst. Growth 244 (2002) 364.

[54] D.B. Yu, D.B. Wang and Z.Y. Meng, J. Mater. Chem. 12 (2002) 403.

[55] J. Lu, Y. Xie, G.A. Du, X.C. Jiang, L.Y. Zhu, X.J. Wang and Y.T. Qian, J. Mater. Chem. 12 (2002) 103.

[56] D.B. Kuang, A.W. Xu, Y.P. Fang, H.Q. Liu, C. Frommen and D. Fenske, Adv. Mater. 15 (2003) 1747.

[57] L.Z. Zhang, J.C. Yu, M.S. Mo, L. Wu, K.W. Kwong and Q. Li, Small 3 (2005) 349.

[58] M.W. Shao, W. Zhang, Z.C. Wu and Y.B. Ni, J. Cryst. Growth 265 (2004) 318.

[59] Q. Xie, Z.P. Liu, M.W. Shao, L.F. Kong, W.C. Yu and Y.T. Qian, J. Cryst. Growth 252 (2003) 570.

[60] C.L. Jiang, W.Q. Zhang, G.F. Zou, L.Q. Xu, W.C. Yu and Y.T. Qian, Mater. Lett. 59 (2005) 1008.

[61] Q.Y. Lu, F. Gao, and D.Y. Zhao, Nano. Lett. 7 (2002) 725.

[62] P. Zhang and L. Gao, J. Mater. Chem. 13 (2003) 2007.

[63] J. Lu, Y. Zhao, N. Chen, and Y. Xie, Chem. Lett. 1 (2003) 30.

[64] J. Yang, S.H. Yu, Z.H. Han, Y.T. Qian, and Y.H. Zhang, J. Solid State Chem. 146 (1999) 387.

[65] Q. Peng, Y.J. Dong, Z.X. Deng, X.M. Sun and Y.D. Li, Inorg. Chem. 40 (2001) 3840.

[66] C.L. Jiang, W.Q. Zhang, G.F. Zou, W.C. Yu and Y.T. Qian, Nanotechnol. 16 (2005) 551.

[67] Q. Peng, Y.J. Dong, Z.X. Deng and Y.D. Li, Inorg. Chem. 41 (2002) 5249.

[68] C.H. An, K.B. Tang, X.M. Liu, F.Q. Li, G.E. Zhou and Y.T. Qian, J. Cryst. Growth 252 (2003) 575.

[69] X.H. Chen and R. Fan, Chem. Mater. 13 (2001) 802.

[70] W.X. Zhang, Y.W. Chen, J.H. Zhan, W.C. Yu, L. Yang, L. Chen and Y.T. Qian, Mater. Sci. Eng. B 79 (2001) 244.

[71] W.X. Zhang, Z.H. Yang, J.H. Zhan, L.Yang, W.C. Yu, G.E. Zhou and Y.T. Qian, Mater. Lett. 47 (2001) 367.

[72] F.A. Ponce and D.P. Bour, Nature 386 (1997) 351.

[73] J. Lu, S. Wei, W.C. Yu, H.B. Zhang and Y.T. Qian, Inorg. Chem. 43 (2004) 4543.

[74] S. Wei, J. Lu, W.C. Yu and Y.T. Qian, J. Appl. Phys. 95 (2004) 3683.

[75] S. Wei, J. Lu, L.L. Zeng, W.C. Yu and Y.T. Qian, Chem. Lett. 2002, 1034.

[76] Y.J. Xiong, Y. Xie, Z.Q. Li, X.X. Li and S.M. Gao, Chem. Eur. J. 10 (2004) 654.

[77] Y.J. Xiong, Y. Xie, Z.Q. Li, X.X. Li and R. Zhang, New J. Chem. 28 (2004) 214.

[78] C.H. An, K.B. Tang, X.M. Liu and Y.T. Qian, Eur. J. Inorg. Chem. 2003, 3250.

[79] B. Gates, B. Mayers, B. Cattle and Y.N. Xia, Adv. Funct. Mater. 12 (2002) 219.

[80] Y.R. Ma, L.M. Qi, J.M. Ma and H.M. Chen, Adv. Mater. 16 (2004) 1023.

[81] M.S. Mo, J.H. Zeng, X.M. Liu, W.C. Yu, S.Y. Zhang and Y.T. Qian, Adv. Mater. 14 (2002) 1658.

[82] Z.P. Liu, S. Li, Y. Yang, Z.K. Hu, S. Peng, J.B. Liang and Y.T. Qian, New J. Chem. 27 (2003) 1748.

[83] Q.Y. Lu, F. Gao and S. Komarneni, Adv. Mater. 16 (2004) 1629.

[84] U.K. Gautam and C.N.R. Rao, J. Mater. Chem. 14 (2004) 2530.

[85] M.H. Huang, S. Mao, H. Feick, H. Yan, Y. Wu, H. Kind, E. Weber, R. Russo and P. Yang, Science 292 (2001) 1897.

[86] C.H. Hung and W.T. Whang, J. Cryst. Growth 268 (2004) 242.

[87] L. Vayssieres, Adv. Mater. 15 (2003) 464.

[88] Q.C. Li, V. Kumar, Y. Li, H.T. Zhang, T.J. Marks and R.P.H. Chang, Chem. Mater. 17 (2005) 1001.

[89] Q. Tang, W.J. Zhou, J.M. Shen, W. Zhang, L.F. Kong and Y.T. Qian, Chem. Commun. 2004, 712.

[90] B. Liu and H.C. Zeng, J. Am. Chem. Soc. 125 (2003) 4430.

[91] H.D. Yu, Z.P. Zhang, M.Y. Han, X.T. Hao and F.R. Zhu, J. Am. Chem. Soc. 127 (2005) 2378.

[92] M.M. Thackeray, Prog. Solid State Chem. 25 (1997) 1.

[93] X. Wang and Y.D. Li, Chem. Commun. 2002, 764.

[94] X. Wang and Y.D. Li, J. Am. Chem. Soc. 124 (2002) 2880.

[95] X. Wang and Y.D. Li, Chem. Eur. J. 9 (2003) 300.

[96] Y.J. Xiong, Y. Xie, Z.Q. Li and C.Z. Wu, Chem. Eur. J. 9 (2003) 1645.

[97] Q. Tang, W.J. Zhou, W. Zhang, S.M. Ou, K. Jiang, W.C. Yu and Y.T. Qian, Cryst. Growth & Design 5 (2005) 147.

[98] Y.P. He, Y.D. Li, J. Yu and Y.T. Qian, Mater. Lett. 40 (1999) 23.

[99] C.X. Guo, M.H. Cao and C.W. Hu, Inorg. Chem. Commun. 7 (2004) 929.

[100] Y.J. Zhu and Y.T. Qian, Mater. Sci. Eng. B 47 (1997) 184.

[101] Q. Yang, K.B. Tang, F. Wang, C.R. Wang and Y.T. Qian, Mater. Lett. 57 (2003) 3508.

[102] X.C. Jiang, Y. Xie, J. Lu, L.Y. Zhu, W. He and Y.T. Qian, Chem. Mater. 13 (2001) 1213.

[103] Z.P. Qiao, Y. Xie and Y.T. Qian, Adv. Mater. 11 (1999) 1512.

[104] R. Giguere, T.L. Bray, S.M. Duncan and G. Majetich, Tetrahedron Lett. 27 (1986) 4945.

[105] X.H. Liao, J.M. Zhu, J.J. Zhu, J.Z. Xu and H.Y. Chen, Chem. Commun. 2001, 937.

[106] D. Chen, K.B. Tang, G.Z. Shen, J. Sheng, Z. Fang, X.M. Liu, H.G. Zheng and Y.T. Qian, Mater. Chem. Phys. 82 (2003) 206.

[107] J.H. Liang, Z.X. Deng, X. Jiang, F.L. Li and Y.D. Li, Inorg. Chem. 41 (2002) 3602.

[108] Y. Jiang and Y.J. Zhu, J. Phys. Chem. B 109 (2005) 4361.

[109] M.W. Shao, F. Xu, Y.Y. Peng, J. Wu, Q. Li, S.Y. Zhang and Y.T. Qian, New J. Chem. 26 (2002) 1440.

[110] Y.J. Zhu, W.W. Wang, R.J. Qi and X.L. Hu, Angew. Chem. Int. Ed. 43 (2004) 1410.

[111] Ultrasounds: It's Chemical, Physical and Biological Effects, Ed. K.S. Suslick, VCH: Weinheim, 1998.

[112] M.M. Mdleleni, T. Hyeon and K.S. Suslick, J. Am. Chem. Soc. 120 (1998) 2301.

[113] Y. Xie, X.W. Zheng, X.C. Jiang, J. Lu and L.Y. Zhu, Inorg. Chem. 41 (2002) 387.

[114] B. Li, Y. Xie, J.X. Huang, Y. Liu and Y.T. Qian, Chem. Mater. 12 (2000) 2614.

[115] J. Liang, X. Jiang, G. Liu, Z. Deng, J. Zhuang, F. Li and Y. Li, Mater. Res. Bull. 38 (2003) 161.

[116] J. Lu, Y. Xie, X.C. Jiang, W. He and G.A. Du, J. Mater. Chem. 11 (2001) 3281.

[117] H. Wang, J.J. Zhu, J.M. Zhu and H.Y. Chen, J. Phys. Chem. B 106 (2002) 3848.

[118] H. Wang, Y.N. Lu, J.J. Zhu and H.Y. Chen, Inorg. Chem. 42 (2003) 6404.

[119] M. Chen, Y. Xie, J.C. Lu, Y.J. Zhu and Y.T. Qian, J. Mater. Chem. 11 (2001) 518.

[120] M.W. Shao, Z.C. Wu, F. Gao, Y. Ye and X.W. Wei, J. Cryst. Growth 260 (2004) 63.

[121] B.T. Mayers, K. Liu, D. Sunderland and Y.N. Xia, Chem. Mater. 15 (2003) 3852.

[122] X.F. Qiu, C. Burda, R.L. Fu, L. Pu, H.Y. Chen and J.J. Zhu, J. Am. Chem. Soc. 126 (2004) 16276.

[123] Y. Sun and Y. Xia, Science 298 (2002) 2176.

[124] R.W.J. Scott, N. Coombs and G.A. Ozin, J. Mater. Chem. 13 (2003) 969.

[125] G. Z. Shen, D. Chen, K. B. Tang, F. Q. Li, Y. T. Qian, Chem. Phys. Lett. 370 (2003) 334.

[126] R.T. Lv, C.B. Cao, H.Z. Zhai, D.Z. Wang, S.Y. Liu and H.S. Zhu, Solid State Commun. 130 (2004) 241.

[127] Y.J. Xiong, Y. Xie, J. Yang, R. Zhang, C.Z. Wu and G.A. Du, J. Mater. Chem. 12 (2002) 3712.

[128] G.Z. Shen, D. Chen, K.B. Tang, X. Jiang and Y.T. Qian, J. Cryst. Growth 252 (2003) 350.

[129] B. Gates, Y.D. Yin and Y.N. Xia, J. Am. Chem. Soc. 122 (2000) 12582.

[130] X.C. Jiang, Y.L. Wang, T. Herricks and Y.N. Xia, J. Mater. Chem. 14 (2004) 695.

[131] Y. Xia, J.A. Rogers, K. Paul and G.M. Whitesides, Chem. Rev. 99 (1999) 1823.

[132] X.C. Jiang, Y. Xie, J. Lu, L.Y. Zhu, W. He and Y.T. Qian, J. Mater. Chem. 11 (2001) 584.

[133] J.P. Xiao, Y. Xie, R. Tang and W. Luo, J. Mater. Chem. 12 (2002) 1148.

[134] Y. Xie, J.X. Huang, B. Li, Y. Liu and Y.T. Qian, Adv. Mater. 12 (2000) 1523.

[135] Y.D. Li, Z.Y. Wang and Y. Ding, Inorg. Chem. 38 (1999) 4737.

[136] W.Z. Wang, Y. Geng, P. Yan, F.Y. Liu, Y. Xie and Y.T. Qian, J. Am. Chem. Soc. 121 (1999) 4062.

[137] Z.H. Wang, X.Y. Chen, M. Zhang and Y.T. Qian, Chem. Lett. 33 (2004) 754.

[138] B. Gates, Y.Y. Wu, Y.D. Yin, P.D. Yang, Y.N. Xia, J. Am. Chem. Soc. 123 (2001) 11500.

[139] B. Gates, B. Mayers, Y.Y. Wu, Y.G. Sun, B. Cattle, P.D. Yang, Y.N. Xia, Adv. Funct. Mater. 12 (2002) 679.

[140] Z.Y. Jiang, Z.X. Xie, X.H. Zhang, R.B. Huang and L.S. Zheng, Chem. Phys. Lett. 378 (2003) 313.

[141] Y. Jiang, Y. Wu, B. Xie, S.Y. Zhang and Y.T. Qian, Nanotechnol. 15 (2004) 283.

[142] W.Z. Wang, Y. Gen, Y.T. Qian, M.R. Ji and X.M. Liu, Adv. Mater. 10 (1998) 1479.

[143] Z.Q. Li, Y. Ding, Y.J. Xiong, Q. Yang and Y. Xie, Chem. Eur. J. 10 (2004) 5823.

[144] Z.P. Liu, Z.K. Hu, J.B. Liang, S. Li, Y. Yang, S. Peng and Y.T. Qian, Langmuir 20 (2004) 214.

CHAPTER 2

Synthetic Architecture of Inorganic Nanomaterials

Hua Chun Zeng

Department of Chemical and Biomolecular Engineering, Faculty of Engineering
National University of Singapore, 10 Kent Ridge Crescent, Singapore 119260

2.1 Introduction

Synthesis and fabrication of nanomaterials have attracted great attention in materials research community over the past decade [1–33]. As prerequisites to lay a foundation for nanoscience and nanotechnology, these novel functional materials have been prepared into various geometrical/morphological structures in addition to chemical compositional tailoring [1–33]. For example, single-component (which has a well-defined chemical composition and formulae) nanomaterials can now be prepared rather routinely into nanotubes, nanorods, nanowires, nanorings, nanodiskettes, nanoribbons, and nanofibers, etc., with many new or existing preparative methods [1–19]. Although they have not been investigated to the same extent, multicomponent nanomaterials such as nanocomposites and supramolecular solid assemblies will undoubtedly represent next challenges for nanomaterials fabrication [20,21]. Among many known techniques [22–43], chemical synthesis under mild reaction conditions has been proven to be an effective approach for this new class of materials. In particular, the synthetic processes can be carried out in solution phase under similar conditions that have been known by chemists for centuries [44], through which numerous new solid compounds were discovered during the evolution of contemporary inorganic or organic chemistry.

To elucidate the above synthetic feasibility for complex inorganic materials, this chapter presents a brief review on recent development of design and synthesis of inorganic nanomaterials under the mild reaction conditions in solution media. Although the materials' functionality is always the first concern for applications, it is now realized that a certain degree of aesthetic expression for the synthesized nanomaterials should also be targeted. Therefore, on the basis of the existing synthetic knowledge, the main objective of this review is to devise a set of general concepts of *synthetic architecture* for constructions of inorganic nanomaterials in a controlled manner [45,46]. Due to space limitation for this review, the works used for demonstration will be quoted mainly from the author's laboratory, although there have been numerous outstanding works reported on the subject area in the literature [1–3,20,39,40,47]. In the following section, we will first look at some general concepts which will be used in this review.

2.1.1 Inorganic nanomaterials

Inorganic materials are generally considered to include metals, alloys, metal salts, carbides, nitrites, oxides, sulfides, and halides etc., according to their chemical compositions [44]. These materials can also be classified according to their functionality into electronic, magnetic, and optical materials, ionic and superconducting materials, structure materials and ceramics, and catalytic and porous materials etc., or simply according to their matter states as crystalline and non-crystalline solids. For small-sized materials, which will be the materials of study in this review, a prefix 'nano' is now commonly added if their spatial scale is in the regime of several to a few hundred nanometers ($1 \text{ nm} = 10^{-9}$ m). As regards nanocomposites, at least one of the solid phases must be in the nanoscale range in their multicomponent material systems. In this chapter, the nanocomposites would also include inorganic–organic nanohybrids [20,21], in which the inorganic phase and organic phase can be arranged either randomly or alternately such as in supramolecular layered solids [43].

2.1.2 Concepts of synthetic architecture

Owing to their intrinsic crystal symmetries, many naturally grown or artificially prepared crystalline inorganic materials exhibit beautiful facets and morphologies. One of these classic examples is magnificent snowflakes formed from gaseous water in cold air; nearly everyone is familiar with this kind of ice morphogenesis [47]. When the crystal sizes of inorganic materials are reduced to nanometer regime, their aesthetic characters still remain (or become even more complex in some cases); the only difference in our aesthetic appreciation is that we have to examine the nanocrystallites under a high-resolution microscope [1–3,20,39,40,47]. This type of appreciation is essentially similar to those for snowflakes, miniature paintings, or ivory micro-sculptures with the aid of a magnifying lens. Apart from their similarity in artistic expression, however, there will be tremendous differences in materials preparation between the macroscopic and nanoscopic materials, which will be the main concern of the present review chapter.

To achieve size control, solution synthesis is a powerful and facile approach, where organic surfactants, liquid crystals, chelating ligands, micelles, and inorganic salts can be used as shape-control limiters or growth-guiding templates. Furthermore, the reactions can be carried out under less demanding conditions, and more than one chemical component can be introduced to one another to form a targeted nanostructure, which is analogous to the process of *synthetic architecture*. In architectural language [49,50], a building or a structure is described as *synthetic* by nature, since it is built from various starting components. Apart from the absolute form of construction, an architectural construct must

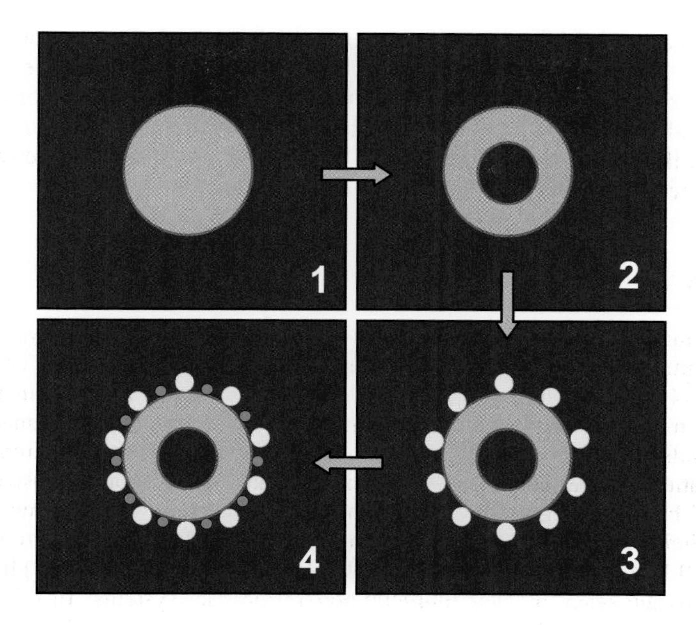

Fig. 2.1. Illustration (cross-sectional view) for the process of synthetic architecture in solution (dark background): (1) forming a spherical aggregation of nanoparticles, (2) hollowing solid center and obtaining an interior space, (3) adding a secondary component (small spheres) to the external surface, and (4) inserting a third component (smaller spheres) to form a new surface phase.

also fit well into the environment that it stands in. In other words, a building and its soundings must be integrated as a whole in a good architectural design, on top of other socio-cultural, political, environmental, and economic concerns and implications. This concept has been recently exploited in our synthesis of inorganic materials [45,46], as schematically illustrated in Fig. 2.1. For example, a spherical colloidal aggregate of nanoparticles synthesized inside a solution may further undergo a hollowing process and result in the creation of an interior space. The shell surface, on the other hand, can be further deposited with a second phase, or even a third surface phase which can be with either the same or different chemical composition. Apparently, from a fundamental viewpoint, the final solid construct produced in this way also increases its structural and chemical complexity and at the same time, possesses certain aesthetic characters (such as highly symmetrical shape, simple round geometry, constant shell wall, surface particles with a up-and-down pattern variation in size, etc.), in addition to its expected material functionalities.

In a wider sense, the above template-free solution methods are somehow analogous to total synthesis of organic compounds or polymerization of copolymers in solution, where one or several functional groups, or monomers can be added to a growing intermediate species sequentially in the synthesis. This similarity adds another tier of emphasis on the prefix word *synthetic* for describing the general architecture of inorganic nanomaterials in solution. However, the difference observed in these analogues is that the primary building units of the nanostructure illustrated in Fig. 2.1 are normally much greater than a single atom or a few atoms that form an organic functional group. In the present context, one more thing we do need to realize is that synthetic architecture concerns constructional design and synthesis of functional nanostructures, and it is not exactly equivalent to self-assembly of nanobuilding units [51–54], although the latter can be certainly utilized in some (or all) of the steps in synthetic architecture. In a simpler description, therefore, nanomaterials can be called as nanostructures when they are prepared with additional structural and compositional designs.

2.2 Methods for Synthetic Architecture

2.2.1 *Single-component nanostructures*

Most of the inorganic nanostructures prepared today are limited only to single-component materials [1–3,47]. In particular, irregularly shaped nanoparticles are still the most common product morphology for the large-scale nanomaterials prepared today. To achieve synthetic control, as mentioned earlier, the nanostructures would be expected to show certain aesthetic expression apart from their basic functionality [45,46]. Figure 2.2 depicts some basic geometric shapes and structures which can be selected in morphological designs for targeted nanostructures. By controlling crystal growth directions, for instance, a resultant nanocrystallite may display crystal isotropy or anisotropy based on its underlying structural properties.

As a first example in this subject, we will use the growth of spinel cobalt oxide Co_3O_4 to demonstrate control of crystal facets in nanomaterials synthesis, and thus the overall shapes of the nanoparticles (i.e. general morphogenesis of crystals). The synthesis of cubical Co_3O_4 powders with a mean size of about 100 nm can be traced back to 1979 with a 'forced hydrolysis' method [55,56]. Interestingly, Co_3O_4 has a cubic crystal-symmetry (space group: $Fd\bar{3}m$; $a_o = 8.084$Å, JCPDS file no. 43–1003), and this intrinsic crystal property can be utilized for further control of crystal morphology. Recently, the formation mechanism of Co_3O_4 nanocubes was investigated with an inorganic salt-assisted method by transforming various cobalt hydroxide and hydroxide nitrate under atmospheric pressure at a temperature below 100°C [57,58]. In particular, β-$Co(OH)_2$ (brucite-like structure), cobalt (II) hydroxide-nitrate $Co^{II}(OH)_{2-x}(NO_3)_x \cdot nH_2O$ ($x \approx 0.2-0.5$, $n \approx 0.1$; also a layered compound), and hydrotalcite-like layered double hydroxide $Co^{II}{}_{1-x}Co^{III}{}_x(OH)_2(NO_3)_x \cdot nH_2O$ ($x \approx 0.26-0.28$, $n \approx 0.3-0.6$) were used as starting solid precursors for the synthesis of Co_3O_4 nanocubes at 95°C with air bubbling [57,58]. The inorganic salt $NaNO_3$ was added to ensure the formation of {100} crystal facets on surfaces of nanocubes. It is believed that the nitrate salt added will reduce the solubility of oxygen and create a salt–(solvent)$_n$ diffusion boundary on the surfaces of oxide nanocubes, which stabilizes the {100} surfaces and increases the crystallinity of the nanocubes [57,58]. Figure 2.3a shows the crystal morphology of Co_3O_4 nanocubes prepared by this method. Using a modified kinetic approach, the size of the Co_3O_4 nanocubes can be tuned within the range of 10–100 nm by changing reaction temperature and time. It is found that the formation of nanocubes proceeds through a 'surface wrapping mechanism' [58], that is, a propagating oxide layer or step is grown parallel to the major surface of a nanocube, as illustrated in Fig. 2.3b. In connection to this growth mechanism, size of the nanocubes can be precisely controlled when the

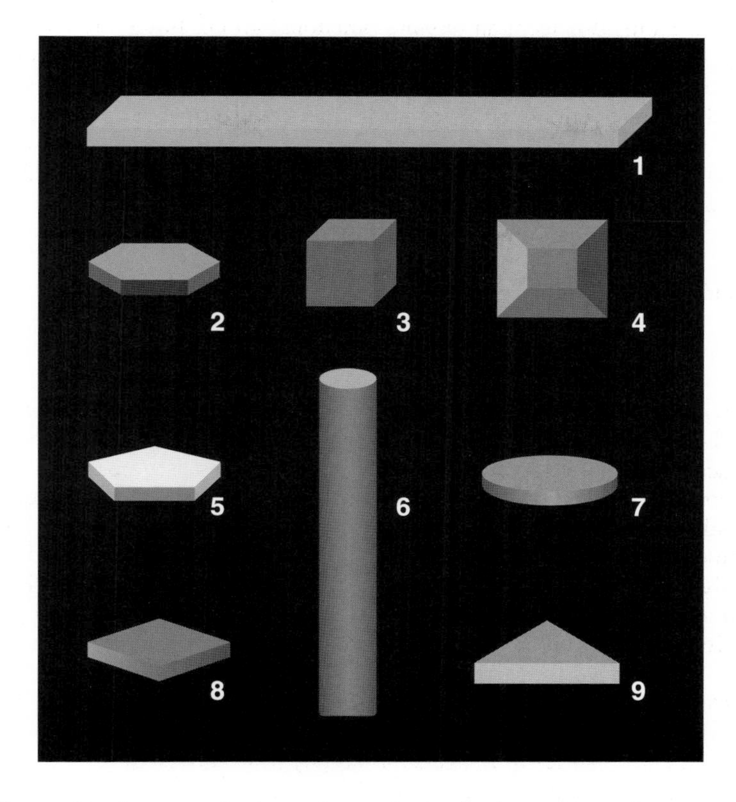

Fig. 2.2. Some of basic geometric shapes for synthesis of primary inorganic nanomaterials with certain aesthetic expression: (1) rectangular, (2) hexagonal, (3) cubic, (4) truncated pyramidal, (5) pentagonal, (6) cylindrical, (7) circular, (8) square, and (9) triangular structures.

supersaturation of reactant nutrients is low, and a linear relationship between the cube size and reaction time has been revealed [58].

To further reduce the crystal size of Co_3O_4 nanocubes, an emulsion method was developed [59]. In this basic solution approach, anionic organic capping (anionic form of an alkylated oleic acid) can be generated from nonionic surfactant polyoxyethylene (20) sorbitan trioleate (Tween-85) while the Tween-85 molecules serve as a micelle formation agent for size control of single crystal Co_3O_4 nanocubes (3–6 nm). It is also found that the nanocubes of Co_3O_4 were formed from related layered hydroxide precursors [$Co_x{}^{III}Co_{1-x}{}^{II}(OH)_2(A)_x \cdot nH_2O$, where A is the anionic form of an alkylated oleic acid; $Co_x{}^{III}Co_{1-x}{}^{II}(OH)_2(NO_3)_x \cdot nH_2O$; and β-$Co(OH)_2$] inside the micellar nanoreactors at 80–95°C [59]. Due to the surface capping formed on the Co_3O_4 nanocubes, two-dimensional superlattices of Co_3O_4 nanocubes in either non-close- or close-packed-arrangement, can be further obtained, as reported in Fig. 2.3c. It should be mentioned that the smaller nanocrystallites of this type are rather reactive; some of them have undergone an 'oriented attachment' process (which refers to the direct attachment of two small crystallites along an identical crystallographic direction using their common crystal planes or facets), through which multiple combinations of the as-prepared nanocubes

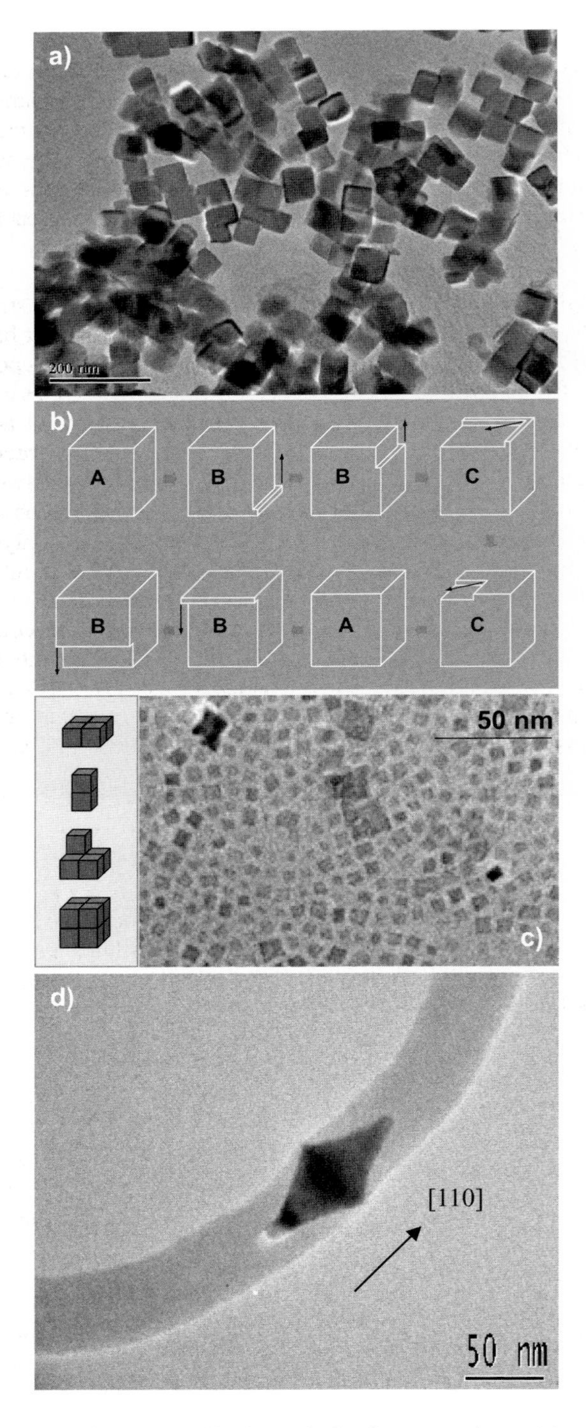

Fig. 2.3. Solution synthesis and application of Co_3O_4 nanocubes: (a) as-prepared Co_3O_4 nanocubes using a salt-assisted approach, (b) a 'surface wrapping' mechanism, (c) as-prepared Co_3O_4 nanocubes prepared with Tween-85, and (d) Co_3O_4 nanocubes used as a model catalyst for a bilateral base growth of carbon nanotubes under chemical vapor deposition condition.

can be achieved. A number of different combination styles has been observed, as illustrated in the inset of Fig. 2.3c. Because the surface capping is abundant in the synthetic medium, only a small population (less than 5%) of the nanocubes can actually break through the surfactant capping and arrive at a neighboring nanocube surface. Nonetheless, the observation made here indicates that the 'oriented attachment' can be utilized to build larger nanostructures if one properly manipulates the surface reactivity of primary building units.

One of the interesting structural features of these nanocubes is their structural isotropy, which may lead to a wide range of potential applications. This point had been recently exploited with carbon nanotubes formation under chemical vapor deposition condition [60]. In particular, the above prepared single-crystal Co_3O_4 nanocubes can be considered as a preshaped freestanding model catalyst. During a reaction with acetylene (C_2H_2) vapor, Co_3O_4 nanocubes are reduced and at the same time, reconstructed into metallic cobalt. It is surprising to note that the resultant metal catalyst has a two-fold symmetry owing to a growth of one-dimensional carbon nanofibers [60], as can be seen in Fig. 2.3d. More importantly, this work indicates that our understanding on catalyst-assisted CVD mechanisms can be acquired, when the shape, size, and crystal orientation of pristine metal catalysts are made known to the chemical reactions. Although it was not an in-situ investigation, the evolutional changes in structure and composition could be studied at different reaction stages using a wide variety of analytical instruments [60]. In general, as demonstrated in this example, any shape-designed nanostructures can be treated as model catalysts for one to pursue mechanistic investigations of heterogeneous catalysis as well as general catalyst-assisted materials synthesis and metal intercalation chemistry.

For crystalline materials with a lower degree of crystal symmetry, low-dimensional nanostructures can be prepared accordingly, where intrinsic crystal structures still play an important role in the controlled crystal growths. For example, zinc oxide (ZnO) has a hexagonal crystal symmetry (Wurtzite-type, space group: $P6_3mc$) and thus a structural anisotropy as expected, can be generated for the product [61,62]. As shown in Fig. 2.4a, indeed, ZnO nanorods along the $\langle 0001 \rangle$-directions have been synthesized with a solutions method [63]. In this work, zinc nitrate ($Zn(NO_3)_2 \cdot 6H_2O$) was dissolved in a water–alcohol–ethylenediamine co-solvent with a high concentration of NaOH. The synthesis was carried out in a Teflon-lined autoclave at 180°C for 20 h [63]. As can be seen, the ZnO nanorods are monodispersed with high crystallinity in the diameter regime of 50 nm. In particular, the fast growing crystal ends are bounded with seven crystallographic facets (10–11), (01–11), (−1011), (0–111), (1–101), (−1101), and (0001). It is found that individual one-dimensional ZnO nanorods can also attach to one another to form a two-dimensional array of nanorods, as shown in Fig. 2.4b, via their smoother prismatic side planes (10–10), (01–10), (−1010), (0–110), (1–100), and (−1100). Potentially the 'oriented attachment' phenomenon observed here may offer a new means to generate freestanding single-crystal sheets of inorganic materials.

To further exploit the above solutions method, the synthesis of ZnO nanorods were carried out at room temperature using the same starting solution precursors [64]:

$$Zn(OH)_2 + 2OH^- = {ZnO_2}^{2-} + 2H_2O \tag{2.1}$$

$$ZnO_2^{2-} + H_2O = ZnO + 2OH^- \tag{2.2}$$

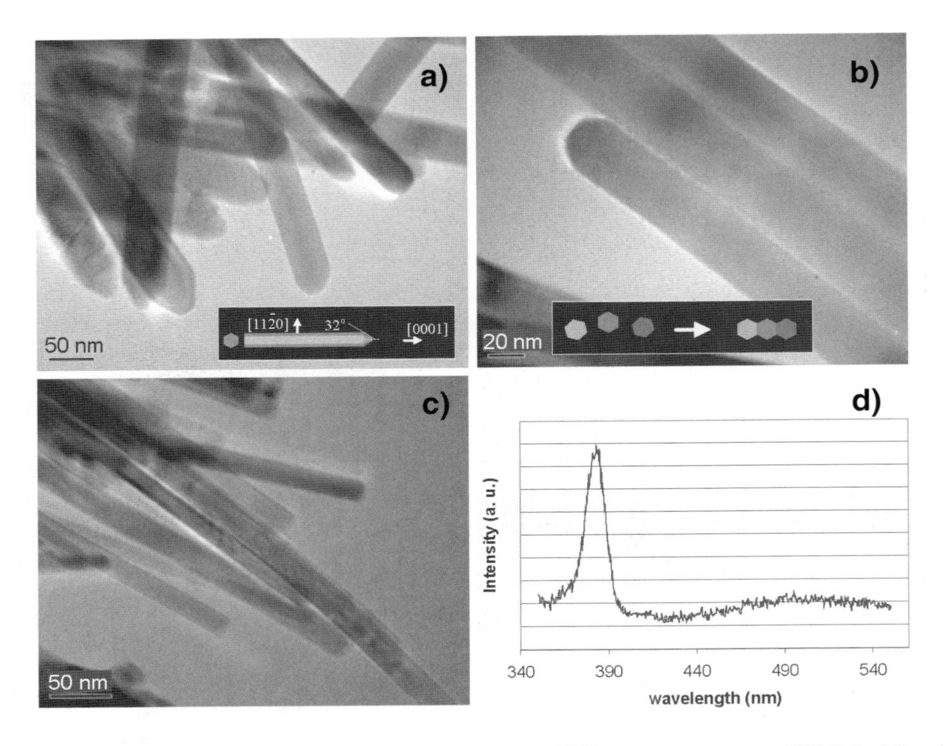

Fig. 2.4. TEM images of ZnO nanorods synthesized at different temperatures: 180°C (a, b) and room temperature (c); insets indicate crystal orientations. A photoluminescence spectrum of the room-temperature synthesized ZnO nanorods is displayed in (d).

At a lower temperature and atmospheric pressure, the growth of ZnO was slowed down and the diameter of the ZnO nanorods was reduced to 10–30 nm regime, while the aspect ratio could be increased up to 50–100 [64], as reported in Fig. 2.4c. The as-prepared ZnO was further characterized with room temperature photoluminescence study. In Fig. 2.4d, the sharp emission band located around 384 nm with a very narrow peak width can be attributed to the recombination of free excitons [64]. It is noted that a normally observed peak in the green region (510–515 nm) is barely observable in the ZnO nanorod sample. This emission is due to the presence of the singly ionized oxygen vacancies (or other point defects in ZnO lattice) [64]. Therefore, the high crystallinity of the nanorods is affirmed despite a large surface-to-volume ratio for this nanomaterial. Furthermore, the observed narrow emission at 384 nm is another indication of the size uniformity of ZnO nanorods [64]. Based on the experimental findings, it is known that the ethylenediamine used in this synthesis works both as surface modifier and zinc carrier in Ostwald ripening. By adjusting synthetic parameters, a variety of hierarchical nanostructures of ZnO can also be prepared. These structures include hexagonally branched, reversed umbrella-type, and cactus-like nanocrystals [64]. It should also be recognized that the method developed offers great synthetic flexibility under biologically relevant conditions (i.e. room temperature and atmospheric pressure).

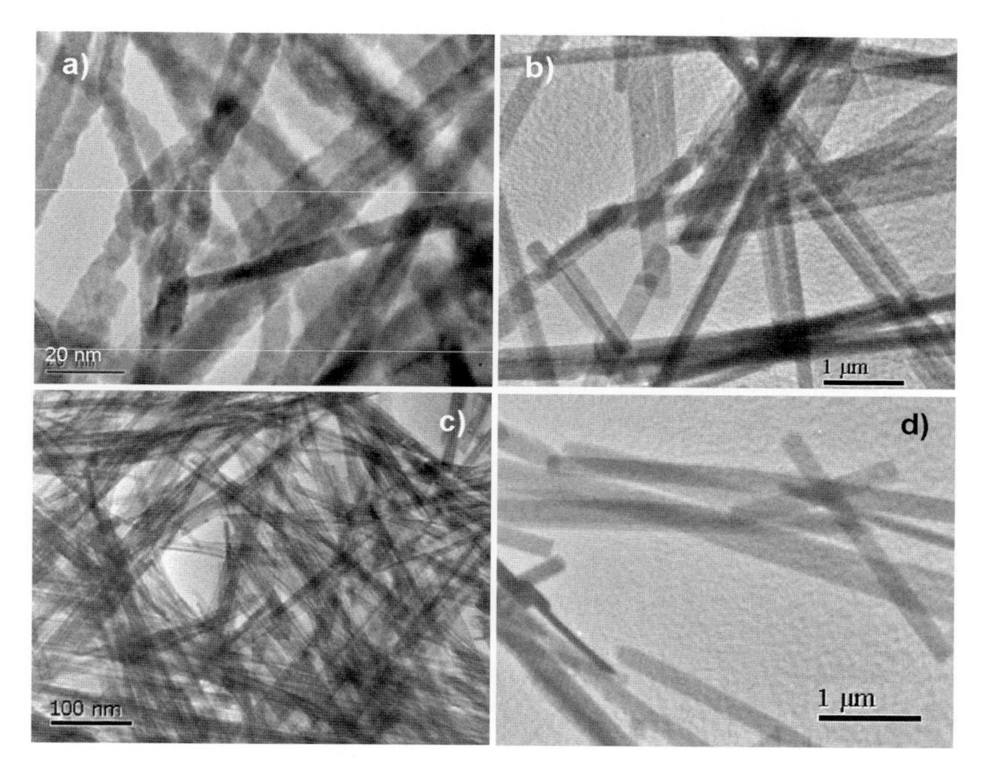

Fig. 2.5. One-dimensional nanostructured materials: (a) CuO nanoribbons, (b) α-MoO_3 nanorods, (c) $W_{18}O_{49}$ nanofibers, and (d) 2H-MoS_2 sulfide nanorods prepared from conversion of the oxide nanorods in (b).

Other low-dimensional nanostructured metal oxides with lower crystal symmetries such as CuO (monoclinic) nanoribbons [65], $W_{18}O_{49}$ (monoclinic) nanofibers [66], and α-MoO_3 (orthorhombic) nanorods [67] etc. can be prepared by various solution routes by manipulating their structural anisotropies with reaction environments. Some of these samples are displayed in Fig. 2.5. Furthermore, it should be mentioned that material conversion from one phase to the other could also be carried out either in solution or gas phase. As shown in Fig. 2.5d, 2H-MoS_2 nanorods were converted from hydrothermally prepared α-MoO_3 nanorods with additional gaseous reactions in a H_2S and H_2 atmosphere at 600°C [67].

It should be mentioned that unidirectional growth is not limited only to crystal systems with a lower symmetry [1–3,47]. In fact, by controlling reaction conditions, unidirectional growth can still be achieved for materials with high symmetries [1–3,47]. For example, a large variety of transition metals have face-centered cubic (fcc) structures, and this class of materials can still be prepared into nanorods, nanowires, and nanoribbons. In Fig. 2.6, a solution synthesis of copper nanowires is reported [68]. The copper nanowires were grown along $\langle 110 \rangle$ directions of the crystal (a similar unidirectional growth can also be observed in Fig. 2.3d for the cobalt cores intercalated inside

Fig. 2.6. Solution synthesis of copper nanowires: (a) an as-prepared Cu nanowire suspension, (b) FESEM image of Cu nanowires, (c) TEM image of a Cu nanowire, and (d) SAED pattern of the nanowire in (c).

carbon nanotubes), as confirmed in Fig. 2.6c and d, noting that metal copper belongs to a space group of $Fm\bar{3}m$. The copper nanowires were synthesized according to the following reaction under the basic condition [68]:

$$2Cu^{2+} + N_2H_4 + 4OH^- \rightarrow 2Cu + N_2 + 4H_2O \tag{2.3}$$

In this synthesis, a high concentration of NaOH is essential to prevent the copper ions from forming copper hydroxide $Cu(OH)_2$. A certain amount of ethylenediamine (EDA) is also indispensable to control the growth direction of copper. With a high concentration of NaOH, for example, the needed EDA will be small while for a lower concentration of NaOH, the amount of EDA should be increased. It is interesting to find that when EDA was overused, the axial one-dimensional growth of copper along $\langle 110 \rangle$ can be entirely terminated to a growth of disk-like copper nanostructure. As confirmed by HRTEM, the copper nanowires prepared via this redox route have high crystal quality, showing constant diameters in the range of 60–160 nm (mostly 90–120 nm) [68]. The nanowires are straight and ultralong, having lengths over 40 μm, which in fact corresponds to an aspect ratio of greater than 350! The directional growth observed here can be attributed to a steric hindrance and charge restriction of various copper complexes in the starting solution (e.g. $[Cu(OH)_4]^{2-}$, $[Cu(EDA)(OH)_2]$, and $[Cu(EDA)_2]^{2+}$ etc.) on a growing copper crystal plane, and to synergetic effects of EDA and NaOH on adsorption and thus deactivation of the grown part of a copper nanostructure [68].

In addition to the above mentioned unidirectional growth, multiple-direction growth can also be attained for high-symmetry crystals. As an example in this topical area, Fig. 2.7a depicts a branching process of Cu_2O along its six principal directions $\langle 100 \rangle$, $\langle 010 \rangle$, and $\langle 001 \rangle$. An immediate result from this process is the development of six octahedral crystals attached to each end of the hexapod (also see Fig. 2.7b) [69]. The formation reaction of Cu_2O solid was carried out with copper nitrate ($Cu(NO_3)_2 \cdot 3H_2O$) and formic acid (a reducing agent) in water–ethanol under hydrothermal conditions (150–200°C for 1.25–5 h) [69]. By changing the ratio of water to ethanol, a full range of novel multipod frameworks of Cu_2O, such as hexapods, octa-pods, and dodeca-pods (two types), can be engineered. As shown in Fig. 2.7a, more importantly, these growth experiments elucidate new organizing schemes for three-dimensional crystal aggregates. For example, the six pieces of Cu_2O crystal octahedra in Fig. 2.7b are viewed to follow a face-centered cubic stacking, while the eight pieces of Cu_2O crystal cubes in Fig. 2.7c are considered to have a simple cubic arrangement [69]. Other even more complicated crystal structures were also prepared with this approach via varying the reactant concentrations and reaction environment. Of course, it should also be recognized that the branching and growth discussed here are not entirely separable during the growths. This conceptual organization scheme provides some means for generation of 'crystals-within-the-crystal' and intra-crystal porosity, as can be seen in the inter-spaces among the primary crystal units.

2.2.2 Multicomponent nanostructures

The previous subsection is primarily focused on single-component inorganic nanostructures in freestanding discrete forms. In many applications, however, nanomaterials have to be prepared into multicomponent composites in their fabrications. For example, the overall material system of an inorganic solid compound and its support at different compositions is considered to be a 'composite' by nature [70]. This consideration can be extended to all thin-film fabrications, where a substrate support used in most cases becomes an integrated part of the final products. In this connection the initial nucleation and subsequent growth processes should be all controllable. This subsection is thus devoted to a review and discussion of synthesis and organization of inorganic nanomaterials either on a solid support or within a solid matrix.

As a first example in this subsection, Fig. 2.8a shows the island formation of anatase TiO_2 (tetragonal, space group: $I4_1/amd$) on glass substrate [71]. The flower-like TiO_2 islands were formed through a liquid phase deposition (LPD) via hydrolysis of a dilute TiF_4 solution at 55–80°C. It was observed that the TiO_2 on this fused silica substrate was first formed as tiny single-crystal nuclei during an initial heterogeneous nucleation that elongated along the $\langle 001 \rangle$-directions with their $\langle 110 \rangle$ axes perpendicular to the glass surface [71]. The elongated crystallites later were grown into the observed flower-like crystal morphology (Fig. 2.8a) due to a significant increase in homogeneous nucleation rate during the continued growth. With a further LPD process, these islands, in either

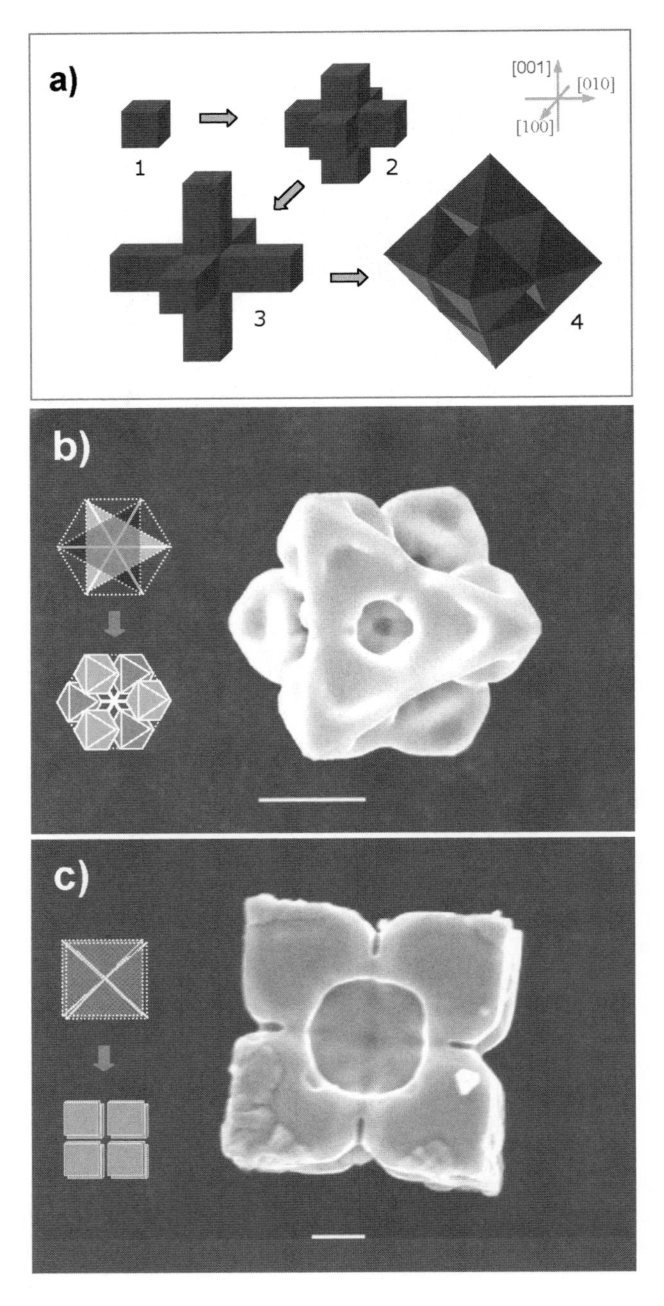

Fig. 2.7. Multiple-directional growths of Cu_2O crystals: (a) an illustration of space-predefined growth (1: a Cu_2O seed, 2: branching formation, 3: growth of the hexapod, and 4: growth of six octahedra at the tips), (b) SEM image of a Cu_2O crystal grown according to (a), and (c) SEM image of a Cu_2O crystal prepared on octa-pod framework. Insets in (b) and (c) indicate the growths of crystal units on the space predefined in their respective multipods.

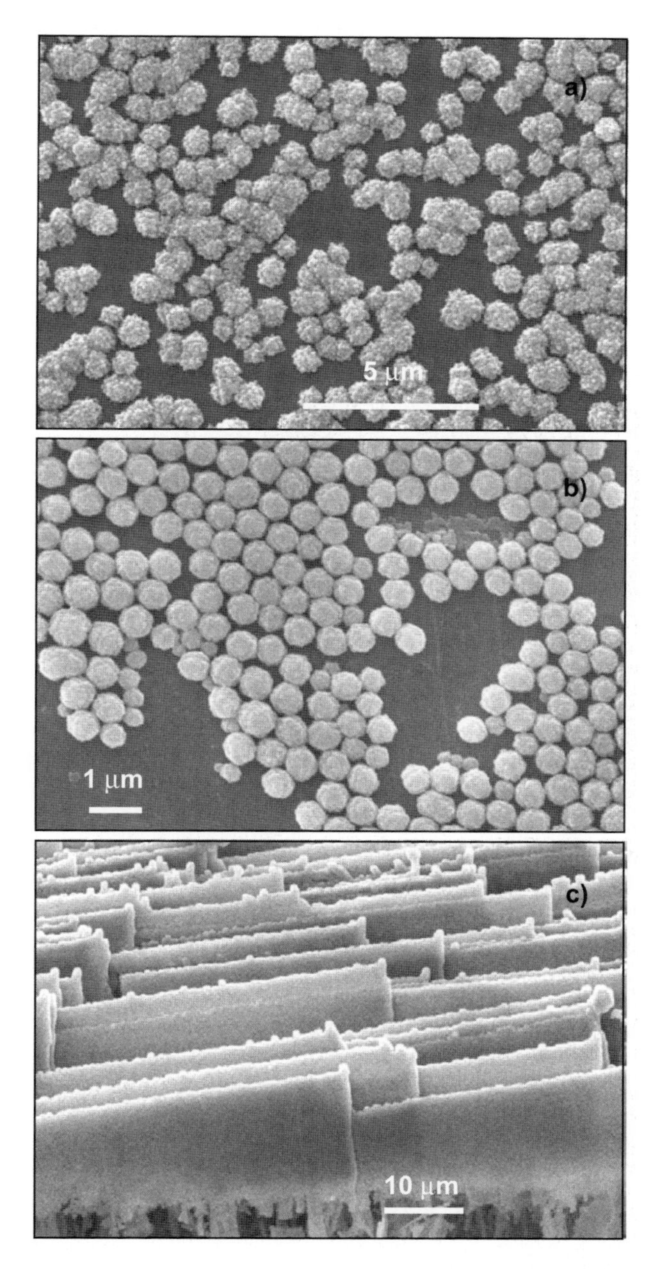

Fig. 2.8. SEM images of supported anatase TiO_2 nanostructures on different substrates: (a) polycrystalline flowerlike islands on silica surface, (b) a hexagonal array of single-crystal anatase TiO_2 nanospheres on the (010) surface of α-MoO_3, and (c) a parallel array of single-crystal anatase TiO_2 platelets standing on the (010) surface of α-MoO_3.

discrete or lined form, could eventually join together, giving away a monolayer planar structure of TiO_2 on the silica substrate. In general, low-pH condition suppresses heterogeneous nucleation and growth while higher pH favors the heterogeneous nucleation and the subsequent planarization of TiO_2 [71]. This example demonstrates that chemical reactions taking place in aqueous phase can be controlled precisely only on the surface of a substrate, and the LPD process thus provides us a great synthetic flexibility for tailor-making surface nanostructures of inorganic materials, noting that the reaction medium and reaction conditions are in fact the least demanding, compared with other common methods for thin film preparation.

To organize inorganic nanostructures on a support, there are two common methods at the present time: (i) surfactant-assisted organization, and (ii) patterned substrate deposition. Using method (i), obtained organizations (i.e. superlattices) of nanomaterials cannot sustain high temperature reaction conditions, whereas in method (ii), surface patterns need to be created via lithographic techniques or other imprinting processes. In the above preparation of surface TiO_2 (Fig. 2.8a), it is noted that the deposited crystallites could be rather random owing to amorphous nature of the glass substrate used. To test the effect of a substrate, single crystals of α-MoO_3 were then selected as supports for the TiO_2 hydrolyzing deposition. Under the hydrothermal condition at 150–200°C, anatase TiO_2 nanospheres in hexagonal arrays can be obtained on the (010) surface of α-MoO_3 single crystal, as displayed in Fig. 2.8b [72]. A number of factors can be attributed to the observed self-aligned growths. Firstly, the low concentration of TiF_4 precursor (in the range of 0.001–0.0133 M) ensured a selective nucleation only on some surface defect sites which is energetically favorable. Secondly, the high degree of crystal perfection avoids random nucleation across an entire crystal plane, and thus prevents the irregular arrangements of nanospheres. And thirdly, the α-MoO_3 support in guiding the advancement of the organized arrays was indispensable. On the basis of experimental evidence from this work, it is found that the HF generated from the hydrolysis of TiF_4 created a localized HF-rich zone in the adjacent areas of TiO_2 spheres, which acted as nucleation sites for new TiO_2 nuclei to grow or to land onto it. In this regard, it seems that the localized HF etchant worked like a chemical 'bulldozer' to pave new expanding paths for a sphere array to advance. The grown TiO_2 nanospheres in the surface arrays are single-crystalline, with the [001] axis perpendicular to the (010) surface of α-MoO_3. The entire growth process can be thus described as a 'growth-cum-assembly' process [72]. This nonlithographic wet assembly scheme may be extendable to fabricate new crystalline composite materials, including organization of surface superstructures using high quality crystal substrate and low solution concentration starting solutions. As a first step toward this end, this work has shown us the possibility of direct organization of oxide superlattices on the simple single crystal planes.

In the above surface deposition, no organic species was present during the hydrolysis of TiF_4. It is interesting to observe that with addition of organic additives to the synthesis, the growth of individual nanospheres can be changed to connected pearl threads which can be further developed into netted or planar overlayers of TiO_2 on the α-MoO_3 support (orthorhombic, space group: *Pbnm*) [73]. When the growth continues, the cracks will develop within the overlayer, resulting in switching the process into a vertical growth of TiO_2 platelets. Some of the anatase platelets are displayed in Fig. 2.8c using the chelating agent ethylenediaminetetraacetic acid disodium salt (EDTA salt,

$C_{10}H_{14}N_2Na_2O_8{\cdot}2H_2O$) [73]. As can be seen, the crystal platelets stand perpendicularly to the (010) plane of α-MoO_3. To explain this observation, it is noted that both the α-MoO_3 and TiO_2 consist of metal-centered oxygen octahedrons MO_6 (M = Mo and Ti). On the one hand, the observed planar growth of the anatase TiO_2 on the orthorhombic α-MoO_3 can be described as a heterogeneous epitaxial process due to the similarity between the two sets of lattice constants for the (001) surface of anatase TiO_2 and the (010) surface of α-MoO_3. On the other hand, the observed cracking can be attributed to a higher degree of lattice mismatch along the [100] (−4.5%) than that along the [001] (+2.4%) of α-MoO_3 [73]. This inter-lattice anisotropy promotes the cracking of the TiO_2 overlayer along the [001] direction of the α-MoO_3, and thus results in the TiO_2 platelets aligned along the same direction. This work should also be applicable to other oxide systems by investigating the interplay between structural resemblance and degree of lattice mismatch.

Another class of the multicomponent nanomaterials is inorganic nanocrystallites supported on or included in metal oxides. For example, nanosized supported hydrotalcite-like compounds $CoAl_x(OH)_{2+2x}(CO_3)_y(NO_3)_{x-2y} \cdot nH_2O$ on commercial γ-Al_2O_3 carrier pallets, noting that the trivalent aluminum ions required to form the hydrotalcite-like compounds were self-released from the surface of the γ-Al_2O_3 support in the solution synthesis [74]. This type of tailor-made biphasic solid precursors (5–7 nm) can be used as nano-catalysts. It has been found that the surface oxides formed after the calcination of samples at 211–217°C are cubic spinel $Co^{II}Co_{2-x}{}^{III}Al_xO_4$ on the alumina support. Catalytic decomposition of greenhouse gas, nitrous oxide (N_2O), had been investigated with this material system, and the highest activity for a high concentration N_2O (30 mol%) decomposition test was about 16 mmol (N_2O)/g·h, which is higher than that of a reported Ru/γ-Al_2O_3 catalyst tested under similar reaction conditions [75]. In addition to the support type of crystallites, hydrotalcite-like compounds can also be included into sol–gel matrices. For example, hydrotalcite-like compounds of $Co_x{}^{III}Co_{1-x}{}^{II}(OH)_2(NO_3)_x \cdot nH_2O$ can be introduced into an organic route derived γ-Al_2O_3 matrix. Furthermore, nanocomposite oxide materials of $Co^{II}Co^{III}_{2-x}Al_xO_4$–$\gamma$-$Al_2O_3$ ($0 \leq x < 1.63$) can be prepared at relatively low temperatures from these biphasic xerogels [76]. It has been found that cubic spinels of Co_3O_4 and $Co^{II}Co_{2-x}{}^{III}Al_xO_4$ are generated sequentially inside the alumina matrices upon the thermal treatments in static air atmosphere. The average spinel crystallites size was in the range of 9–13 nm at 350°C to 31–38 nm at 800°C, while the specific surface area of the oxide composites decreased from 277–363 m^2/g at 350°C to 184–224 m^2/g at 800°C [76]. In relation to this type of preparation of inorganic nanocomposites, thermal processes of volatile metal oxides as a secondary phase in a nanocrystalline oxide matrix can also be investigated. For instance, volatile nanocrystalline RuO_2 had been introduced to sol–gel-derived γ-Al_2O_3 matrices [77]. At 300–400°C, nanocrystallites (2 nm) of γ-Al_2O_3 were formed in the presence of ruthenium. Single phase RuO_2 was formed from its precursor compound (ruthenium red) at 400°C. The γ-Al_2O_3 matrix is converted to α-Al_2O_3 at *ca* 1000°C [77]. This transformation was promoted by the nanosized RuO_2 crystallites preformed in the alumina matrix.

A final class of the multicomponent nanomaterials within the review scope of this subsection should be looked at is inorganic–organic nanohybrids. As an example, single crystals of MoO_3(4,4′-bipyridyl)$_{0.5}$ were prepared with from a hydrothermal synthesis with α-MoO_3 and 4,4′-bipyridine ($C_{10}H_8N_2$) at 150°C for 72 h [78]. It should be indicated that the high degree of separation of individual molecular metal-oxide sheets (inorganic phase, e.g. MoO_3) can be realized with organic ligands or intercalants (organic phase, e.g. 4,4′-bipyridine in the present case) in this type of supramolecular solid assemblies. In the architecture of this nano-hybrid, molecular sheets of MoO_3 and 4,4′-bipyridine ligands are arranged alternately into a layered structure. Two important applications of this compound have been elucidated. In the first application, the prepared MoO_3(4,4′-bipyridyl)$_{0.5}$ was converted to H_xMoO_3(4,4′-bipyridyl)$_{0.5}$ ($x = 0.5$), and finally to MoS_2(4,4′-bipyridyl)$_{0.2}$ in a stream of H_2S and H_2 at 100–250°C. Apparently, this gas–solid reaction provides a new means to circumvent general synthetic difficulties in fabrication of MoS_2-related inorganic–organic nanohybrids. In the second application, the single crystal MoO_3(4,4′-bipyridyl)$_{0.5}$ was used as a solid precursor to form metal-oxide nanostructures. It had been found that the molecular MoO_3 sheets condensed after removal of 4,4′-bipyridine intercalants at 300–450°C [79]. The thermal process can generate organized nanoplatelets of β-MoO_3 and α-MoO_3, depending on the heating temperature adopted.

2.2.3 Creation of interior spaces for nanostructures

Self-assembly of nano-building units is a powerful approach to form larger organized conformations and geometrical architectures for device applications. To meet new challenges in nanoscience and nanotechnology, an interior space for the resultant micro- and nanostructures are further required in many new applications. The interior ‘nanospace’ of these nanostructures, when coupled with chemical and physical functionalities of boundary materials, creates both aesthetic beauty and scientific attractions. For instance, in addition to the general core-shell nanostructures, there has been an increasing interest in the fabrication of hollow inorganic nanostructures owing to their important applications in optical, electronic, magnetic, catalytic, and sensing devices ranging from photonic crystals, to drug-delivery carriers, and nano-reactors etc. This subsection will thus be concentrated on a review of synthetic architecture of inorganic nanostructures through which interior spaces with architectural design and aesthetic expression can be attained.

Among many existing methods, templating is a most commonly used approach for the preparation of inorganic nanomaterials with hollow interiors. The templates used in synthesis can be broadly divided into two categories: soft and hard templates. Examples of the soft templates can be found in the syntheses of mesoporous silica MCM-41 and SBA-15 and other porous materials prepared with molecular-imprints, liquid crystals, micelles, vesicles, gas bubbles, and ionic liquids [17,80–83]. Examples of hard templates can be seen in the syntheses of nanotubes, nanorods, and nanowires using anodic aluminum oxide (AAO) and polycarbonate (PC) membranes [84–91]. The inorganic nanomaterials can be grown inside the interior space (e.g. channels of AAO) as well as on the exterior surface (e.g. on the surface a nanorod) of a template. In all cases, templates must be removed after synthesis, which could become a difficult task in many preparations of this type [40,92,93].

Although the templating approach has been widely adopted, and it has been the subject of many excellent reviews, one should also recognize that nonlithographic preparation of hard template used in the template-assisted approach itself is already posed as a difficult task. Future development of nanomaterials may depend on our ability to come up with other alternatives, such as template-free syntheses [94–102]. Though these methods are less commonly used at present, they are complementary to the existing templating methods. Some details of these newly developed template-free methods are discussed below.

The following summarizes a few methods regarding the creation of interior spaces for inorganic nanostructures, which includes Ostwald ripening and Kirkendall diffusion for solid evacuation, and oriented attachment for construction of three-dimensional hollow structures. More importantly, chemical reactions of the three basic methods can be conducted in solution media without prefabricated templates or supports [94–104]. Thus these techniques, as well as many others, may form a methodical toolbox for synthetic architecture of inorganic nanomaterials in solution phases, analogous to the total synthesis of organic compounds.

The first method discussed is Ostwald ripening. The physical phenomenon utilized herein has been known for more than a century [105,106], which states 'the growth of larger crystals from those of smaller size which have a higher solubility than the larger ones' (*IUPAC Compendium of Chemical Terminology*, 2nd edition, 1977). As a first example for the research in this area, preparation of hollow anatase TiO_2 nanoparticles had been recently demonstrated. For a spherical crystallite aggregate of TiO_2 formed by the hydrolysis of a TiF_4 dilute solution, crystallite density and size across the radius of this aggregate may not be identical. As a result of Ostwald ripening, a dissolution and re-growth process will then take place, and smaller or less densely packed crystallite will be removed in certain locations of this aggregate. This solid evacuation process is shown in Fig. 2.9a and b [98], from which we understand that the pristine crystallites located in the central area were either smaller or less packed prior to the Ostwald ripening process. Similar to this result, the solid hollowing is further demonstrated in the formation of Cu_2O nanospheres, in which a redox reaction was further coupled into the solid state conversion of CuO to Cu_2O [99]. In this work, the CuO nanocrystallites were formed in a copper nitrate salt solution [$Cu(NO_3)_2 \cdot 3H_2O$ salt was dissolved in the organic solvent *N,N*-dimethylformamide (DMF)], which were then converted into Cu_2O spherical aggregates, followed by a solid evacuation for central space via a prolonged treatment of Ostwald ripening in the same reaction liquors (Fig. 2.9c) [99]. The generality of this approach had also been elucidated in a recent synthetic investigation of symmetric and asymmetric Ostwald ripening, through which homogeneous symmetric ZnS core-shells (Fig. 2.9d) and asymmetric Co_3O_4 core-shells can also be prepared [100]. Since different materials in different reaction environments lead to different types of aggregation, and the forms of aggregation determine the final vacant space, more complex interior spaces can also be obtained with this solution method.

The above hollowing processes are in fact based on simple matter relocation. In this regard, any chemical or physical phenomena which lead to mass migration and/or relocation should also be utilizable for creating interior spaces for nanomaterials. Recently, the Kirkendall effect [107–109], which is a well-known physical phenomenon in metallurgy, had been introduced to nanoscale synthesis of inorganic materials [101].

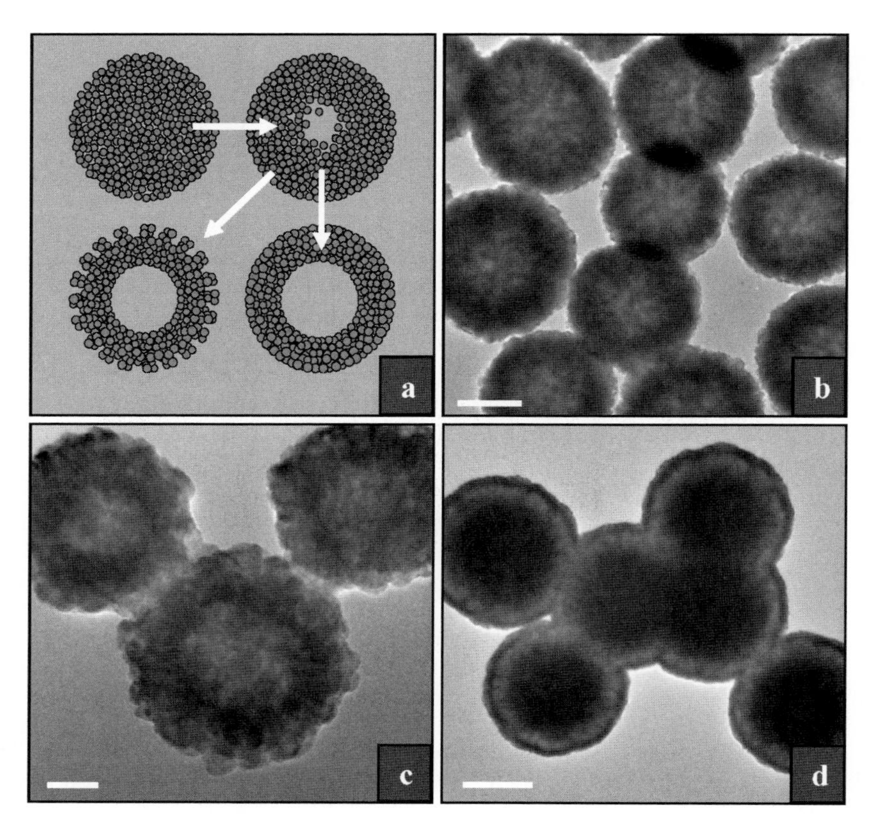

Fig. 2.9. (a) Schematic illustrations of Ostwald hollowing process. (b) TEM image of anatase TiO_2 hollow spheres. (c) TEM image of Cu_2O hollow spheres. (d) TEM image of ZnS core–shell structures with inter-space between core and shell. Scale bars in (b) to (d) are 500 nm, 50 nm, and 500 nm respectively.

Conventionally, the Kirkendall effect describes comparative diffusive migrations among different atomic species in metals and alloys under thermally activated conditions. This effect can be understood by looking at the fact that a zinc atom diffuses into the copper faster than copper diffuses into brass in a brass–copper interface due to their difference in atomic diffusivities [107–109]. Owing to the mass relocation, porosity can be thus obtained in the lower-melting component side of the diffusion couple. Nanoscale Kirkendall effect had been recently illustrated with the formations of Co_3S_4 and Co_9S_8 by reacting metallic cobalt nanoparticles with sulfur-containing compounds [101]. It was also demonstrated that yolk-shell nanostructures of CoO–Pt yolk shells were formed by oxidizing composite metal core-shells of Pt@Co in a stream of O_2/Ar gas mixture at 182°C [101]. The Kirkendall type diffusion had been recently extended to synthesis-cum-organization of nanomaterials. As a first example in this area, synthetic architecture of ZnO nanostructures was investigated with metallic zinc powders (in spheres) at different diameters, as indicated in Fig. 2.10a. In particular, the following two chemical reactions described additional chemical reactions taking place in the solid–solution

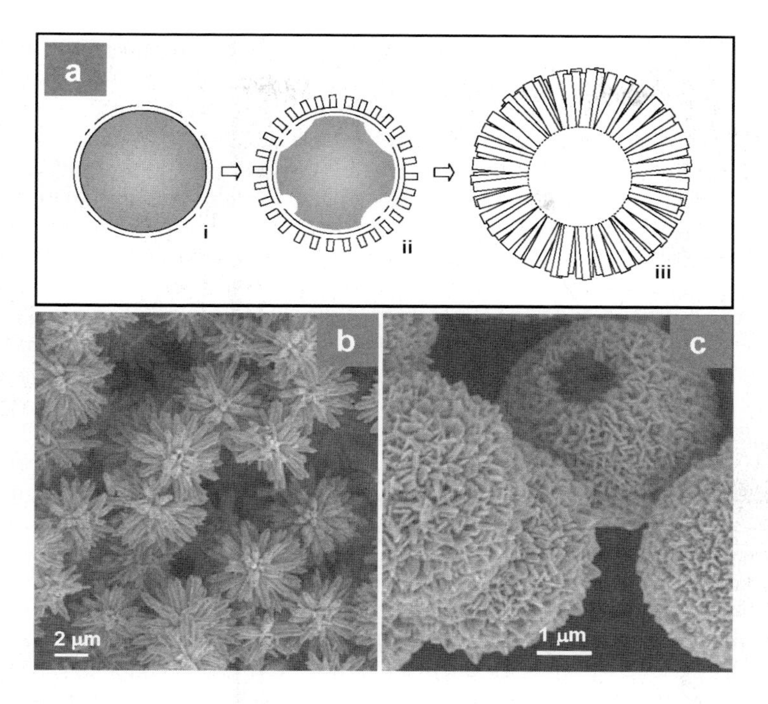

Fig. 2.10. (a) A schematic illustration of Kirkendall diffusion in forming hollow structures of ZnO: (i) starting zinc particle, (ii) surface nucleation and initial hollowing, and (iii) continued growth and complete core evacuation. (b) SEM image of ZnO dandelions formed with small zinc particles. (c) SEM image of hollow ZnO structures formed from larger zinc particles.

interface [102]:

$$Zn + 2OH^- \rightarrow ZnO_2{}^{2-} + H_2 \tag{2.4}$$

$$ZnO_2{}^{2-} + H_2O \leftrightarrow ZnO + 2OH^- \tag{2.5}$$

One important finding from this work is that in addition to the synthesis of hollow crystals, the Kirkendall-type diffusion can also be utilized for nanomaterials organization when additional interfacial chemical reactions are introduced. ZnO single crystalline nanorods (Fig. 2.10b) or nanoplatelets (Fig. 2.10c) can be formed and aligned simultaneously into dandelion-like arrays from metallic particles in a basic aqueous solution. The central parts of the final ZnO products are empty because zinc atoms in the precursor cores undergo an out-diffusion. As the shape and size of ZnO nanorods can be controlled, the Kirkendall-type diffusion is now thought to be a new means for synthetic architecture of hollow inorganic nanomaterials.

In the above two methods, Ostwald ripening and Kirkendall-type diffusion, the solid evacuation processes are somehow analogous to *sculpturing* or *caving* process. On the other hand, 'oriented attachment' is a truly *building* process from smaller primary units. In the latter process, small crystallites (such as colloidal suspensions) attach to each other through their suitable crystallographic planes [110, 111]. As a result of this combination, multidimensional-size multiplications can be attained. Recent investigations in this area were started with one- and two-dimensional attachments of TiO_2 and ZnO nanocrystals [110,112]. As indicated earlier, nanocrystals with well facets easily undergo this process, as can be seen in the Co_3O_4 nanocubes reported in Fig. 2.3c and the ZnO nanorods in Fig. 2.4b [59,63]. Apart from the length or size multiplication, 'oriented attachment' is also applicable to generate three-dimensional structures with interior space, using some lower-dimensional building blocks. As an example, oriented aggregated CuO nanoribbons can be synthesized in a copper nitrate solution ($Cu(NO_3)_2 \cdot 3H_2O$), together with ethanol, ammonia, and NaOH at 60–180°C [103]. It is found that the basic shape of the CuO crystal aggregates is rhombic, which allows a geometry-limiting construction of puffy CuO spheres [103], as reported in Fig. 2.11a. Quite interestingly, these dandelion-like structures were formed through a donut-like intermediate assembly, as rhombic-shaped structures can easily give rise to curved spherical surface [103]. In addition to the two-dimensional aggregation observed in CuO rhombic strips, nanocrystallites can also attach to one another into two-dimensional crystal sheets that can then turn into three-dimensional structures. This possibility was very recently demonstrated with self-construction of hollow SnO_2 octahedra [104]. In Fig. 2.11b, the polyhedrons of SnO_2 were prepared in a SnF_2 solution, together with water, 2-propanol, and ethylenediamine under 'one-pot' reaction conditions [104]. It is also revealed that the hollow octahedra were grown by a plane-by-plane process, starting with round nanocrystallites (3–5 nm). The hollow structures can be as small as 150 nm and as large as several micrometers (μm) with wall thickness in the range of 20–200 nm. Apparently, the presence of ethylenediamine, in combination with water–alcohol cosolvent, is a key to obtain small crystallite size as well as to stabilize their sheet formation. On the basis of these works, it is believed that there is great potential for 'oriented attachment' to be developed into a truly self-assembling methodology for the construction of inorganic nanomaterials.

2.2.4 *Synthetic architecture of complex nanostructures*

As reviewed in the previous subsections, numerous synthetic strategies have been in place for us to carry out investigations for even more complex inorganic nanostructures. In this subsection, we will use three recently reported syntheses to illustrate some important aspects of basic synthetic architecture for fabrication of complex nanostructures:

(a) *Structural control*: At the present time, many functional inorganic materials are prepared according to their intrinsic structural anisotropies. Control of structure, shape, and chirality of individual inorganic nanostructures is prevailingly achieved via *organic* capping assisted methods, which use various organic surfactants, liquid crystals, ligands, and chiral supramolecular assemblies to guide the growth directions. However, *inorganic* capping has been so far rarely touched in this area. In Fig. 2.12a, therefore, we will introduce an entirely inorganic approach, which is capable of controlling the structure and shape of complex nanostructures of α-MoO_3, in addition to incorporating a secondary phase [15].

In this synthetic scheme, the intrinsic structural anisotropy is utilized to control one-dimensional growth of α-MoO_3 nanorods with a hydrothermal synthesis [15], where the

Fig. 2.11. Oriented attachments in formation of hollow structures: (a) SEM image of a CuO dandelion (with a hollow interior) which is formed from smaller CuO nanoribbons. (b) FESEM image of a SnO_2 hollow octahedron which is constructed from tiny SnO_2 nanocrystallites.

growth rates are ranged in the following hierarchical order: $r_{001} \gg r_{100} \gg r_{010}$ [15]. This process involved an acidification of ammonium heptamolybdate tetrahydrate ($(NH_4)_6Mo_7O_{24} \cdot 4H_2O$) with nitric acid and formation of an intermediate crystal compound $((NH_4)_2O)_{0.0866} \cdot MoO_3 \cdot 0.23H_2O$ [67]. If a crystal of α-MoO_3 is capped with TiO_2 in the two $\langle 001 \rangle$ ends, the growth of α-MoO_3 nanorods in $\langle 001 \rangle$ is ceased and the growth in $\langle 100 \rangle$ is then promoted [15]. However, when the crystal along [100] exceeds the breadth of the TiO_2 nanocrystal caps, the growth along the $\langle 001 \rangle$ can be resumed,

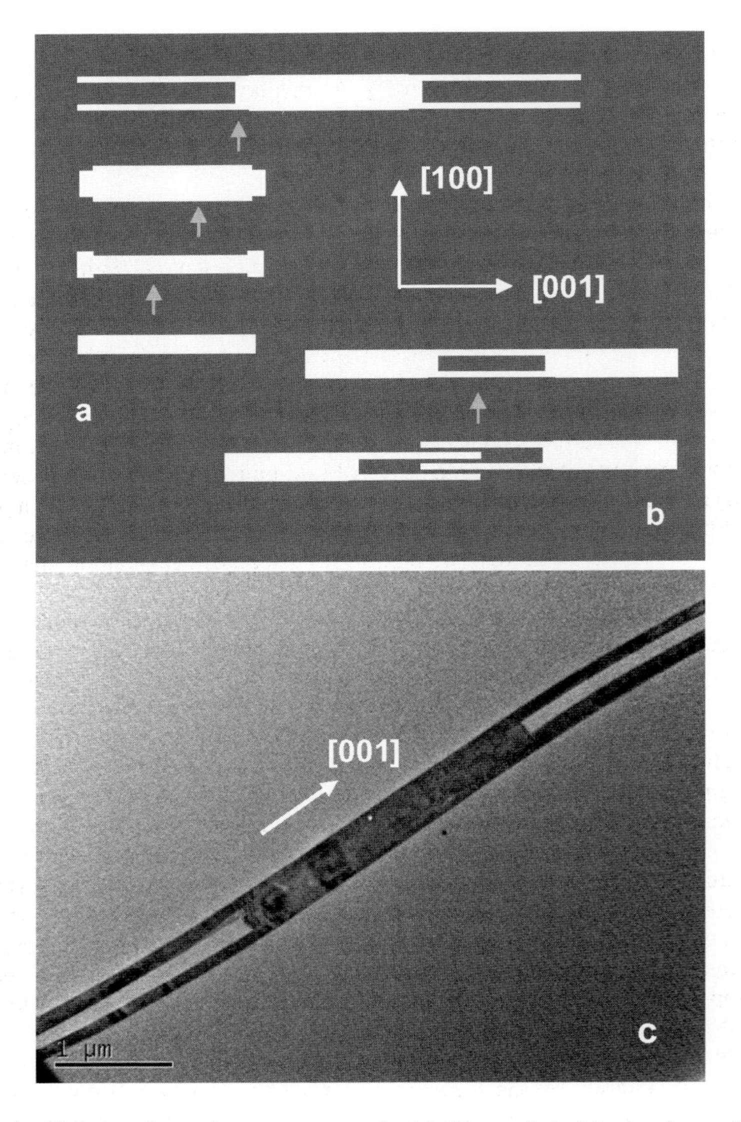

Fig. 2.12. Architecture of complex nanostructured α-MoO_3 nanofork: (a) route for synthetic architecture, (b) self-assembly of two nanostructures, and (c) TEM image of a double-ended α-MoO_3 nanofork. In (a) and (b), anatase TiO_2 capping is indicated in white, and α-MoO_3 nanoforks in light grey.

as the growth rate of ⟨001⟩ is much faster than that of ⟨100⟩. This structural manipulation will produce complex nanostructures of α-MoO_3, i.e. double-ended fork-like morphology, as depicted in Fig. 2.12a. Some intermediate crystals with anatase TiO_2 capping in this process. The epitaxial growth of TiO_2 on the α-MoO_3 ends as well as crystal orientations were confirmed with SAED investigations, which indicate that the *a*-axes of

TiO_2 crystals are parallel to the *a*- and *c*-axes of α-MoO_3 respectively, while the *c*-axis of TiO_2 is parallel to the *b*-axis of α-MoO_3. As displayed in Fig. 2.12c, double-ended α-MoO_3 nanoforks are formed when the growth along ⟨001⟩ is resumed; the desired product of this type can be as high as 90% [15].

α-MoO_3 is a typical layered structure material with two-dimensional 'molecular' sheets stacking along ⟨010⟩ directions by van der Waals interactions [113–115]. In this regard, these surfaces possess a strong tendency for external bonding. Indeed, it has been observed that the four arms of a nanofork can be combined in an inward manner, creating a new type of artificial morphology [15]. A number of combinations are noted. For example, as shown in Fig. 2.12b, four different arms approach each other from opposite directions, giving away a central space in the combined nanostructure. On the other hand, the main α-MoO_3 nanorod stems can simply stack together, while the secondary arms are freely dangling; this kind of final products can be considered as nanobrushes. In a certain sense, a four-armed nanofork is thought as a basic building unit in the observed one-dimensional expansion, which is essentially similar to the 'monomer' units in a polymeric molecule. Further investigation using XPS indicates that there is no atomic interdiffusion among Mo and Ti, owing to their significant difference in oxidation states [15]. As a whole, therefore, the final nanoforks are in fact a nanocomposite of MoO_3–TiO_2 in which TiO_2 is a secondary phase. Through this fabrication of complex oxide composite, more importantly, it is demonstrated that additional structural control on geometrical complexity can also be attained when introducing a minor material phase to the first.

(b) *Compositional control*: The above work on fabrication of MoO_3–TiO_2 composite nanoforks elucidates a basis for synthetic architecture of complex inorganic nanostructures through an entire solution synthesis. At present, there have been two new types of inorganic nanocomposites in addition to conventional 'matrix-plus-a-secondary-phase' composites [21]. In the first type of materials, two or more phases of nanoparticles are physically mixed. Among the second type of materials, nanocomposites are normally prepared into core–shell structures with spherical, belt-like, or rod-like composite units [45]. Apparently, the second type of nanocomposites allows a better mixing or more intimate contact among the different material phases. Concerning the fabrication of the second type of composites, solution approach can be considered, as we have seen in the synthesis of α-MoO_3–TiO_2 nanoforks [15]. Nonetheless, a number of issues related to the synthetic chemistry must be further considered. First of all, intrinsic chemical reactivity and lattice compatibility among various components should be addressed. Second, interactions among the forming materials and reaction environments should be investigated, as the reaction conditions may cause dissolution or conversion of existing phases. And third, the relationship between the composition and structural integration of composites should be further revealed, because both factors determine their ultimate performance. Herein we will use some step-by-step syntheses for the type II complex nanocomposites in order to address these fundamental issues.

As a backbone of these new composites, a degradable protonated pentatitanate ($H_2Ti_5O_{11}\cdot H_2O$) was first synthesized into a belt-like morphology through a hydrothermal route at 180°C followed by an ion exchange reaction at room temperature [45,116–128]. Figure 2.13a shows the nanobelts of $H_2Ti_5O_{11}\cdot H_2O$ synthesized. Taking these nanobelts as substrate, tiny anatase TiO_2 nanorods can be deposited on the surfaces

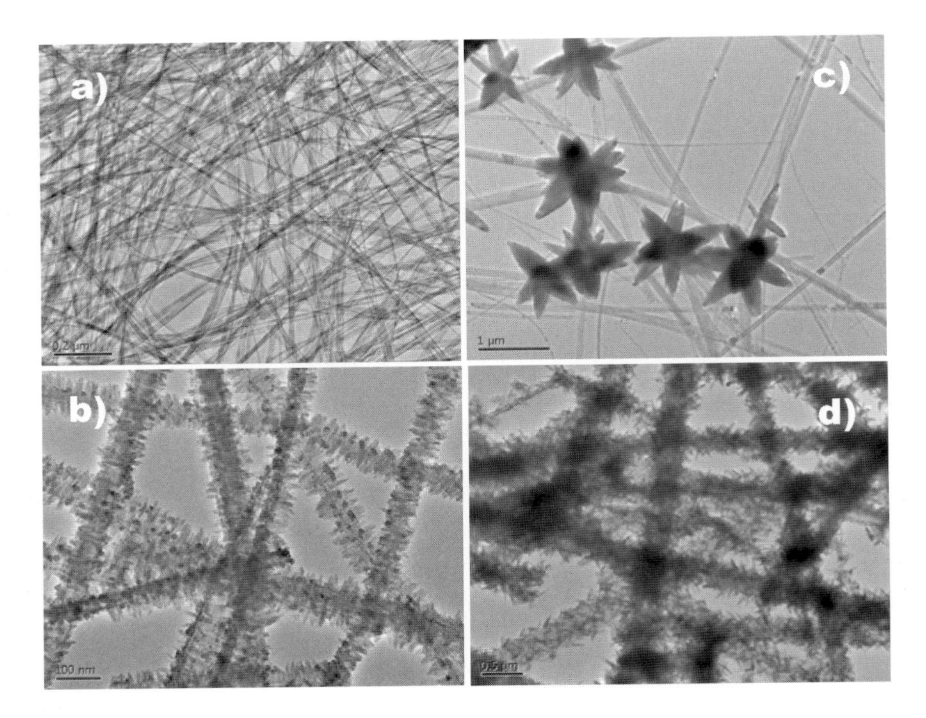

Fig. 2.13. Synthetic architecture of metal oxide nanocomposites: (a) starting $H_2Ti_5O_{11}{\cdot}H_2O$ nanobelts, (b) small TiO_2 nanorods on $H_2Ti_5O_{11}{\cdot}H_2O$ nanobelts, (c) ZnO nanoflowers on $H_2Ti_5O_{11}{\cdot}H_2O$ nanobelts, and (d) small ZnO nanorods on larger TiO_2 nanorods which were converted from $H_2Ti_5O_{11}{\cdot}H_2O$ nanobelts.

of nanobelts in a dilute aqueous solution of TiF_4 at 50°C (Fig. 2.13b) [45]. The diameter and length of the prepared TiO_2 nanorods on $H_2Ti_5O_{11}{\cdot}H_2O$ support can be controlled by varying the reaction time. Under the hydrothermal conditions, however, the prepared TiO_2–$H_2Ti_5O_{11}{\cdot}H_2O$ composites will be changed entirely to an anatase phase TiO_2 during which the substrate $H_2Ti_5O_{11}{\cdot}H_2O$ is dissolved or deposited onto the TiO_2 nanorods formed previously, resulting in a pure phase TiO_2 nanostructures [45]. This observation indicates that the substrate used in a synthesis may be degradable after serving as a template. By choosing a suitable template, hollow interior of a newly grown phase may be created [45]. On the other hand, ZnO nanoflowers, which usually have multiple rods, can also be grown on the surfaces of $H_2Ti_5O_{11}{\cdot}H_2O$ nanobelts, and ZnO–$H_2Ti_5O_{11}{\cdot}H_2O$ composites are thus obtained (Fig. 2.13c). This result shows that a new secondary phase can be introduced to a primary phase in a discrete manner and with proper morphological design.

Furthermore, ZnO can also be added onto the TiO_2–$H_2Ti_5O_{11}{\cdot}H_2O$ and TiO_2 reported earlier, from which ZnO–TiO_2–$H_2Ti_5O_{11}{\cdot}H_2O$ and ZnO–TiO_2 (Fig. 2.13d) can be fabricated [45]. The morphologies of these crystal composites reveal that when the compositional complexity of a nanocomposite increases, surface structures, crystal facets, and the lattice match and mismatch would then have to be investigated carefully, as they determine the overall structural aspects of the nanocomposites.

As shown in the above syntheses, compositionally complex inorganic nanostructures can be prepared with wet chemical approaches. In addition to their multiple solid phases, this class of nanocomposites belongs to lightweight materials since all the prepared nanocomposites display protruding structural features. Their puffy surface structures can generate certain interconnectivity among the nanostructures while keeping them at fixed distances, especially for the primary phase in the central part of the nanocomposites. More importantly, the above syntheses demonstrate the possibility of adding a new component in a one-at-a-time manner to an existing materials system. Thus, the main findings from this step-by-step process may serve as a general guide for compositional control in synthetic architecture of complex nanocomposites.

(c) *Structural and compositional control*: In the above two cases, we see that the structure and composition are two important aspects of nanocomposites. Although it is not possible to totally decouple them, very often, one of them may carry a higher weight than the other in a specific synthetic architecture, as discussed in the above two cases respectively. In the following discussion, we look at an even more complicated case where both structural control and compositional control play a more or less similar role in the design and synthesis of complex nanostructure.

Figure 2.14a presents a process flowchart in the synthesis of Au–TiO_2 nanocomposites that can be used as nanoreactors [129]: (1) preparation of colloidal solution of gold nanoparticles, which will be used as cores (metallic Au catalyst), (2) deposition of anatase TiO_2 nanoparticles on the Au core surface in solution and formation of Au–TiO_2 core–shell structure, (3) solid evacuation of TiO_2 crystallites located in the central part of the core–shell structures, which creates an interior space, and (4) tuning metallic Au cores to a desired size, that is, the growing Au particles inside TiO_2 hollow spheres, and (5) entirely filled TiO_2 spheres, where Au has fully occupied the interior space previously created in step (3). This proposed process has been in fact realized recently with a total synthetic architecture in solution reactions. In this synthesis, the metallic gold cores were prepared in a low concentration $HAuCl_4$ and sodium citrate solution at room temperature (step 1). Anatase TiO_2 was then deposited on gold surface by hydrolyzing a dilute TiF_4 solution (step 2), followed by the generation of central cavity under hydrothermal conditions at 180°C (step 3). Further control on the Au particle size inside TiO_2 hollow spheres can be found in the literature report (steps 4 and 5) [129]. Figure 2.14b displays some TEM images of the Au–TiO_2 nanocomposite synthesized using the above process. As the central TiO_2 crystallites were removed through a solution Ostwald ripening process [98], it is understandable that there must be intercrystallite interstitials between the central cavity and outer solution space [129]. Indeed, these communicable channels are observable in these TEM images as white contrast areas.

In view of their unique shell structure and catalytic cores, there are many potential applications for these novel nanostructures. For example, Fig. 2.14c illustrates one of the possible applications. Among the molecules with different sizes, only those permeable to the shell channels can have chances to access gold metallic cores and thus to be able

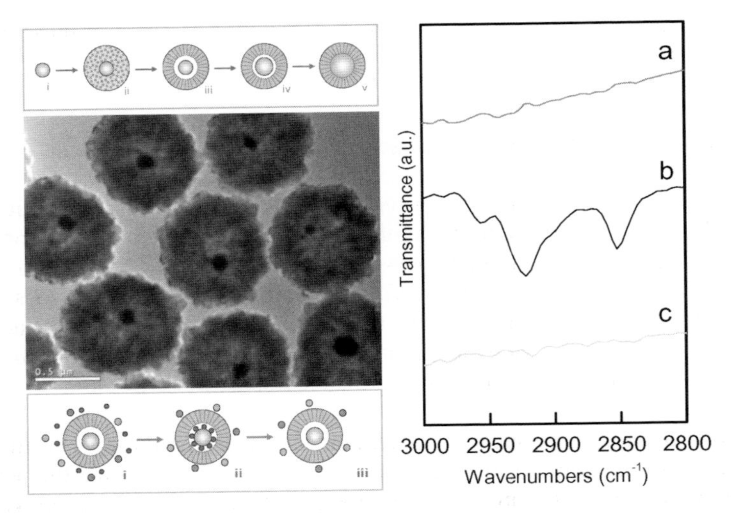

Fig. 2.14. Synthetic architecture of complex core-shell nanocomposites: the flowchart of synthesis (top inset); the product morphology in the stage (iii) of the top inset (TEM image); operating principle of nanoreactors (bottom inset): (i) Au-TiO_2 nanoreactors together with molecules at various sizes, (ii) only small-sized molecules can reach central metal catalyst, and (iii) photoinduced desorption and/or photocatalytic decomposition of the anchored molecules. FTIR spectra on the right: (a) clean nanoreactors, (b) after introducing $C_{12}H_{25}SH$ molecules, and (c) after photoinduced desorption and/or photocatalytic decomposition.

to modify core catalyst or to undergo metal-catalyzed reactions. This is further illustrated in the FTIR spectrum of Fig. 2.14 for the Au–TiO_2 sample modified with an aliphatic thiol ($C_{12}H_{25}SH$). As can be seen, the –SH ends of $C_{12}H_{25}SH$ were anchored on the gold surface, while the alkane ends are pointing toward the aqueous phase, from which surface hydrophobicity is attained [129]. Similarly, by introducing a carboxyl group to the thiol molecules, hydrophilicity is also attainable. Photocatalysis reactions have been further conducted in this type of nanoreactors. Indeed, the adsorbed thiolate anions had undergone photocatalytic reactions under UV irradiation (254 nm; optical energy band gap of $TiO_2 \approx 3.2$ eV); they were either desorbed and/or decomposed, as revealed by FTIR and XPS techniques [129]. It is understood that the holes generated in the valance band of TiO_2 are responsible for the oxidation of organic molecules while the electrons promoted to the conduction band may transfer to the Au core (near the Fermi level) and induce the observed desorption. Based on these investigations, it is now known that further chemical modification and size control of channels may lead to more effective screening of chemical species into the nanoreactors. On the other hand, further reduction in size of catalyst cores and better control of surface active sites of catalyst (not limited to metal only) will undoubtedly provide higher activity and selectivity for the heterogeneous catalytic reactions. It is also important to recognize that this type of photosensitized nanoreactors possesses excellent vehicle for the energy input into specific parts of a reactor without heating non-reacting chemical reactants. For example, the photoenergy (such as energetic electrons generated) converted is only directed to those adsorbed on the catalyst surface.

2.3 Conclusion and Future Directions

In summary, based on the above review for simple single component and multicomponent nanomaterials, we can see a significant increase in structural complexity when compositional complexity increases. In this regard, architectural consideration and design for the nanomaterials have to be investigated along the line of materials synthesis. With more stringent future compositional and structural requirements imposed, a certain degree of aesthetic expression for nanomaterials will become naturally desirable in addition to compositional and structural concerns on the targeted materials.

It is noted that most inorganic nanomaterials synthesized at the present time are only single-component. Therefore, multicomponent materials will become the next immediate research area to meet more challenging applications. Through the above examples, it is also clear that wet chemical synthesis is a versatile approach for the preparation of complex inorganic nanostructures. In this connection, the search for new organizing schemes and synthetic methods should be continued. Furthermore, chemistry of liquid–solid interface should be investigated in a greater depth. This will involve investigations of chemical reactivity of crystal plane under a set of reaction parameters (such as chemical additives used, reaction medium, pH value, pressure, and temperature, etc), and of size- and direction-controlled growths of nanostructures in solution. Future research endeavor should be continuously devoted to, though not limited to, wet chemical synthesis of nanostructures. A number of important issues in this area are needed to be further addressed: (i) chemical compatibility among the different solid components; (ii) structural (lattice) compatibility among different solid phases in a designed nanocomposite; (iii) surface chemistry of adsorbed ligands and surfactants on studied materials; (iv) chemical stability of resultant nanostructures in their forming environments; and (v) sustainability of prepared nanomaterials under working conditions. All these factors will determine our material selection and synthetic strategy, and they deserve fuller investigations in future synthetic architecture.

As a long-term research effort in this area, traditional and contemporary concepts of general architecture should be exploited and rejuvenated constantly in our design and synthesis of inorganic materials. An important note to be mentioned is that unlike the macroscopic construction for a building, which always starts from the Earth's surface (even for suspended roofs in any constructs), nanoscopic constructions can be started from all directions since nanostructures can freely rotate in solution media. It has been witnessed that new technologies have generated dramatic impacts on new architecture over the past century. The impact of technological change on buildings and structures can be seen everywhere, from high-tech skyscrapers, long-span suspension bridges and airport terminals, to monorails for magnetically propelled bullet trains. Similarly, new synthetic methods, though they have been accumulated rapidly over the past two decades, should be further explored and devised. In any case, synthetic architecture should be taken into consideration and the aesthetic expression for final products is no longer a luxury in our materials design and synthesis.

Acknowledgments

Thanks are directed to R. Xu, J. Feng, H.G. Yang, Y. Chang, B. Liu, X.W. Lou, J.J. Teo, M.L. Lyn, S.M. Liu, L.H. Liu, Y. Wang, and J.Y. Lee for their contributions to the work reviewed in this article. The author gratefully acknowledges the research funding supported by the National University of Singapore and the Ministry of Education, Singapore.

References

[1] C.N.R. Rao and A.K.J. Cheetham, Mater. Chem. 11 (2001) 2887.

[2] C.N.R. Rao, G.U. Kulkarni, P.J. Thomas and P.P. Edwards, Chem. Eur. J. 8 (2002) 29.

[3] G.R. Patzke, F. Krumeich and R. Nesper, Angew. Chem. Int. Ed. 41 (2002) 2446.

[4] F. Caruso and R.A. Caruso and H. Möhwald, Science 282 (1998) 1111.

[5] M.P. Zach, K.H. Ng and R.M. Penner, Science 290 (2000) 2120.

[6] K.P. Velikov, C.G. Christova, R.P.A. Dullens and A. van Blaaderen, Science 296 (2002) 106.

[7] H. Cölfen and S. Mann, Angew. Chem. Int. Ed. 42 (2003) 2350.

[8] V.R. Thalladi and G.M. Whitesides, J. Am. Chem. Soc. 124 (2002) 3520.

[9] S. Kobayashi, N. Hamasaki, M. Suzuki, M. Kimura, H. Shirai and K. Hanabusa, J. Am. Chem. Soc. 124 (2002) 6550.

[10] S. Park, J.-H. Lim, S.-W. Chung and C.A. Mirkin, Science 303 (2004) 348.

[11] N.I. Kovtyukhova and T.E. Mallouk, Chem. Eur. J. 8 (2002) 4355.

[12] M.P. Pileni, J. Phys. Chem. B 105 (2001) 3358.

[13] A. Sugawara, T. Ishii and T. Kato, Angew. Chem. Int. Ed. 42 (2003) 5299.

[14] J.T. Sampanthar and H.C. Zeng, J. Am. Chem. Soc. 124 (2002) 6668.

[15] X.W. Lou and H.C. Zeng, J. Am. Chem. Soc. 125 (2003) 2697.

[16] F. Krumeich, H.-J. Muhr, M. Niederberger, F. Bieri, B. Schnyder and R. Nesper, J. Am. Chem. Soc. 121 (1999) 8324.

[17] J.Y. Ying, C.P. Mehnert and M.S. Wong, Angew. Chem. Int. Ed. 1999, 38, 56.

[18] J.M. Thomas, Angew. Chem. Int. Ed. 38 (1999) 3588.

[19] M.E. Spahr, P. Bitterli, R. Nesper, M. Müller, F. Krumeich and H.-U. Nissen, Angew. Chem. Int. Ed. 37 (1998) 1263.

[20] K.J.C. van Bommel, A. Friggeri and S. Shinkai, Angew. Chem. Int. Ed. 42 (2003) 980.

[21] H.C. Zeng, in Handbook of Organic-Inorganic Hybrid Materials and Nanocomposites, Vol. 2: Nanocomposites, American Scientific Publishers: New York, 2003; Ch. 4, pp. 151–180.

[22] H. Zeng, J. Li, J.P. Liu, Z.L. Wang and S. Sun, Nature 420 (2002) 395.

[23] F.X. Redl, K.-S. Cho, C.B. Murray and S. O'Brien, Nature 423 (2003) 968.

[24] B.H. Hong, S.C. Bae, C.-W. Lee, S. Jeong and K.S. Kim, Science 294 (2001) 348.

[25] L.J. Lauhon, M.S. Gudiksen, D. Wang and C.M. Lieber, Nature 420 (2002) 57.

[26] J. Sloan, M. Terrones, S. Nufer, S. Friedrichs, S.R. Bailey, H.-G. Woo, M. Rühle, J.L. Hutchison and M.L.H. Green, J. Am. Chem. Soc. 124 (2002) 2116.

[27] M. Wilson, Nano Lett. 4 (2004) 299.

[28] Y. Wu, R. Fan and P. Yang, Nano Lett. 2 (2002) 83.

[29] R. He, M. Law, R. Fan, F. Kim and P. Yang, Nano Lett. 2 (2002) 1109.

[30] M.S. Gudiksen, L.J. Lauhon, J. Wang, D.C. Smith and C.M. Lieber, Nature 415 (2002) 617.

[31] J. Pyun and K. Matyjaszewski, Chem. Mater. 13 (2001) 3436–3448.

[32] A.R. Urbach, J.C. Love, M.G. Prentiss, G.M. Whitesides, J. Am. Chem. Soc. 125 (2003) 12704.

[33] T. Mokari, E. Rothenberg, I. Popov, R. Costi and U. Banin, Science 304 (2004) 1787.

[34] D. Wu, X. Ge, Z. Zhang, M. Wang and S. Zhang, Langmuir 20 (2004) 5192.

[35] M. Danek, K.F. Jensen, C.B. Murray and M.G. Bawendi, Chem. Mater. 8 (1996) 173.

[36] L. Manna, E.C. Scher, L.-S. Li and A.P. Alivisatos, J. Am. Chem. Soc. 124 (2002) 7136.

[37] J. Hwang, B. Min, J. S. Lee, K. Keem, K. Cho, M.-Y. Sung, M.-S. Lee and S. Kim, Adv. Mater. 16 (2004) 422.

[38] C.J. Barrelet, Y. Wu, D.C. Bell and C.M. Lieber, J. Am. Chem. Soc. 125 (2003) 11498.

[39] C. Burda, X. Chen, R. Narayanan and M.A. El-Says, Chem. Rev. 105 (2005) 1025.

[40] J.C. Love, L.A. Estroff, J.K. Kriebel, R.G. Nuzzo and G.M. Whitesides, Chem. Rev. 105 (2005) 1103.

[41] A.T. Bell, Science 299 (2003) 1688.

[42] M. Valden, X. Lai and D.W. Goodman, Science 281 (1998) 1647.

[43] H.C. Zeng, in The Dekker Encyclopedia of Nanoscience and Nanotechnology; Marcel Dekker: New York, 2004; pp. 2539–2550.

[44] F.A. Cotton and G. Wilkinson, Advanced Inorganic Chemistry; 6th Ed. (with a chapter on boron by R. Grimes); John Willey & Sons: New York, 1999.

[45] H.G. Yang and H.C. Zeng, J. Am. Chem. Soc. 127 (2005) 270.

[46] H.C. Zeng, J. Mater. Chem. 16 (2006) 649.

[47] Y. Xia, P. Yang, Y. Sun, Y. Wu, B. Mayers, B. Gates, Y. Yin, F. Kim and H. Yan, Adv. Mater. 15 (2003) 353.

[48] K.G. Libbrecht, Engineering & Sci. (2001) 10.

[49] J. Glancey, 20th Century Architecture: the structures that shaped the century, Overlook Press: Woodstock, New York, 1999.

[50] P. Guedes, The Macmillan Encyclopedia of Architecture and Technological Change, Macmillan: London, 1979.

[51] G.M. Whitesides and B. Grzybowski, Science 295 (2002) 2418.

[52] J.-M. Lehn, Science 295 (2002) 2400.

[53] D.N. Reinhoudt and M. Crego-Calama, Science 295 (2002) 2403.

[54] O. Ikkala and G. ten Brinke, Science 295 (2002) 2407.

[55] T. Sugimoto and E. Matijević, J. Inorg. Nucl. Chem. 41 (1979) 165.

[56] E. Matijević, Chem. Mater. 4 (1993) 412.

[57] R. Xu and H.C. Zeng, J. Phys. Chem. B 107 (2003) 926.

[58] J. Feng and H.C. Zeng, Chem. Mater. 15 (2003) 2829.

[59] R. Xu and H.C. Zeng, Langmuir 20 (2004) 9780.

[60] J. Feng and H.C. Zeng, J. Phys. Chem. B 109 (2005) in press.

[61] Z.W. Pan, Z.R. Dai and Z.L. Wang, Science 291 (2001) 1947.

[62] P. Gao and Z.L. Wang, J. Phys. Chem. B 106 (2002) 12653.

[63] B. Liu and H.C. Zeng, J. Am. Chem. Soc. 125 (2003) 4430.

[64] B. Liu and H.C. Zeng, Langmuir 20 (2004) 4196.

[65] Y. Chang and H.C. Zeng, Cryst. Growth Design 4 (2004) 397.

[66] X.W. Lou and H.C. Zeng, Inorg. Chem. 42 (2003) 6169.

[67] X.W. Lou and H.C. Zeng, Chem. Mater. 14 (2002) 4781.

[68] Y. Chang, M.L. Lye and H.C. Zeng, Langmuir 21 (2005) 3746.

[69] Y. Chang and H.C. Zeng, Cryst. Growth Design 4 (2004) 273.

[70] T. Ishikawa, H. Yamaoka, Y. Harada, T. Fujii and T. Nagasawa, Nature 416 (2002) 64.

[71] H.G. Yang and H.C. Zeng, J. Phys. Chem. B 107 (2003) 12244.

[72] H.G. Yang and H.C. Zeng, Chem. Mater. 15 (2003) 3113.

[73] H.G. Yang and H.C. Zeng, J. Phys. Chem. B 108 (2004) 819.

[74] R. Xu and H.C. Zeng, Chem. Mater. 13 (2001) 297.

[75] H.C. Zeng and X.Y. Pang, Appl. Catal. B 13 (1997) 113.

[76] J.T. Sampanthar and H.C. Zeng, Chem. Mater. 13 (2001) 4722.

[77] L. Ji, J. Lin and H.C. Zeng, Chem. Mater. 13 (2001) 2403.

[78] X.M. Wei and H.C. Zeng, Chem. Mater. 15 (2003) 433.

[79] X.M. Wei and H.C. Zeng, J. Phys. Chem. B 107 (2003) 2619.

[80] C.T. Kresge, M.E. Leonowicz, W.J. Roth, J.C. Vartuli and J.S. Beck, Nature 359 (1992) 710.

[81] K. Moller and T. Bein, Chem. Mater. 10 (1998) 2950.

[82] R.A. Caruso, Angew. Chem. Int. Ed. 43 (2004) 2746.

[83] H.G. Yang and H.C. Zeng, Angew. Chem. Int. Ed. 43 (2004) 5206.

[84] C.R. Martin, Science 266 (1994) 1961.

[85] B.B. Lakshmi, P.K. Dorhout and C.R. Martin, Chem. Mater. 9 (1997) 857.

[86] D.T. Mitchell, S.B. Lee, L. Trofin, N.C. Li, T.K. Nevanen, H. Soderlund and C.R. Martin, J. Am. Chem. Soc. 124 (2002) 11864.

[87] Z. Liang, A.S. Susha, A. Yu and F. Caruso, Adv. Mater. 15 (2003) 1849.

[88] J. Chen, Z.L. Tao and S.L. Li, J. Am. Chem. Soc. 126 (2004) 3060.

[89] M. Lahav, T. Sehayek, A. Vaskevich and I. Rubinstein, Angew. Chem. Int. Ed. 42 (2003) 5576.

[90] M. Steinhart, R.B. Wehrspohn, U. Gösele and J.H. Wendorff, Angew. Chem. Int. Ed. 43 (2004) 1334.

[91] B. Liu and H.C. Zeng, J. Phys. Chem. B 108 (2004) 5867.

[92] S.M. Liu, L.M. Gan, L.H. Liu, W.D. Zhang and H.C. Zeng, Chem. Mater. 14 (2002) 1391.

[93] Y. Wang, J.Y. Lee and H.C. Zeng, Chem. Mater. 17 (2005) 3899.

[94] T. Nakashima and N. Kimizuka, J. Am. Chem. Soc. 125 (2003) 6386.

[95] C.-W. Guo, Y. Cao, S.-H. Xie, W.-L. Dai and K.-N. Fan, Chem. Commun. (2003) 700.

[96] H.J. Hah, J.S. Kim, B.J. Jeon, S.M. Koo and Y.E. Lee, Chem. Commun. (2003) 1712.

[97] P. Afanasiev and I. Bezverkhy, J. Phys. Chem. B 107 (2003) 2678.

[98] H.G. Yang and H.C. Zeng, J. Phys. Chem. B 108 (2004) 3492.

[99] Y. Chang, J.J. Teo and H.C. Zeng, Langmuir 21 (2005) 1074.

[100] B. Liu and H.C. Zeng, Small 1 (2005) 566.

[101] Y. Yin, R.M. Rioux, C.K. Erdonmez, S. Hughes, G.A. Somorjai and A.P. Alivisatos, Science 304 (2004) 711.

[102] B. Liu and H.C. Zeng, J. Am. Chem. Soc. 126 (2004) 16744.

[103] B. Liu and H.C. Zeng, J. Am. Chem. Soc. 126 (2004) 8124.

[104] H.G. Yang and H.C. Zeng, Angew. Chem. Int. Ed. 43 (2004) 5930.

[105] W. Ostwald, Z. Phys. Chem. 22 (1897) 289.

[106] W. Ostwald, Z. Phys. Chem. 34 (1900) 495.

[107] E. Kirkendall, L. Thomassen and C. Upthegrove, Trans. AIME 133 (1939) 186.

[108] E.O. Kirkendall, Trans. AIME 147 (1942) 104.

[109] A.D. Smigelskas and E. O. Kirkendall, Trans. AIME 171 (1947) 130.

[110] R.L. Penn and J.F. Banfield, Science 281 (1998) 969.

[111] J.F. Banfield, S.A. Welch, H.Z. Zhang, T.T. Ebert and R.L. Penn, Science 289 (2000) 751.

[112] C. Pacholski, A. Kornowski and H. Weller, Angew. Chem. Int. Ed. 41 (2002) 1188.

[113] Z.Y. Hsu and H.C. Zeng, J. Phys. Chem. B 104 (2000) 11891.

[114] H.C. Zeng, W.K. Ng, L.H. Cheong, F. Xie and R. Xu, J. Phys. Chem. B 105 (2001) 7178.

[115] H.C. Zeng, F. Xie, K.C. Wong and K.A.R. Mitchell, Chem. Mater. 14 (2002) 1788.

[116] Q. Chen, W. Zhou, G. Du and L. Peng, Adv. Mater. 14 (2002) 1208.

[117] T. Sasaki, Y. Komatsu and Y. Fujiki, Chem. Mater. 4 (1992) 894.

[118] H. Izawa, S. Kikkawa and M. Koizumi, J. Phys. Chem. B 86 (1982) 5023.

[119] Y. Zhu, H. Li, Y. Koltypin, Y.R. Hacohen and A. Gedanken, Chem. Commun. (2001) 2616.

[120] S. Yin, S. Uchida, Y. Fujishiro, M. Aki and T. Sato, J. Mater. Chem. 9 (1999) 1191.

[121] T. Sasaki, M. Watanabe, Y. Fujiki and Y. Kitami, Chem. Mater. 6 (1994) 1749.

[122] T. Sasaki, F. Izumi and M. Watanabe, Chem. Mater. 8 (1996) 777.

[123] T. Sasaki, M. Watanabe, Y. Komatsu and Y. Fujiki, Inorg. Chem. 24 (1985) 2265.

[124] Z.R. Tian, J.A. Voigt, J. Liu, B. Mckenzie and H. Xu, J. Am. Chem. Soc. 125 (2003) 12384.

[125] X. Sun and Y. Li, Chem. Eur. J. 9 (2003) 2229.

[126] H. Zhu, X. Gao, Y. Lan, D. Song, Y. Xi and J. Zhao, J. Am. Chem. Soc. 126 (2004) 8380.

[127] S. Cheng and T. Wang, Inorg. Chem. 28 (1989) 1283.

[128] M.W. Anderson and J. Klinowski, Inorg. Chem. 29 (1990) 3260.

[129] J. Li and H.C. Zeng, Angew. Chem. Int. Ed. 44 (2005) 4342.

CHAPTER 3

Self-Assembled Monolayer on Silicon

Hiroyuki Sugimura
Department of Materials Science and Engineering
Kyoto University
Sakyo, Kyoto 606-8501, Japan

3.1 Introduction

Self-assembling, in which minute elements such as atoms, molecules, and clusters are spontaneously organized into an ordered array of the elements, is a key process of the bottom–up nanotechnology [1–4]. One of the promising material processes on the basis of self-assembly is the fabrication of organic thin films with a monomolecular thickness. Although it has been well-known that some types of organic molecules adsorb on a particular substrate and form a monolayer since several decades [5–7], such types of organic monolayers have recently been named as self-assembled monolayers (SAMs). As schematically illustrated in Fig. 3.1, when a particular substrate is immersed in a solution of precursor molecules which have a chemical reactivity to the substrate surface, the molecules chemisorb on the surface with their reactive sites facing the surface. Most of the SAM precursors have a long alkyl chain or an aromatic ring so that some types of intermolecular interactions, e.g. van der Waals, hydrophobic, and π-electron interactions, work between the chemisorbed molecules. Consequently, the molecules are attracted to each other so as to be closely packed and to form a thin and uniform film with a monomolecular thickness. Due to immobilization of the molecules to the substrate through chemical bondings and the presence of intermolecular attractive interactions, SAMs are more stable mechanically, chemically, and thermodynamically compared with similar monolayers fabricated by the Langmuir–Blodgett technique.

Once a SAM is formed on a substrate, its surface is entirely covered with organic molecules. Since there is no room on the surface to be further adsorbed with molecules, the SAM growth stops automatically at this stage. The thickness of the SAM is determined with the length of the precursor molecules and an adsorption angle of the molecules. There is no need for the precise control of process conditions in order to

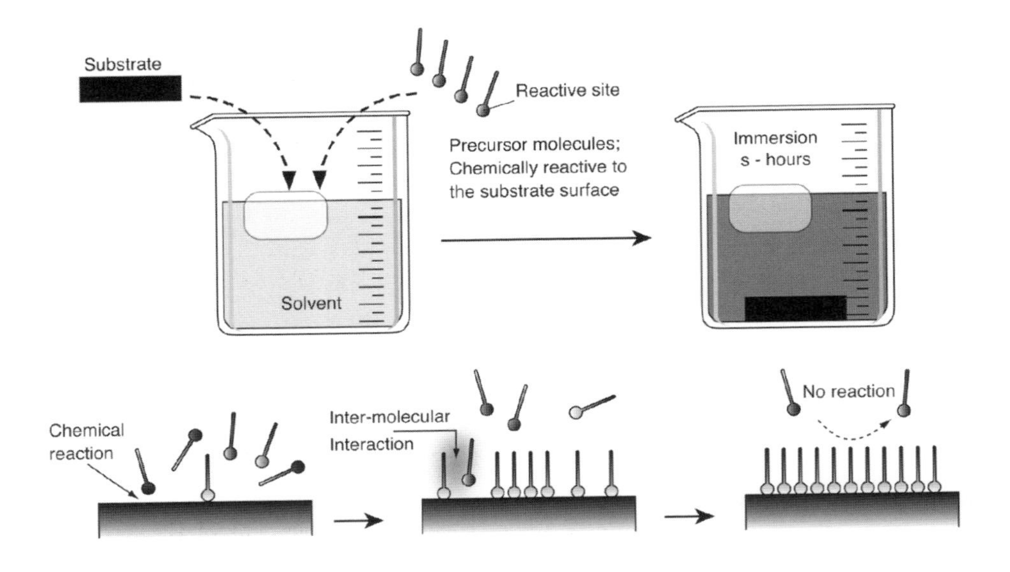

Fig. 3.1. Preparation procedure for self-assembled monolayer (SAM).

fabricate molecular level ultrathin films with their thicknesses in the range of 1–2 nm. Besides thickness, the presence of a SAM on a substrate alters surface physical and chemical properties being markedly different from those of the bare substrate.

However, the formation of SAMs depends on unique chemical reactions between a substrate and organic molecules. Hence, specific pairs of a substrate and a precursor are required to fabricate SAMs. Typical examples for these SAM formation pairs are summarized in Table 3.1. Among various materials, Si is most important in micro- and nano electronics and mechatronics. Thus, SAMs on Si are of special interest in order to integrate Si micro- and nanodevices with organic molecules. As described in Section 3.2, there are two major methods for preparing SAMs on Si.

3.2 Self-Assembled Monolayers on Si

3.2.1 SAM formation on oxide-covered Si through the silane coupling chemistry

A molecule consisting of one Si atom connected with four functional groups, SiX_4, is named as 'silane.' A molecule in which at least one of these four functional groups are substituted with organic functional groups, i.e. SiR_nX_{4-n}, is organosilane. Organosilane molecules react with hydroxyl groups on an oxide surface so as to be fixed on the surface as illustrated in Fig. 3.2. This surface modification chemistry has been practically used as the silane coupling reaction for preparing organic layers on inorganic material surfaces [8]. Sagiv and co-workers have reported, for the first time, that SAMs could be formed from organosilane molecules with one long alkyl chain in each of the molecules [9,10].

Table 3.1. Pairs of precursors and substrates.

Precursor molecules	Substrate materials
Organosulfurs alkylthiol: R–SH dialkyldisulfide: RS–SR′ thioisocyanide: R–SCN etc.	Metals/compound semiconductors Au, Ag, Cu, Pt, Pd, Hg, Fe, GaAs, InP
Fatty acid R–COOH	Oxide Al_2O_3, AgO, CuO
Phosphonic acid $R–PO_3H_2$	Oxide ZrO_2, TiO_2, Al_2O_3, Ta_2O_5, etc.
Organosilanes: $R–SiX_3$ ($X = Cl, OCH_3, OC_2H_5$)	Oxide glass, mica, SiO_2, SnO_2, GeO_2, ZrO_2, TiO_2, Al_2O_3, ITO, PZT, etc.
Unsaturated hydrocarbons alkene, alkyne: $R–CH = CH_2$, $R–C{\equiv}H$ Alchols, Aldehydes R–OH, R–CHO	Silicon hydrogen-terminated Si: Si–H halogen-terminated Si: Si–X (X = Cl, Br, I)

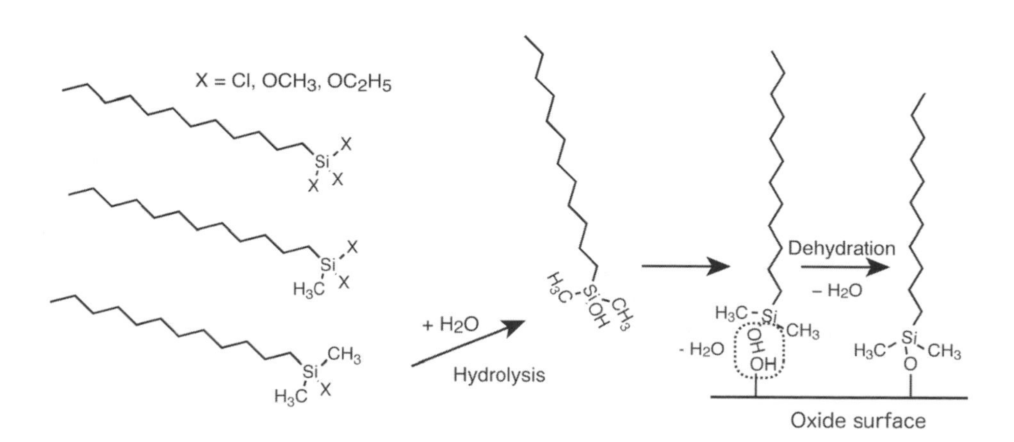

Fig. 3.2. Organosilane SAMs.

On a Si substrate, such a SAM can be formed as well by covering the substrate with its oxide.

As shown in Fig. 3.2, a trace amount of water is necessary to form organosilane SAMs. Halogen or alcohoxy groups in an organosilane molecule are converted to hydroxyl (–OH) groups by hydrolysis. Dehydration reaction between these silanol (Si–OH) sites in the molecule with –OH groups on the surface oxide of a Si substrate immobilizes the molecules on the oxide through siloxane (Si–O–Si) bonds.

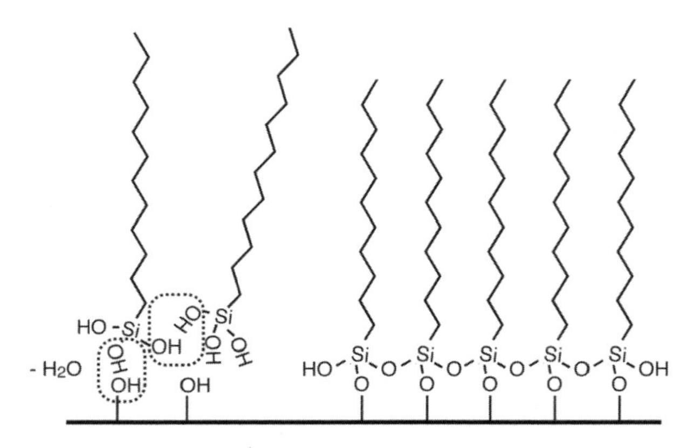

Fig. 3.3. SAM formation from tri-functional organosilane molecules.

There are three types of organosilane precursor molecules which have one, two, or three reactive sites. When a SAM is formed from organosilane molecules with a single reactive site, the molecular density assembled into the SAM remains to be relatively low due to steric hindrance between methyl ($-CH_3$) groups of adjacent molecules. On the other hand, when a SAM is formed from organosilane molecules each of which has three reactive sites, its growth behavior is complicated somewhat. Since two Si–OH groups remain in the organosilane molecule after that is immobilized on the substrate, the molecule is further linked with adjacent organosilane molecules through Si–O–Si bonds as illustrated in Fig. 3.3. This type of organosilane SAM is more closely packed and stable mechanically, chemically, and thermally due to the siloxane network laterally connecting molecules in the SAM in addition to the chemical attachment to the substrate.

3.2.2 SAMs directly bonded to Si

If a SAM can be formed on a semiconductor surface, it is expected to be a springboard for integrating functionalities of semiconductors and organic molecules. Although the silane-coupling method described in Section 3.2.1 is powerful and indispensable for the preparation of SAMs on Si, the organosilane SAMs have a disadvantage from the viewpoint of electronic applications. Such SAMs require the presence of a thin oxide layer, namely, a very good insulator, of 1–2 nm in thickness at least on Si substrates. Therefore, in this case, we can utilize electronic functions of the SAMs only in the situation inserting the insulator between SAM and Si. An alternative technology is needed in order to form SAMs on Si without its surface oxide.

Such a process technique has been reported in 1993 by Linford and Chidsey [11] and, thereafter, extended markedly by them and other researchers [12]. In general, Si radicals are first formed by extracting hydrogen atoms from a hydrogen-terminated Si surface or halogen atoms from a halogen-terminated Si surface, usually by the use of a reaction initiator, thermal excitation, or photo irradiation. Some types of organic molecules are covalently immobilized on Si through a reaction with the Si radicals. As shown in Fig. 3.4,

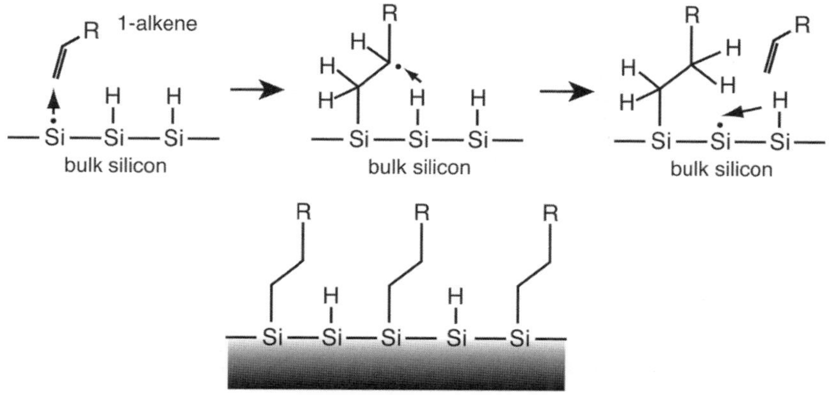

Fig. 3.4. SAM formation on Si through the radical reaction.

for example, an 1-alkene molecule reacts with a mono-hydrated Si(111) surface and is connected to the surface through a Si–C bond. The remaining C radical extracts a H atom from an adjacent Si–H resulting in the formation of a Si radical again. The chain reaction proceeds by repeating these reactions. Steric hindrance between the organic molecules limits the replacement degree of H with R to be about 50% at most [13].

3.3 Vapor Phase Growth of Organosilane SAM

3.3.1 SAM formation at the vapor–solid interface

Organosilane SAMs are formed on OH-bearing oxide surfaces through the chemical reaction of organosilane molecules with the –OH sites. The SAMs are usually prepared at the liquid–solid interface by simply immersing a substrate in a solution of precursor molecules [10,14]. Besides this liquid phase process, the vapor phase process [15–24] is also promising particularly because it has no need for the use of a large amount of solvents which is necessary in the liquid phase processes. Furthermore, particulate deposits of aggregated organosilane molecules, which frequently degrades the quality of the SAMs, is expected to be fewer in the vapor phase method than in the liquid phase method, since such aggregated molecules have lower vapor pressures and are rarely vaporized. Thus, the vapor phase method is considered to be practically convenient, although at present the method is not widely applicable since it depends on whether a precursor can vaporize or not, and the molecular ordering of vapor-grown SAMs is inferior to that of liquid-grown SAMs [23].

3.3.2 Preparation procedure

Organosilane SAMs were formed through a simple method described elsewhere [21]. Here, we demonstrate results on three types of precursors, i.e. n-octadecyltrimethoxysilane [ODS: $H_3C(CH_2)_{17}Si(OCH_3)_3$], n-(6-aminohexyl)aminopropyltrimethoxysilane

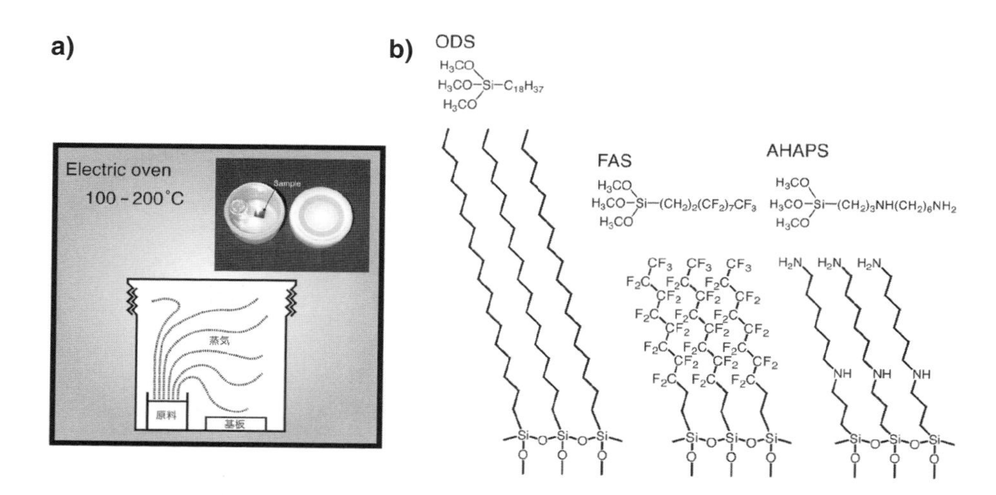

Fig. 3.5. a) Organosilane vapor treatment in a closed vessel, b) Chemical structures of precursor molecules and corresponding SAMs.

[AHAPS: $H_2N(CH_2)_6NH(CH_2)_3Si(OCH_3)_3$], and fluoroalkylsilane (FAS), heptadeca-fluoro-1,1,2,2-tetrahydro-decyl-1-trimethoxysilane [$F_3C(CF_2)_7(CH_2)_2Si(OCH_3)_3$]. The chemical structures of these precursors and their corresponding SAMs are shown in Fig. 3.5b. A photocleaned SiO_2/Si plate was placed together with a glass cup filled with organosilane liquid in a Teflon^{TM} container. When ODS and FAS were employed, the container was sealed with a cap and placed in an oven maintained at 150°C. In the case of AHAPS, organosilane liquid was diluted with toluene under dry N_2 atmosphere in order to avoid gelation of AHAPS through polymerization. A lower temperature of 100°C was employed for AHAPS in order to minimize polymerization of AHAPS in toluene. Subsequently, each of the samples treated with AHAPS as sonicated for 20 min in dehydrated ethanol and dehydrated toluene in succession. Then, the samples were further sonicated in NaOH (1 mM) and HNO_3 (1 mM) in order to remove excessively adsorbed AHAPS molecules. Finally, the samples were rinsed with Milli-Q water and were blown dry with a N_2 gas stream.

3.3.3 Vapor-grown organosilane SAMs

Organosilane molecules in a vapor or liquid phase react with –OH groups on an oxide surface resulting in the formation of a SAM as illustrated in Fig. 3.3. Figure 3.5a and b follow the formation of ODS-, FAS-, and AHAPS–SAMs. When a SiO_2/Si substrate is treated with ODS or FAS, it becomes hydrophobic as shown in Fig. 3.2. The water contact angles of the ODS- and FAS-treated substrates increase with an increase in reaction time at the initial stage. However, they hardly increase even when the process is prolonged for more than 3 and 1 h, in the cases of ODS- and FAS-treated substrates, respectively. The water contact angles of the ODS- and FAS-treated substrates reached 105° and 112°C, respectively. X-ray photoelectron spectroscopy XPS-C1s spectra of the deposited films are shown in Fig. 3.7. The spectrum of the film prepared from ODS (Fig. 3.7a) consists mostly, a single peak centered at 285.0 eV, indicating that a hydrocarbon film corresponding to its precursor was formed. On the other hand, the spectrum of the film prepared from FAS (Fig. 3.7b) could be resolved into six features, centered at binding energies of 283.5–283.6, 285.0, 286.6–286.7, 290.5, 291.7–291.9, and 294.1 eV. These components correspond to Si–C, C–C, C–O, $-CF_2-CH_2-$, $-CF_2-CF_2-$, and CF_3-CF_2- groups, respectively. The film deposited from FAS is a fluorocarbon film which is more hydrophobic than the hydrocarbon film.

As clearly demonstrated in Fig. 3.6b, that is in a graph of the films' thicknesses as estimated by ellipsometry, the thicknesses of the ODS and FAS films increase and stop to increase similarly. A film of 2 nm thickness was grown on the ODS-treated substrate at a reaction time of longer than 3 h, while a film of 1.1 nm thickness was formed on the substrate treated with FAS for longer than 1 h. These thicknesses of the deposited films are shorter than the lengths of the corresponding precursor molecules, which are 2.35 and 1.34 nm for ODS and FAS, respectively. Both the deposited films are thus considered to be monomolecular layers composed of packed molecules inclined more than 30° to

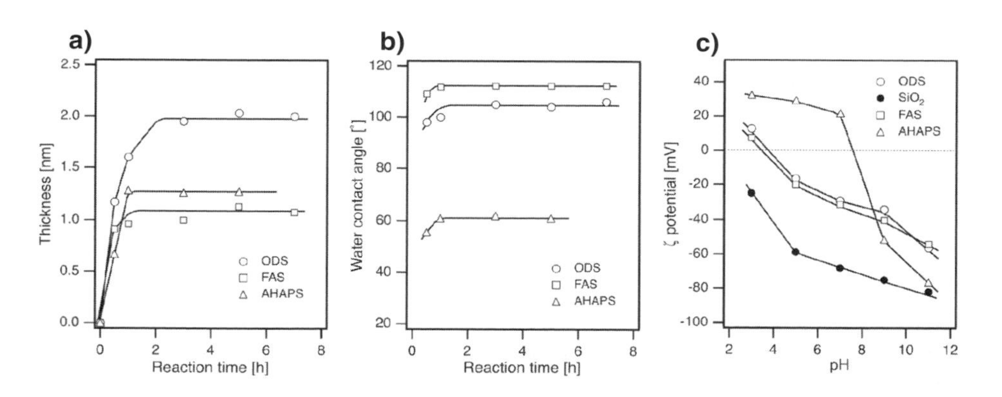

Fig. 3.6. Organosilane SAMs grown by the vapor phase method. a) Thickness, b) water contact angle, and c) ζ potential.

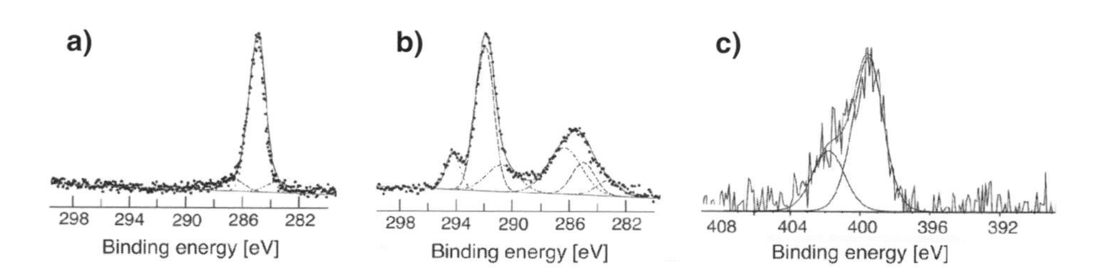

Fig. 3.7. XPS spectra of the organosilane SAMs. a) C1s spectrum of ODS–SAM, b) C1s spectrum of FAS–SAM, and c) N1s spectrum of AHAPS–SAM.

the normal. ODS has a vapor pressure of 2 Torr at 150°C, while the vapor pressure of FAS is 1 Torr at 86°C. Thus, FAS is expected to show a higher vapor pressure than ODS at our preparation temperature of 150°C. This is the reason why ODS takes longer than FAS to form a monolayer.

Besides ODS and FAS, AHAPS forms a monolayer as well. Its thicknesses reach 1.3 nm at a reaction time of 1 h and remain unchanged even when the reaction time is extended up to 5 h. However, unlike the other SAMs prepared from ODS and FAS, the AHAPS–SAM formation was not reproducible when it was conducted without the sonication in the organic solvents and in the ionic solutions. Since an amino group in the aminosilane molecule, i.e. $-NH_2$ or –NH–, may form a hydrogen or ionic bond with a methoxysilane group or its hydrolyzed form, i.e. $SiOCH_3$ or SiOH, respectively, in another aminosilane molecule, AHAPS molecules are thought to form aggregates and to be further adsorbed on the AHAPS–SAM surface. Indeed, thicknesses of the AHAPS deposits prior to the sonication were sometimes 2–3 times greater than the true thickness of the AHAPS–SAM. A considerable amount of AHAPS molecules was thought to be adsorbed on the SAM surface. The thickness of the AHAPS–SAM, i.e. 1.3 nm, is slightly smaller than the calculated molecular length of 1.48 nm. The adsorbed AHAPS molecules form a monolayer but probably inclined about 25° to normal. Although simply physisorbed AHAPS molecules can be removed by the sonication in the organic solvents, the chemisorbed AHAPS molecules through hydrogen and ionic bondings still remain. This is the reason why the sonication in the ionic solutions is needed.

Water contact angles of the AHAPS-treated substrates also increased with the reaction time, as indicated by open triangles in Fig. 3.6a. The atomic ratio of nitrogen to carbon (N/C) of the AHAPS–SAM was estimated to be about 0.17 from its XPS. This is slightly smaller than that of the chemical formula of AHAPS molecule (N/C = 0.19) probably due to adventitious carbon contaminants on its surface. As shown in Fig. 3.8, an N1s XPS peak consists of at least two chemical components with binding energies at 399.6 and 400.9 eV. The former is assigned to –NH– and $-NH_2$ groups while the latter corresponds

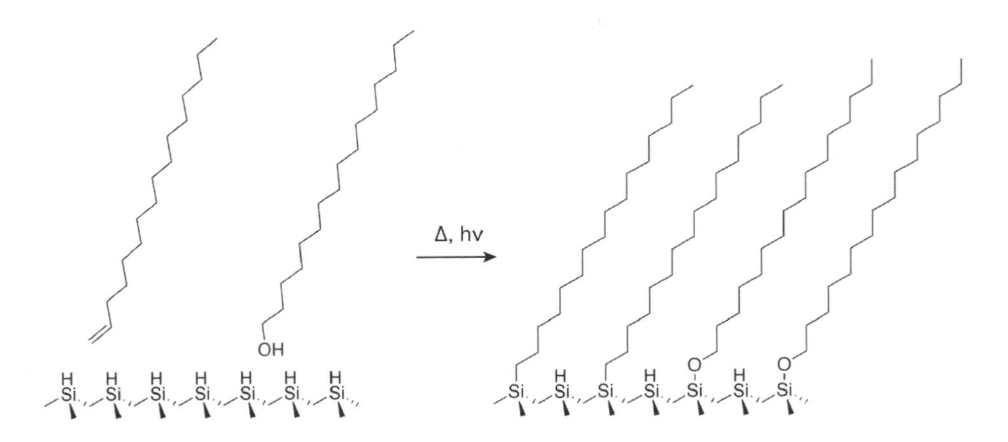

Fig. 3.8. SAM formation on SiH from 1-hexadecene and 1-hexadecanol.

to the protonated amino groups. The AHAPS–SAM is found to be protonated partially, probably due to washing in the acidic solution in the preparation process.

Figure 3.6c depicts the variation in ζ-potentials of the ODS–SAM (open circle), FAS–SAM (open square), AHAPS–SAM (open triangle) covered, and uncovered SiO_2/Si substrates (closed circle). These ζ-potentials were measured at a temperature of 298 K with an electrophoretic light scattering spectrophotometer, as described in detail elsewhere [25]. A solution containing 1 mM KCl as supporting electrolyte was used, adjusting its pH over the range of 3–11 by adding HCl or NaOH. Each ζ-potential was estimated from the average of the three values measured. The error of the ζ-potentials was about $\pm$ 5 mV. In the pH range of 3–11, the SiO_2/Si substrate shows negative ζ-potentials of ca. 25–82 mV due to partial ionization of the surface silanol groups (SiOH) to $-SiO^-$, similar to silica particles [26,27]. Although there are no isoelectric point (IEP) in this pH range, by extrapolating the potential curve, the IEP of the SiO_2/Si substrate is estimated to be pH 2.0 coinciding with the reported IEP of silica [28].

The pH dependence of the ζ-potentials for the ODS–, and FAS–SAM on the SiO_2/Si substrates is significantly different from those of the naked SiO_2/Si substrate. These ζ-potentials vs. pH curves are nearly identical in shape and magnitude in the entire pH range. From the potential curves, the negative ζ-potentials of the ODS– and FAS–SAM-covered samples are approximately 35–65% lower in magnitude than those of the naked SiO_2/Si substrate. This is attributable to the reduction in the number of silanol groups on the SiO_2/Si surface since they are consumed through the covalent bonding at the SAM–SiO_2/Si interface. Furthermore, the IEPs of the ODS– and FAS–SAM-covered samples are estimated to be pH 3.5–4.0. This IEP value is higher than the IEP of the naked SiO_2/Si substrate, i.e. pH 2.0, and are almost same with IEPs of polyethylene and polytetrafluoroethylene plates, whose surfaces are terminated with $-CH_2-$ and $-CF_2-$ groups, respectively [27]. On the other hand, the AHAPS–SAM-covered sample shows an IEP of pH 7.5–8.0. Its ζ-potentials are positive below pH 7.0. Under such acidic conditions, amino groups on the AHAPS–SAM are considered to be protonated to $-NH_3^+$. On the contrary, under basic conditions, the amino groups are probably converted to $-NH^-$ or $-NH_3O^-$ due to deprotonation or attachment of hydroxyl anions, resulting in large negative ζ-potentials, as shown in Fig. 3.6c.

3.4 SAMs Covalently Bonded on Si

3.4.1 Chemical reactions of hydrogen-terminated Si surface

As described in Section 3.2.2, SAMs covalently bonded to bulk Si can be formed on the basis of chemical reactions of hydrogen-terminated Si (Si–H) surfaces with some types of organic molecules. Thermal activation, by which Si–H bonds on a substrate surface are thermally excited and dissociated typically at a temperature higher than 100°C, are most commonly employed [29]. Si radicals, i.e. dangling bonds, are consequently formed on the surface so as to act as reaction sites with organic molecules. In addition to molecules containing a carbon–carbon double or triple bond such as alkenes and alkynes [29,30], primary alcohol and aldehyde molecules are reported to react as well resulting in the formation of SAMs with Si–O–R forms [31].

Besides the thermal method, photoactivation of Si–H by irradiating with ultraviolet (UV) light less than 400 nm in wavelength has been successfully applied to form SAMs from alkenes, aldehydes, etc. [32–34]. This photochemical process has some advantages over the thermal process. One is that the photoprocess can be conducted at room temperature. Thus, the method is applicable to SAM formation of thermally unstable molecules. Another is the capability in micropatterning. SAMs were successfully grown only in a selected area on a Si surface which it had been irradiated with UV light through a photomask [32]. Recently, a photoactivation process using visible light at 488 nm has been employed in order to form SAMs on Si [35]. In this case, electron-hole pairs generated by excitation of the substrate Si are considered to promote chemical reactions, since photon energy of the visible light is too low to dissociate Si–H. The visible light process is certainly useful in the case of a precursor unstable with UV irradiation.

3.4.2 Hexadecyl and hexadecyloxy monolayers on Si

In this section, actual examples of SAMs covalently-bonded to bulk Si are described. Cleaned Si(111) substrates (phosphorus-doped n-type wafers with a resisitivity of 10–50 Ωcm) were first etched in 5% HF for 5 min at room temperature. Next, the substrates were treated in 40% NH_4F for 30 s at a temperature of 80°C. Through these treatments, a surface oxide on each of the Si substrates was removed and the substrate surfaces became terminated with hydrogen. Due to this hydrogen termination, the substrate surfaces became hydrophobic hence their water contact angles increased to 80–85° from those of surface oxide etching. The surface was initially covered with hydroxylated oxide showing a water contact angle of almost 0°. Some of the Si–H substrates were treated in 1-hexadecene ($C_{16}H_{32}$) liquid, and the other ones were processed in a solution of 1-hexadecanol ($C_{16}H_{33}OH$) in mesitylene at a concentration of 200 mM. Before and throughout the reaction, a stream of nitrogen was bubbled into the liquid or the solution in order to purge oxygen from the entire reaction system and to suppress the Si–H surfaces from being oxidized. For UV excitation, a super high-pressure Hg lamp was used without filtering. Thus, in addition to its main radiation peaks at 365, 405, and 436 nm, the irradiated light contains some amounts of visible light. However, as reported in [32], UV components less than 380 nm in wavelength plays a central role in the activation of Si–H. A white light from a Xe lamp in the wavelength range from 400 to 800 nm was used for experiments that activated with visible light.

As schematically illustrated in Fig. 3.8, the 1-alkene or primary alcohol molecules react with a Si–H surface to form corresponding hexadecyl (HD–SAM, Si–$C_{16}H_{33}$) or hexadecyloxy (HDO–SAM, Si–O–$C_{16}H_{33}$) monolayer, respectively. Table 3.2 summarizes water contact angles of HD– and HDO–SAM surfaces prepared by the thermal, UV, and visible activation processes. The substrate surfaces become further hydrophobic to show water contact angles more than 105°, since those have been terminated with methyl (–CH_3) groups owing to the alkyl or alkoxy monolayer grown on each of the substrates. Figure 3.9 shows topographic images of the thermally grown HD– and HDO–SAM surfaces acquired by AFM in the contact mode. Even after forming SAMs of about 2 nm thick, structures of Si(111) surface consisting of flat terraces separated with a monoatomic step of near 0.3-nm high are clearly retained.

Table 3.2. Water contact angles of HD– and HDO–SAMs.

SAM	Activation	Temperature (°C)	Reaction time (h)	Water contact angle (°)
HD	Thermal	150	2	109
	UV (660 $mWcm^{-2}$)	R.T.	10	107
	Visible (15 $mWcm^{-2}$)	R.T.	15	106
HDO	Thermal	150	2	108

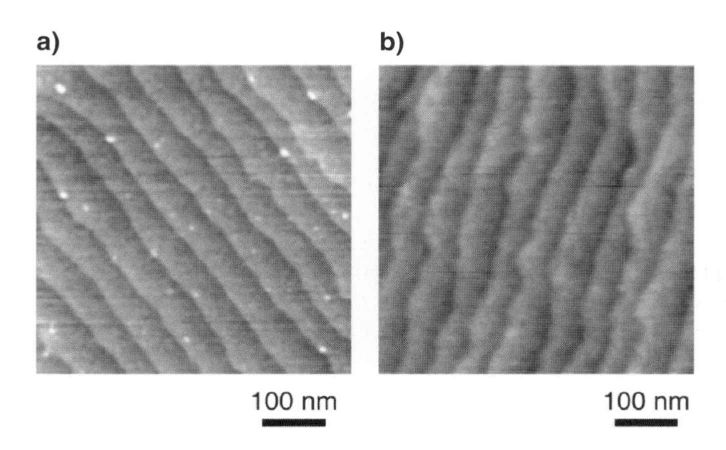

Fig. 3.9. Topographic images acquired by AFM. a) HD–SAM covered Si(111) surface. b) HDO–SAM covered Si(111) surface.

As shown in Fig. 3.10a, an XPS-C1s spectrum for the HD–SAM on Si prepared by the thermal activation, the carbon signal corresponds to alkyl chains has been detected. The thickness of HD–SAM was in the range of 2.0–2.4 nm as estimated by ellipsometry. Figure 3.10b shows XPS-Si2p spectra of HD-SAM/Si, Si–H, and Si covered with an oxide layer of about 2 nm thick. There are two peaks corresponding to Si^{4+} (oxidized Si) and Si^{0} centered near 103 and 99 eV, respectively, in the spectrum of the oxide-covered Si indicated by a broken curve. Such an oxidized Si peak is not present in the spectrum of the Si–H sample as indicated by a dotted curve. As confirmed by the spectrum of the HD–SAM sample indicated as a solid curve in Fig. 3.10b, there is no oxidized Si on this sample as well. This is an evidence that the monolayer is attached to bulk Si without an interfacial oxide layer, on the contrary to organosilane SAMs.

As reported in [13], almost 50% of Si–H groups are considered to remain at the monolayer–Si interface. If these Si–H groups are exposed to air, they gradually dissociate and the underlying Si surface oxidizes, since Si–H is in a metastable state. Nevertheless, the XPS-Si2p spectrum of the HD–SAM sample shows the absence of Si oxide. This result indicates that the monolayer formed on Si is tightly packed and the remaining Si–H groups are effectively shielded. It was confirmed as well that there was no Si oxide on the other SAM samples photochemically prepared from 1-hexadecene and thermally prepared from 1-hexadecanol shown in Table 3.2.

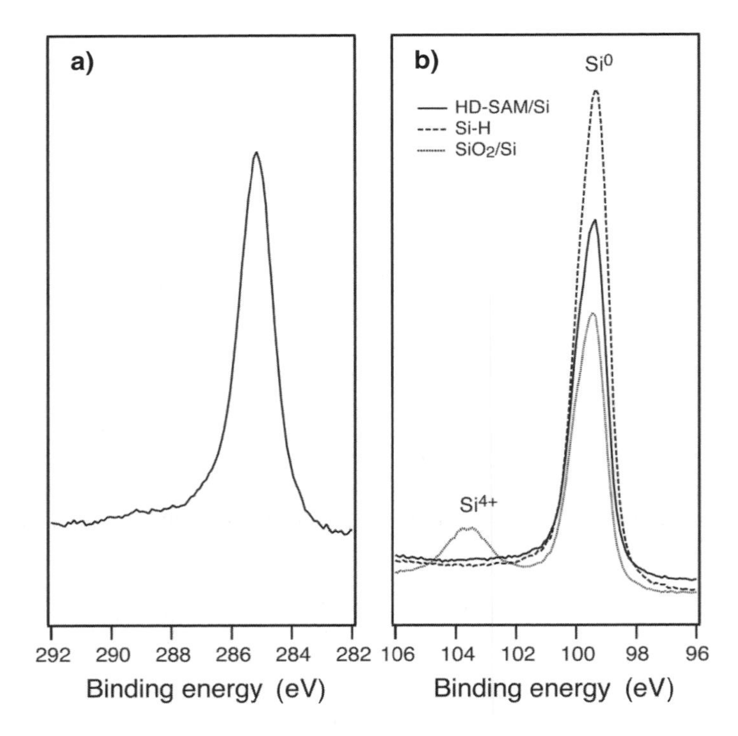

Fig. 3.10. a) XPS-C1s spectrum of a HD–SAM/Si surface. b) XPS-Si2p spectra of HD–SAM/Si, Si–H, and SiO_2/Si surfaces.

3.5 SAMs on Si in Micro–Nano Patterning

3.5.1 SAMs as ultra thin resist films for patterning

For advanced applications of SAMs in microdevices with mechanical, electronic, chemical, and biological functions, micro–nano structuring technologies for the SAMs are of primary importance [36–41]. Among various patterning methods, photolithography has been commonly employed, since the technique is most practical mainly due to its high throughput.

Besides photolithography, other lithographic methods including electron beam irradiation [42–47], ion beam etching [48], neutral atomic beam exposure [49–51], X-ray lithography [52–56] have been used in order to fabricate micro–nanopatterns on various types of SAMs. Furthermore, a unique method, the so-called μ-contact printing, in which micropatterns are printed on a solid substrate using a microstructured silicone rubber as a stamp and a solution of precursor molecules as an ink, has been developed and improved rapidly in the last decade [37,38,57–60]. Another promising nanopatterning method is nanolithography based on scanning probe microscopy, as described in Section 3.5.5.

3.5.2 Photopatterning of organosilane SAMs

Photopatterning of SAMs on Si was reported for the first time by the research group in Naval Research Laboratory, USA [36, 61–64]. They employed excimer lasers of 248- and 192-nm wavelengths in order to induce particular photochemical reactions of organosilane SAMs, for example, dissociation of Si–C bonds through excitation of the adjacent aromatic rings [61] as illustrated in Fig. 3.11a. This method has been extended to deactivate $-NH_2$ groups and oxidise of $-CH_2Cl$ groups. We have also reported that the photochemical conversion of surface SH and –SS– groups to SO_3H groups using an excimer laser at 248 nm is as illustrated in Fig. 3.11b [65].

These approaches are successful for the fabrication of micropatterns on organosilane SAMs. The photochemical reactions governing the processes depend on the particular functional groups, such as phenyl, amino, and mercapto groups, and hence these methods are not applicable generally to other organosilane monolayers, including alkylsilane and fluoroalkylsilane SAMs, in spite of the fact that these SAMs are frequently used for surface modification. We have developed a promising way applicable to any type of organic thin films [66–72].

This method is based on the use of vacuum ultraviolet (VUV) light at a wavelength of 172 nm radiation from an excimer lamp. Why a VUV light of 172 nm wavelength is used? One crucial advantage is its high photon energy of 7.2 eV. Such high-energy photons can excite a variety of organic molecules which has no photosensitivity to UV light with a longer wavelength usually used in photolithography. Furthermore, the VUV light of 172 nm wavelength dissociates oxygen molecules into two oxygen atoms in the singlet and triplet states, (O(1D) and O(3P), respectively) as described in Eq. 3.1 [73].

$$O_2 + h\nu \rightarrow O(ID) + O(3P) \qquad (3.1)$$

These oxygen atoms, particularly O(1D), have oxidative reactivities strong enough to oxidize alkyl and fluoroalkyl chains, which are hardly decomposed by the irradiation with VUV light at 172 nm alone, VUV-micropatterning rates of SAMs are distinctly accelerated through the simultaneous VUV excitation of the SAMs and oxygen molecules existing on and around the SAM surfaces.

Several types of excimer lamps radiating VUV light with a wavelength shorter than 172 nm have been developed. Such short-wavelength VUV lights are more favorable for

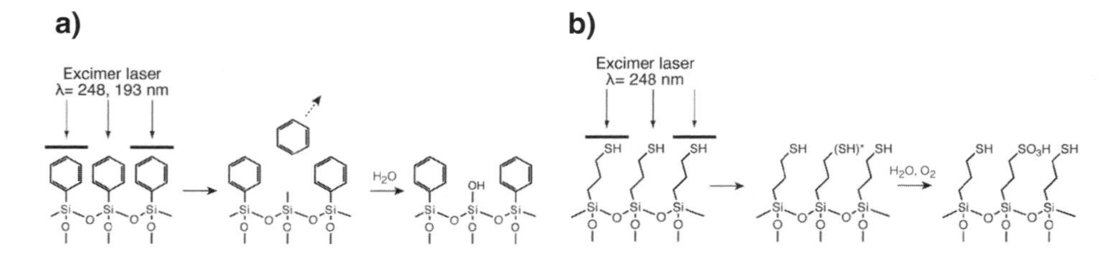

Fig. 3.11. Examples of UV exposure. a) Dissociation of phenylsilane SAM through the excitation of aromatic rings. b) Photochemical conversion of SH to SO_3H.

promoting surface modification reactions of organic monolayers, as they can induce direct excitations of C–C and C–H bonds. Nevertheless, VUV light of 172 nm is advantageous in respect of practical applications. At 172 nm, quartz has an adequate transparency so as to be applicable as a material for optical parts. Quartz-made optical parts including photomasks are frequently used in common photolithographic processes and are readily available. This is the second advantage of VUV lithography at 172 nm.

As schematically illustrated in Fig. 3.12a, a sample (a SiO_2/Si substrate covered with ODS–SAM prepared by the vapor phase method, as described in Section 3.3) was micropatterned by a mask-contact photolithography. A sample along with a photomask was located in a vacuum chamber, the pressure of which was controlled by introducing

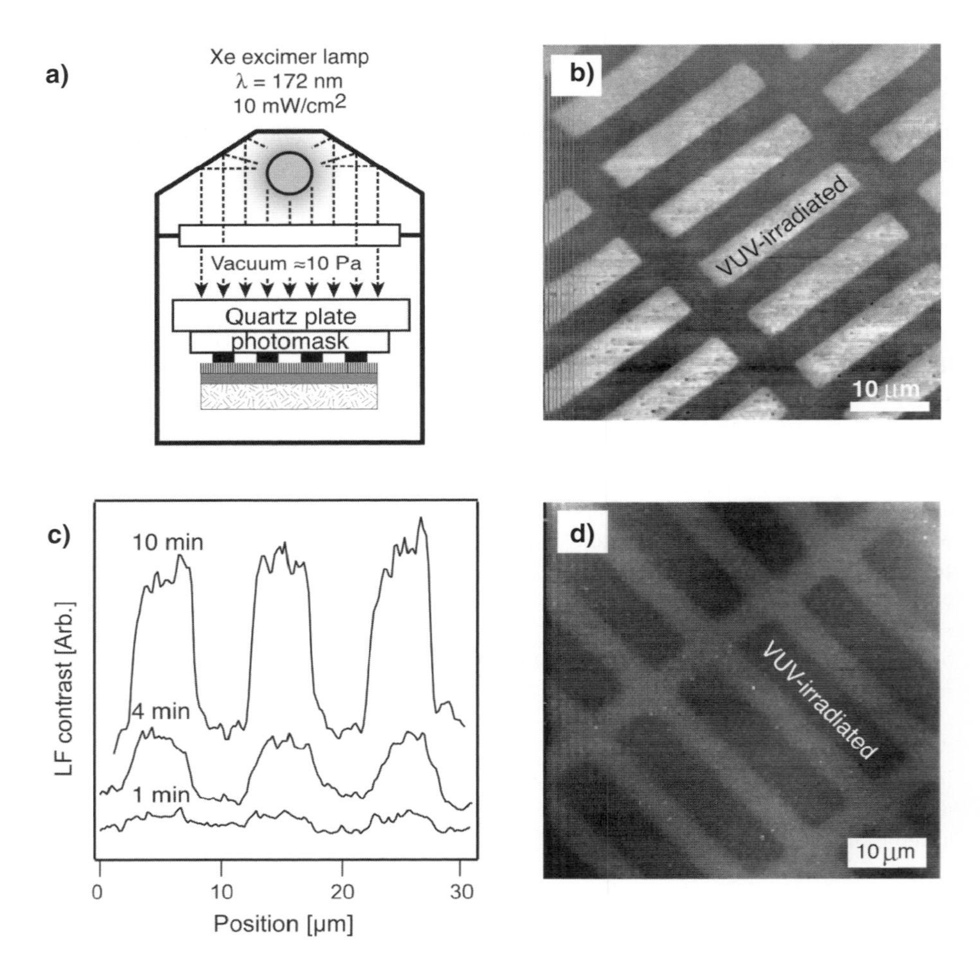

Fig. 3.12. VUV lithography. a) VUV micropatterning apparatus. b) LFM image of ODS–SAM irradiated for 10 min through the photomask. c) Cross sections of LF images for the samples irradiated for 1, 4, and 10 min. d) Topographic AFM image of the ODS–SAM irradiated for 20 min.

air through a variable leak valve. A weight of 10-mm thick quartz glass plate was placed on the photomask in order to attain a satisfactory contact between the mask and the SAM surface.

Figure 3.12b shows a lateral force microscope (LFM) image of the sample irradiated for 10 min with VUV light through the photomask. This VUV irradiation was performed at a pressure of 10 Pa. The image consists of two distinct regions which are bright 5 μm × 25 μm rectangular features and a dark surrounding area. These bright features correspond to the VUV-irradiated regions that possess a higher friction coefficient than that of the unirradiated ODS–SAM. Such an image contrast becomes more distinct with an increase in irradiation time. As shown in Fig. 3.12c, at an irradiation for 1 min, an LFM contrast is faint, while at irradiation periods of 4 and 10 min, the contrasts become ca. 6 and 40 times greater, respectively. The LFM contrast increased slowly when VUV irradiation was further prolonged and finally remained unchanged when irradiated for more than 15 min, at which ODS–SAM was considered to be almost removed as expected from other experimental results such as water contact angle measurements, through X-ray photoelectron spectroscopy (XPS), and infra-red reflection absorption spectroscopy. Indeed, as confirmed by a topographic AFM image shown in Fig. 3.12d, regions VUV-irradiated for 20 min are recessed about 1.5–2.0 nm. This depth corresponds to the thickness of ODS–SAM.

3.5.3 Photopatterning of the hexadecyloxy SAM

The VUV-patterning method described in Section 3.5.2 is applicable to SAMs covalently bonded to Si as well [75–77]. In this section, VUV-patterning of HDO–SAM is described. First, the VUV-degradation chemistry of HDO-SAM was studied based on XPS. Figure 3.13a–c shows XPS-C1s, -O1s, and -Si2p spectra, respectively, of the monolayer-covered Si samples prior to VUV-irradiation and VUV-irradiated for 120 and 480 s at a pressure of 10^3 Pa. By the VUV-irradiation at 10^3 Pa for 120 s, the amount of carbon on the sample decreases. On the contrary, the amount of oxygen on the sample increases. As indicated, in the binding energy range between 287 and 289 eV of the XPS-C1s spectrum, polar functional groups containing oxygen have been formed, hence the sample surface became hydrophilic. Owing to the polar functional groups themselves and an increased amount of water molecules on the hydrophilic surface, the O1s signal intensity increased.

As can be seen in Fig. 3.13c, the XPS-Si2p spectrum of the monolayer without VUV-irradiation has only a single peak near 100 eV, indicating that no oxide is formed between the Si substrate and the monolayer. At an irradiation time of 120 s, no oxide is still present at the monolayer–substrate interface. The Si2p signal from the substrate increases, since the monolayer thickness has decreased. Finally, at an irradiation time of 480 s where the monolayer has been considered to be completely decomposed, the C1s signal further decreases. This intensity is almost equal to a C1s signal caused by a contamination layer, which always adsorbs on the sample surface before brought into the measurement chamber of XPS. Thus, the alkyl part of the monolayer has been recognized to be almost removed form the substrate. It is noteworthy that an additional peak centered around 103 eV appears in the XPS–Si2p spectrum. Furthermore, the O1s signal

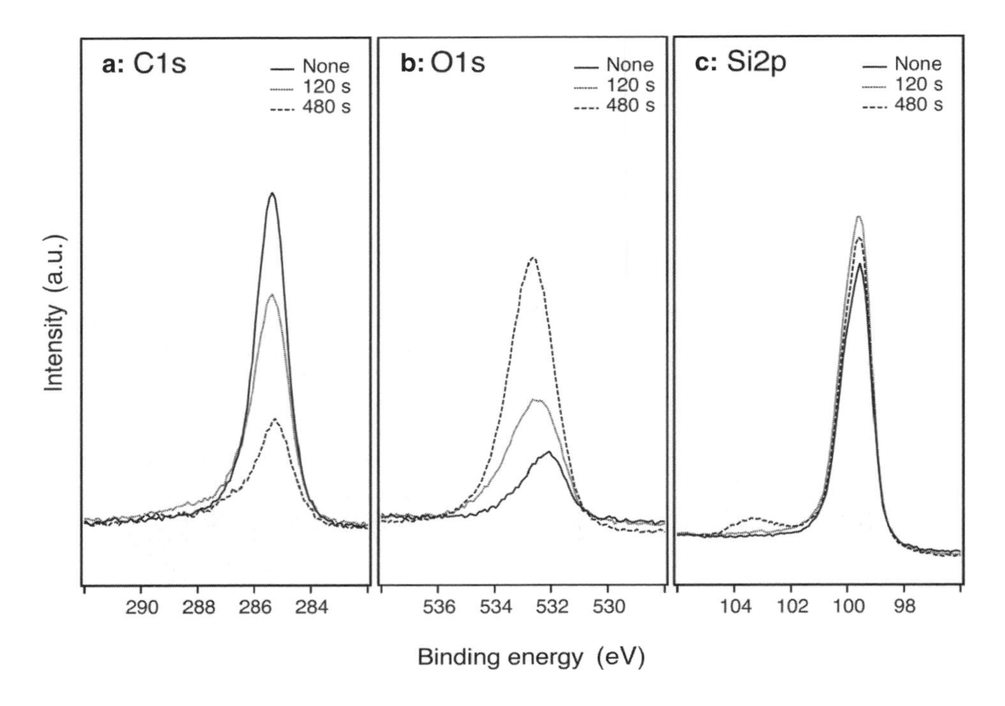

Fig. 3.13. XPS-C1s, O1s and Si2p spectra of HDO–SAM surfaces unirradiated and irradiated with VUV light for 120 or 480 s.

increases markedly. These results indicate that silicon oxide is formed on the sample surface. The oxide thickness was estimated to be about 1.5 nm by ellipsometry.

Two of the microfabrication processes based on VUV-photolithography and HDO–SAM are demonstrated here. As schematically illustrated in Fig. 3.14a, the monolayer was VUV-photoetched and a Si oxide layer grown in the VUV-irradiated regions. One of these VUV-patterned samples was treated with HF. The oxide layer grown on the VUV-irradiated region is etched, while the monolayer is resistant to HF. Thus, as illustrated in Fig. 3.14b, dimples are formed on the sample surface through the selective etching of the oxide. As shown in the AFM topographic image, the VUV-irradiated region is distinctly depressed from the surrounding area where it is covered with the monolayer. The monolayer has successfully worked as an etching mask. Such a resistivity to HF etching is a characteristic feature of organic monolayers covalently attached to Si substrates [78], on the contrary to organosilane monolayers fixed on Si with insertion of an oxide layer. For example, ODS–SAM on oxide-covered Si could resist only for a few seconds or even less in 2% HF solution [79].

In the next step, the VUV-patterned sample was treated with ODS by the vapor phase method described in Section 3.3, in order to form an organosilane monolayer on the oxide surface fabricated by the VUV exposure. The details of this organosilane monolayer growth by the vapor phase method has been reported in Section 3.3. The oxide surface

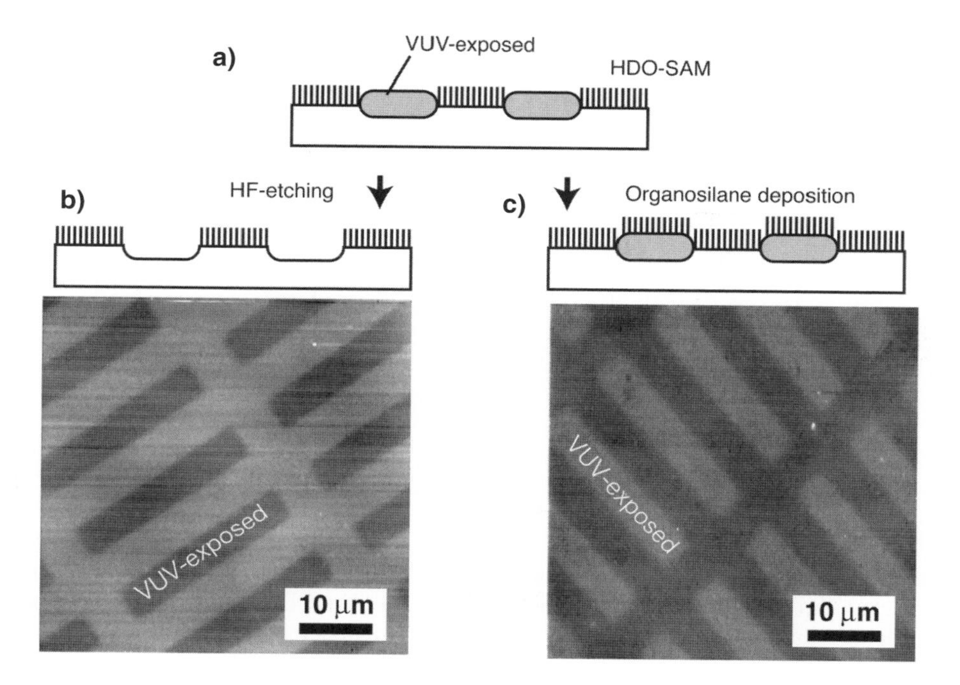

Fig. 3.14. VUV-photolithography process. a) A micropatterned HDO–SAM/Si sample. VUV-patterning was conducted at an exposure time for 480 s and a camber pressure of 10^3 Pa. b) Etching in 5% HF solution for 5 min. c) organosilane deposition on the VUV-grown oxide surface.

has an affinity to the silane coupling chemistry with organosilane molecules, while the monolayer-covered surface does not react with the organosilane molecules. Accordingly, an organosilane monolayer is formed selectively on the VUV-grown oxide surface, as illustrated in Fig. 3.14c. The AFM topographic image indicates that the VUV-exposed region protrudes from the region covered with HDO–SAM.

3.5.4 VUV lithography system

The VUV-induced degradation rates of the SAMs are governed mainly by two factors. The first is the intensity of VUV light at the SAM surface. The second is the amount of oxygen supplied to the surface. In the present case as depicted in Fig. 3.12a, it is difficult to satisfy these two factors simultaneously since the VUV light intensity decreases when the chamber pressure is increased in order to supply large amount of oxygen molecules. Hence, we have constructed a new VUV lithography system as schematically illustrated in Fig. 3.15 [80,81]. In this system, a sample is placed in air in order to supply large amount of oxygen, while the space between the photomask and the lamp window is purged with nitrogen to avoid the absorption of VUV light with oxygen molecules, i.e. the photomask works as a separation wall between the nitrogen and air atmospheres. Furthermore, a proximity gap between the photomask and the sample is controlled precisely at an accuracy of 0.1 μm using a mechanical stage.

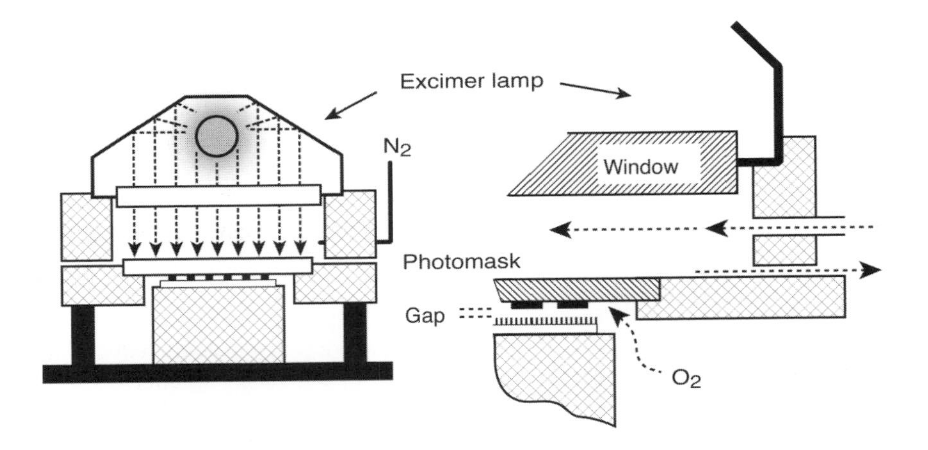

Fig. 3.15. VUV exposure system.

The performance of this VUV-exposure system was characterized by measuring water contact angles of VUV-irradiated ODS–SAM samples. The ODS–SAM surface becomes hydrophilic due to VUV irradiation, since polar functional groups have been formed through oxidation of the alkyl chains. Finally, at a certain VUV dose, its surface is completely wetted with water showing a water contact angle of almost 0°, when all the ODS molecules had been decomposed and removed so that the underlying SiO_2 surface had appeared. As shown in Fig. 3.16a, a VUV dose of about 8 J/cm^2 is required in order

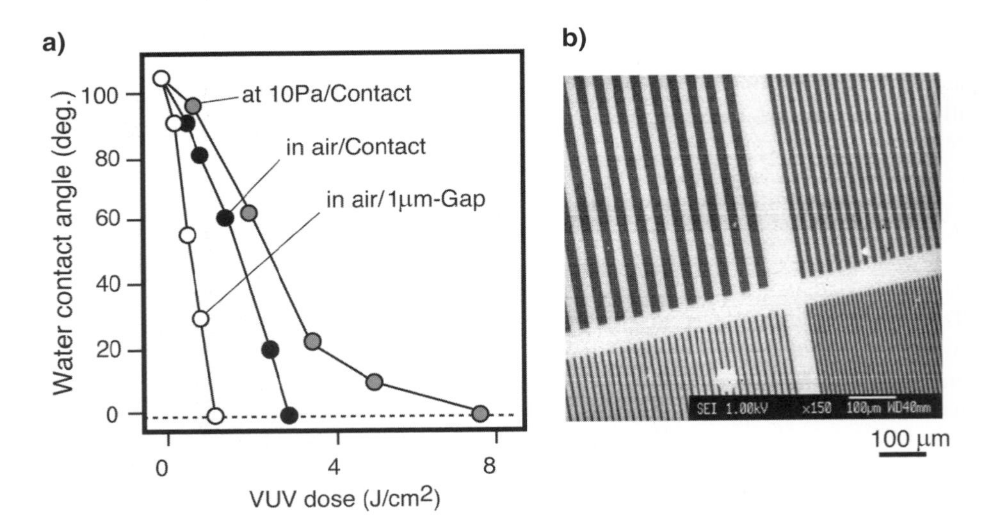

Fig. 3.16. The performance of the VUV exposure system. a) Changes in water contact angles of ODS–SAM covered samples. b) FE–SEM image of an patterned ODS–SAM acquired at an electron energy of 1 keV.

to fabricate a micropattern on ODS–SAM at a pressure of 10 Pa using the apparatus shown in Fig. 3.12a, which is similar to the result shown in Fig. 3.12d. On the contrary, a required dose for micropatterning using the VUV system is only 3 J/cm^2 or less as indicated by the closed circles in Fig. 3.16a. The dose is further reduced down to near 1 J/cm^2 as indicated by the open circles, when a gap of 1 μm was located between the photomask and the sample. An example of a printed pattern on the ODS–SAM sample is shown in Fig. 3.16b. Fine lines of 5–20 μm wide were successfully transferred from the photomask. These patterns were observed by a field-emission scanning electron microscope (FE-SEM) based on the difference in secondary electron emission efficiencies between ODS–SAM and SiO_2 [82].

The VUV-exposure system enabled faster patterning rates. However, it requires several minutes to decompose ODS–SAM entirely. It is necessary for an alternative idea to further speed up patterning rates without increasing the power of the VUV light source. We have proposed and demonstrated the micro-photo-conversion patterning in which only the top surface parts of a monolayer are photochemically modified instead of decomposing an entire monolayer [83]. As illustrated in Fig. 3.17a, an organosilane SAM prepared from p-chloromethylphenyltrimethoxysilane (CMPhS–SAM) was used as a sample.

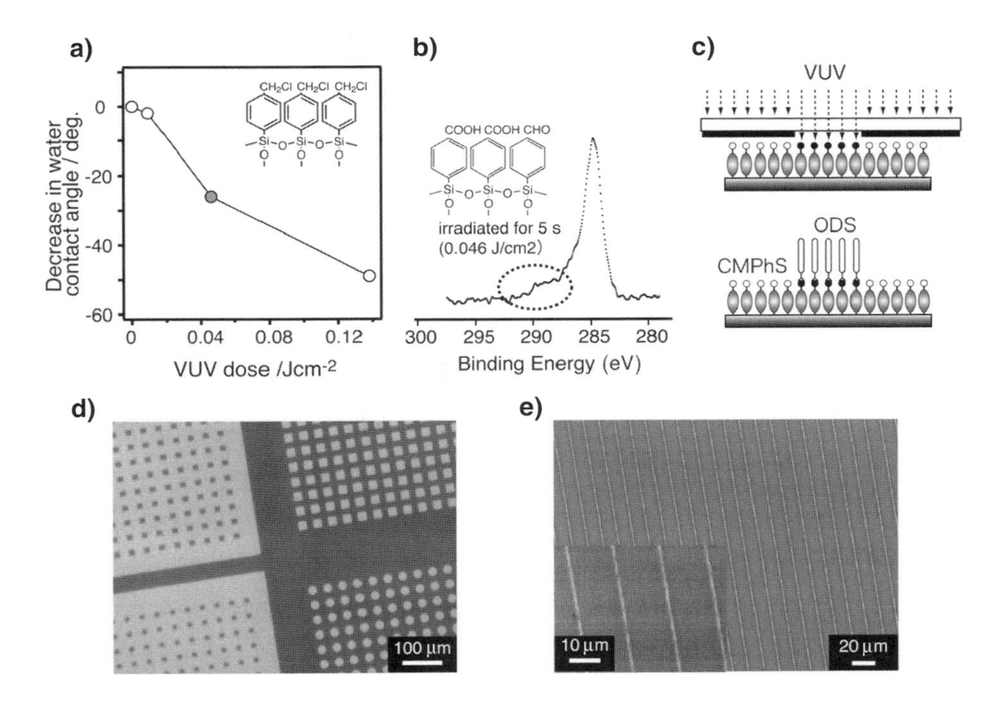

Fig. 3.17. VUV-exposure of CMPhS–SAM. a) Decrease in water contact angle due to VUV-irradiation. b) XPS-C1s spectrum of the CMPhS–SAM irradiated for 5 s. c) VUV-patterning of CMPhS–SAM followed by ODS treatment. d), and e) FE-SEM images of the ODS/CMPhS microstructures.

The water contact angle of the CMPhS–SAM surface decreased due to VUV irradiation from its initial value of 76° before irradiation to approximately 0° at 60 s (dose = 0.56 J/cm^2). In contrast, it took 1800 s (dose = 15.1 J/cm^2) to decompose the CMPhS–SAM to the extent that its water contact angle became approximately 0°, when it was irradiated under a reduced pressure of 10 Pa using the apparatus of Fig. 3.12a. Figure 3.17b shows an XPS-C1s spectrum of CMPhS–SAM in an intermediate state before its complete decomposition, i.e. irradiation for 5 s (dose = 0.046 J/cm^2). This VUV irradiated CMPhS–SAM surface became slightly hydrophilic and showed a water contact angle of approximately 50°, probably because some amounts of polar functional groups had been formed photochemically as shown in the spectrum as a distinct shoulder at binding energies ranging from 287 to 290 eV. This should correspond to –COOH and –CHO groups.

Such an oxidized CMPhS surface covered with the polar functional groups is expected to have an affinity for the silane coupling reaction. To confirm this assumption, we exposed a VUV-patterned CMPhS–SAM sample to ODS vapor. ODS molecules were assumed to be deposited selectively on the VUV-irradiated region and, consequently, to form a micropatterned ODS–SAM. Figures 3.17d and e shows FE-SEM images of a binary CMPhS/ODS coplanar structure. Dark and bright regions correspond to the regions terminated with CMPhS and ODS, respectively. As we have reported, secondary electron emission from organosilane SAMs is strongly dependent on their molecular species. The CMPhS–SAM was darker than the ODS–SAM in FE-SEM images as a function of the electron affinity of the SAMs [82]. Therefore, the results shown in Figs. 3.17d and e clearly demonstrate that ODS molecules have been deposited selectively on the VUV-irradiated CMPhS–SAM surface. At present, a spatial resolution near 1 μm was achieved as demonstrated in Fig. 3.17e.

3.5.5 *Scanning probe nanopatterning of SAMs on Si*

Modern microscopic technologies, which enable to obtain enlarged images of minute objects, are closely related to lithographic technologies, in which mechanisms for generation and replication of micro- and nanometer-scale patterns are required. Scanning probe microscopy (SPM) is a powerful means to investigate material surfaces with high spatial resolutions at nanometer scales and, in favorable cases, at atomic and molecular scales. Accordingly, SPM technologies have attracted much attention as nanoprocessing tools [84–91].

In order to apply SPM to nanolithography, the optimum resist material is a key factor. Among a variety of materials, organic thin films are practically important and have been frequently applied as patterning materials for lithography. Thus, the patterning of organic thin films by SPM is of special importance. However, in order to attain high spatial resolution in nanometer scale, resist films must be prepared in a thin and uniform layer. Furthermore, the films must be compatible with pattern transfer processes particularly with chemical etching. SAMs fulfill the requirements for high-resolution resist films including thickness, uniformity, patternability, and compatibility to various pattern transfer processes so as to be the most promising candidate [92–98].

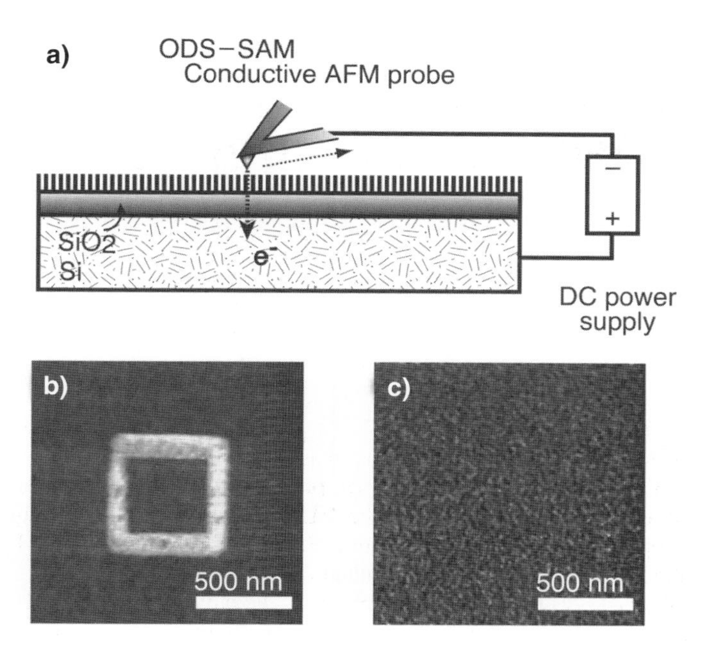

Fig. 3.18. Current injecting AFM lithography using ODS–SAM as a resist film. The probe was scanned at a speed of 0.1 μm/s while being pressed to the sample surface at a load force of 3 nN. A bias voltage of 10 V was applied between the probe and the substrate Si. a) Schematic illustration. b) LFM image of the ODS–SAM/Si sample current-injected in air. c) LFM image of the ODS–SAM/Si sample current-injected in vacuum at a pressure of 1.5×10^{-6} Torr.

An ODS–SAM/Si sample surface was modified using an AFM as illustrated in Fig. 3.18a. In order to inject current into the sample, a DC bias voltage was applied between the conductive AFM probe and the substrate Si which served as cathode and anode, respectively. An electrically conductive probe (a heavily doped Si probe) was used. Figure 3.18b shows an LFM image of a patterned ODS–SAM/Si sample surface. The square feature showing a higher lateral force corresponds to the region where the current was injected from an AFM probe. From a topographic AFM image of this sample, the probe-scanned region was confirmed to protrude a few nanometer from the surrounding ODS–SAM/Si surface where the current was not injected. On the contrary, there are no features detectable by LFM on the ODS–SAM/Si sample scanned and current injected in vacuum, although the same bias and load force were applied. Mechanical scratching is not a mechanism of the surface modification, since the ODS–SAM/Si is so robust that no damages were induced even at a load force of 600 nN [99]. We, thus, conclude that the ODS–SAM/Si sample was modified through electrochemical reactions proceeding in the adsorbed water column formed at the probe–sample junction as similarly to scanning probe anodization [100]. Due to these reactions, the organic molecules consisting of the ODS monolayer were anodically degraded and finally decomposed. Consequently, a lateral force contrast between the modified and unmodified regions was generated. In addition, anodization of the substrate Si occurred simultaneously so that the current-injected region protruded.

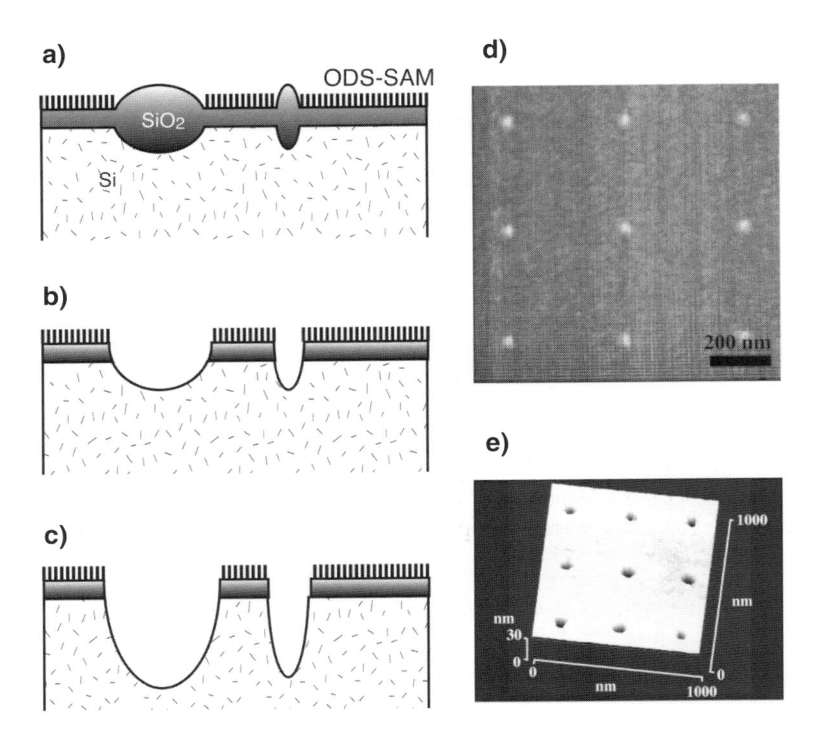

Fig. 3.19. Pattern transfer via chemical etching. a) Patterned ODS–SAM/Si at a bias voltage of 10 V. b) Oxide etching with HF (0.1%, 10 min, R.T.). c) Si etching in a solution of $NH_4F{:}H_2O_2{:}H_2O = 10{:}3{:}100$ (weight ratio), 1 min, R.T.). d) LFM image of a patterned sample. e) Topographic image of the etched sample.

Here, we demonstrate a pattern transfer process in which a pattern drawn on an ODS–SAM/Si sample is transferred into its substrate Si via chemical etching (Fig. 3.19). First, the sample patterned by AFM is treated in a HF solution. An LFM image of the patterned sample surface is shown in Fig. 3.19d. Dots of 30 nm diameter have been formed. Due to the HF etching, oxide in the region modified by AFM is etched as illustrated in Fig. 3.19b. Next, the sample is further etched in a mixed solution of ammonium fluoride and hydrogen peroxide. The substrate Si is selectively etched in this solution as illustrated in Fig. 3.19c. Indeed, as demonstrated in Fig. 3.19e, nanoholes of 50 nm diameter and 30 nm depth are formed on the sample.

3.5.6 Reversible nanochemical conversion

Among the various principles behind scanning probe lithography, local oxidation as described in Section 3.5.5 is the most promising way. Based on electrochemistry, local oxidation proceeds in a minute water column formed between the sample and the tip of an SPM [100]. In order to achieve reversible chemical nanopatterning based on SPM, both oxidation and reduction reactions must be manipulated.

Here, we demonstrate a reversible surface modification of a SAM surface using SPM by manipulating both electrochemical oxidation and reduction reactions in a controlled manner [101,102]. We employed a SAM prepared from p-aminophenyl-trimethoxysilane (APhS, $H_2N(C_6H_4)Si\ (OCH_3)_3$) as a sample material. Figure 3.20a shows the chemical structures of the APhS molecule and the corresponding SAM. Surface-modification experiments were conducted in air using an AFM with a gold-coated Si cantilever. A DC bias was applied to the Si substrate, while the cantilever grounded, in order to induce electrochemical reactions at the probe–sample junction. Kelvin-probe force microscopy (KFM) imaging was conducted using the same AFM and probe used to modify the surface.

Figure 3.20c and d show the surface-potential images of the probe-scanned APhS–SAM surfaces. The probe-scanned regions have different surface potentials compared to the as-prepared APhS–SAM surface. At a negative sample bias voltage, a bright square with a higher surface potential than the as-prepared APhS–SAM was formed, while a dark square with a lower surface potential was formed at the positive sample bias voltage. It is noteworthy that there were no apparent topographic changes in the probe-scanned regions.

The surface potentials of organic thin films are reported to be governed by their polarization states [103–106]. In an aqueous solution, the amino (NH_2) groups can be

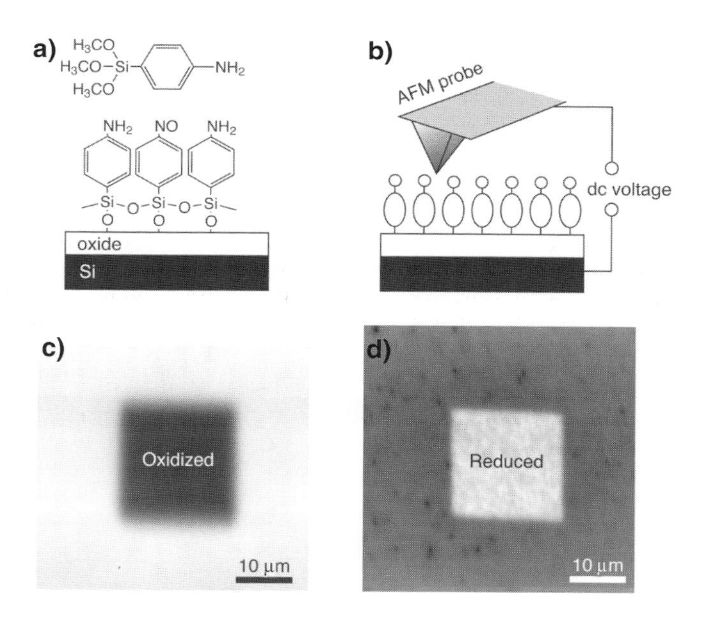

Fig. 3.20. Sample preparation and surface modification. a) Chemical structures of APhS molecule and the corresponding SAM. A partially oxidized APhS–SAM (0.6 nm thick) was prepared on Si substrates (100 orientation, n-type, 4–6 Ωcm) covered with an 2-nm thick oxide layer. b) Schematic illustration of the surface modification procedure. KFM images of surface-modified samples at dc biases of c) +3 and d) −3 V.

electrochemically oxidized into nitroso (NO) groups and the NO groups can be electrochemically reduced to NH_2 groups. Thus, the reduced and oxidized states of the APhS–SAM fabricated in this study are considered to have NH_2-terminated and NO-terminated surfaces, respectively. The NO-terminated SAM surface was assumed to be negatively polarized since the NO groups attract electrons from the aromatic rings that consist of the SAM. Therefore, the surface potential of the SAM shifts toward the negative direction when terminated with NO groups. This is the reason that the probe-oxidized area showed lower surface potentials than the surrounding as-prepared SAM surfaces (Fig. 3.20c). However, it was assumed that the NH_2-terminated surface is positively polarized, since the NH_2 groups supply electrons to the aromatic rings. The NH_2-terminated surface, i.e. the reduced state of the APhS–SAM, should show a higher surface potential than the as-prepared surface (Fig. 3.20d). Considering the results that both reducing and oxidizing modifications could be conducted on the as-prepared APhS–SAM surface, it was most likely to be partially oxidized as schematically shown in Fig. 3.20a. Namely, a portion of NH_2 groups had been oxidized to NO groups during the preparation process in which APhS molecules were heated up to be 100°C in air [101].

A writing and erasing experiment based on these oxidation and reduction mechanisms was conducted as shown in Fig. 3.21a. In this experiment, a 10-μm square region on the as-prepared APhS–SAM was initially reduced at a bias of −2 V. This reduced region appears as a bright square feature in the KFM image as shown in Fig. 3.21b. Next, six oxidized dots were fabricated in this initialized area under the writing conditions as indicated in Fig. 3.21a. As can be seen in the surface potential image of Fig. 3.21b,

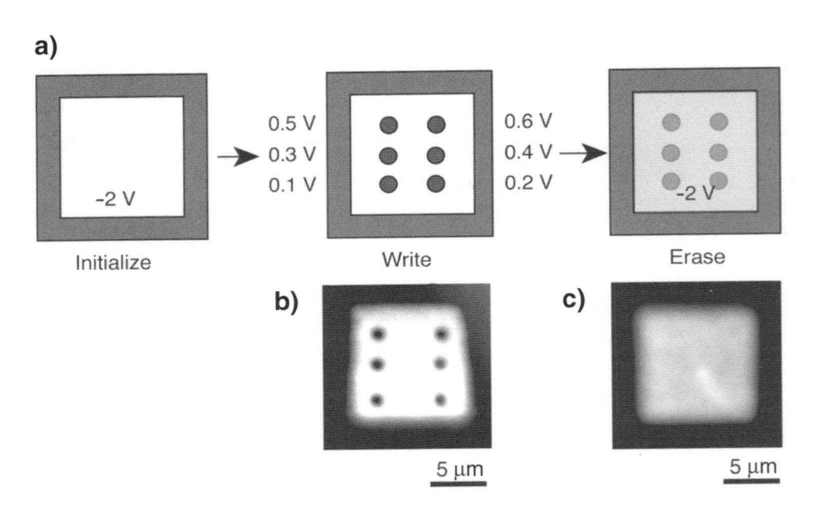

Fig. 3.21. Reversible nanochemical conversion. a) Initializing, writing and erasing. b) KFM image of an initialized 10 μm square at a dc bias of −2 V and six dots written in the initialized square by oxidation with 1 sec pulses. Each peak voltage is indicated in the figure. c) KFM image of the same area with that of Fig. 3.21b acquired after erasing with a reductive over-scanning at a dc bias of −2 V.

the oxidized dots are detected as dark dots with 40–50 mV lower potential than the initialized surface. Then, the same 10-μm square region was reduced. The KFM image of Fig. 3.21c indicates that the oxidized dots were successfully erased. This writing and erasing process can be repeated several times.

3.6 Some Electrical Properties of SAMs on Si

3.6.1 Surface potentials of organosilane SAMs

SAMs markedly alter surface properties of substrates, including the electrical properties. For example, organosilane SAMs have been applied to control characteristics of organic field-effect transistors through the modification of the interface between an organic semiconductor layer and a gate oxide film [107]. The knowledge on dipole moments of SAMs, consequently, surface potentials of SAMs well known to be closely related to dipole moments of the molecules consisting of the SAMs [108–110], is of primary importance for such an application.

Five types of SAMs, i.e. 3,3,3-trifluoropropyltrimethoxysilane [TFPS, $CF_3(CF_2)_2Si(OCH_3)_3$] in addition to ODS, FAS, CMPhS, and AHAPS as depicted in Figure 3.22a–e were studied. Samples for surface potential measurements were fabricated by the

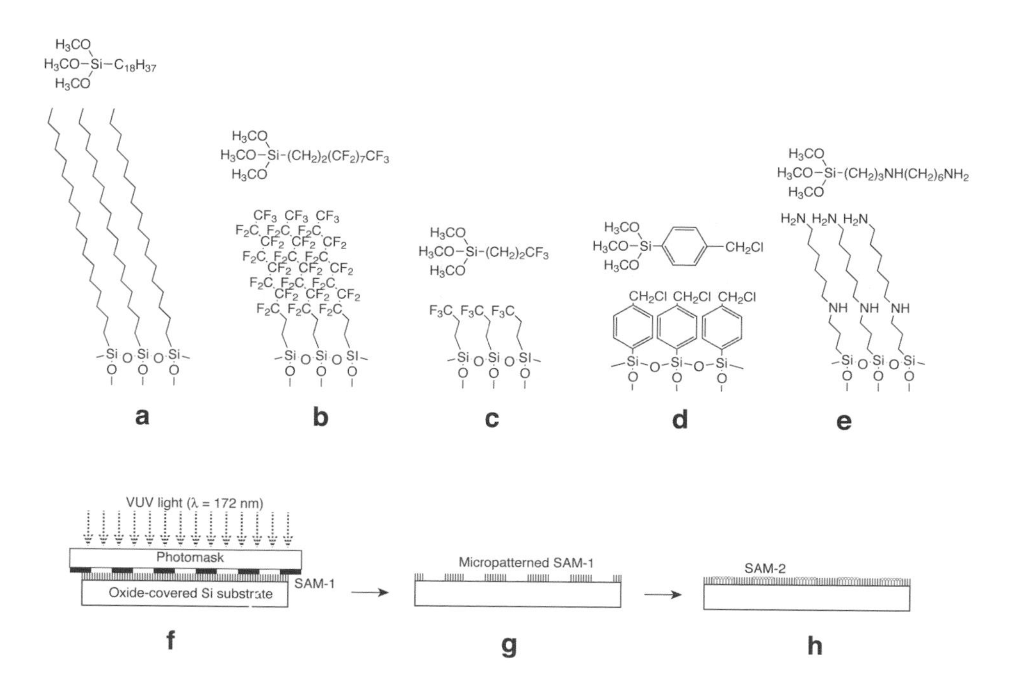

Fig. 3.22. Chemical structures of SAMs and their precursors a) ODS–SAM, b) FAS–SAM, c) TFPS–SAM, d) CMPhS–SAM and f) AHAPS–SAM. f–h) The procedure for sample preparation.

procedures as shown in Figure 3.22f–h. First, an ODS–SAM was prepared on Si substrates (n-type, 4–6 Ω cm) by the vapor phase method. Next, the substrate covered with ODS–SAM was micropatterned by VUV-lithography as described in Section 3.5.3. In the VUV-irradiated regions, the SAM was selectively removed so that the underlying SiO_2 layer was exposed. Finally, the VUV-patterned ODS–SAM sample was treated with a different precursor. Since such a photochemically exposed oxide surface was most likely to be terminated with OH groups, it has an affinity to organosilane molecules. The second SAM (SAM-2) consisting of FAS, TFPS, CMPhS, or AHAPS area-selectively formed confining to the VUV-irradiated pattern. In order to obtain surface potential differences between ODS–SAM and the other SAMs, the coplanar FAS/ODS, TFPS/ODS, CMPhS/ODS, and AHAPS/ODS microstructures are observed by KFM.

Figures 3.23a–d show KFM images of the FAS/ODS, TFPS/ODS, CMPhS/ODS, and AHAPS/ODS coplanar microstructures, respectively. Rectangular features of 5 μm × 25 μm correspond to the regions covered with a SAM other than ODS–SAM, while the surrounding area is covered with ODS–SAM. As clearly seen in Figure 3.23a–c, the regions covered with FAS–SAM, TFPS–SAM, and CMPhS–SAM show lower surface potentials than ODS–SAM. On the contrary, the region covered with AHAPS–SAM, as shown in Fig. 3.23d, possesses a higher surface potential than ODS–SAM. The potential

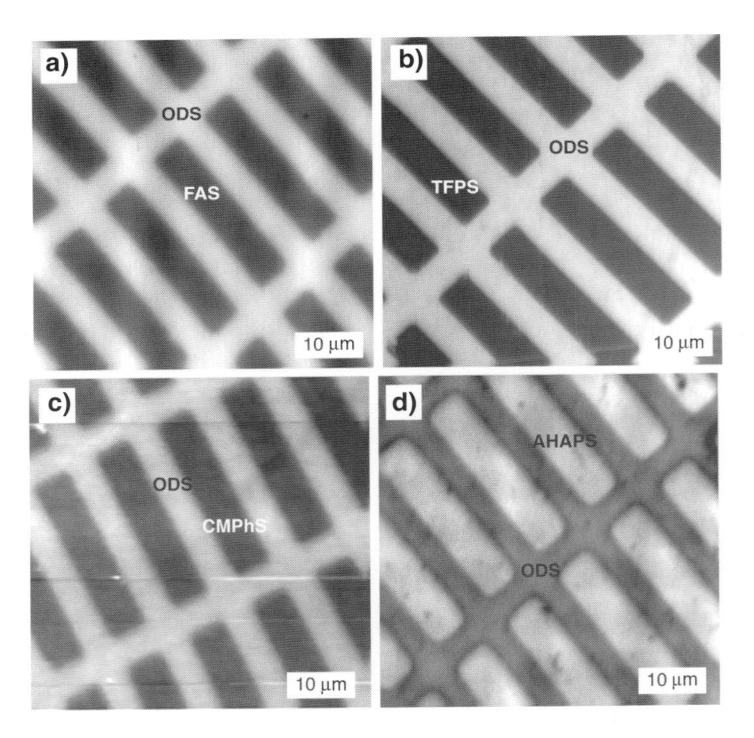

Fig. 3.23. Surface potential images of a) FAS/ODS, b) TFPS/ODS, c) CMPhS/ODS, and d) AHAPS/ODS.

contrasts of the regions covered with FAS–SAM, TFPS–SAM, CMPhS–SAM, and AHAPS–SAM with reference to ODS–SAM are ca. −180, −150, −30, and +50 mV, respectively. The main advantage of ODS–SAM as a reference is its hydrophobicity. Since the amount of adsorbed water, which affects measured surface potentials significantly [111,112], on ODS–SAM is small, surface potentials are reliably measured.

Here, we discuss surface potential contrasts between ODS–SAM and others. A surface potential of a SAM on a Si substrate is expressed by Eq. 3.2,

$$V_{\mathrm{SAM}} = -\frac{(\phi_{\mathrm{subst}} - \phi_{\mathrm{tip}})}{e} + \frac{\mu}{A\varepsilon_{\mathrm{SAM}}\varepsilon_0} + \alpha \tag{3.2}$$

where ϕ_{subst} and ϕ_{tip} are work functions of the Si substrate and the KFM tip, respectively, e is the electric charge, μ is the net dipole moment directed normally to the substrate surface, A is the area occupied by each molecule, and $\varepsilon_{\mathrm{SAM}}$ and ε_0 are the permittivity of the SAM and free space, respectively. Equation 3.1 is consists of three terms: the first is $-(\phi_{\mathrm{subst}} - \phi_{\mathrm{tip}})/e$, which represents the contact potential difference between the Si substrate and KFM tip, the second is $\mu/A\varepsilon_{\mathrm{SAM}}\varepsilon_0$, which represents the dipole moment of an organic thin film derived from Helmholtz equation, and the third is α which corresponds to a potential generated by trapped charges in the SiO_2 layer. The surface potential difference between the regions covered with ODS–SAM and another SAM is obtained by Eq. 3.3,

$$V_{\mathrm{SAM}} - V_{\mathrm{ODS}} = \frac{M_{\mathrm{SAM}}}{A_{\mathrm{SAM}}\varepsilon_{\mathrm{SAM}}\varepsilon_0} - \frac{\mu_{\mathrm{ODS}}}{A_{\mathrm{ODS}}\varepsilon_{\mathrm{ODS}}\varepsilon_0} \tag{3.3}$$

where the first and third terms of Eq. 3.2 do not remain. We assumed that differences in A and ε between ODS–SAM and the other SAMs are not so significant, since the SAMs are laterally connected with an identical Si–O–Si network as shown in Fig. 3.3 and are formed mainly from hydrocarbons. Thus, the potential contrast is considered to be primarily governed by the difference in dipole moment.

Dipole moments of ODS, FAS, TFPS, CMPhS, and AHAPS molecules were calculated using a slightly simplified model for each molecule in which each of the three methoxy groups is replaced with a hydrogen atom, since the methoxy groups in the precursor molecules do not remain in the SAMs. The modeled ODS, FAS, TFPS, CMPhS, and AHAPS molecule have dipole moments of 2.35, 3.41, 2.91, 2.40, and 1.04 debye, respectively. These dipole moments incline 26.5, 25.3, 3.6, 55.1, and 12.4°, respectively, with respect to each of the molecular chains. Vertical components of the dipole moments for the model ODS and AHAPS molecules are positive, namely, the dipole moments are directing from the bottom to the top, while those for the other model molecules, i.e. FAS, TFPS, and CMPhS are negative due to electron negativities of fluorine and chlorine atoms existing in their head groups. The dipole moments of the model FAS, TFPS, CMPhS, and AHAPS molecules were compared with that of the ODS model with regard to their vertical components. The estimated vertical dipole moment differences

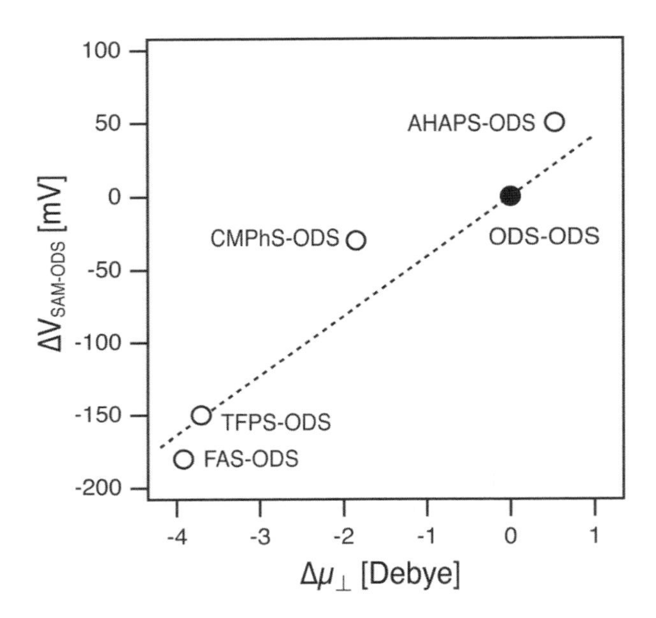

Fig. 3.24. Relation between $\Delta\mu\perp$ and ΔV.

($\Delta\mu\perp$) are as follows:

$$\begin{aligned}&\Delta\mu\perp(\text{FAS–ODS})(-3.92\,\text{D}) < \Delta\mu\perp(\text{TFPS–ODS})(-3.71\,\text{D})\\&\quad < \Delta\mu\perp(\text{CMPhS–ODS})(-1.85\,\text{D}) < \Delta\mu\perp(\text{ODS–ODS})(0\,\text{D})\\&\quad < \Delta\mu\perp(\text{AHAPS–ODS})(+0.53\,\text{D})\end{aligned} \tag{3.4}$$

The order and signs of $\Delta\mu\perp$ qualitatively agree with the order and signs of the surface potential contrasts (ΔV) as shown in Fig. 3.24. The line indicated in Fig. 3.24 is the relation derived from Eq. 3.3 with the assumption that A_{SAM} and ε_{SAM} are identical for all the SAMs. One plausible reason for the discrepancy between the experimental and calculated results is the molecular orientation in SAM, since the above discussion is based on the assumption that molecules in a SAM are aligned perpendicular to the substrate surface. It is known that SAMs are frequently constructed from inclined molecules [1]. The difference in A_{SAM} is also responsible. For example, TFPS–SAM incompletely covered a substrate compared with FAS–SAM. Since the molecular length of TFPS is much shorter than that of FAS, intermolecular interactions, which are necessary to form a densely packed monolayer, is so weak that TFPS–SAM is loosely packed.

3.6.2 Conductive and electroactive SAMs covalently bonded to Si

When an unsaturated hydrocarbon molecule having a carbon–carbon double bond reacts with a hydrogen-terminated Si surface, the molecule is attached to the substrate with a interfacial chemical structure of $\equiv$Si–CH_2–CH_2–, as shown in Fig. 3.4. Consequently,

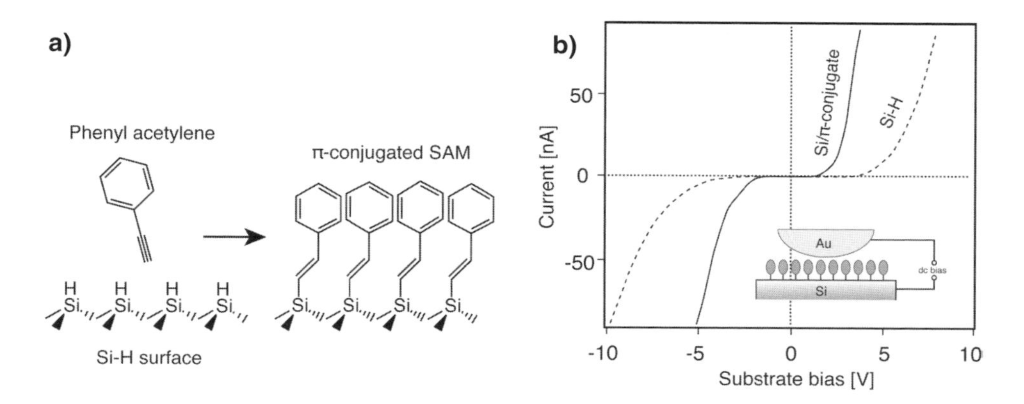

Fig. 3.25. SAM on Si prepared from phenylacetylene. a) Chemical strcutrures of phenylacetylene and the corresponding SAM. b) I–V characteristics of the junctions formed under an Au-coated AFM probe with the SAM-covered Si or Si–H surfaces.

the SAM formed entirely consists of σ bondings and is essentially an electrical insulator. On the contrary, in the case of a molecule containing a carbon–carbon triple bond, a SAM with an interfacial structure of $\equiv Si{-}CH_2{=}CH_2{-}$ is formed. For example, phenylacetylene (PA) molecules form a π-conjugated SAM which is expected to be electrically conductive as illustrated in Fig. 3.25a.

This π-conjugated SAM was actually formed on a Si(111)–H surface through a chemical reaction of PA at a temperature of 180°C [76]. Fig. 3.25b shows a current–voltage (I–V) characteristic of a junction between the π-conjugated SAM/Si sample and a gold-coated conductive probe of AFM in addition to an I–V curve of a Si(111)–H surface as a control. The junction current on the SAM surface distinctly flows at lower substrate voltages than that on the Si– and H surface. These results mean that a contact resistance between Au and π-conjugated SAM is lower than that between Au and Si– and H. It has been actually demonstrated that the π-conjugated SAM is electrically conductive to some extent and electrical properties of Si surfaces can be controlled by preparing a direct-bonding SAM on the surfaces.

Redox-active molecules have an ability to store and release electric charges reversibly. SAMs of such redox-active molecules attached to silicon [112–115] are, thus, expected to be applied to solid-state memory devices. Ferrocene has appropriate properties, such as redox activity, reversibility in redox reactions, and durability in repetition of the reactions. Thus, vinylferrocene (VFc) molecules were attached to a Si(111) surface through Si–C bonds by immersing a Si(111)–H substrate to a bath of 10-mM VFc dissolved in mesitylene at 150°C for 15 h as schematically illustrated in Fig. 3.26a.

A cyclic voltammogram for the Si(111) electrode functionalized with VFc was measured in a 0.1 M $HClO_4$ solution under a nitrogen purged atmosphere as shown in Fig. 3.26b. Oxidation and reduction peaks are observed at potentials around 0.36 and 0.3 V vs. Ag/AgCl, respectively. This means that ferrocenyl groups in the VFc molecules attached to Si undergo the oxidation/reduction reactions.

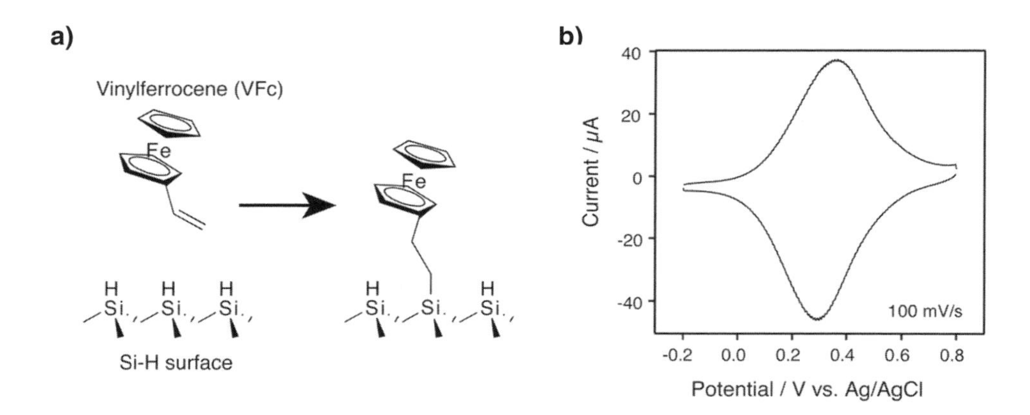

Fig. 3.26. Immobilization of vinylferrocene molecules on Si. a) Chemical reaction scheme. b) Cyclic-voltammogram of a VFc-treated Si electrode.

3.7 Summary and Conclusion

Besides a brief introduction to general aspects of SAMs formed on various solid substrates, preparation and properties of SAMs on Si were reviewed in this chapter. Two of the major methods for the fabrication of SAMs on Si were described. One is the silane coupling chemistry, i.e. the chemical reactions between organosilane molecules and hydroxyl groups on a surface oxide layer of Si. Thus, an interfacial oxide layer between the SAM and Si substrate is substantially present. The other method is based on the chemical reactions of unsaturated hydrocarbon, alcohol, and aldehyde with Si–H activated by thermal- or photo-excitation. By this method, the organic molecules are directly attached to bulk Si through Si–C or Si–O–C bonds. Furthermore, micro- and nanopatterning of the SAMs on Si, based on VUV-light irradiation and current-injecting AFM, respectively, were described. Finally, some of the electric properties of the SAMs, i.e. surface potential, contact resistance, and redox activity, were demonstrated.

The growth of thin films via chemisorptions of organic molecules onto solid surfaces has been a well-known phenomenon since 1940s. However, recent advances in science and technology, particularly in surface characterization and analyzing methods, have elucidated fundamental and practical aspects of SAMs, for example, their structures in molecular scales, detailed growth mechanisms, and practically useful functionalities. Consequently, the research and development on SAM and its application have been markedly pushed forward. The self-assembling approach to fabricate ultra thin films using molecules as building blocks will become a crucial technology increasingly as one of the bottom–up nanotechnologies in surface and interface engineering with a variety of practical applications.

References

[1] A. Ulman, Chem. Rev. 96 (1996) 1533.

[2] Frank Schreiber, Prog. Surf. Sci. 65 (200) 151.

[3] S. Flink, F.C.J.M. van Veggrl and D.N. Reinhoudt, Adv. Mater. 12 (2000) 1315.

[4] J.C. Love, L.A. Estroff, J.K. Kriebel, R.G. Nuzzo and G.M. Whitesides, Chem. Rev. 105 (2005) 1103.

[5] W.C. Biegelow, D.L. Pickett and W.A. Zisman, J. Colloid Sci. 1 (1946) 513.

[6] L.O. Brockway and J. Karle, J. Colloid Sci. 2 (1947) 277.

[7] E.G. Shafrin and W.A. Zisman, J. Colloid Sci. 4 (1949) 571.

[8] E.P. Plueddemann, Silane Coupling Agents; 2nd Ed. Plenum Press: New York and London, 1991.

[9] E.E. Polymeropoulos and J. Sagiv, J. Chem. Phys. 69 (1978) 1836.

[10] J. Sagiv, J. Am. Chem. Soc. 102 (1980) 92.

[11] M.R. Linford and C.E.D. Chidsey, J. Am. Chem. Soc. 115 (1993) 12631.

[12] J.M. Buriak, Chem. Rev. 102 (2002) 1271.

[13] A.B. Sieval, B. van den Hout, H. Zuilhof and E.J.R. Sudhoelter, Langmuir 16 (2000) 2987.

[14] S.R. Wasserman, Y.-T. Tao and G.M. Whitesides, Langmuir 5 (1989) 1074.

[15] U. Jonsson, G. Olofsson, M. Malmqvist and I. Ronnberg, Thin Solid Films 124 (1985) 117.

[16] H. Tada and H. Nagayama, Langmuir 11 (1995) 136.

[17] P.W. Hoffmann, M. Stelzle and J.F. Rabolt, Langmuir 13 (1997) 1877.

[18] H. Sugimura and N. Nakagiri, J. Photopolym. Sci. Technol. 10 (1997) 661.

[19] A.Y. Fadeev and T.J. McCarthy, Langmuir 15 (1999) 3759.

[20] A. Hozumi, K. Ushiyama, H. Sugimura and O. Takai, Langmuir 15 (1999) 7600.

[21] H. Sugimura, A. Hozumi, T. Kameyama and O. Takai, Surf. Interf. Anal. 34 (2002) 550.

[22] G.A. Husseini, J. Peacock, A. Sathyapalan, L.W. Zilch, M.C. Asplund, E.T. Sevy and M. R. Linford, Langmuir 19 (2003) 5169.

[23] T. Koga, M. Morita, H. Ishida, H. Yakabe, S. Sasaki, O. Sakata, H. Otsuka and A. Takahara, Langmuir 21 (2005) 905.

[24] G.-Y. Jung, Z. Li, W. Wu, Y. Chen, D.L. Olynick, S.-Y. Wang, W.M. Tong and R.S. Williams, Langmuir 21 (2005) 1158.

[25] A. Hozumi, H. Sugimura, Y. Yokogawa, T. Kameyama and O. Takai, Colloids and Surfaces A 182 (2001) 257.

[26] M.D. Butterworth, R. Corradi, J. Johal, S.F. Lascelles, S. Maeda and S.P. Armes, J. Colloid Interface Sci. 174 (1995) 510.

[27] J.W. Goodwin, R.S. Harbron and P.A. Reynolds, Colloid Polym. Sci. 268 (1990) 766.

[28] G.A. Parks, Chem. Rev. 65 (1965) 177.

[29] M.R. Linford, P. Fenter, P.M. Eisenberger and C.E.D. Chidsey, J. Am. Chem. Soc. 117 (1995) 3145.

[30] A.B. Sieval, R. Opitz, H.P.A. Maas, M.G. Schoeman, G. Meijer, F. J. Vergeldt, H. Zuilhof and E.J.R. Sudhlter, Langmuir 16 (2000) 10359.

[31] R. Boukherroub, S. Morin, P. Sharpe, D.D.M. Wayner and P. Allongue, Langmuir 16 (2000) 7429.

[32] F. Effenberger, G. Götz, B. Bidlingmaier and M. Wezstein, Angew. Chem. Int. Ed. 37 (1998) 2462.

[33] R. Boukherroub, S. Morin, F. Bensebaa and D.D.M. Wayner, Langmuir 11 (1999) 3831.

[34] R.L. Cicero, M.R. Linford and C.E.D. Chidsey, Langmuir 16 (2000) 5688.

[35] Q.-Y. Sun, L.C.P.M. de Smet, B. van Lagen, A. Wright, H. Zuilhof and E.J.R. Sudhölter, Angew. Chem. Int. Ed. 43 (2004) 1352.

[36] W.J. Dressick and J .M. Calvert, Jpn. J. Appl. Phys. 32 (1993) 5829.

[37] A. Kumar, N.L. Abbott, E. Kim, H.A. Biebuyck and G.M. Whitesides, Acc. Chem. Res. 28 (1995) 219.

[38] Y. Xia and G.M. Whitesides, Annu. Rev. Mater. Sci. 28 153 (1998).

[39] H. Sugimura, T. Hanji, O. Takai, T. Masuda and H. Misawa, Electrochim. Acta 47 (2001) 103.

[40] S. Krämer, R.R. Fuierer and C.B. Gorman, Chem. Rev. 103 (2003) 4367.

[41] R.K. Smith, P.A. Lewis and P.S. Weiss, Prog. Surf. Sci. 75 (2004) 1.

[42] R.C. Tiberio, H.G. Craighead, M. Lercel, T. Lau, C.W. Sheen and D.L. Allara, Appl. Phys. Lett. 62 (1993) 476.

[43] J.A.M. Sondag–Huethorst, H.R.J. van Helleputte and L.G.J. Fokkink, Appl. Phys. Lett. 64 (1994) 285.

[44] N. Mino, S. Ozaki, K. Ogawa and M. Hatada, Thin Solid Films 243 (1994) 374.

[45] M.J. Lercel. H.G. Craighead, A.N. Parikh, K. Seshadri and D.L. Allara, Appl. Phys. Lett. 68 (1996) 1504.

[46] R. Hild, C. David, H.U. Muller, B. Volkel, D.R. Kayser and M. Grunze, Langmuir 14 (1998) 342.

[47] W. Geyer, V. Stadler, W. Eck, M. Zharnikov, A. Gölzhäuser and M. Grunze, Appl. Phys. Lett. 75 (1999) 2401.

[48] P.C. Rieke, B.J. Tarasevich, L.L. Wood, M.H. Engelhard, D.R. Baer, G.E. Fryxell, C.M. John, D.A. Laken and M.C. Jaehnig, Langmuir 10 (1994) 619.

[49] K. Berggren, A. Bard, J.L. Wilbur, J.D. Gillaspy, A.G. Helg, J.J. McClelland, S.L. Rolston, W.D. Phillips, M. Prentiss and G.M. Whitesides, Science 269 (1995) 1255.

[50] R. Younkin, K.K. Berggren, K.S. Johnson, M. Prentiss, D.C. Ralph and G.M. Whitesides, Appl. Phys. Lett. 71 (1997) 1261.

[51] S.B. Hill, C.A. Haich, F.B. Dunning, and G.K. Walters, J.J. McClelland, R.J. Celotta, H. G. Craighead, J. Han and D.M. Tanenbaum, J. Vac. Sci. Technol. B 17 (1999) 1087.

[52] X.M. Yang, R.D. Peters, T.K. Kim and P.F. Nealey, J. Vac. Sci. Technol. B 17 (1999) 3203.

[53] T.K. Kim, X.M. Yang, R.D. Peters, B.H. Sohn and P.l F. Nealey, J. Phys. Chem. 104 (2000) 7403.

[54] M. Zharnikov and M. Grunze, J. Vac. Sci. Technol. B 20 (2002) 1793.

[55] Y.-H. La, Y.J. Jung, H.J. Kim, T.-H. Kang, K. Ihm, K.-J. Kim, B. Kim and J.W. Park, Langmuir 19 (2003) 4390.

[56] R. Klauser, M.-L. Huang, S.-C. Wang, C.-H. Chen, T.J. Chuang, A. Terfort and M. Zharnikov, Langmuir 20 (2004) 2050.

[57] A. Kumar, H.A. Biebuyck, N.L. Abbott and G.M. Whitesides, J. Am. Chem. Soc. 114 (1992) 9188.

[58] A. Kumar, H.A. Biebuyck and G.M. Whitesides, Lanmguir 10 (1994) 1498.

[59] A. Kumar and G.M. Whitesides, Science 263 (1994) 60.

[60] Y. Xia, M. Mrksich, E. Kim and G.M. Whitesides, J. Am. Chem. Soc. 117 (1995) 9576.

[61] C.S. Dulcey, J.H. Georger, V. Krauthamer, D.A. Stenger, T.L. Fare and J.M. Calvert, Science 252 (1991) 551.

[62] D.A. Stenger, J.H. Georger, C.S. Dulcey, J.J. Hickman, A.S. Rudolph, T.B. Nielsen, S.M. McCort, and J.M. Calvert, J. Am. Chem. Soc. 114 (1992) 8435.

[63] C.S. Dulcey, J.H. Georger, M.-S. Chen, S.W. McElvany, C.E. O'Ferrall, V.I. Benezra and J.M. Calvert, Langmuir 12 (1996) 1638.

[64] S.L. Brandow, M.-S. Chen, R. Aggarwal, C.S. Dulcey, J.M. Calvert and W.J. Dressick, Langmuir 15 (1999) 5429.

[65] N. Ichinose, H. Sugimura, T. Uchida, N. Shimo and H. Masuhara, Chem. Lett. (1961) (1993).

[66] H. Sugimura and N. Nakagiri, Jpn. J. Appl. Phys. 36 (1997) L968.

[67] H. Sugimura and N. Nakagiri, Appl. Phys. A 66 (1998) S427.

[68] H. Sugimura, K. Ushiyama, A. Hozumi and O. Takai, Langmuir 16 (2000) 885.

[69] H. Sugimura, T. Shimizu and O. Takai, J. Photopolym. Sci. Technol. 13 (2000) 69.

[70] H. Sugimura, K. Hayashi, Y. Amano, O. Takai and A. Hozumi, J. Vac. Sci. Technol. A 19 (2001) 1261.

[71] H. Sugimura, N. Saito, Y. Ishida, I. Ikeda, K. Hayashi and O. Takai, J. Vac. Sci. Technol. A 22 (2004) 1428.

[72] H. Sugimura, L. Hong and K.-H. Lee, Jpn. J. Appl. Phys. 44 (2005) 5185.

[73] R.P. Roland, M. Bolle and R.W. Anderson, Chem. Mater. 13 (2001) 2493.

[74] L. Hong, H. Sugimura, O. Takai, N. Nakagiri and M. Okada, Jpn. J. Appl. Phys. 42 (2003) L394.

[75] N. Saito, Y. Kadoya, K. Hayashi, H. Sugimura and O. Takai, Jpn. J. Appl. Phys. 42 (2003) 2534.

[76] N. Saito, K. Hayashi, H. Sugimura and O. Takai, Langmuir 19 (2003) 10632.

[77] H. Sugimura, K.-H. Lee, H. Sano and R. Toyokawa, Colloids and Surfaces A, submitted.

[78] N. Saito, K. Hayashi, S. Yoda, H. Sugimura and O. Takai, Surf. Sci. 532-535 (2003) 970.

[79] H. Sugimura, T. Hanji, O. Takai, K. Fukuda and H. Misawa, Materials Research Society Symposium Proceedings, 584 (2000) 163.

[80] H. Sugimura, K. Hayashi, N. Saito, L. Hong, O. Takai, A. Hozumi, N. Nakagiri and M. Okada, Trans. Mater. Res. Soc. Japan 27 (2002) 545.

[81] L. Hong, H. Sugimura, O. Takai, N. Nakagiri and M. Okada, Jpn. J. Appl. Phys 42 (2003) L394.

[82] N. Saito, Y. Wu, K. Hayashi, H. Sugimura and O. Takai, J. Phys. Chem. B 107 (2003) 664.

[83] H. Sugimura, L. Hong and K.-H. Lee, Jpn. J. Appl. Phys. 44 (2005) 5185.

[84] H. Rohrer, Jpn. J. Appl. Phys.32 (1993) 1335.

[85] P. Avoris, Acc. Chem. Res. 28 (1995) 95.

[86] H. Sugimura and N. Nakagiri, Jpn. J. Appl. Phys. 34 (1995) 3406.

[87] R.M. Nyffebegger and R.M. Penner, Chem. Rev. 97 (1997) 1195.

[88] C.F. Quate, Surf. Sci. 386 (1997) 259.

[89] H.T. Soh, K.W. Guarini and C.F. Quate, Scanning Probe Lithography, Kluwer Academic Publishers: Boston, 2001.

[90] K. Sattler, Jpn. J. Appl. Phys. 42 (2003) 4825.

[91] D. Wouters and U.S. Schubert, Angew. Chem. Int. Ed. 43 (2004) 2480.

[92] C.R.K. Marrian, F.K. Perkins, S.L. Brandow, T.S. Koloski, E.A. Dobisz and J.M. Calvert, Appl. Phys. Lett. 64 (1994) 390.

[93] F.K. Perkins, E.A. Dobisz, S.L. Brandow, T.S. Koloski, J.M. Calvert, K.W. Rhee, J. E. Kosakowski and C.R.K. Marrian, J. Vac. Sci. Technol. B 12 (1994) 3725.

[94] H. Sugimura and N. Nakagir, Langmuir 11 (1995) 3623.

[95] H. Sugimura and N. Nakagir, J. Vac. Sci. Technol. A 14 (1996) 1223.

[96] H. Sugimura, K. Okiguchi and N. Nakagiri, Jpn. J. Appl. Phys. 35 (1996) 3749.

[97] H. Sugimura, K. Okiguchi and N. Nakagiri and M. Miyashita, J. Vac. Sci. Technol. B 14 (1996) 4140.

[98] H. Sugimura, T. Hanji, K. Hayashi and O. Takai, Ultramicroscopy, 91(2002) 221.

[99] K. Hayashi, H. Sugimura and O. Takai, Jpn. J. Appl. Phys. 40 (2001) 4344.

[100] H. Sugimura, T. Uchida, N. Kitamaura and H. Masuhara, J. Phys. Chem. 98 (1994) 4352.

[101] H. Sugimura, N. Saito, S.-H. Lee and O. Takai, J. Vac. Sci. Technol. B 22 (2004) L44.

[102] H. Sugimura, Jpn. J. Appl. Phys. 43 (2004) 4477.

[103] M. Fujihira, Ann. Rev. Mater. Sci. 28 (1999) 353.

[104] X. Chen, H. Yamada, T. Horiuchi and K. Matsushige, Jpn. J. Appl. Phys. 38 (1999) 3932.

[105] J. Lü, E. Delamarche, L. Eng, R. Bennewits, E. Myer and H-J. Güntherodt, Langmuir 15 (1999) 8184.

[106] H. Sugimura, K. Hayashi, N. Saito, N. Nakagiri and O. Takai, Appl. Surf. Sci. 188 (2002) 403.

[107] S. Kobayashi, T. Nishikawa, T. Takenobu, S. Mori, T. Shimoda, T. Mitanai, H. Shimotani, N. Yoshimoto, S. Ogawa and Y. Iwasa, Nature Materials 3 (2004) 317.

[108] H. Sugimura, K. Hayashi, N. Saito, N. Nakagiri and O. Takai, Appl. Surf. Sci. 188 (2002) 403.

[109] K. Hayashi, N. Saito, H. Sugimura, O. Takai and N. Nakagiri, Langmuir 18 (2002) 7469.

[110] O. Gershevitz, C.N. Sukenik, J. Ghabboun and D. Cahen, J. Am. Chem. Soc. 125 (2003) 4730.

[111] S. Ono, M. Takeuchi and T. Takahashi, Appl. Phys. Lett. 78 (2001) 1086.

[112] H. Sugimura, Y. Ishida, K. Hayashi, O. Takai and N. Nakagiri, Appl. Phys. Lett. 80 (2002) 1459.

[113] Peter Kruse, Erin R. Johnson, Gino A. DiLabio and Robert A. Wolkow, Nano Lett. 2 (2002) 807.

[114] K.M. Roth, A.A. Yasseri, Z. Liu, R.B. Dabke, V. Malinovskii, K.-H. Schweikart, L. Yu, H. Tiznado, F. Zaera, J.S. Lindsey, W.G. Kuhr and D.F. Bocian, J. Am. Chem. Soc. 125 (2003) 505.

[115] Q. Li, G. Mathur, M. Homsi, S. Surthi and V. Misra, Appl. Phys. Lett. 81 (2002) 1494.

[116] R. Zanoni, F. Cattaruzza, C. Coluzza, E.A. Dalchiele, F. Decker, G. Di Santo, A. Flamini, L. Funari and A.G. Marrani, Surf. Sci. 575 (2005) 260.

CHAPTER 4

Fabrication and Structural Characterization of Ultrathin Nanoscale Wires and Particles

Ning Wang and Kwok Kwong Fung

Department of Physics
The Hong Kong University of Science and Technology
Hong Kong, China

4.1 Introduction

The tremendous interest in nanoscale structures such as quantum dots (0-dimension) and wires (quasi-one-dimension (1D)) stems from their size-dependent optical, electrical, and mechanical properties as well as potential applications in nanoscale electronic and optoelectronic devices. In the physics of nanoscale 0D and 1D semiconductors, quantum effects are expected to play an increasingly prominent role [1]. As predicted theoretically, quantum wires have shown interesting transport properties and may have applications in high-performance devices. This is because the current transport in a quantum wire is easily influenced by nearby charges, just like the gate in a field effect transistor. For sufficiently small width of the 1D nanostructure, this can be extended even down to the single-charge limit.

In addition to new studies in physics, much effort has been devoted to fabricating high-quality semiconductor nanostructures by employing different techniques [2–13]. The most popular technique used to fabricate semiconductor artificial structures with feature sizes in sub-100 nm range is nanofabrication (electrons, ions or ultraviolet lithography) [2,3]. 1D nanostructures are generally fabricated on the surface of a substrate by tedious processes of lithography such as photoresist removal, chemical or ion-beam etching, and surface passivation. For semiconductor nanostructures, etching processes can lead to significant surface structure damage, and thus introduce surface states to the nanostructure. This damage may not be serious for the structures in the micrometer range. However, nanostructures with dimensions in the range of nanometers are very sensitive to the surface states or impurities induced by the fabrication processes. 1D nanostructures formed 'naturally' (also called self-organized growth) without the aid of ex situ techniques such as chemical etching are desirable not only in fundamental research but also in future device design and fabrication.

Various methods have been used to realize 1D nanostructures by self-organized techniques. For example, a stepped surface can be used as a template to grow nanowires along the steps [4]. The deposited atoms can be absorbed at the step edges forming nanowire arrays. Metal catalytic growth (also called the vapor–liquid–solid (VLS) [5] growth) is another important technique widely used in growing 1D nanostructures. VLS growth can produce freestanding 1D nanostructures with fully controlled diameters. It offers a number of advantages over other techniques; for example, the nucleation sites and the sizes of the nanostructures can be well controlled by the pre-formed metal catalysts. Since the 1960s, Si and other semiconductor whiskers grown from the VLS reaction [5,6] have been extensively studied. In the VLS reaction, Au particles on Si substrates are used as the mediating solvent since Au and Si form a eutectic alloy with a relatively low melting temperature. At a certain high temperature, Au–Si alloy droplets form on the substrate. Si in the vapor phase diffuses into the liquid alloy droplet and bonds to the solid Si at the liquid–solid interface resulting in 1D crystal growth. Since 1997, new techniques such as laser-assisted chemical vapor deposition (CVD) [7,8,9], oxide-assisted CVD (without metal catalyst) [10], thermal CVD [11], and metal-catalyzed molecular beam epitaxy (MBE) [12,13] and chemical beam epitaxy (CBE) [14] techniques have been developed to fabricate semiconductor 1D nanostructures. However, how to fabricate desired 1D semiconductor nanostructures with well-controlled atomic structures (e.g. chemical composition, density of defects, doping state, and ohmic contacts) is still a challenging issue for materials scientists. 1D semiconductor nanostructures produced by conventional synthesis techniques, such as CVD or laser ablation have to be purified and manipulated in order to measure their properties. For electrical property measurement, surface chemicals (often introduced during growth or by chemical etching) always influence the quality of the contacts between the nanostructure and the electrodes. Moreover, it is difficult to control the doping state and doping site in 1D nanostructures by most conventional synthesis techniques. CBE and MBE, however, provide an ideal clean growth environment; moreover, the atomic structure and doping state can be well controlled. For example, using Au-catalyzed CBE, high quality 1D heterostructure (InAs/InP and 40 nm in diameter) nanowires have been fabricated and characterized [14]. The InP barrier layer with a thickness of 1.5 nm and good interface structure can be precisely controlled.

The discovery of carbon nanotubes [15] by direct high-resolution transmission electron microscopy (HRTEM) imaging was one of the most important events in the history of electron microscopy. Transmission electron microscopy is a technique for characterizing materials down to the atomic limit. It has revolutionized the understanding of materials in 20th century by correlating the material properties to the processing and structures down to atomic levels. At the atomic level, the structure of a crystal is the organization of atoms regularly arranged with respect to one another. In other words, it depends mainly on the arrangements of the atoms, ions, or molecules that make up the crystals and the bonding forces between them. It has been known that the microstructure of a crystal is often correlated with its physical properties, such as electrical resistance, optical refractivity, elastic modulus, and magnetic property. Most engineering materials consist of imperfect crystals. Particularly, the existence of defects plays the most important role in enhancing the mechanical properties of solid materials. Structural characterization and determination of solid materials is an indispensable task not only for materials research and development but also for materials engineering. Nowadays, modern electron microscopes provide more powerful methods to perform diffraction and even directly reveal atomic structure in crystals. Most nanostructures which show quantum effects are in the nanometer range. HRTEM has become a routine tool for the characterization of the atomic structure of nanomaterials and the in situ study of their physical properties.

In this chapter, several novel methods for nanomaterial formation (the VLS, laser-assisted growth, oxide-assisted growth, and metal-catalyzed MBE techniques) are described in detail. HRTEM and electron diffraction characterization of the nucleation and growth mechanism of 1D nanomaterials are also presented. Later, some interesting properties of ultrathin semiconductor 1D nanostructures, such as the size-dependence of growth direction and photothermal effect are illustrated as well. For metallic nanoparticles, complications arise from the surface oxide films. The structure of the surface film depends on how the nanoparticles are passivated, i.e. oxidized. In general, the oxide films are polycrystalline rather than amorphous. This means that the oxide films are not able to provide adequate protection to the metal core from further oxidation. Under conditions that remain unclear, epitaxial oxide films on metals are obtained. According to Caberra and Mott[70], an epitaxial oxide film of thickness 2–3 nm is able to protect the metal from further oxidation at room temperature. It is critical to eliminate the grain boundaries of the oxide film in order to provide protection to the metal. Furthermore, the physical properties of the epitaxial oxide films are very different from those of the metal. It is expected that the physical properties of the metallic nanoparticles will be seriously affected, particularly for particles of a few nm in size. Another critical issue in the study of metallic nanoparticles is the size distribution. The physical properties will not be manifested in nanoparticles with a wide size distribution. Fabrication of mono-sized nanoparticles is therefore very important. We have observed novel magnetic properties in mono-sized epitaxial γ-Fe_2O_3-coated nanoparticles of iron, but not in epitaxial γ-Fe_2O_3-coated nanoparticles with a wide size distribution. The stability of the epitaxial oxide films at temperatures above room temperature is scientifically interesting and technologically important. We have obtained encouraging results showing the self-limiting growth of chromium oxide on chromium nanoparticles when they are heated by electron illumination in a TEM.

4.2 Novel Methods of 1D Nanomaterial Formation

4.2.1 The classical VLS technique

In 1964, Wagner and Ellis [5] first reported the novel unidirectional growth phenomenon of Si whiskers from vapor phase deposition. The principle of Si whisker growth is schematically shown in Fig. 4.1a. Au particles deposited on the surface of a Si substrate react with Si to form Au–Si alloy droplets at a certain temperature. The structure of Au–Si alloy in solid solution is as shown in the Au–Si phase diagram in Fig. 4.1c. The melting point of Au–Si alloy at the eutectic point is only about 363°C. For the vapor source of the mixture of $SiCl_4$ and H_2, it is known that a reaction between them may happen at

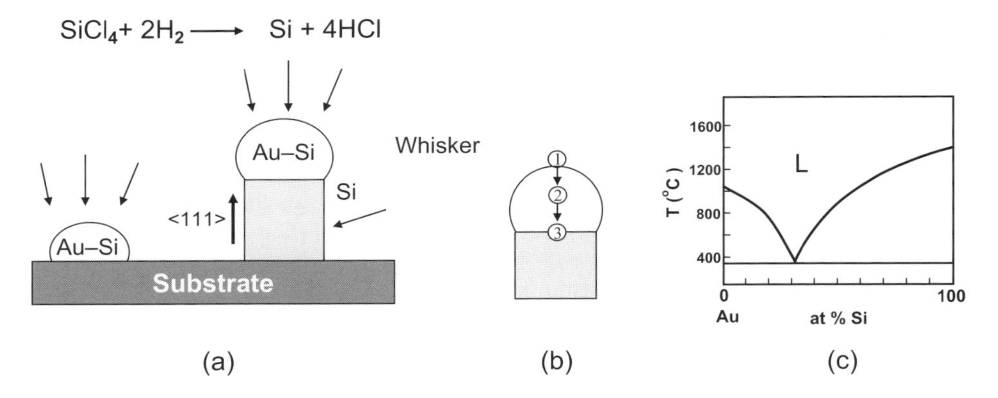

Fig. 4.1. Schematic illustration of Si whisker growth from the vapor phase via Au–Si catalytic droplets. (a) Au–Si droplet and the whisker growth; (b) A whisker tip; (c) Au–Si phase diagram.

temperatures above 800°C. Below this temperature, almost no deposition of Si occurs on the surface of the substrate [6]. However, at a temperature above 363°C, Au particles can easily form Si–Au eutectic droplets on the substrate surface. Due to the catalytic effect, the reduction reaction occurs at the droplets. The Au–Si droplets absorb Si from the vapor phase resulting in supersaturated alloy droplets. Since the melting point of Si is high (1414°C), Si atoms in the droplets precipitate and bond at the liquid (droplet)–solid (Si) interface, and the liquid droplet rises from the original Si substrate surface. This process shown by the path 1→2→3 in Fig. 4.1b finally results in the unidirectional growth. Such a unidirectional growth is also the consequence of anisotropy in solid–liquid interface energy. This growth process is called vapor–liquid–solid (VLS) mechanism because the vapor, liquid, and solid phases are involved. The main feature of the VLS process is its low activation energy compared to the normal vapor–solid process. The whiskers grow only at the areas seeded by the Au–Si particles. The whisker diameter is mainly determined by the droplet size.

The kinetics on the VLS whisker growth has been well studied [6]. For Si whiskers, a typical kinetic experimental result is the growth rate dependence on the diameter of the whisker. The thicker the whisker diameter, the faster is its growth rate. Depending on the driving force (the chemical potential difference), Si whiskers with small diameters (<0.1 μm) grow very slowly. There is a critical diameter at which whisker growth stops completely. This growth phenomenon is attributed to the well-known Gibbs–Thomson effect, i.e. the decrease of supersaturation as a function of the whisker diameter [6]:

$$\frac{\Delta\mu}{k\mathrm{T}} = \frac{\Delta\mu_0}{k\mathrm{T}} - \frac{4\alpha\Omega}{k\mathrm{T}} \cdot \frac{1}{d}, \tag{4.1}$$

or

$$\Delta\mu = \Delta\mu_0 - 4\alpha\Omega/d, \tag{4.2}$$

where $\Delta\mu$ (also the driving force for whisker growth) is the effective difference between the chemical potentials of Si in the vapor phase and in the whisker. $\Delta\mu_0$ is the chemical potential difference for the plane boundary case, i.e. the whisker diameter $d \rightarrow \infty$. Ω is the atomic volume of Si; α the specific free energy of the whisker surface. Obviously, there is a critical diameter for the whisker growth at $\Delta\mu = 0$. Whiskers with diameters smaller than the critical diameter will stop growing. Thick whiskers grow faster than narrow ones [6]. At the thermodynamic equilibrium state, the stability of a liquid droplet depends on the degree of supersaturation. The whisker growth rate as a function of the driving force (supersaturation $\Delta\mu/kT$) has been determined experimentally. However, recent studies have shown that although Eq. (4.1) can be used to predict the VLS growth of most whiskers, it is not sufficient to describe the VLS growth because the droplet size may not be the same as that of the whisker, and the binary alloy nature of the droplet should be considered [16].

Under isothermal condition, the main growth direction of Si whiskers is $\langle 111 \rangle$ on Si (111) substrates, i.e. perpendicular to the substrate surfaces. Si whiskers are dislocation-free. It is believed that this particular growth direction arises because the solid–liquid interface is a single (111) plane, and this plane is very stable during the VLS growth. Twinning structure and twin-dendrites (or branched whiskers) have been frequently observed in the whiskers. Ribbon-like whiskers often coexist and show $\langle 111 \rangle$ or $\langle 112 \rangle$ growth direction [17]. Their crystalline structure is generally perfect, though steps and facets occur on the whisker surfaces. Dislocations or other crystalline defects are not essential for the growth of the whiskers in the VLS process. Although the VLS mechanism discussed above has been well acknowledged, the real absorption, reaction, and diffusion processes of Si atoms through the catalyst are complicated in different experimental conditions and material systems. As proposed by Wang *et al.* [18], Si atoms may easily diffuse along the surface of the liquid droplet to the liquid (droplet)–solid interface, where Si atoms deposit. Recently, a systematical investigation of in-situ HRTEM carried out by Howe *et al.* [19] reveals that the metal droplets on Si substrates are in a partially molten state. The surface and interface regions are liquid, while the cores of the droplets are in solid state. It is therefore suggested that diffusion of Si atoms along the droplet surface should be a more favorable mass transport path for the whisker growth. Alternatively, a model based on surface diffusion has been proposed recently [20]. Unlike the classical VLS catalytic growth, it is proposed that atoms may deposit on the whole surfaces (on the whiskers and the substrate) and then diffuse along the surface of the whisker to the liquid–solid interface at the catalyst tip.

In different semiconductor material systems, whiskers with similar morphology and structures have been investigated. A variety of whisker forms is also obtained, for example, GaP whiskers [21] display rotational twins around their $\langle 111 \rangle$ growth axes, while GaAs [22,23] whiskers could grow in the form of wurtzite structure. Many other examples of semiconductor and oxide whiskers are given in [6,58,59].

4.2.2 Laser-assisted catalytic growth

Since the mid-1990s, ultrathin semiconductor nanowires (or nanowhiskers) have aroused great interest because of their interesting properties, e.g. quantum size effects and their potential application as sensitive detectors or nanodevices. Many efforts have been made to synthesize Si nanowires with diameters smaller than 100 nm by employing different methods. Of particular interest is the laser ablation of metal-containing

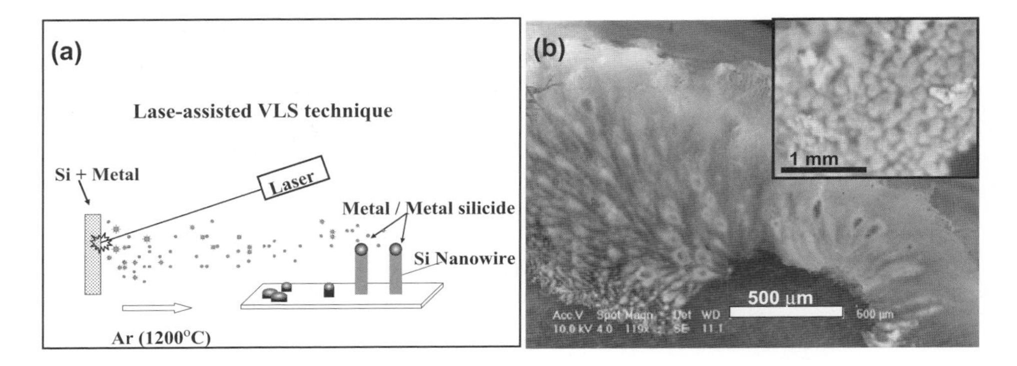

Fig. 4.2. (a) Experimental setup for the synthesis of Si nanowires by laser ablation. (b) SEM image of Si nanowires formed by laser ablation. The insert is an image from an optical microscope.

semiconductor targets [7,8,24], by which bulk-quantity semiconductor nanowires can be readily obtained. Compared to the classical VLS, nanoparticles of metal or metal silicide in large quantity are rather easy to obtain from the high-temperature induced by laser ablation. Assisted by laser ablation, these nanoparticles act as the critical catalyst for the nucleation and growth of nanowires.

Figure 4.2a shows schematically the experimental set up of the laser-ablation technique [24]. The experiment in [24] was carried out using a high power excimer laser (248 nm, 10 Hz, 400 mJ/pulse) to ablate the target in an evacuated quartz tube containing Ar (~500 torr). The temperature around the target was about 1200°C. The target used was highly pure Si powder mixed with metals (e.g. Fe, Ni, or Co). The laser beam (1×3 mm^2) was focused on the target surface. Si nanowire products (sponge-like, dark yellow in color) (Fig. 4.2b) formed on the Si substrate or the inner wall of the quartz tube near the water-cooled finger after 1–2 h laser ablation. The temperature of the area around the quartz tube where the nanowire grew was approximately 900–1000°C.

In the laser-ablation process, the metal catalysts are envisioned as clusters. They are in liquid state and serve as the energetically favored reaction sites for absorption of the reactant. They are also the nucleation sites for crystallization when supersaturated. Then, preferential 1D growth occurs with the presence of reactant. Si nanowires obtained by ablating a metal-containing (1–3% at.) Si powder target are extremely long and straight. The typical diameters of the nanowires are 20–50 nm. There is a metal catalyst at the tip of each nanowire (Fig. 4.3a). As identified by electron diffraction and HRTEM (Fig. 4.3b), the axis of most nanowires are along $\langle 111 \rangle$ direction. Defects such as stacking faults can be easily found in the nanowires. An amorphous thin shell always occurs on each Si nanowire. The growth mechanism has been extrapolated from the classical VLS reaction [7,8,24]. This is because Fe (or other metal catalysts) mixed in the targets can easily form Fe-silicides at a high temperature induced by laser ablation. Compared to the classical VLS, the major difference is that the reaction during the laser-ablation process is not under equilibrium conditions. Therefore, small-size metal catalysts in the range of 10–30 nm can easily form.

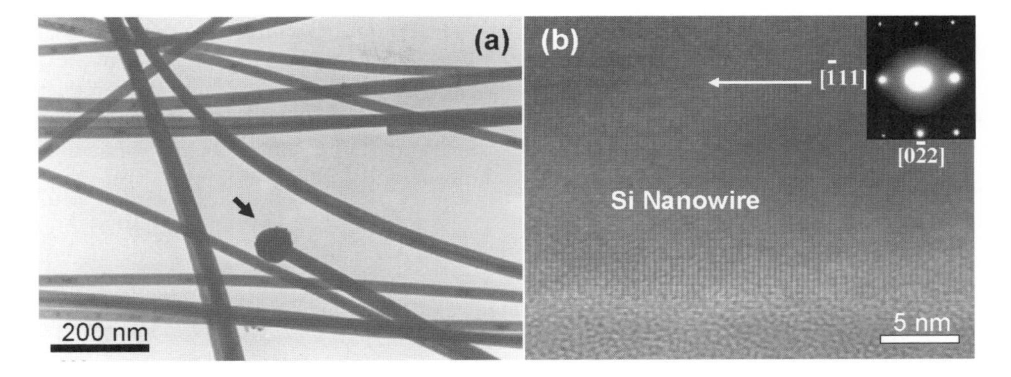

Fig. 4.3. (a) TEM image of Si nanowires synthesized by laser ablation technique. The arrow indicates the metal catalyst on the tip. (b) HRTEM image showing the ⟨111⟩ growth direction of the nanowire.

4.2.3 Oxide-assisted growth

In 1998, Wang *et al.* [10,25] suggested that a metal catalyst was unnecessary during the formation of Si nanowires by laser ablation. Instead, SiO_2 has been found to be an effective catalyst which largely enhanced Si nanowire growth. A model called Si-oxide-assisted growth was therefore proposed which was evidenced by the experiment not only in Si nanowire growth but also in Ge [26], III–V [27] and II–VI [28] semiconductor nanowire growth.

Si nanowires from Si-oxide-assisted growth are highly pure. Their diameters are uniform as shown in Fig. 4.4a. HRTEM investigations show that the defects and silicon oxide outer layers existing at the nanowire tips may play important roles for the formation and growth of Si nanowires. The growth direction of most Si nanowires synthesized by this method are along the ⟨112⟩ direction. The experimental setup for the Si-oxide-assisted

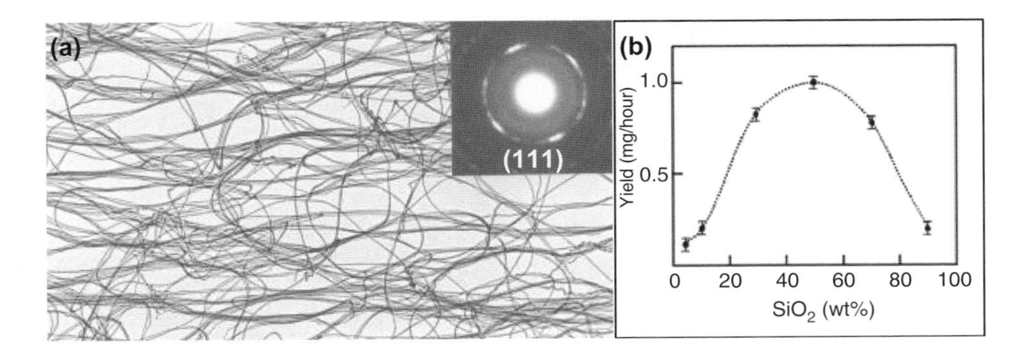

Fig. 4.4. (a) Si nanowires synthesized by the oxide-assisted growth. (b)Yield of Si nanowires vs. the percentage of SiO_2 in the target.

growth is very similar to that of the laser ablation technique. The target is a mixture of Si and SiO_2. No metal catalyst is needed. As demonstrated in Fig. 4.4b, the presence of SiO_2 in the powder target significantly enhances Si nanowire growth. The Si nanowire product obtained using a powder target composed of 50% SiO_2 and 50% Si is 30 times greater than the amount generated by using a metal containing target [10]. The mechanism of the oxide-assisted growth is described by the following reactions [25,29].

$$\mathrm{Si\,(solid)} + \mathrm{SiO_2\,(solid)} \xrightarrow{\textit{high temperature}} \mathrm{2SiO\,(gas)} \tag{4.3}$$

$$\mathrm{2SiO\,(gas)} \xrightarrow{\textit{low temperature}} \mathrm{Si\,(solid)} + \mathrm{SiO_2\,(solid)} \tag{4.4}$$

$$\mathrm{Si}_x\mathrm{O\,(gas)} \xrightarrow{\textit{low temperature}} \mathrm{Si}_{x-1}\,\mathrm{(solid)} + \mathrm{SiO\,(solid)}\ (x > 1) \tag{4.5}$$

By laser ablation, Si reacts with SiO_2 to form SiO or Si_xO ($x > 1$) vapor phase. This vapor phase is the key factor for the nucleation and growth of Si nanowires because it deposits on the substrate and decomposes into nanoparticles at a relatively low temperature of 900–1000°C. These nanoparticles are the nuclei of Si nanowires as discussed later in detail.

The model of oxide-assisted growth has been proven by a simple experiment [30] as shown in Fig. 4.5, which was carried out by simply sealing pure SiO powder or the mixture of Si and SiO_2 (1:1) in an evacuated (vacuum < 10 torr) quartz tube and then inserting the tube into a preheated furnace (1250–1300°C). No catalyst or special ambient gas was needed. Put one end of the tube outside the furnace in order to generate a temperature gradient between the source material and the nanowire formation zone. After 20–30 min, a high yield of sponge-like Si nanowire product formed on the cooler parts of the tube where the temperature was about 800–1000°C. A similar experiment was performed in 1950 [31,32]. Unfortunately, Si nanowires, labeled as '*light brown loose material*' in [31], were overlooked at the time.

4.2.4 Metal-catalyzed molecular beam or chemical beam epitaxy growth

One of the challenging issues in producing 1D nanostructured materials is how to fabricate and control the morphology of as-grown nanostructures or to make nanostructures with well-organized arrangement. The VLS method has been widely used to grow 1D nanostructures because the nucleation sites and the sizes of the 1D nanostructures

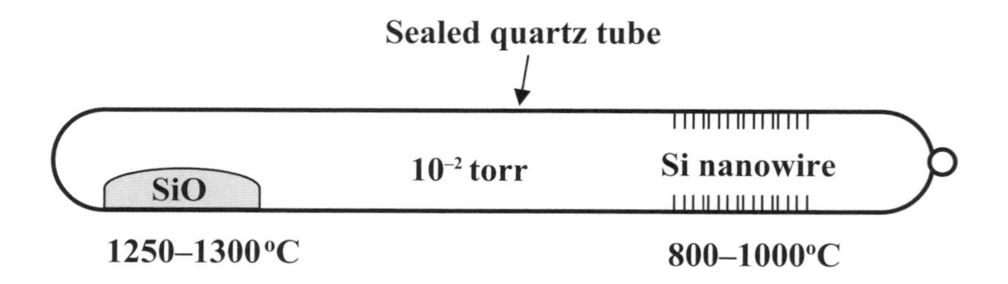

Fig. 4.5. Experimental setup for the fabrication of Si nanowires from SiO powder.

can be well controlled by the pre-formed metal catalysts. Since 2001, MEB and CBE techniques have been employed to synthesize Si [33], II–V [13], and III–V [12,14,34] compound semiconductor nanowires. The morphology and structure of these nanowires could be fully controlled at atomic scale. Using electron beam lithography (EBL), nano-sized Au catalyst patterns can be easily prepared on a substrate. Figure 4.6 shows the flowchart of the EBL for the preparation of Au-catalyst for MBE or CBE nanowire growth. Under certain growth conditions, nanowires may grow perpendicularly to the substrate surface at the seeded position forming nanowire arrays (Fig. 4.7).

For Au-catalyzed MBE growth of II–VI (ZnSe or ZnS) nanowires [13], for example, the synthesis is carried out using a ZnSe compound-source effusion cell at temperatures above 500°C. Au nanoparticles are in a molten state at this temperature. The growth rate is about 0.1 nm/s. The growth temperature is a very important factor for the formation of high-quality ZnSe nanowires. On the one hand, the deposition of ZnSe

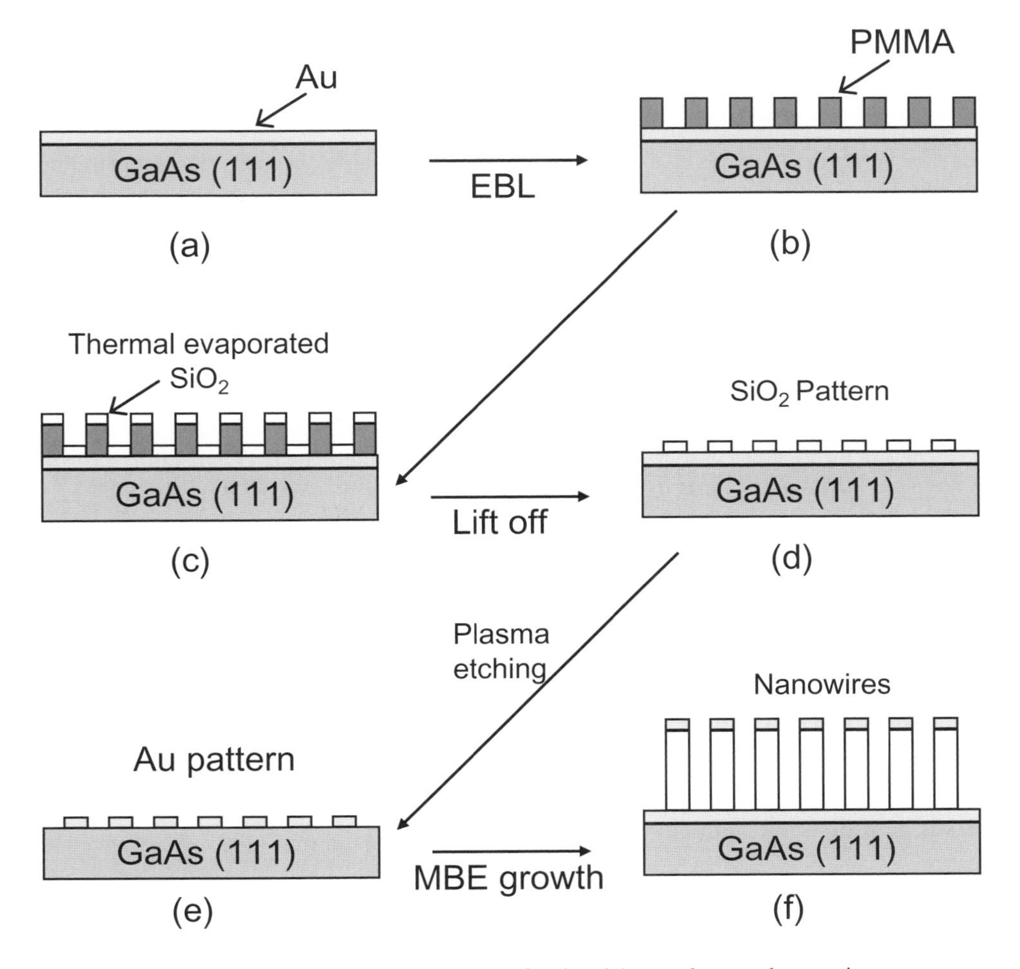

Fig. 4.6. The flow chart for the fabrication of ordered Au catalysts and nanowire arrays.

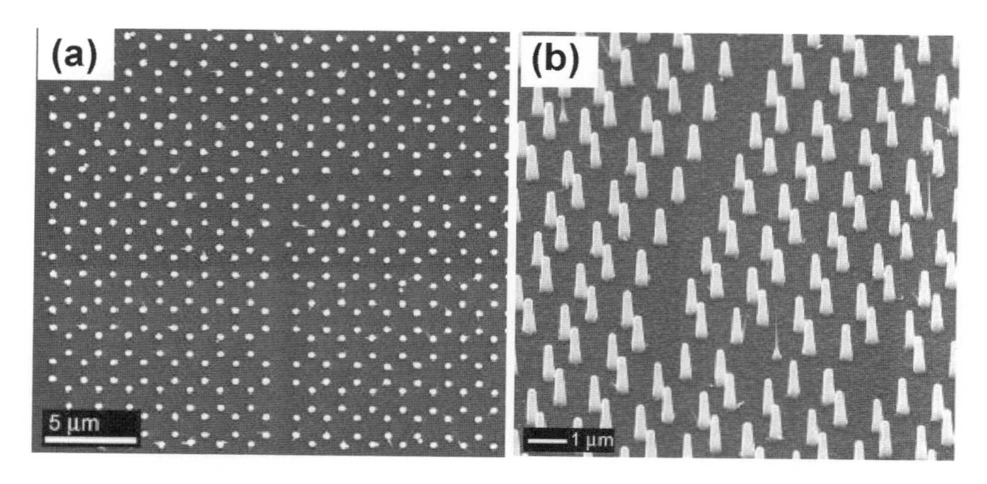

Fig. 4.7. (a) Au catalyst prepared by EBL. (b) InP nanowire arrays formed by CBE (Courtesy by American Chemical Society [34]).

on the substrate (e.g. GaP or GaAs) is restrained because the substrate temperature is substantially higher than 300°C. Therefore, no ZnSe deposition occurs on the fresh surface of the substrate. On the other hand, a certain high temperature is needed for the formation of the eutectic Au–ZnSe alloy droplets on the substrate in order to catalyze the growth of ZnSe nanowires epitaxially on the substrate. Due to the surface melting effect, the growth of ZnSe nanowires is possible at a much lower temperature of about 400°C. The deposition of ZnSe on the substrate surface at a low temperature is significant, and the quality of ZnSe nanowires is poor compared with that grown at a higher temperature. These nanowires contain a high density of defects e.g. stacking faults. However, a too-high growth temperature may result in the coarsening of Au catalyst and a low growth rate, and in turn lead to non-uniform diameters of ZnSe nanowires. The resulting growth rate of the nanowires is mainly determined by the ZnSe flux at a fixed temperature.

The growth direction of thick nanowires with diameters greater than 30 nm can be relatively easily controlled. In many cases, however, defects may cause the change of the nanowire growth direction. For ultrathin nanowires, their initial growth direction is mainly influenced by the diameters or the sizes of the catalysts. The solid–liquid interface structure at the tip of the nanowire is in fact the most critical factor influencing the nanowire growth direction. The size-dependent growth direction of ZnSe nanowires will be discussed later based on the estimation of the surface and interface energies of ZnSe nanowire nuclei.

4.3 Ultrathin Semiconductor Nanowire Growth

4.3.1 Nucleation and growth mechanisms

The growth mechanisms of the classical VLS have been systematically studied since the 1960s [6]. Though the diameters of the nanowires synthesized by new technologies are

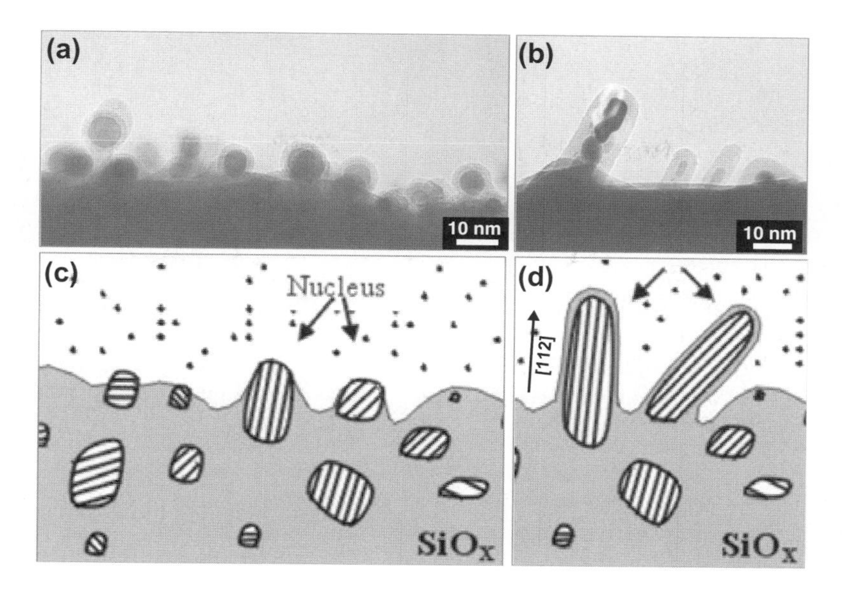

Fig. 4.8. (a) Si nanoparticles first precipitate from the Si-oxide matrix. (b) The nanoparticles in a preferred orientation (e.g. ⟨112⟩) grow fast and form nanowires. (c) and (d) The model for the nucleation and initial growth of Si nanowires from Si-oxide.

much smaller than that of Si whiskers, the VLS mechanism can still be extrapolated to explain nanowire growth phenomena in most cases. However, ultrathin nanowires from different techniques show distinct growth behaviors. It is therefore necessary to study the nucleation and growth mechanisms of ultrathin nanowires at the atomic level using modern electron microscopy. One of the novel growth mechanisms, the oxide-assisted growth will be first discussed in detail in comparison with the VLS technique.

The mechanism of the oxide-assisted nanowire growth is described by the reactions in Eqs. 4.3, 4.4, and 4.5. The vapor phase of SiO and Si_xO ($x > 1$) generated by the thermal effect (thermal evaporation or laser ablation) is the key factor. The nucleation of Si nanoparticles has been observed to occur on the substrate due to the decomposition or precipitation of Si oxide [25,29]. All nuclei are clad by shells of silicon oxide. The precipitation, nucleation, and growth of Si nanowires always occur at the area near the colder region, which suggests that the temperature gradient provides the external driving force. The formation of Si nanowire nuclei at the initial growth stage is revealed in Fig.4.8a [25], in which some nanoparticles pile up on the matrix surface soon after the start of the deposition of SiO. Notably, the nanowire nuclei that stand separately with their growth direction (⟨112⟩ direction) normal to the substrate surface undergo fast growth (Fig. 4.8b). Each nucleus consists of a Si crystalline core and an amorphous (silicon oxide) outer layer. Figure 4.8c and d are the model for the nucleation and initial growth of Si nanowires from Si oxide. Different from the VLS growth, no metal catalysts or impurities exist at the tips of the nuclei. The key point for the formation of nanowires is the fast growth of the nuclei in the ⟨112⟩ orientation of the Si cores. Therefore, only

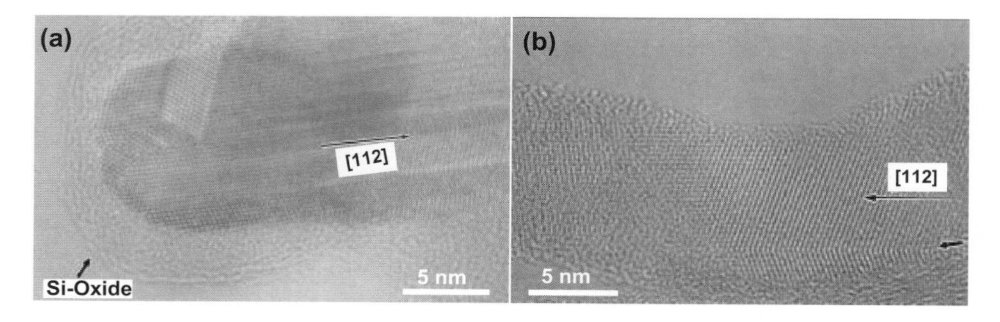

Fig. 4.9. (a) The tip structure of an individual Si nanowire from oxide-assisted growth. (b) Si nanowires grown along the ⟨112⟩ direction. Defects such as micro-twins (marked by the arrow) are along the wire axis.

those nuclei with their growth direction (⟨112⟩ direction) normal to the substrate surface undergo fast growth.

Figure 4.9a illustrates the typical tip structure of Si nanowires from the oxide-assisted growth. Most tips are round and covered by a relatively thick Si oxide layer (2–3 nm). A high density of stacking faults and micro-twins exists in the Si crystal core near the tip. Most stacking-faults and micro-twins are along the axis of the nanowire in ⟨112⟩ direction (see also Fig. 4.9b). These defects are believed to enhance the growth of the nanowires.

Si nanowires from the oxide-assisted growth are mainly determined by the following factors: (1) The catalytic effect of the Si_xO ($x > 1$) layer on the nanowire tips. It has been well known that the surface melting temperatures of nanoparticles can be much lower than that of their bulk materials, for example, the difference between the melting temperatures of Au nanoparticles (2 nm) and Au bulk material is about 400°C [35]. The materials at the Si nanowire tips (similar to the case of nanoparticles) may be in or near their molten states. This enhances atomic absorption, diffusion, and deposition. (2) The SiO_2 component in the shells, which is formed from the decomposition of SiO, retards the lateral growth of nanowires. (3) The main defects in Si nanowires are stacking faults along the nanowire growth direction of ⟨112⟩, which normally contain easy-moving 1/6[112] partial dislocations, and microtwins. The presence of these defects at the tip areas should result in the fast growth of Si nanowires since dislocations are known to play a very important role in crystal growth and (4) the charging effect at the tips may enhance the absorption of Si source [36].

For the VLS growth, metal catalyst should form initially and dictate the one-dimensional growth. On Si substrates, it is known that Au particles react with Si to form simple Au–Si solid solution. For compound semiconductors, for example GaAs, the structure and chemical composition of Au catalysts are complicated. Figures 4.10a and d show the Au particles formed on a GaAs (111) surface after the annealing treatment at 530°C. Figure 4.10a is an enlarged image showing the typical hexagonal shape of Au catalysts in which 2D moiré patterns are clearly visible. The selected-area electron

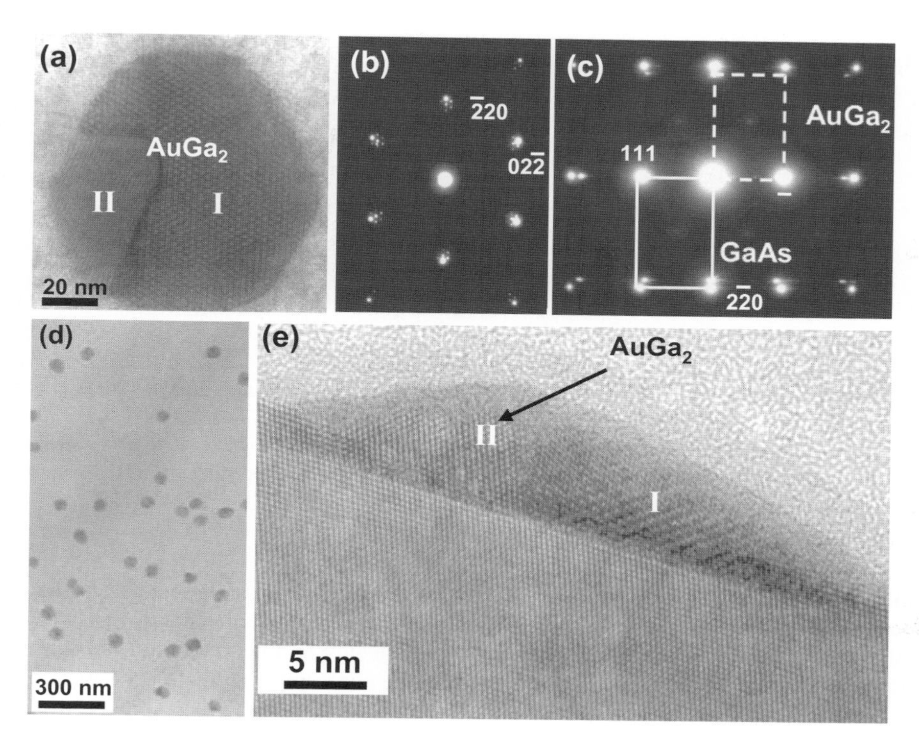

Fig. 4.10. (a) Plan-view TEM image of Au catalysts formed on GaAs substrate surface by the annealing treatment. (b) and (c) SAED patterns taken along the [111] and [11] zone axes of the catalyst, respectively. (d) Morphology of Au catalyst. (e) Cross-sectional HRTEM image (along the [110] direction) of the catalyst formed on the substrate.

diffraction (SAED) pattern (Fig. 4.10b) taken from such a particle illustrates clearly the strong diffractions of GaAs (along the [111] zone axis) surrounded by satellite spots. The structure of the catalysts have been identified to be $AuGa_2$ (FCC, space group Fm3m, lattice parameter $a = 0.6073$ nm) by electron diffraction and TEM image simulation [37]. The moiré fringes are actually due to the overlap between GaAs substrate and $AuGa_2$ particles. The $AuGa_2$ catalysts are single crystalline if their sizes are small. Two grains may form (marked by I and II in Fig. 4.10a) in a large catalyst. As shown in the SAED patterns in Fig. 10b (along the [111] zone axis) and Fig. 4.10c (along the [112] zone axis), only one $AuGa_2$ grain epitaxially forms on the substrate with orientation relations of $[100]_{GaAs}//[100]_{AuGa_2}$ and $[010]_{GaAs}//[010]_{AuGa_2}$.

For Au–GaAs system, it was observed that all catalysts reacted with the substrate during the annealing treatment [37]. The sharp interface between the catalyst and GaAs substrate has moved toward the substrate compared to the neighboring substrate surface. Figure 4.10e is the cross-sectional view of an individual $AuGa_2$ catalyst. The orientation relations between $AuGa_2$ grain II and the substrate agree well with the SAED results. The chemical composition of the catalysts was characterized using electron energy-loss spectroscopy (EELS) and X-ray energy dispersive spectroscopy (EDS), and the results

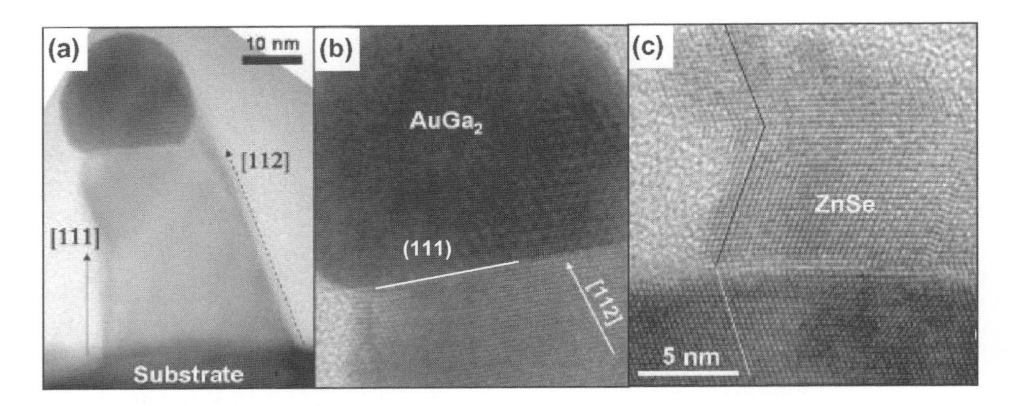

Fig. 4.11. (a) Initial growth of a ZnSe nanowire on the GaAs (111) substrate. (b) The tip structure. (c) The interface structure at the substrate. Twining structure frequently occurs in the nanowire. The solid lines indicate the {111} planes in twining relation.

indicated that the catalysts consisted of Au and Ga, and As was not detected in the catalysts. The interface of the catalyst at the substrate (about 7.4% mismatch) was (111) at which interfacial dislocations occurred. It was interesting to note that only $AuGa_2$ binary alloy (not Ga–As–Au ternary alloy) resulted by the annealing treatment, and arsenic did not participate in the nanowire growth. The reaction of the catalysts can be described as

$$\text{GaAs (solid)} + \text{Au (solid)} \rightarrow \text{AuGa}_2 \text{ (solid)} + 2\text{As (gas)} \tag{4.6}$$

In this reaction, arsenic is extracted from the substrate during the formation of $AuGa_2$ alloy. Then, arsenic may diffuse out of the catalyst surface and evaporate [38]. A similar reaction occurred when Au catalysts formed on the surfaces of ZnSe buffer layers. Au reacted with ZnSe to form Zn–Au alloy droplets and Se evaporated.

For the VLS growth, nanowires of different materials show very similar growth behaviors. As an example, Fig. 4.11a illustrates the initial growth of a ZnSe nanowire on the GaAs (111) substrate. The interfaces at the substrate and at the catalyst (Fig. 4.11b) are always {111}. Though the growth of the nanowire is due to the stacking of ZnSe layer by layer perpendicularly to the substrate and the interfaces are {111}, the resulting nanowire axis may not be perpendicular to the substrate. The ⟨112⟩ growth direction frequently appears as shown in Fig. 4.11a and c. It turns out that the growth direction of ultrathin nanowires largely relies on the diameter of the catalyst [37] as discussed below.

4.3.2 Growth direction of semiconductor nanowires

Since the properties of ultrathin nanowires are very sensitive to impurities, structural defects, and crystal orientation, much effort has been devoted to improving the synthesis process and the quality of semiconductor nanowires by employing different techniques. The lattice orientation in semiconductor nanowires is important because it may affect their

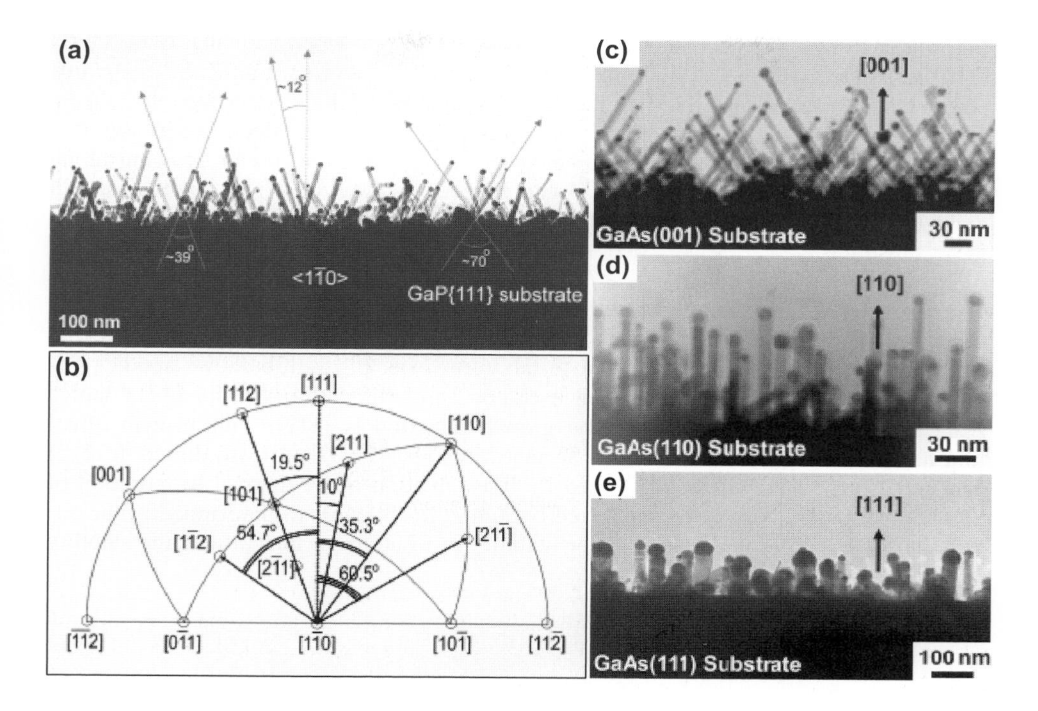

Fig. 4.12. (a) TEM image of the cleaved specimen taken with the electron beam nearly parallel to the [1$\bar{1}$0] direction of the GaP(111) substrate. ZnSe nanowires inclined mainly along four directions at ±19° and ±35° as indicated by the arrows. (b) The [1$\bar{1}$0] pole stereographic projection diagram of cubic crystals. Some poles of $\langle 110 \rangle$ and $\langle 112 \rangle$ are indicated. (c), (d) and (e) are ZnSe nanowires grown on GaAs (001), (110) and (111) substrate surfaces, respectively.

optical and transport properties [39,40]. For semiconductor nanowires, several growth directions have been frequently observed, such as $\langle 111 \rangle$ [7,33], $\langle 112 \rangle$ [10,13,41], and $\langle 110 \rangle$ [39,42,43]. It is known that most thick nanowires grow along the $\langle 111 \rangle$ direction. Ultrathin nanowires, for example ZnSe (or ZnS) nanowires, grow epitaxially on GaP or GaAs (111) substrates. These ZnSe nanowires contain few defects. However, they may grow along $\langle 111 \rangle$, $\langle 112 \rangle$ or $\langle 110 \rangle$ directions on the same substrate. This feature can be seen in Fig. 4.12a. ZnSe nanowires are inclined mainly along four directions, i.e. at ±19° or ±12° (along $\langle 112 \rangle$ direction) and ±35° (along $\langle 110 \rangle$ direction) to the normal of the GaP (111) surface, respectively.

Figures 4.12c–e illustrate the typical morphologies of ZnSe nanowires grown on GaAs (001), (110), and (111) substrates, respectively. Figure 4.12c was recorded with the electron beam nearly parallel to the [110] direction of the GaAs (001) substrate. These ultrathin ZnSe nanowires grew mainly along two directions inclined approximately ±35° to the normal of the substrate surface. When observed along the [110] direction, the nanowires displayed the same inclination. Obviously, these nanowires grew along four $\langle 110 \rangle$ directions of the substrate. ZnSe nanowires with diameters smaller than 10 nm grew only along $\langle 110 \rangle$ direction on different substrate. This can also be seen in

Fig. 4.12d, in which almost all ultrathin ZnSe nanowires grow perpendicular to the GaAs (110) substrate surface, i.e. along the [110] direction. In addition, thin ZnSe nanowires (diameter <20 nm) grown by MBE technique contain few defects. However, thick ZnSe nanowires (diameter >30 nm) always have stacking faults. It is interesting to note that most thick ZnSe nanowires with diameters greater than 30 nm prefer growing along ⟨111⟩ direction on GaAs (111) (Fig. 4.12e) or (001) substrates. ZnSe nanowires with diameters of 10–20 nm or even smaller may grow along the ⟨110⟩ or ⟨112⟩ direction on these substrates.

The microstructures of ZnSe nanowires are revealed in Fig. 4.13. The axis or the growth direction of the nanowire shown in Fig. 4.13a is along the ⟨111⟩ direction. Stacking faults and twinning are perpendicular to the wires axis. Fig. 4.13b shows another single crystalline nanowire lying flat on a carbon supporting film with its {111} lattice plane parallel to its side-surface. The growth direction is ⟨112⟩. This growth direction has been frequently observed in Si nanowires [10,41]. The growth direction ⟨112⟩ and ⟨110⟩ of Si nanowires synthesized by laser ablation or by thermal evaporation of SiO source materials has been analyzed by Tan *et al.* [44,46]. In their model, four criteria have been considered, such as the stability of Si atoms at the surface, the stability

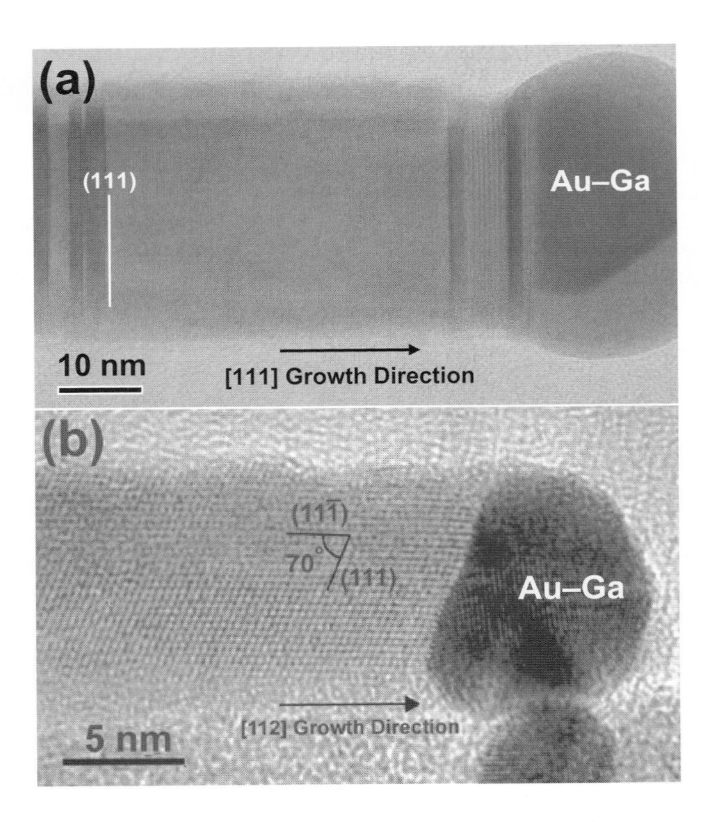

Fig. 4.13. HRTEM images showing ZnSe nanowires grown along (a) [111] and (b) [112] directions.

of Si {111} surface, the stepped Si {111} surface growth, and the dislocation effect. With these criteria, it is concluded that ⟨112⟩ and ⟨110⟩ are the preferred growth directions, and that ⟨111⟩ and ⟨110⟩ are not. For the VLS growth, due to the presence of Au catalyst at the tip of the nanowires, the interface and surface energies play important roles for the nucleation and growth direction of different sizes of nanowires as discussed below.

The interfaces for ZnSe nanowires grown along ⟨111⟩, ⟨112⟩, and ⟨110⟩ on GaAs (111) substrate were always {111} since the nanowires epitaxially formed on the substrate. At the tips of ZnSe nanowires, the interfaces between Au–Ga catalysts and ZnSe nanowires with different diameters were found to be also {111}. These features are illustrated in Fig. 4.14a–c. This is obviously due to the fact that {111} interfaces have the

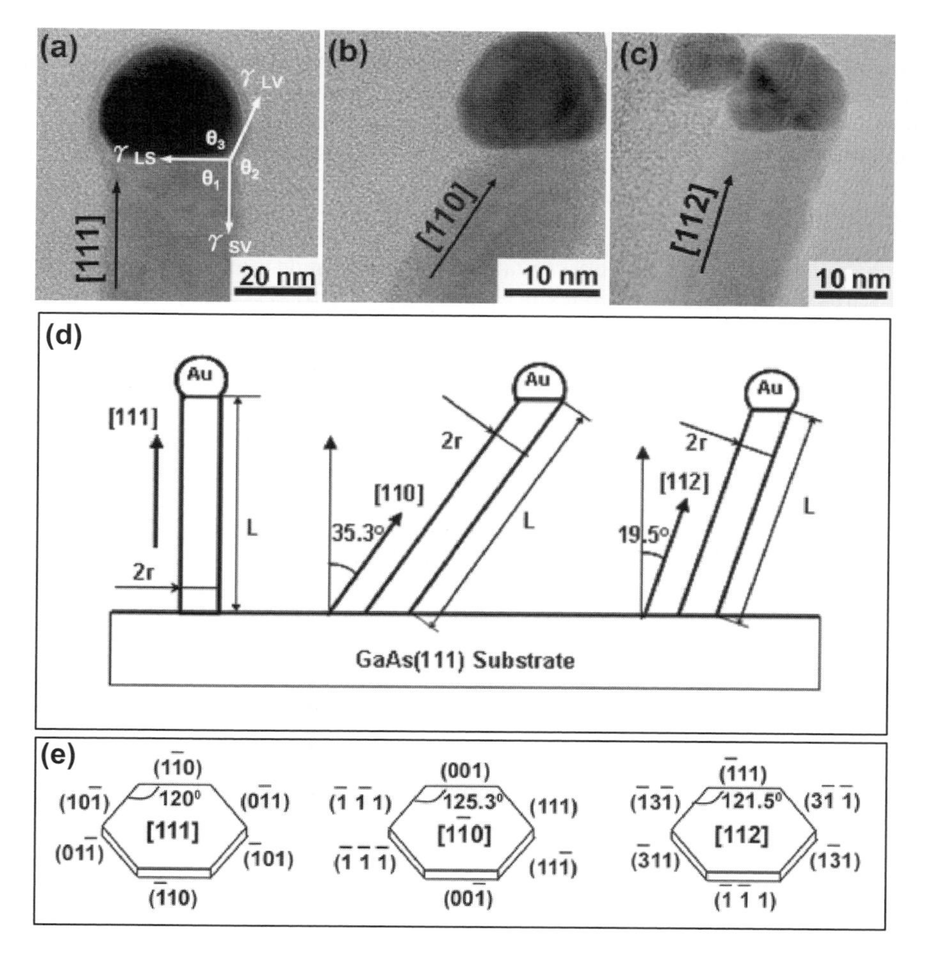

Fig. 4.14. The structures of ZnSe nanowire tips grown along (a) [111], (b) [110], and (c) [112] directions and (d) their models. (e) Models of the nanowire nuclei grown along [111], [110], and [112] directions. There are two interfaces and six side faces for each nucleus.

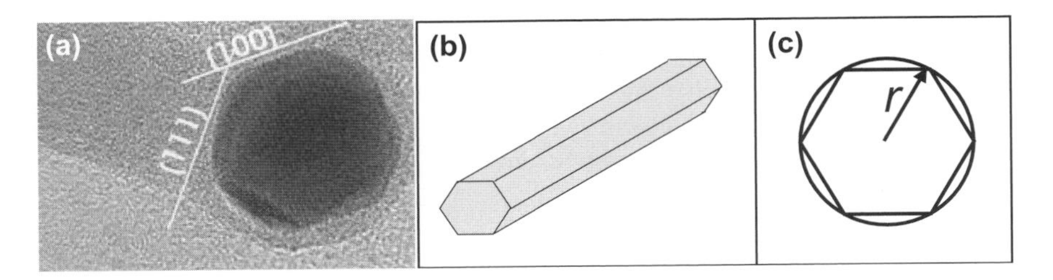

Fig. 4.15. (a) Top view of a nanowire tip with Au catalyst. The shape of the catalyst/nanowire section is close to a hexagon. (b) The hexagonal shape approximation. (c) The cross-section of a nanowire.

lowest interface energy. The solid–liquid interface structure together with the size of the catalyst or the diameter of the nanowire at the tip of the nanowire is the most critical factor influencing the nanowire growth direction. It is believed that ZnSe nanowire growth is driven by the minimum state of the total system energy of the nanowire. The size-dependent growth direction of ZnSe nanowires can be understood by this criterion.

To estimate the total energy of a nanowire, we consider the nucleus of a ZnSe nanowire as a column with two plane interfaces and a cylindrical side surface. Since the surface-to-volume ratio is high, and the bulk crystal energy is independent of nanowire orientation, only surface and interface energies need to be considered. For nano-sized crystals, a cylindrical side surface generally consists of low-energy surfaces and steps. As shown in Fig. 4.15, the hexagonal shape is a moderately good approximation to describe the cross section of a nanocatalyst or a nanowire [45]. For the three growth directions (⟨111⟩, ⟨112⟩, and ⟨110⟩), the most possible side surfaces are schematically shown in Fig. 4.14e. The initially formed nanowires should be hexagonal disks. The side surfaces can be {111}, {110}, {113}, and {100} (the energetically most favorable surfaces with low surface energies) [44]. The top and bottom surfaces are interfaces with the catalyst and the substrate, respectively. Thus, the total energy of a nanowire nucleus is:

$$F = \sum A\gamma_s + E_e \tag{4.7}$$

Here, $A\gamma_s$ and E_e denote the surface/interface energy and edge/step energies, respectively. A is the area of the surface/interface. For the hexagonal shape approximation, E_e can be considered as a fixed term for different oriented nanowires. Therefore, only surface and interface energies need to be considered, and they are the major factors affecting the growth direction of ZnSe nanowires. For a certain volume, nanowires preferentially grow in the direction minimizing their total interface/surface energy. For a nanowire with radius r and length L, the total surface/interface energy is given by

$$F = 2\pi r L\gamma_{SV} + \pi r^2\gamma_{LS} + \pi r^2\gamma_{SS} \tag{4.8}$$

where γ_{SV} is the solid–vapor interface energy (or surface energy) and γ_{LS} is the liquid–solid (L–S) interface energy between Au catalyst and ZnSe. γ_{SS} is the interface energy between ZnSe and the substrate. Since ZnSe nanowire epitaxially formed on the substrate, γ_{SS} is small compared to γ_{SV} and γ_{LS}. It will not be counted in the present model. Dividing F by the nanowire volume, the surface/interface energy per volume f is expressed as

$$f = \frac{2\gamma_{SV}}{r} + \frac{\gamma_{LS}}{L} \tag{4.9}$$

The surface energies of the major facets in ZnSe (111), (100), (311), and (110) normalized with respect to $\gamma_{(\bar{1}\bar{1}\bar{1})}$ are determined to be 1.18, 1.10, 1.16, and 1.19, i.e. $\gamma_{(\bar{1}\bar{1}\bar{1})} < \gamma_{\{100\}} < \gamma_{\{311\}} < \gamma_{\{110\}}$ [37]. Then, the absolute values of these surface energies can be calculated using $\gamma_{(\bar{1}\bar{1}\bar{1})} = 0.563\ \mathrm{Jm}^{-2}$ [47]. The interface energy γ_{SL} was determined based on the Young–Dupre equation,

$$\frac{\gamma_{LV}}{\sin\theta_1} = \frac{\gamma_{LS}}{\sin\theta_2} = \frac{\gamma_{SV}}{\sin\theta_3}, \tag{4.10}$$

which represents the equilibrium condition of the surface and interface energies. Here, θ_1, θ_2, and θ_3 are the contact angles as depicted in Fig. 4.14a. Because $\gamma_{(\bar{1}\bar{1}\bar{1})}$ equals 0.563 J/m^2 [47], γ_{LS} and γ_{LV} are determined to be approximately 0.33 J/m^2 and 0.58 J/m^2, respectively.

For the $\langle 111\rangle$ oriented nanowire with a hexagonal cross section, there are six $\{110\}$-type side surfaces and two $\{111\}$ interfaces. We assume that the nanowire cross section is an inscribed hexagon of the circular cross section of the nanowire with a diameter r (see Fig. 4.15c), which can also be considered as the diameter of the circumscribed circle of the hexagon. The radius r of the $\langle 111\rangle$ oriented nanowire is equal to the width of each $\{110\}$ facet. Its surface/interface energy is

$$f_{\langle 111\rangle} = \frac{2\gamma_{(110)}}{r} + \frac{\gamma_{(\bar{1}\bar{1}\bar{1})SL}}{L} \tag{4.11}$$

For the $\langle 110\rangle$ oriented nanowire, the side surfaces consist of four $\{111\}$-type facets and two $\{100\}$-type facets. According to the Wulff construction, the cross section should not be a regular hexagon with equal sides because there are two large $\{\bar{1}\bar{1}\bar{1}\}$ facets and two small $\{111\}$ facets for polar cubic semiconductor materials. To simplify the calculation, $\bar{\gamma}_{\{111\}}$ (the average surface energy of the polar $(\bar{1}\bar{1}\bar{1})/(111)$ pair planes) is introduced, and then the four $\{111\}$ planes are equivalent. $\bar{\gamma}_{\{111\}}$ is estimated based on the relative proportion of $\gamma_{(111)}/\gamma_{(\bar{1}\bar{1}\bar{1})}$. We further define a geometrical parameter q to represent the percentage of $\{111\}$ facets among the nanowire circumference. Then, the surface/interface energy for $\langle 110\rangle$ oriented nanowire is

$$f_{\langle 110\rangle} = \frac{2[q\bar{\gamma}_{\{111\}} + (1-q)\gamma_{(100)}]}{r} + \frac{\gamma_{(\bar{1}\bar{1}\bar{1})SL}}{L\cos 35.3^0} \tag{4.12}$$

35.3° is the incline angle of the $\langle 110\rangle$ nanowire (see Fig. 4.15). Similarly, the $\langle 112\rangle$ oriented nanowire consists of a pair of polar side surfaces $(\bar{1}\bar{1}\bar{1})/(111)$ and four $\{311\}$

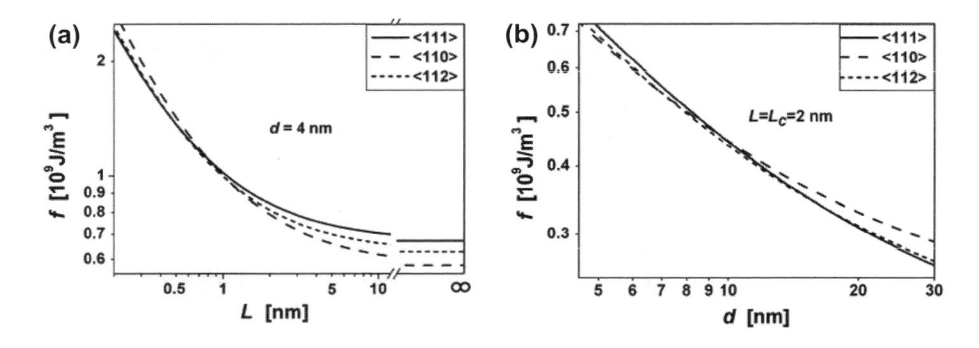

Fig. 4.16. (a) Nanowire surface/interface energy (f) vs. the nanowire length (L) for ZnSe nanowires with a fixed diameter ($d = 2$ nm). (b) Nanowire surface/interface energy (f) vs. the nanowire diameter at the critical length ($L_c = 2$ nm).

side surfaces. Its surface/interface energy can be expressed as

$$f_{\langle 112\rangle} = \frac{2[p\bar{\gamma}_{\{111\}} + (1-p)\gamma_{(311)}]}{r} + \frac{\gamma_{(\bar{1}\bar{1}\bar{1})SL}}{L\cos 19.5^0} \tag{4.13}$$

where p is the geometrical parameter and 19.5° is the incline angle of $\langle 112\rangle$ nanowires (see Fig. 4.14d).

The estimated nanowire surface/interface energies for different oriented ZnSe nanowires are shown in Fig. 4.16. The changes of the surface/interface energy (f) vs. the nanowire length (L) indicate that $\langle 111\rangle$ ultrathin nuclei are always at the minimum energy state and therefore $\langle 111\rangle$ is the preferred growth direction when the nanowire length is shorter than a certain value. This is because in the initial stage of nucleation, the side surface area is relatively insignificant. The total surface energy of a ZnSe nanowire is mainly determined by the $\{111\}$ interfaces. When the length of the nanowire nucleus exceeds a certain value, $\langle 111\rangle$ growth may not offer the minimum energy state.

At the tip of the nanowire, there is a terminal growth zone in the order of several monolayers in which ZnSe atoms are nearly in the molten state during growth. The critical thickness (L_c) can be regarded as the width of the L–S interface at the tips of the nanowires. Howe *et al.* [19] have studied the dynamic behavior of the L–S interface in $\{111\}$ Si–Al nanocrystal growth and revealed that the atoms in the L–S region were in a partially molten state. The width of this region was about 1.9 nm (about 6 stacking layers of Si). The L–S interface width (or the critical thickness) in ZnSe nanowires is estimated to be about 2 nm (6 stacking layers of ZnSe). Within this zone, the change of nanowire growth direction can be triggered in order to favor a lower energy state. The thickness of this terminal zone is defined as L_c (the critical thickness).

Then, the diameter of the catalyst is critical. If the diameter of the catalyst is large, the contribution of the two interfaces in the total system energy will be always dominant when the length is shorter than L_c. In this case, the nanowire will keep on growing along the ⟨111⟩ direction even though its final energy is higher compared to that in the other growth direction. For ZnSe nanowires catalyzed by small-diameter Au catalysts, within the critical thickness, the contribution from the side surfaces will be largely increased. As a consequence, this triggers the transition of nanowire growth direction to reach a lower energy state, i.e. either along ⟨112⟩ or ⟨110⟩. For ZnSe nanowires with ultrasmall diameters, ⟨110⟩ growth will be more energetically preferable (see Fig. 4.16a).

Figure 4.16b illustrates the change of f vs. the nanowire diameter d at the critical thickness $L_C = 2$ nm. For ZnSe nanowires with small diameters ($d < 7$ nm), ⟨110⟩ will be the most preferable growth direction. Medium-sized ZnSe nanowires (7 nm $< d <$ 17 nm) prefer ⟨110⟩ and ⟨112⟩ growth direction. ⟨111⟩ is the most preferable growth direction for ZnSe nanowires with large diameters ($d >$ 17 nm). It should be noted that the quantitative determination of the diameters at which the growth direction transition may occur largely relies on the surface and interface energies of the nanowire. The real surface and interface energies should depend on many factors, e.g. the impurities in the nanowire and concentration of the vapor source materials.

4.3.3 Photothermal effect in ultrathin semiconductor nanowires

Nanomaterials have shown unusual and often unexpected properties because of their unique structures and small dimensions. For example, Si nanoclusters showed interesting quantum confinement properties of light-emitting due to their small size effect [48,49]. A typical photoluminescence (PL) spectrum from Si nanowires covers the range of 400–800 nm which strongly depends on the oxidation state of the nanowires and the oxide shells [29,50]. ZnO and GaN nanowires have shown strong nanowire lasing or quantum lasing [9,51]. III–V semiconductor heterostructure nanowire devices also showed interesting I–V characteristics [52]. Among these interesting properties, some unexpected phenomena have been uncovered. One of these is the enhanced photothermal and photoacoustic effects. Si nanowires have a strong ability to confine photo energy from visible light. Si nanowires ignited in air and exhibited a large optoacoustic effect when exposed to a conventional photographic flash. Due to the enhanced photothermal effect, the microstructure of the Si nanowires was significantly changed upon flashing [53].

The as-prepared Si nanowire samples are very stable in air. Because of the thin oxide shells formed on those Si nanowires, the nanowire samples can be stored in air for years. A gas flame cannot ignite Si nanowires. But they may be slowly oxidized in the flame, forming silica nanowires. However, when exposed to a camera flash at a short range (3 cm), Si nanowires may ignite and fiercely burn down in air (see Fig. 4.17a). The light power needed for the ignition in [53] was 0.1–0.2 J/cm^2. The pulse duration was about 5 ms. It is obvious that Si nanowires absorbed and confined the energy from the flash, resulting in a sufficiently high temperature and hence leading to instantaneously fierce oxidation. The ignition was found to start at certain areas of the sample and then the fierce burning propagated the entire sample. The remaining materials consisted of various forms of SiO_2 nanostructures, e.g. particles, wires, and tubes (see Fig. 4.17b,d,e). When flashing the Si nanowires with low light power, a large audible acoustic wave, i.e. optoacoustic effect was observed. Large optoacoustic effects, resulting from the absorption of the

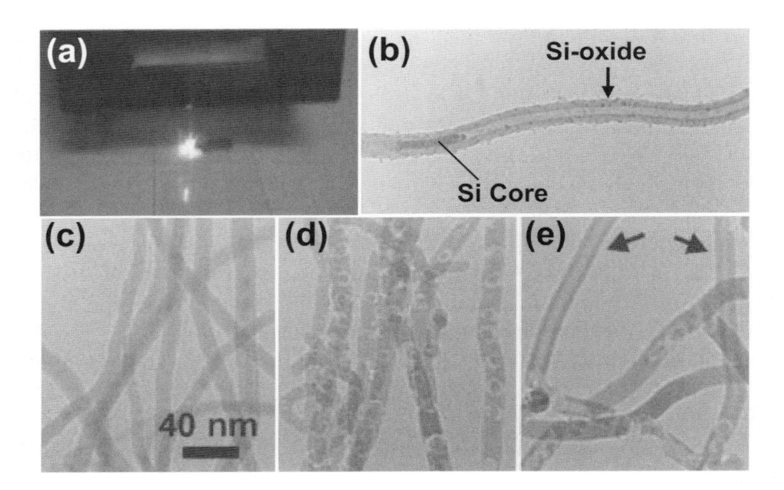

Fig. 4.17. (a) Burning of Si nanowires soon after flashing. (b) Si nanowires ignite like blast fuses resulting in Si-oxide tube structure. (c) The original Si nanowires. (d) Si nanowires change to Si nanoparticles after flashing in Ar. (e) Si-oxide nanotubes formed by flashing the Si nanowires in vacuum.

incident light, occurred also in Ge and B nanowires. However, TiO_2, ZnO, and SiO_2 nanowires (all transparent to visible light) did not show obvious optoacoustic effect.

When the flashing experiment was performed in inert gases, no ignition was observed. Flashing Si nanowires in Ar or He, resulted in structural transformation from Si nanowires to Si nanoparticles. As illustrated in Fig. 4.17d, Si nanoparticles were embedded in the Si-oxide nanowires. It is assumed that the Si cores instantaneously reached the melting state after flashing and subsequently segregated into crystalline Si nanoparticles in milliseconds. Therefore, the temperature within the Si cores must be at least ~1500°C (Si melting point 1414°C). However, Si nanowires with diameters greater than 40 nm can not be easily ignited.

A similar photothermal effect has been observed in single-walled carbon nanotubes (SWNTs) [54]. SWNTs burned way in air after flashing, forming mainly CO_2 gas. It was hypothesized that the ignition in SWNTs might arise from the photophysical effect at interfaces between the metal catalysts and SWNTs [55]. The associated large photoacoustic effect caused by the absorption of the flashlight on the SWNTs was suggested to be due to the expansion and contraction of the trapped gases in the tubes [54]. Si nanowires, however, showed distinct photoeffect because no gas was trapped in Si or other semiconductor nanowires, e.g. Ge and B (no oxide shell) nanowires. The large photoeffect is another unusual phenomenon in Si nanowires. Flashlight emulates sunlight and mainly consists of visible light. The question raised is why Si nanowires captured so much energy from light?

Some recent theoretical considerations may provide better understanding of the enhanced photothermal effect in Si nanowires. The increase of the absorption coefficient of a material means that the dielectric constant of the material may increase. Due to the quantum conferment effect in Si nanowires, Nishio *et al.* [56] report that the imaginary part of the dielectric constant of Si in the core of a Si nanowire is larger than that of the rest of the Si atoms in the nanowires. Another work reported by Ding *et al.* [57] reveals that a photothermal heating does occur when the diameter of the nanowire is about 20 nm. The optical ignition of Si nanowires can be explained entirely within the domain of classical Maxwell theory. This analysis is based on the assumption that the nanowire is a blackbody though the diameter of the nanowire is much smaller compared with visible wavelengths. Then, the energy it absorbs per unit length is proportional to its radius (r), while the number of atoms that shares the absorbed energy $\sim r^2$ per unit length. The temperature rise should then be $\sim 1/r$. Small diameter nanowires should be easily heated up compared with large ones. It is shown that Si nanowires can have an absorption cross section that is even larger than the geometric cross section because of the resonance which is unique to the cylindrical geometry of the nanowires. The total scattering cross section per unit length of the cylinder nanowire is given by [57]

$$\sigma_{\text{scatt}} = \frac{4}{k_0} \sum_{m=-M}^{M} |\mathrm{D}_m|^2 \tag{4.14}$$

The total absorption cross section per unit length is

$$\sigma_{\text{absorb}} = -\frac{4}{k_0} \sum_{m=-M}^{M} [\mathrm{Re}(\mathrm{D}_m) + |\mathrm{D}_m|^2] \tag{4.15}$$

where k_0 is the wave vector for the light in air and D_m is the scattering coefficient. The temperature rise ΔT of a Si nanowire induced by the absorption of the incident light from the photographic flash is

$$\Delta T = \frac{I_0 \sigma_{\text{absorb}}}{\rho C_p V}, \tag{4.16}$$

where I_0 is the intensity of the incident light, ρ is the density of Si, C_{p} is the specific heat of Si, and V is the volume of the Si nanowire per unit length. It is found that the resonant absorption constitutes one of the most important elements in nanowire light absorption and heating [57].

In Fig. 4.18, the temperature increase induced by the absorption of incident light is plotted as a function of the nanowire diameter. The highest temperature rise ΔT is attained at the diameter $d = 20$ nm. ΔT decreases rapidly if $d > 20$ nm. The large photo energy confinement effect in Si nanowires is unusual. This effect indicates also applications such as in smart ignition systems, self-destructing systems for electronic devices, nanosensors, and nanostructural or nanophase reformation. Several other properties of metallic and semiconductor nanowires are given in reviews and books [58,59].

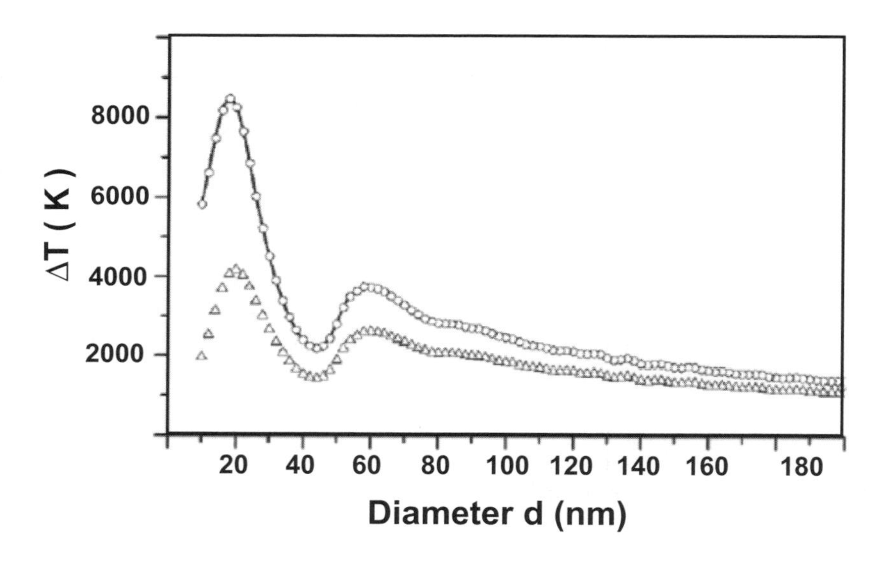

Fig. 4.18. The temperature rise ΔT vs. the nanowire diameter. The solid line indicates the idea case that the photo energy absorbed from the flash is totally converted to heat. The open circles are the result if the nanowire is coated with 2 nm of silica. The triangles indicate the result calculated by considering the heat dissipation (Courtesy by C. T. Chan [57]).

4.4 Epitaxial-oxide-passivated Metallic Nanoparticles

Metallic oxides, rather than metals, are found in nature because the oxides are thermodynamically more stable than the metals. Metals are obtained from their oxides by reduction, i.e. the removal of oxygen. There is a tendency for the metals to be oxidized in the atmosphere. Consequently, the surfaces of metals are always covered by a thin layer of oxide. The surface oxide layer tends to protect the metal from further oxidation. The successful fabrication of nanoparticles (NPs) of metals depends critically on their passivation in an atmosphere of low oxygen concentration in which a surface oxide shell or film is formed to protect the metal core from further oxidation. The oxide shell is usually a polycrystalline film which is not a very effective protective passive layer because the grain boundaries between the crystalline grains provide readily accessible channels for the diffusion of oxygen into the metal core. The protection against further oxidation would be effective if the grain boundaries are eliminated. This is the case in silicon where the surface oxide film is amorphous. Amorphous oxide enclosing Si is clearly evident in Figs. 4.3 and 4.9 above. Furthermore, when the amorphous silicon dioxide is damaged, the film can re-heal itself spontaneously and instantaneously. Indeed the success of the silicon technology is dependent on the self-healing of the surface silicon dioxide. But passivated NPs of metals are not so stable and not so effective in keeping out the oxygen so much so that they are generally regarded as inflammable and have to be stored in vacuum or in an inert gas atmosphere. Remarkably stable NPs of iron have been fabricated in Tianjin University [60]. These NPs remain unchanged in air and in water for several years. It turns out that these NPs are passivated by epitaxial oxide films.

In the early days, metallic NPs were fabricated by thermal gas evaporation and condensation in an argon atmosphere. The metallic NPs were characterized by TEM. The early studies of NPs (mainly in Japan) are summarized by Uyeda [61]. The high-resolution imaging capability of modern TEM facilitates the determination of the atomic structure of the oxide films when the NPs are oriented along major crystallographic directions. The tilting of a NP quickly to the desired crystallographic orientation in the TEM is the most demanding part of the experiment.

The size of the metallic NPs fabricated by gas condensation of plasma evaporated vapor in Tianjin University ranges from about 10 to 400 nm. In order to study the physical properties of the NPs, particles with a narrow size distribution have to be fabricated. This has been accomplished by using a plasma–gas-condensation-type deposition system in HKUST in which a size selector of collimated apertures is placed between the evaporation source and the deposition chamber [62]. The metal target is sputtered by high power to create a plasma in a vacuum chamber filled with Ar gas of 0.66 torr. The sputtered metal atoms nucleate and form small clusters in the Ar gas. NPs formed are extracted by pressure difference through a series of collimated apertures and are collected in a chamber with a pressure of 10^{-5} torr. The NPs are then passivated by filling the deposition chamber slowly with air. NPs are dispersed on porous carbon films on copper grid for study in a JEM-2010 transmission electron microscope.

4.4.1 NPs with a wide size distribution

NPs of iron and chromium The size distribution of the NPs of iron and chromium fabricated in Tiianjin University ranges from about 20–400 nm and 10–200 nm, respectively. It was found by Uyeda and coworkers that the body-centred-cubic (bcc) iron and chromium NPs are {100} truncated rhombic dodecahedra with 12 {110} facets and 6 {100} facets [63–65]. NPs of iron are 0–60% truncated while those of chromium are 60–100% truncated. The degree of truncation has been defined as the ratio of the truncated part of an edge to the full length of the edge of the rhombic dodecahedron. 100% truncation gives a cube. We have found that NPs of iron are mostly 60% truncated while partially truncated NPs and nanocubes of chromium have been observed.

Most NPs of iron do not show obvious polyhedral outlines. This is not surprising since a randomly oriented eighteen-faceted polyhedron with rounded edges and corners would appear to be spherical. But when viewed along a high symmetry low index zone axis orientation, truncated polyhedral shapes with rounded edges and vertices are readily discernible. Along a [001] zone axis, the outline of a typical 60% truncated NP is a rounded octagon. However, 30% truncated NPs, which were only observed occasionally, give a squarish outline in a [001] zone axis which is readily recognized. A typically truncated NP of iron with a rounded octagonal cross section together with its corresponding [001] electron diffraction pattern are shown in Fig. 4.19. The NP is 60% truncated with large {100} facets. The diffraction pattern has been obtained by focusing a convergent electron beam on the nanoparticle using a small condenser aperture. Thus the convergent beam electron diffraction is obtained from the iron core and the oxide films on the top and bottom facets only. The diffraction pattern does not contain diffraction spots from oxide films on other facets. Oxide diffraction spots appear at the centers of the square array of

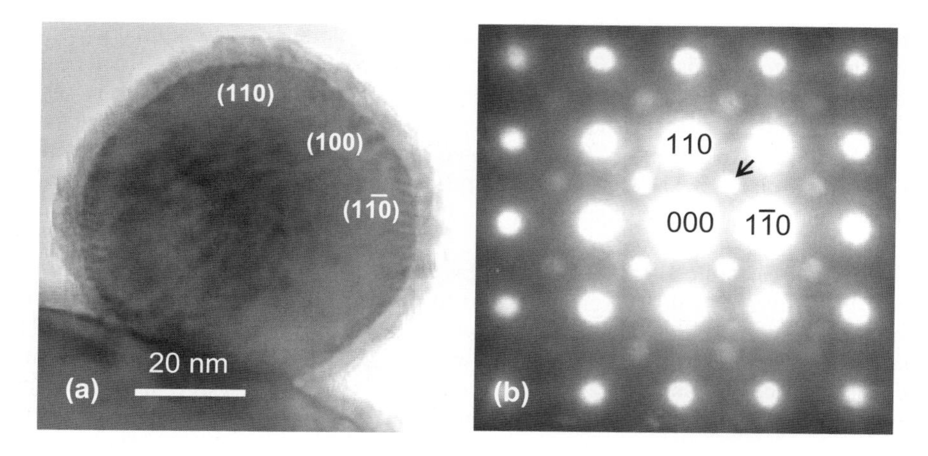

Fig. 4.19. [001] image (a) and diffraction pattern (b) of a 60% truncated nanoparticle of iron, showing the presence of epitaxial oxide on the large {100} and small {110} iron facets. The oxide spots are at the centers of the square array of iron spots, the arrowed spot is a 220 oxide spot.

iron diffraction spots. The oxide spot indicated by a downward pointing arrow has been identified to be a 220 spot. The orientation relationship between the oxide and iron is therefore [66]:

$$(001)_{\text{oxide}}//(001)_{\text{Fe}},\ [110]_{\text{oxide}}//[100]_{\text{Fe}} \tag{4.17}$$

There is a rotation of 45° about the vertical [001] cubic axis between the oxide film and iron. This is the well-known Bain relationship between face-centered cubic (fcc) and bcc crystals. X-ray photoelectron spectroscopy (XPS) study of nanoparticles of iron has shown the presence of Fe^{3+} only [66]. The oxide is therefore γ-Fe_2O_3, a cubic spinel ($a = 0.83515$ m). There is a 3% compressive mismatch in the oxide film relative to the iron core.

A [001] HRTEM image of the nanoparticle is shown in Fig. 4.20. The thickness of the oxide film is 4 nm. The two sets of intersecting γ-Fe_2O_3 {111} fringes in the oxide film on the (100) iron facet are highlighted by dashed and solid lines, while the {111} fringes on the (0$\bar{1}$0) facet are highlighted by solid and dotted lines. The oxide films on the (100) and (0$\bar{1}$0) facets are in ⟨011⟩ orientation. Here the 45° rotation between the oxide films and iron particle is about the horizontal ⟨110⟩ cubic axes. The sets of {111} fringes highlighted by solid lines on the (100) and (0$\bar{1}$0) facets extend laterally and link up into bent concave fringes on the (1$\bar{1}$0) facet. The continuous rotation of the {111} fringes on the (0$\bar{1}$0) facet relative to the {111} fringes on the (100) and (0$\bar{1}$0) facets is accommodated by bending. When the laterally extending oxide layers from the (100) and (0$\bar{1}$0) facets meet on the (1$\bar{1}$0) facet, twofold rotational twin domains, or more accurately rotational twin-like domains, as shown by the dashed and dotted {111} fringes in Fig. 4.20, are formed. It would be twin-like if the fringes highlighted by solid lines are not bent. Rotational twin-like domains are also visible in area 1 on

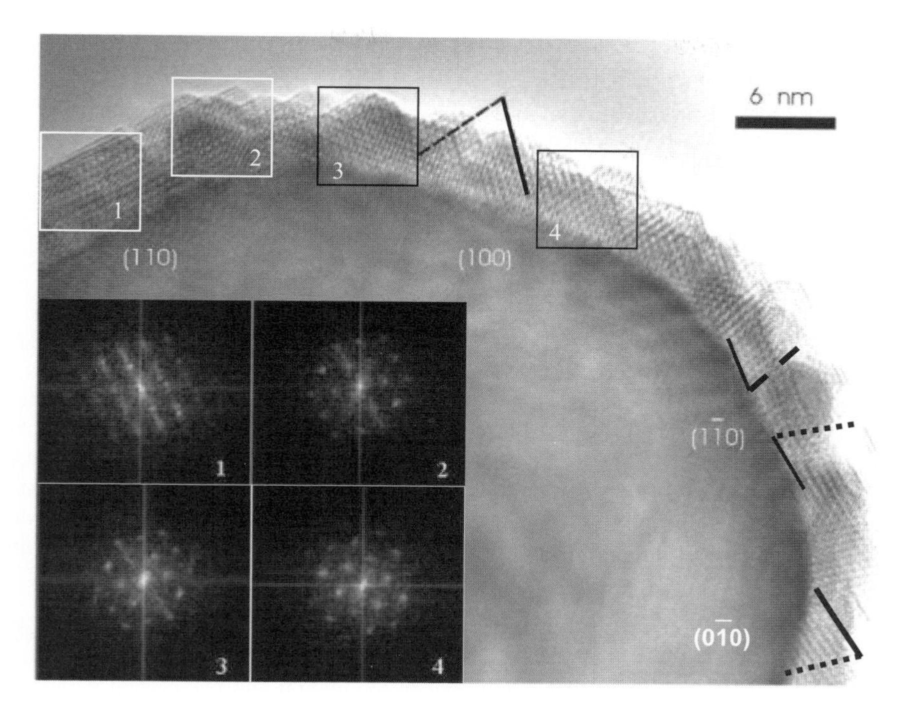

Fig. 4.20. [001] HRTEM image of the 60% truncated nanoparticle of iron in Fig. 1. The {111} fringes of the γ-Fe_2O_3 films on the (100), (1$\bar{1}$0) and (0$\bar{1}$0) facets are highlighted as shown. One set of {111} fringes from the (100) iron facet extend laterally to join up with a set of {111} fringes from (0$\bar{1}$0) facet on the (1$\bar{1}$0) facet. The remaining sets of {111} fringes are related as in rotational twins. Diffractograms obtained by Fourier transforming the enclosed images in areas 1, 2, 3 and 4 in the inset confirm the ⟨011⟩ orientation of the γ-Fe_2O_3 films on the iron facets.

the (110) facet. Diffractograms obtained by Fourier transforming the images in area 1 and 2 are shown in the inset. The diffractogram corresponding to area 2 confirms the ⟨011⟩ orientation of the oxide layer. Non-collinear twin-like spots are visible in the diffractogram of area 1. The resultant epitaxial orientation relationship on the {110} facets is

$$(111)_{\text{oxide}}//(110)_{\text{Fe}},\ [1\bar{1}0]_{\text{oxide}}//[001]_{\text{Fe}}, \tag{4.18}$$

which is the Nishiyama–Wassermann relationship between fcc and bcc crystals [67]. The variation of intensity in the {111} and {220} oxide fringes on the {100} facet is clearly evident in Fig. 4.20. {022} oxide fringes are dominant in areas 3 while {111} fringes are dominant in area 4. Diffractograms obtained from areas 3 and 4 show that both areas are oriented along ⟨011⟩ and there is no signs of off-axis misalignment.

We have studied NPs of chromium fabricated in Tianjin University by electron diffraction, HRTEM and XPS [68,69]. It was found that NPs of chromium have sharp vertices, thus when viewed along the [001] direction, squares, and octagons are clearly seen.

Squares correspond to 100% truncated nanocubes with {100}, facets while octagons correspond to partially truncated NPs with {100} and {110} facets. Epitaxial oxide films on {100} facets are fcc structured Cr_2O_3 with a lattice constant of 0.407 nm. Cr^{3+}, as in the case of Fe^{3+}, was established by XPS. Epitaxial oxide films on {110} facets are rhombohedral α–Cr_2O_3. Thus the structure of the epitaxial oxide film is different on different crystal facets. This behavior has also been observed in 30% truncated NPs of iron where the {110} facets are considerably larger than the {100} facets. Whereas the epitaxial films on {100} facets are fcc γ-Fe_2O_3, double layers of rhombohedral α–Fe_2O_3 on top of a very thin imperfect γ-Fe_2O_3 are found on {110} facets.

A [001] HRTEM image of a relatively small nano-cube of Cr is shown in Fig. 4.21. Lattice fringes are clearly visible in the oxide films and Cr cube. The plane spacings of oxide fringes can readily be deduced by reference to {110} Cr fringes. The thickness of the oxide film is deduced to be slightly less than 4 nm and the size of the Cr cube, 28 nm. The thickness of the oxide films observed in all the NPs of chromium we have studied is about 4 nm. Diffractograms obtained from the chromium core and the oxide layers show that the chromium core is oriented along [001] while the fcc Cr_2O_3 oxide is oriented along [011]. This establishes the epitaxial relationship between the oxide and the chromium nanoparticle as

$$(100)_{\text{oxide}}//(100)_{\text{Cr}},\ [01\bar{1}]_{\text{oxide}}//[001]_{\text{Cr}}. \tag{4.19}$$

This is the well-known Bain relationship between fcc crystal and bcc crystal which has been observed in epitaxial γ-Fe_2O_3 on NPs of iron. Here the 022 oxide spacings are matched to the 200 chromium spacing on the {100} facets.

A [001] HRTEM image of a 70% truncated NP of chromium is shown in Fig. 4.22. It can be seen that the oxide film on {100} facets are the same as those of Fig. 4.21. But the oxide film on the (110) facet is quite different. It is inferred from the diffractogram that the oxide is rhombohedral α–Cr_2O_3.

According to the theory of oxidation of Caberra and Mott [70], an epitaxial oxide film of 3 nm gives adequate protection against further oxidation at room temperature. The 4 nm epitaxial oxide films we have observed on NPs of iron and chromium would therefore protect the NPs against further oxidation at temperature above room temperature.

NPs of cobalt, nickel, and aluminium NPs of fcc metals are generally {100} truncated octahedra with 8 {111} facets and 6 {100} facets [64]. NPs of nickel and cobalt fabricated in Tianjin University are indeed {100} truncated octahedral enclosed by epitaxial oxide films. NPs of aluminium are spherical with an amorphous film of Al_2O_3. Like amorphous SiO_2 on silicon, amorphous Al_2O_3 on aluminium NPs provides excellent protection against further oxidation. We have studied oxide films on NPs of cobalt and nickel [71]. Epitaxial NiO(CoO) hillocks on the {111} and {100} facets of the truncated octahedral NPs of nickel (cobalt) have been directly observed by HRTEM. These nanometer tetrahedral hillocks (2.4–3.6 nm) form a rough shell enclosing the metal core. The epitaxial relationship of the oxide hillock layer on the {111} facets is

$$(\bar{1}11)_{\text{NiO}}//(\bar{1}11)_{\text{Ni}},\ [110]_{\text{NiO}}//[110]_{\text{Ni}}. \tag{4.20}$$

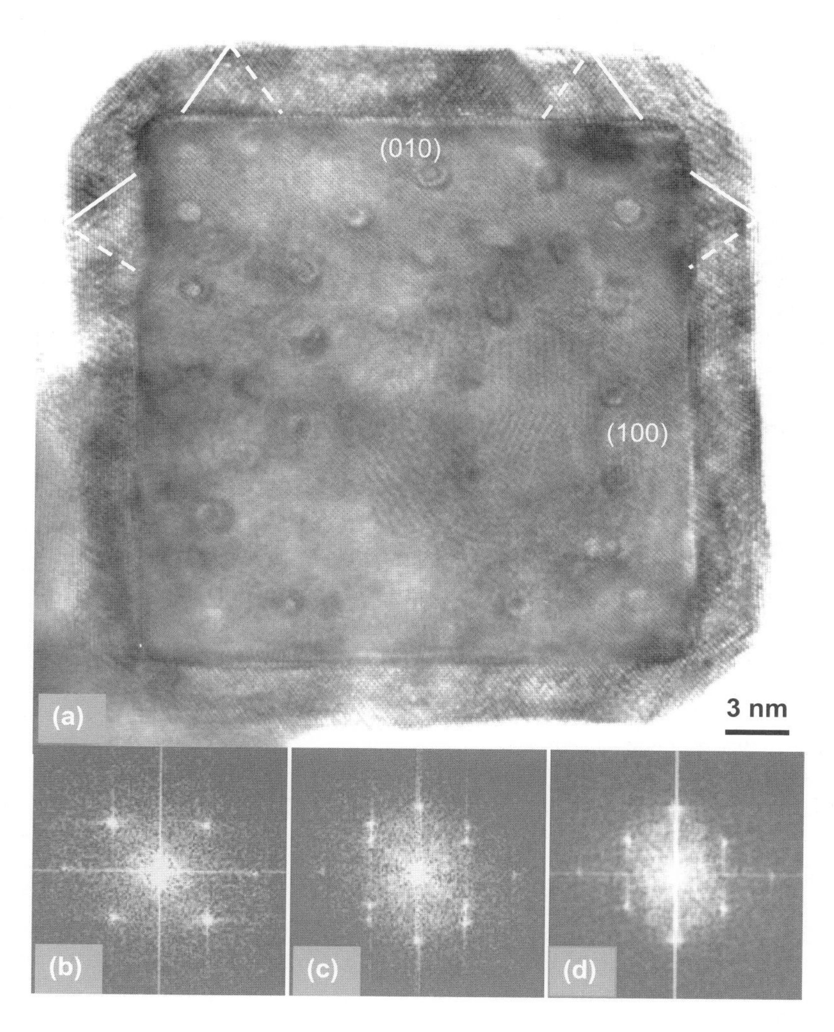

Fig. 4.21. [001] HRTEM image of a nano-cube of Cr. {111} fringes of oxide films on the (100), (010) and (100) facets are highlighted as shown. Moire fringes and electron beam induced damage spots are clearly visible. (b)-(d) Diffractograms obtained by Fourier transforming the lattice fringes of Cr and the oxide film. The orientation of Cr [001] and the cubic oxide [011] are confirmed in (b) and (d). The oxide metal epitaxy is shown in (c).

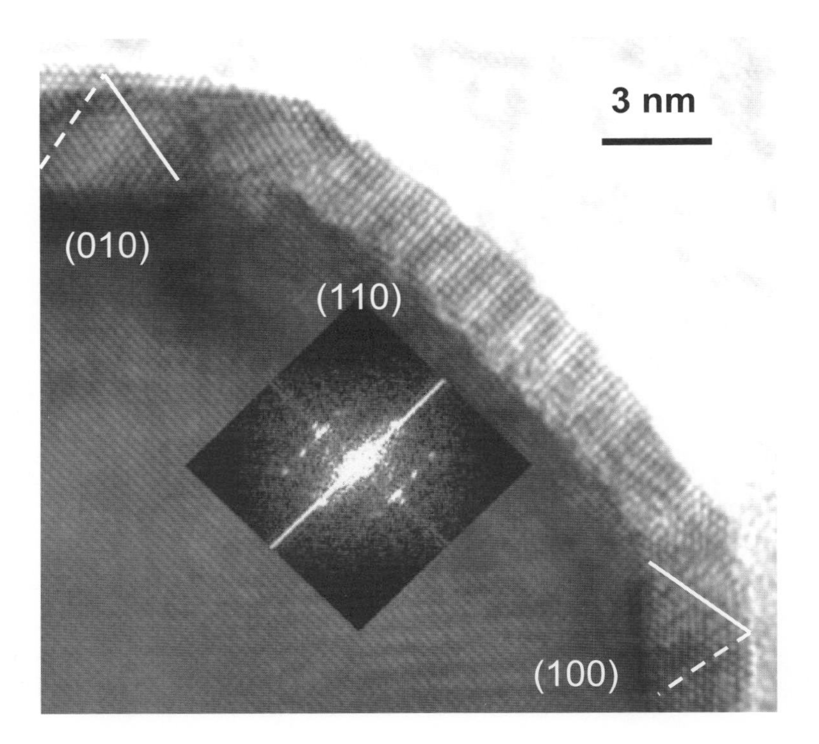

Fig. 4.22. [001] HRTEM image showing the lattice fringes of the oxide films on the (100), (010) and (110) facets of a 70% truncated NP of chromium. The {111} fringes of the cubic oxide films on {100} facets are highlighted. The diffractogram is from the oxide on the (110) facet.

The epitaxy on the {100} facets is

$$(111)_{NiO}//(001)_{Ni},\ [110]_{NiO}//[110]_{Ni}. \tag{4.21}$$

The epitaxial relationships are the same as those of NiO on bulk Ni {111} and {100} surfaces. The nominal mismatch at the NiO/Ni {111} facet is 18.7%, far exceeding the misfit limit for pseudomorhpic growth. The hillock morphology is the result of island growth mode of tetrahedral NiO on Ni [72]. The formation of hillocks leads to the relaxation of the compressive misfit stress in NiO [73]. The compressively stressed rough epitaxial oxide shell provides effective protection against further oxidation.

4.4.2 NPs with a narrow size distribution

NPs of cobalt and iron fabricated in HKUST are also passivated by epitaxial oxides of CoO and γ-Fe_2O_3. Because these NPs have a very narrow size distribution, the influence of the oxide shell on magnetic properties of the metals can be studied. We have succeeded in observing the exchange bias [74] arising from the interfacial exchange coupling between the ferromagnetic metal core with the antiferromagnetic oxide shell

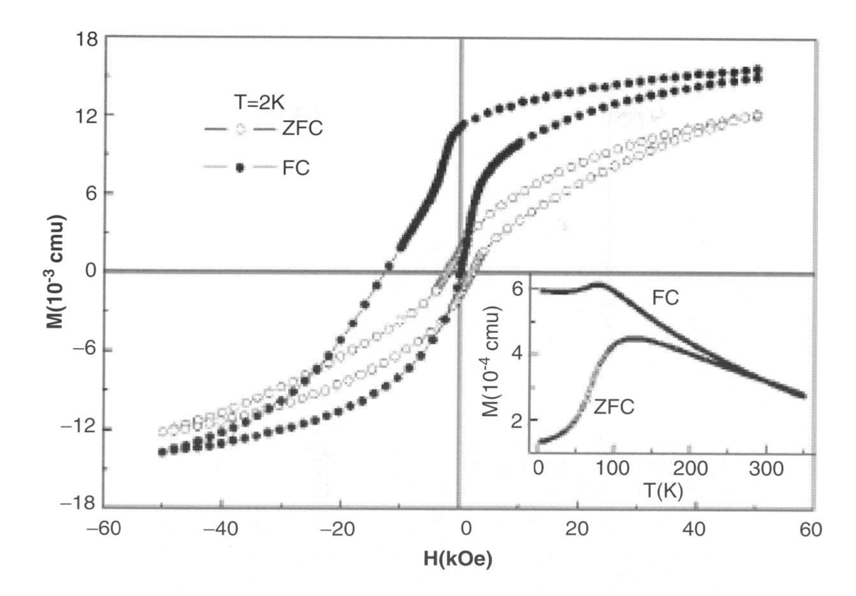

Fig. 4.23. The ZFC and 50 kOe FC magnetic hysteresis loops at 2 K. Note the substantial horizontal and vertical shifts in the FC loop. The high field irreversibility up to 50 kOe is clearly evident in both the ZFC and FC loops. The inset shows the magnetization curves in a 0.1 kOe field.

in Co–CoO [62]. The magnetization and magnetic hysteresis loops of NPs of iron with a narrow size distribution, a 5-nm core enclosed by a 3-nm oxide shell, were measured at different temperatures under zero-field-cooled (ZFC) and field-cooled (FC) conditions from 350 K in a 50 kOe field [75]. The loops measured at 2 K are shown in Fig. 4.23. The magnetization curves are shown in the inset. The ZFC curve peak at a blocking temperature of 126 K. The broad peak in the ZFC is attributed to the broad energy barrier distribution from exchange coupling and interparticle interactions due to the existence of spin-glass-like (SGL) phase [76] at the interfaces of Fe/γ-Fe_2O_3. High field irreversibility at 50 kOe is clearly shown by the open loops in both the ZFC and FC cases. This is interpreted as due to the existence of SGL phase. On the other hand, the coercive field has been substantially enhanced from 2.4 kOe in the ZFC loop to 6.4 kOe in the FC loop. A giant exchange bias of 6.3 kOe, much larger than values reported previously, has been obtained from the asymmetrical FC loop. Exchange bias is the macroscopic manifestation of microscopic exchange interaction. The giant exchange bias in Fe/γ-Fe_2O_3 results from the exchange coupling between frozen spins of SGL phase in the oxide shell and the reversible spin in the iron core and oxide shell. An interesting characteristic of exchange bias is its strong dependence on the thermal and magnetic histories. Consecutive hysteresis loops measured after field cooling show that the exchange bias fields decrease with ordinal measurement cycles, which is known as the training effect [77]. We have observed training effect of exchange bias in γ-Fe_2O_3-coated Fe NPs [78]. Both the exchange bias and training effect are explained in a modified Stoner–Wohlfarth model [79]. The horizontal and vertical FC hysteresis shifts are associated with frozen spins in γ-Fe_2O_3, the configuration of which changes with the field cycling during the hysteresis loop measurement.

4.4.3 Stability of passivated epitaxial NPs

Damage induced by electron beam heating in the TEM has been noted in NPs of chromium (Fig. 4.21). We have also found that the thickness of the epitaxial oxide increases substantially when the NPs are illuminated by an intense electron beam for a long period, say half an hour. The growth of the oxide film is due to the outward diffusion of metal which results in vacancies in the metal core. The condensation of vacancies results in the formation of voids in the metal core. Indeed, voids have been observed

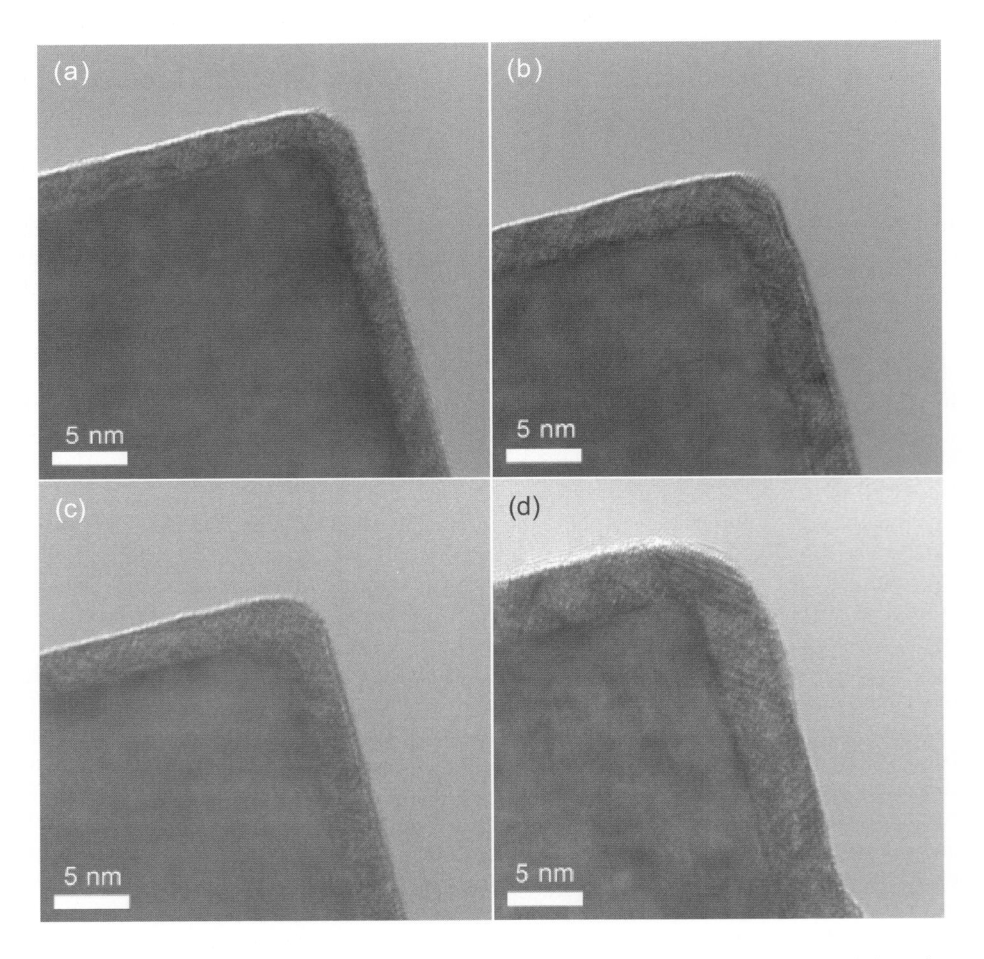

Fig. 4.24. [001] HRTEM images showing the increase of oxide thickness when a NP of chromium is illuminated by an intense electron beam in a TEM. The illuminated times and illumination levels of each of the images are given in Fig. 4.25.

in NPs of iron and chromium [80]. It turns out that the thickness of the 'as-passivated' Cr_2O_3 film is not 4 nm as shown in Fig. 4.21. When the NP is tilted to the [001] direction in a weak electron beam, the thickness of Cr_2O_3 film is slightly thicker than 2 nm and there is no induced damages in the NPs. A word of caution is order as it is important not to illuminate a NP with an intense electron beam and it is preferable to tilt the NP to the deisred crystallographic direction in a short time, say a few minutes.

We have taken HRTEM images of a passivated NP of chromium to monitor the thickness increase in the oxide film while illuminating the NP with different levels of electron beam illumination in the JEM 2010 TEM for more than two hours [81]. Figure 4.24 shows HRTEM images taken after 1200, 2760, 4440, and 6300 s. Accurate oxide thickness can be obtained from the lattice fringes. Oxide thickness after being illuminated for different time intervals are plotted in Fig. 4.25. We note the Cr_2O_3–Cr interface remain intact as the Cr_2O_3 increases in thickness. The continued epitaxial growth of oxide on chromium takes place in an ambient pressure of about 10^{-6} torr in the TEM. According to the Caberra–Mott theory of oxidation, epitaxial oxide layer of 2–3 nm provides adequate protection at room temperature to the metal below from further oxidation. The initial thickness of the oxide is slightly greater than 2 nm. It increases and saturates at about 3.2 nm. Thus the growth of the oxide is self-limiting if the electron beam intensity is increased by one order of magnitude, from 0.5 pA/cm^2 to 5 pA/cm^2. If the beam intensity is again increased by an order of magnitude, the growth of the oxide film resumes and linear growth takes place to a thickness of more than 5 nm. It is quite remarkable that the thickness of the oxide film is more than doubled and yet the epitaxy of the oxide is not changed. We interpret

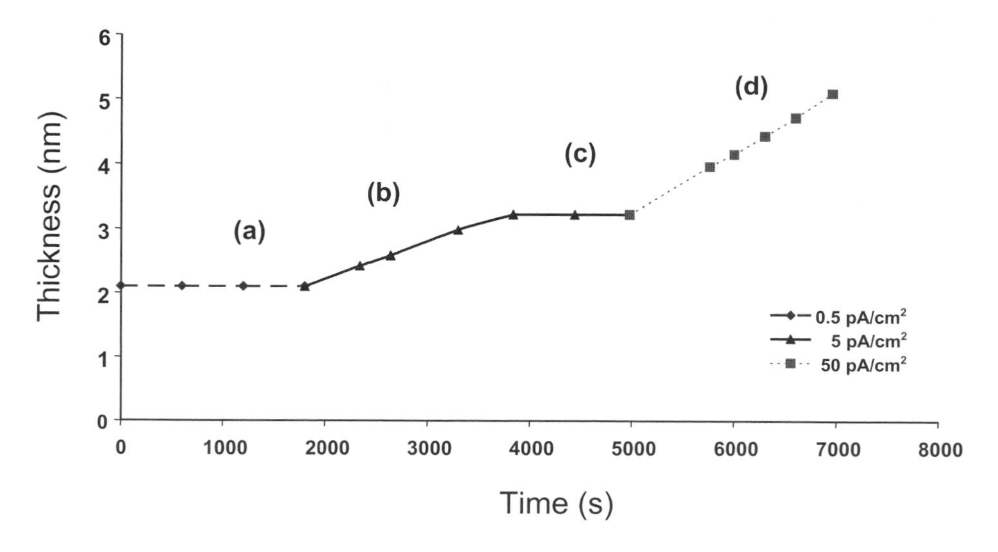

Fig. 4.25. Increasing thickness in passivated Cr_2O_3 oxide on a NP of chromium as a function of time and electron beam illumination intensity. There is no change in thickness in half an hour when the illumination is low. For intermediate illumination, self-limiting increase in thickness is evident. Under high illumination, oxide thickness increases linearly. The alphabetic labels refer to the HRTEM images in Fig. 4.24 taken at the locations shown.

this result as the continuation of the epitaxial growth of the oxide from the previous self-limiting passivation at a higher temperature arising from strong electron illumination. The self-limiting growth of Cr_2O_3 means that the oxide film is stable with moderate temperature increase in a very low oxygen ($<10^{-6}$ torr) environment. The implication of this is that the epitaxial oxide film is quite robust since it is able to passivate the NP at temperature above room temperature. The determination of the highest temperature for the self-limiting growth of epitaxial oxide would establish the temperature below which NPs are protected by self-limiting passivation by epitaxial oxide.

References

[1] H. Sakaki, Jap. J. Appl. Phys. 19 (1980) L735.

[2] E.A. Dobisz, F.A. Bout, and C.R.K. Mrrian: in Nanomaterials: Synthesis, Properties and Applications, Eds. A.S. Eddlstein and R.C. Cammarata, Institute of Physics, Bristol, 1996.

[3] M.A. Reed, J.N. Randall, R.J. Aggarwal, R.J. Matyi, T.M. Moore and A.E. Wetsel, Phys. Rev. Lett. 60 (1988) 535.

[4] T.M. Jung, S.M. Prokes and R. Kaplan, J. Vac. Sci. Technol. A12 (1994) 1838.

[5] R.S. Wagner and W.C. Ellis, Appl. Phys. Lett. 4 (1964) 89.

[6] E.I. Givargizov, Eds., Highly Anisotropic Crystal, D. Reidel Pub. Co., Tokyo, 1987.

[7] A.M. Morales and C.M. Lieber, Science, 279 (1998) 208.

[8] D.P. Yu, Z.G. Bai, Y. Ding, Q.L. Hang, H.Z. Zhang, J.J. Wang, Y.H. Zou, W. Qian, G.C. Xiong, H.T. Zhou and S.Q. Feng, Appl. Phys. Lett. 72 (1998) 3458.

[9] P. Yang, MRS Bulletin 30 (2005) 85.

[10] N. Wang, Y.F. Zhang, Y.H. Tang, C.S. Lee and S.T. Lee, Appl. Phys. Lett. 73 (1998) 3902.

[11] Z.W. Pen, Z.R. Dai and Z.L. Wang, Science 291 (2001) 1947.

[12] Z.H. Wu, X.Y. Mei, D. Kim, M. Blumin and H.E. Ruda, Appl. Phys. Lett. 81 (2002) 5177.

[13] Y.F. Chan, X.F. Duan, S.K. Chan, I.K. Sou, X.X. Zhang and N. Wang, Appl. Phys. Lett. 83 (2003) 2665.

[14] M.T. Björk, B.J. Ohlsson, T. Sass, A.I. Persson, C. Thelander, M.H. Magnusson, K. Deppert, L.R. Wallenberg and L. Samuelson, Appl. Phys. Lett. 80 (2002) 1058.

[15] S. Iijima, Nature (London) 354 (1991) 56.

[16] T.Y. Tan, N. Li and U. Goesele, Appl. Phys. Lett. 83 (2003) 1199.

[17] R.S. Wagner, Whisker Technology, Ed. A.P. Levitt, John Willey & Sons Inc., 1971, p47.

[18] H. Wang and G.S. Fischman, J. Appl. Phys. 76 (1994) 1557.

[19] J.M. Howe and H. Saka, MRS Bulletin 29 (2004) 951.

[20] L. Schubert, P. Werner and N.D. Zakharov, Appl. Phys. Lett. 84 (2004) 4968.

[21] M. Gershenzo and R.M. Mikulyak, J. Electrochem. Soc. 108 (1961) 548.

[22] E.N. Laverko, V.M. Marakhonov and S.M. Polyyakov, Sov. Phys. Crystallogr. 10 (1966) 611.

[23] K. Hiruma, T. Katsuyama, K. Ogawa, G.P. Morgan, M. Koguchi and H. Kakibayashi, Appl. Phys. Lett. 59 (1991) 431.

[24] Y.F. Zhang, Y.H. Zhang, N. Wang, D.P. Yu, C.S. Lee, I. Bello and S.T. Lee, Appl. Phys. Lett. 72 (1998) 1835.

[25] N. Wang, Y.H. Tang, Y.F. Zhang, C.S. Lee and S.T. Lee, Phys. Rev. B58 (1998) R16024.

[26] Y.F. Zhang, Y.H. Tang, N. Wang, C.S. Lee, I. Bello and S.T. Lee, Phys. Rev. B61 (2000) 4518.

[27] W.S. Shi, Y.F. Zheng, N. Wang, C.S. Lee and S.T. Lee, Adv. Mater. 13 (2001) 591.

[28] B.D. Yao, Y.F. Chan and N. Wang, Appl. Phys. Lett. 81 (2002) 757.

[29] N. Wang, Y.H. Tang, Y.F. Zhang, C.S. Lee and S.T. Lee, Chem. Phys. Lett. 299 (1999) 237.

[30] N. Wang, B.D. Yao, Y.F. Chan and X.Y. Zhang, Nano Lett. 3 (2003) 475.

[31] G. Hass, J. Am. Ceramic Soc. 33 (1950) 353.

[32] L. Holland, Ed., Vacuum Deposition of Thin Films; Chapman and Hall Ltd., London, 1966, p485.

[33] V. Schmidt, S. Senz and U. Goesele, Appl. Phys. A80 (2005) 445.

[34] T. Martensson, P. Carlberg, M. Borgstrom, L. Montelius, W. Seifert and L. Samuelson, Nano Lett. 4 (2004) 699.

[35] Ph. Buffat and J.P. Borel, Phys. Rev. A13 (1976) 2287.

[36] S.W. Cheng and H.F. Cheung, Appl. Phys. Lett. 85 (2004) 5709.

[37] Y. Cai, S.K. Chan, I.K. Sou, Y.F. Chan, D.S. Su and N. Wang, Adv. Mater. 18 (2006) 109.

[38] M. Borgström, K. Deppert, L. Samuelson and W. Seifert, J. Crystal Growth 260 (2004) 18.

[39] J.D. Holmes, K.P. Johnston, R.C. Doty and B.S. Korgel, Science 287 (2000) 1471.

[40] M.S. Gudiksen, L.J. Lauhon, J. Wang, D.C. Smith and C.M. Lieber, Nature (London) 415 (2002) 617.

[41] N. Ozaki, Y. Ohno and S. Takeda, Appl. Phys. Lett. 73 (1998) 3700.

[42] Y. Cui, L.J. Lauhon, M.S. Gudiksen, J. Wang and C.M. Lieber, Appl. Phys. Lett. 78 (2001) 2214.

[43] Z.H. Wu, X. Mei, D. Kim, M. Blumin, H.E. Ruda, J.O. Liu and K.L. Kavanagh, Appl. Phys. Lett. 83 (2003) 3368.

[44] T.Y. Tan, S.T. Lee and U. Goesele, Appl. Phys. A74 (2002) 423.

[45] Y. Zhao and B.I. Yakobson, Phys. Rev. Lett. 91 (2003) 35501.

[46] T.Y. Tan, N. Li and U. Gösele, Appl. Phys. A78 (2004) 519.

[47] S.B. Zhang , S.H. Wei, Phys. Rev. Lett. 92 (2004) 86102.

[48] M. Zacharias, J. Heitmann, R. Scholz, U. Kahler, M. Schmidt and J. Bläsing, Appl. Phys. Lett. 80 (2002) 661.

[49] J. Valenta, R. Juhasz and J. Linnros, Appl. Phys. Lett. 80 (2002) 1070.

[50] D.P. Yu , Z.G. Bai, J.J. Wang, Y.H. Zou , W. Qian, J.S. Fu, H.Z. Zhang, Y. Ding, G.C. Xiong, L.P. You, J. Xu and S.Q. Feng, Phys. Rev. B59 (1999) R2498.

[51] M.H. Huang, S. Mao, H. Feick, H. Yan, Y. Wu, H. Kind, E. Weber, R. Russo and P. Yang, Science 292 (2001) 1897.

[52] L. Samuelson, Materialstoday, Oct. 2003, p22.

[53] N. Wang, B.D. Yao, Y.F. Chan and X.Y. Zhang, Nano Letters, 3 (2003) 475.

[54] P.M. Ajayan, M. Terrones, A. de la Guardia, V. Huc, N. Grobert, B.Q. Wei, H. Lezec, G. Ramanath and T.W. Ebbesen, Science 296 (2002) 705.

[55] B. Bockarth, J.K. Johson, D.S. Sholl, B. Howard, C. Matranga, W. Shi and D. Sorescu, Science 297 (2002) 192.

[56] K. Nishio, J. Koga, H. Ohtani, T. Yamaguchi and F. Yonezawa, J. Non-cryst. Solid. 293 (2001) 705.

[57] G.H. Ding, C.T. Chan, Z.Q. Zhang and P. Sheng, Phys. Rev. B71 (2005) 20532.

[58] M. Law, J. Goldberger and P. Yang, Annual Review of Materials Research 34 (2004) 83.

[59] Z.L. Wang, Ed., Nanowires and Nanobelts - Materials, Properties and Devices Metal and Semiconductor Nanowires, Vol. I and II, Kluwer Academic Publishers, 2003.

[60] K.K. Fung, B. Qin and X.X. Zhang, Mater. Sci. Eng. A286 (2000) 135.

[61] R. Uyeda, Prog. Mater. Sci. 35 (1991) 1.

[62] G.H. Wen, R.K. Zheng, K.K. Fung and X.X. Zhang, J. Magn. Magn. Mater. 270 (2004) 407.

[63] K. Kimoto and I. Nishida, Jpn. J. Appl. Phys. 6 (1967) 1047.

[64] T. Hayashi, T. Ohno, S. Uatsuya and R. Uyeda, Jpn. J. Appl. Phys. 16 (1977) 705.

[65] Y. Saito, K. Mihama and R. Uyeda, Jpn. J. Appl. Phys. 19 (1980) 1603.

[66] Y.S. Kwok, X.X. Zhang, B. Qin and K.K. Fung, Appl. Phys. Lett. 77 (2000) 3971.

[67] K.W. Andrews, D.J. Dyson and S.R. Keown, 'Interpretation of Electron Diffraction Patterns' Hilger, London, 1971.

[68] J.C. Rao, X.X. Zhang, B. Qin and K.K. Fung, Philos. Mag. Lett. 83 (2003) 395.

[69] J.C. Rao, X.X. Zhang, B. Qin and K.K. Fung, Ultramicroscopy 98 (2004) 231.

[70] N. Caberra and N.F. Mott, Rep. Prog. Phys. 12 (1948) 184.

[71] Y.S. Kwok, X.X. Zhang, B. Qin and K.K. Fung, J. Appl. Phys. 89 (2001) 3061.

[72] P.H. Holloway and J.B. Hudson, Surf. Sci. 43 (1974) 123.

[73] F.H.M. d'Heurle, Int. Mater. Rev. 34 (1989) 53.

[74] W.H. Meiklejohn and C.P. Bean, Phys. Rev. 102 (1956) 1413.

[75] W.H. Meiklejohn and C.P. Bean, Phys. Rev. 105 (1957) 904.

[76] R.K. Zheng, G.H. Wen, K.K. Fung and X.X. Zhang, J. Appl. Phys. 95 (2004) 5244

[77] F. Bodker, S. Morup and S. Linderoth, Phys. Rev. Lett. 72 (1994) 282.

[78] B. Martinez, X. Obradors, L. Balcells, A. Rouanet and C. Monty, Phys. Rev. Lett. 80 (1998) 181.

[79] R.K. Zheng, G.H. Wen, K.K. Fung and X.X. Zhang, Phys. Rev. B69 (2004) 214431.

[80] E.C. Stoner and E.P. Wohlfarth, Philos. Trans. R. Soc. London, A240 (1948) 599.

[81] Chun Man Chan, M.Phil thesis, Department of Physics, HKUST, 2003.

CHAPTER 5

Synthesis and Characterization of Various One-Dimensional Semiconducting Nanostructures

Jeunghee Park

Department of Chemistry
Korea University
Jochiwon 339-700, Korea

5.1 Introduction

Since the discovery of carbon nanotubes (CNTs), one-dimensional (1D) nanostructures have attracted much attention as well-defined building blocks to fabricate nanoscale electronic, and optoelectronic devices [1–5]. The formation of various 1D semiconductor nanostructures is believed to be of importance in tailoring the optical, electronic, electrical, magnetic, and chemical properties of 1D nanostructures. The diverse 1D nanostructures include extensively the heterostructures having modulated compositions and interfaces, i.e. nanowire heterojunctions [6–9], superlattices [10–12], nanotapes [13], biaxial nanowires [14], and coaxial nanocables [15–43]. Indeed, they have already demonstrated the great potential in the nanodevice applications such as coaxial-gated transistors and laser diode [5.11(b), 5.16, and 5.21(c)]. However, fabrication of the nanostructures into the nanodevices needs essentially the detailed information on the structural properties, electronic structure, and magneto-optical properties.

Herein, we explain the synthesis of various semiconductor 1D nanostructures; GaN, ZnO, GaP, SiC, Si_3N_4, and BN nanostructure, via chemical vapor deposition, sublimation, or pyrolysis method, performed in our research group. The structure and composition were completely analyzed by scanning electron microscopy (SEM), high-resolution transmission electron microscopy (TEM), electron diffraction (ED), energy dispersive X-ray spectroscopy (EDS), electron energy-loss spectroscopy (EELS), X-ray diffraction (XRD), and Raman spectroscopy. The electronic structure of nanostructures has also been examined by synchrotron X-ray photoelectron spectroscopy (XPS) and angle-resolved X-ray absorption near edge structure (XANES) spectroscopy. We intensively investigated distinctive structural, optical, and magnetic properties of these nanostructures and that of the corresponding bulk and nanoparticles.

5.2 Experimental

All syntheses were performed using electrically heated furnaces. The Si or alumina substrates were deposited with various catalytic nanoparticles, using an ethanol solution of $FeCl_2 \cdot 4H_2O$ (99%, Aldrich), or $NiCl_2 \cdot 6H_2O$ (Aldrich, 99.99%), or $HAuCl_4 \cdot 3H_2O$ (98+%, Sigma), and placed in a quartz boat located inside a quartz tube reactor. The substrate was positioned either on the top of the source boat or at a distance of about 10–15 cm from the source. The source was sublimated under the Ar flow with a rate of 500 standard cubic centimeters per minute (sccm). The temperature of the source was set at 500–1200°C.

The structure, composition, and magnetic properties were thoroughly investigated by SEM, TEM, high-voltage TEM (HVEM) using 1.25 MeV, scanning TEM (STEM) elemental mapping, EDX, EELS (GATAN GIF-2000) attached to TEM (TECNAI G^2), and powder X-ray diffraction (XRD, Philips X'PERT MPD). Raman spectroscopy (Jobin–Ivon JY U1000 or Renishaw 1000) was measured using 514.5 nm line of argon ion laser. Room temperature cathodoluminescence (CL, Gatan MonoCL2) measurement was performed at an acceleration voltage of 15 kV. The photoluminescence (PL) measurement was carried out using a He–Cd laser ($\lambda = 325$ nm) as the excitation source. Magnetic properties were measured by superconducting quantum interference (SQUID, Quantum Design) magnetometer.

High-resolution XRD pattern was measured using the 8C2 beam line of the Pohang Light Source (PLS) with monochromatic radiation ($\lambda = 1.54520$ Å). X-ray photoelectron measurements were performed at the U7 and U10 beam lines of the Pohang Light Source (PLS) [44]. The XPS data were collected using the photon flux of $4–7 \times 10^{11}$ (photons/s/200 mA). The experiment was performed in an ultra-high vacuum chamber with a base pressure $\leq 5 \times 10^{-10}$. The photoelectrons emitted from the surface of the sample were collected and their energy was analyzed with an electron energy analyzer (Physical Electronics: Model PHI 3057 with a 16-channel detector). The analyzer was located at 55° from the surface normal. The angle-resolved X-ray absorption near edge structure (XANES) measurement was performed at the U7 beam line of the Pohang Light Source (PLS). The spectral resolving power ($E/\Delta E$) of the incident photon is about 5000 at 400 eV. The angle of incident X-ray beam to the sample plane is tuned from 15 to 90°. All spectra were taken in a total electron yield mode recording the sample current at room temperature. The measurement was performed in an ultra-high vacuum chamber with a base pressure $\leq 7 \times 10^{-10}$ mbar. All spectra were taken in a total electron yield mode recording the sample current at room temperature. The photon energy was calibrated by the second peak in the π^* transition of N_2 gas to be 401.1 eV as a reference. To eliminate the effect of incident beam intensity fluctuations and monochromator absorption features, all spectra were normalized by a reference signal from an Au mesh with 90% transmission.

5.3 Results and Discussion

5.3.1 GaN nanostrucures

Gallium nitride (GaN) material has been the subject of intense research for UV or blue emitters, detectors, high-speed field-effect transistors, and high-temperature microelectronic devices [45–48]. Lately, one-dimensional GaN nanostructure, GaN nanowire, has attracted much attention because of its potential for new visible and UV optoelectronic applications [49–51]. It is also expected that the GaN nanowires would be ideally suited to understand the role of dimensionality and size in optical, electrical, mechanical, and magnetic properties. Many research groups have developed various synthesis methods such as carbon nanotube-confined reaction [52], arc discharge [53], laser ablation [54], sublimation [55], pyrolysis [56], and chemical vapor deposition (CVD) [57–61]. In this section, we report the accomplished CVD synthesis of various GaN nanostructures with uniformly controlled morphology, enabling the investigation of the structural and optical properties.

Strained GaN nanowires. Highly strained GaN nanowires were synthesized by the CVD method using the reaction of Ga metal and GaN powder mixture with NH_3 [62]. It has a uniform diameter of 25 nm and the length of 20–40 μm. Iron nanoparticles with the diameter of 8–10 nm produced from the photodissociation of iron pentacarbonyl were used as catalysts. HRTEM images and SAED patterns reveal that the GaN nanowires have single crystalline wurtzite structure with a few stacking faults (Fig. 5.1). The peaks of XRD shift to a higher angle from those of the GaN epilayer and powders (Fig. 5.2). Raman spectrum has been measured using 514.5 nm line of argon ion laser, showing the peak shift to the lower frequency and the peak broadening relative to those of the epilayer. The PL of the GaN nanowires has been studied as a function of temperature by 325 nm line of He–Cd laser. A strong broad PL band is found in the range 2.9–3.6 eV. Almost no yellow band appears even at room temperature (Fig. 5.3). The peak of the

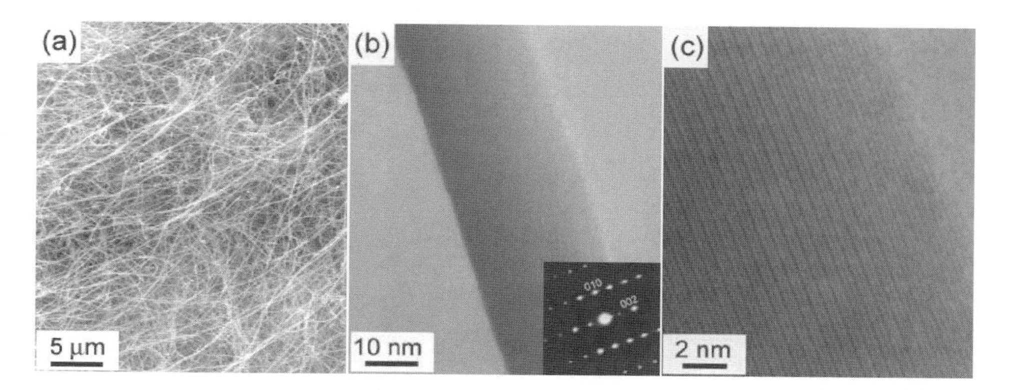

Fig. 5.1. (a) SEM micrograph for the GaN nanowires grown on the catalyst-deposited silicon substrate using a reaction of Ga/GaN and NH_3. (b) HRTEM images and its corresponding electron diffraction (inset) of a GaN nanowire whose growth direction is [010]. (c) A magnified view of (b) showing nearly defect-free (001) planes. [62]

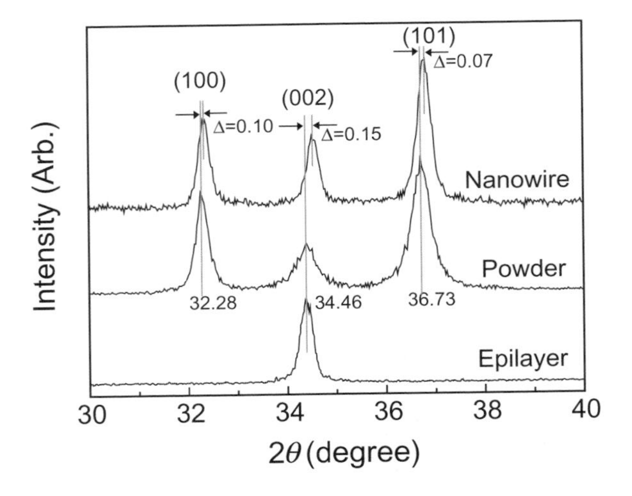

Fig. 5.2. XRD pattern taken from the GaN nanowires, epilayer, and powders, showing the higher-angle shift of the (100), (002), and (101) peaks for the nanowires. [62]

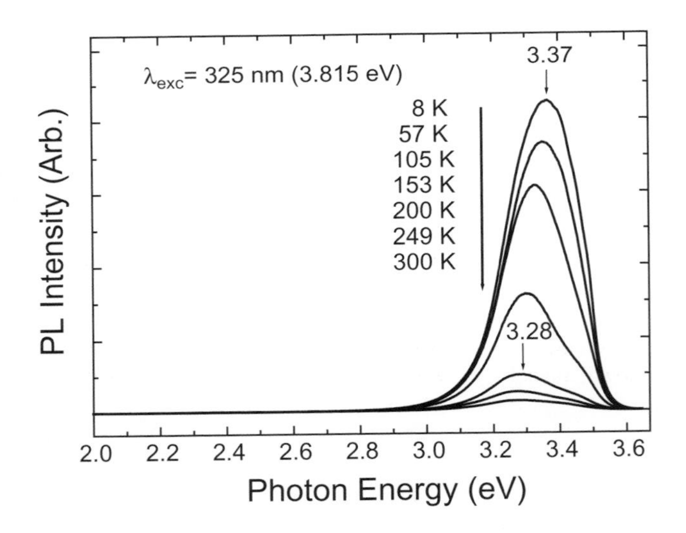

Fig. 5.3. PL spectrum of the GaN nanowires as a function of temperature. The exciton laser is 325 nm He–Cd laser. [62]

PL measured at 8 K is positioned at 3.37 eV, which is of lower energy than the bound exciton peak of the epilayer by 0.08 eV. The thermal quenching is less significant than that of the epilayer.

The shift of XRD peaks to higher angles indicates that the separations of the neighboring lattice planes along the growth direction are shorter than those of bulk GaN. The nanowire

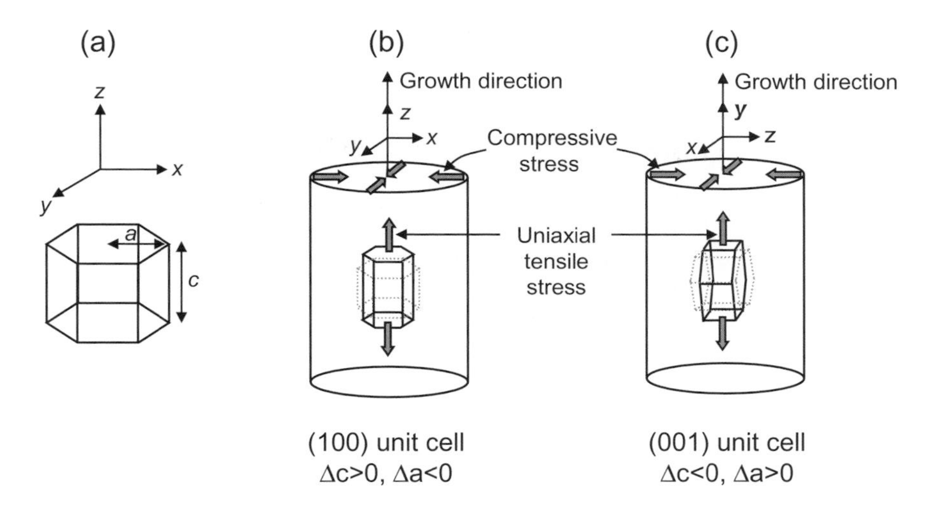

Fig. 5.4. Schematic diagrams for (a) a hexagonal unit cell with the lattice constants a and c, (b) the (100) unit cell that z-axis is aligned to the growth direction of a nanowire, and (c) the (001) unit cell that z-axis is perpendicular to the growth direction of a nanowire. [62]

would experience the biaxial compressive stress perpendicularly to the growth direction and the induced uniaxial tensile stress along the growth direction. The analysis of XRD data suggest that the strongest compressive strain could occur in the unit cells in which the [001] direction is parallel to the growth direction and the strongest tensile strain could occur in the unit cells in which the [001] direction is perpendicular to the growth direction (Fig. 5.4). We estimated the band gap shifts for the strongest compressive and tensile strains. The decrease of the band gap due to the strongest tensile stress is more significant than the increase due to the strongest compressive stress. The lower frequency shift of Raman scattering peaks are also related with the decrease of the lattice constants due to the strains. The shift of the E_2(high) peak reveals the reduction of lattice constant c, which also made it possible to estimate the lower energy shift of the band gap due to the tensile stress. The broad PL of the nanowires could originate from the recombination of bound excitons. The strong room temperature PL would be indicative of the strains inside the nanowires. The reduction of the band gap due to the tensile stresses would lead to the shift of the PL to a lower energy. The broad energy range of the PL results from various strengths of stresses in the differently oriented unit cells along the growth direction. The highest and lowest energy of the PL are consistent with the shifts of band gap calculated using the XRD and Raman data.

Triangle GaN nanowires. We synthesized the GaN nanowires on the Au-deposited alumina substrate by thermal CVD reaction of Ga/GaN with NH_3 [63]. The diameter is 40–70 nm with an average value of 50 nm and the length is about 20 μm. SEM and TEM images reveal that most of the nanowires are single-crystalline wurtzite structured GaN crystals grown with the [010] direction (Fig. 5.5). Remarkably, the cross section is nearly an equilateral triangle. The negligible shift of the XRD peaks and the E_2(high) Raman peak from those of the GaN powder suggests that the lattice constants of the GaN

Fig. 5.5. (a) SEM micrograph of the high-density GaN nanowires homogeneously grown on the substrate. (b) A magnified SEM image reveals the triangle cross section and the smooth surface. (c) TEM image showing the tip of the nanowires. [63]

nanowires would be the same as those of the bulk. The temperature-dependent PL shows a strong emission in the energy range of 2.75–3.6 eV. The PL at 8 K consists of a dominant peak at 3.478 eV and a series of peaks at 3.30, 3.22, and 3.14 eV. The 3.478 eV peak probably corresponds to the I_2 line, responsible for the recombination of excitons bound to neutral donors, and the series at the lower energy can be assigned to a zero phonon line and its phonon replicas of the free-to-bound emission. The I_2 peak shows a slight blueshift and asymmetric broadening from that of the stress-free GaN epilayers, which could be explained by the higher carrier concentration. The PL of the GaN nanowires reveals negligible energy shift of the bandgap, confirming the stress-free properties of the present GaN nanowires.

GaN nanobelts. The GaN nanobelts were grown on the Fe-deposited alumina substrate by a catalyst-assisted CVD reaction of Ga, GaN, and B_2O_3 mixture with NH_3 [64]. Fe and B_2O_3 were used as catalysts for the high-yield synthesis of GaN nanobelt. The Fe-deposited substrates were prepared using the ethanol solution containing Fe salt. NH_3 flowed at a rate of 200 sccm through a furnace with a heating zone maintained at 1100°C. The widths of the GaN nanobelts are 200–300 nm and their lengths are 10–20 μm (Fig. 5.6a). Most of the nanobelts have a triangle tip in which the angle between end edges is 60°, and non-uniform thickness due to the thick side edges (Fig. 5.6b). The shape is approximately symmetric with respect to the growth direction. The thickness of thin belt plane is about 1/10 of the width. The nanobelts consist of single-crystalline wurtzite crystal with the [010] direction perpendicular to the growth direction. The EELS analysis confirms only gallium and nitrogen components and the distinctive shape of the nanobelts.

Another typed GaN nanobelts were grown on the Ni-deposited silicon substrate by a catalyst-assisted CVD reaction of Ga, GaN, and B_2O_3 mixture with NH_3 [65]. The use of Ni and B_2O_3 as catalysts yields the formation of high-density single-crystalline wurtzite structured GaN nanobelts. The widths of nanobelts are in the range of 100 nm–1 μm, the ratio of thickness to width is about 1/10, and the lengths are up to a few tens of μm. Majority of the nanobelts have uneven thickness, jagged side edges, and

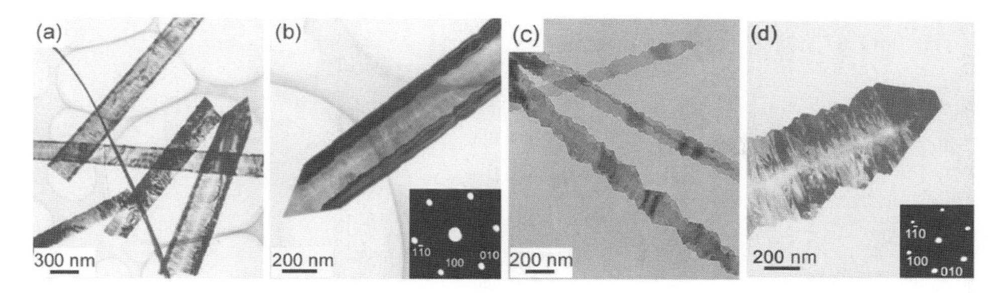

Fig. 5.6. (a) TEM image showing a general morphology of the GaN nanobelts grown Fe-deposited alumina substrates. (b) TEM image shows that single-crystalline nanobelt has a triangle tip and thick side edge. The inset is the corresponding SAED pattern, showing that the [010] direction is perpendicular to the growth direction. [64] (c) TEM image showing a general morphology of another typed GaN nanobelts, grown on the Ni-deposited silicon substrate. (d) A typical GaN nanobelt, showing a sharp tip, jagged side edges, and a thin center part along the growth direction. The inset is the corresponding SAED pattern, recording along the ⟨001⟩ zone axis. The [100] direction is parallel to the growth direction. [65]

an approximately symmetric shape with respect to the belt axis, resembling seaweeds (Fig. 5.6c). The growth directions of nanobelts are not the same, but the [001] direction is perpendicular to the belt plane (Fig. 5.6d). The XRD and EELS analyses confirm the highly crystalline wurtzite structure without boron component over the entire nanobelt.

GaN nanosaws. A new form of 1D GaN nanostructures was grown with high density, on the Ni-deposited alumina substrate by a catalyst-assisted CVD reaction of Ga, Ga_2O_3, and B_2O_3 mixture with NH_3 [66]. The temperature of Ga source was set at 1200°C and that of the substrate was approximately 1000°C. The nanomaterials are flat like a belt, with the saw teeth at one side (Fig. 5.7a). Therefore they can be called as *nanosaws*. The length of the nanosaws is up to 1 mm, the average width is in the range of 100 nm–1 μm, and the ratio of thickness to the average width is about 1/10. They have single-crystalline wurtzite GaN crystals. The TEM images and ED pattern analysis reveal that the [011] direction of hexagonal unit cell is parallel to the growth direction and its [001] direction is perpendicular to the edge of saw teeth (Fig. 5.7b). The edges of the saw tooth are at an angle of 100–110°. Schematic diagram showing the alignment of the hexagonal unit cells in the nanosaw is displayed in Fig. 5.7c. The XRD confirms the highly crystalline wurtzite structure of the nanosaws. It shows the strongest (110) peaks originated from the lattice planes parallel to the saw plane. The EELS data suggests an existence of thin amorphous B outerlayers on the nanosaws. The room temperature CL spectrum consists of intensive band-edge emission centered at 3.46 eV and a negligible yellow band, showing excellent optical properties of the GaN nanosaws.

GaN nanotrees. Tree-like GaN nanowires were grown on the Fe-deposited alumina substrate by the CVD reaction of Ga, GaN, and B_2O_3 mixture with NH_3 [67]. Most of the nanowires have diameters of 5–10 nm and lengths of 40–50 μm (Fig. 5.8a). The nanowires with the larger diameter exhibit the zigzag configuration due to the stacked hexagonal-cones along the axis, in which case they have the appearance of

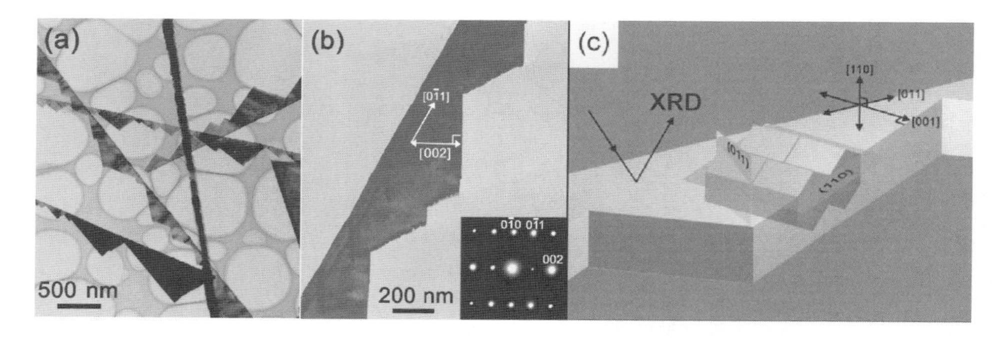

Fig. 5.7. (a) TEM image showing a typical morphology of the GaN nanosaws. The angle of saw tooth is 100–110 degrees. (b) HRTEM and its corresponding SAED pattern (inset) show that the [011] direction is parallel to the growth direction and the [002] direction is perpendicular to the edge of saw tooth. (c) A schematic diagram showing the alignment of the hexagonal unit cells in the nanosaw. [66]

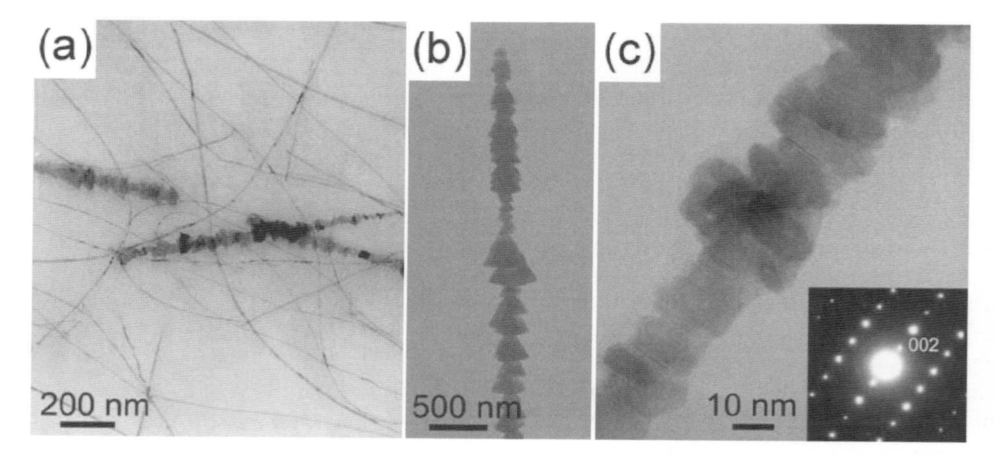

Fig. 5.8. (a) TEM image shows that thin nanowires with a diameter of 5–10 nm occupy over 90% of total nanowires. (b) The GaN nanotree with the diameter of 100–300 nm exhibits a zigzag feature of the surface. (c) A HRTEM image for the GaN nanotree with a diameter of 20–30 nm. The inset is the corresponding SAED pattern, revealing the single-crystalline wurtzite structure grown with the [001] direction. [67]

a 'tree' (Fig. 5.8b). So, we refer to them as 'nanotrees'. TEM, ED, and EELS reveal the single-crystalline wurtzite structure with a few stacking faults and the amorphous B outerlayers. Remarkably, all nanotrees have the same growth direction, [001] (Fig. 5.8c). The XRD pattern supports the uniform [001] growth direction of the nanotrees. The XRD and Raman spectrum analyses reveal no significant strains inside the nanowires. The B_2O_3 would assist to grow the GaN nanotrees with uniform [001] growth direction.

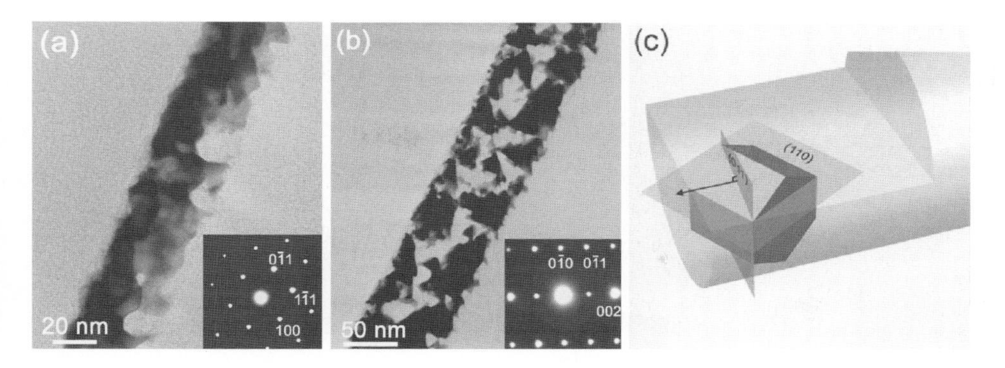

Fig. 5.9. (a) HRTEM image and its corresponding SAED pattern of a porous GaN nanowire, showing the single-crystalline wurtzite structure with the [011] growth direction. (b) HRTEM image and the SAED pattern of another porous GaN nanowire. The growth direction is also [011]. (c) A schematic diagram for the alignment of the hexagonal unit cells in the porous nanowire. [68]

Porous GaN nanowires. We reported a large-scale synthesis of porous structured GaN nanowires on the alumina substrates directly by a thermal CVD using the reaction of Ga/C-containing ball-milled Ga_2O_3 and B_2O_3 powder mixture with NH_3 [68]. The alumina substrates were deposited with Fe/Ni nanoparticles. The temperature of the source was 1200°C and that of the substrates was 950°C. The diameter of porous nanowires is 10–80 nm with an average value of 40 nm and the length is about 1 mm. The size of pores is 5–20 nm. The HRTEM images and ED patterns reveal that the GaN nanowires have wurtzite single-crystalline structure with few stacking faults (Fig. 5.9a and b). The growth direction is uniformly [011]. The porous GaN crystals are partially surrounded with the nearly amorphous layers. A schematic diagram for the alignment of the hexagonal unit cells in the nanowire is shown in Fig. 5.9c. The EELS reveals that these outerlayers are composed of B, C, and N. The XRD pattern confirms the highly crystalline wurtzite GaN crystals and the [011] growth direction. The GaN nanowires exhibit a broad PL in the range of 2.1–3.6 eV by the excitation of the 325 nm line from a He-Cd laser. The position of the PL band has been interpreted in terms of the strains. The emission from the surface defects at the pores may contribute in the broadening of the PL band down to 2.1 eV. Our results are useful in addressing the physical properties depending on the size and structure of the nanostructured material.

GaN–BCN nanocables. The graphitic BCN-layer-coated GaN nanowires were synthesized in a large quantity by thermal CVD method [69]. The mixture of Ga, ball-milled Ga_2O_3 and B_2O_3 powders, and multiwalled CNTs reacts under NH_3 atmosphere. The temperature of the source was maintained at 1100°C and the nanowires were grown on the Fe nanoparticles deposited-alumina substrates at 950°C. The light gray-colored wool-like nanowires were produced with a visible thickness. The average diameter is 30 nm and the length is up to 1 mm. The TEM, ED, and ELLS analysis show that less than 20 graphitic sheets coat the single-crystalline wurtzite-structured GaN nanowires and the growth direction of GaN nanowires is not uniform. Figure 5.10 corresponds to TEM images for a GaN–BCN nanocable whose growth direction of GaN is (002).

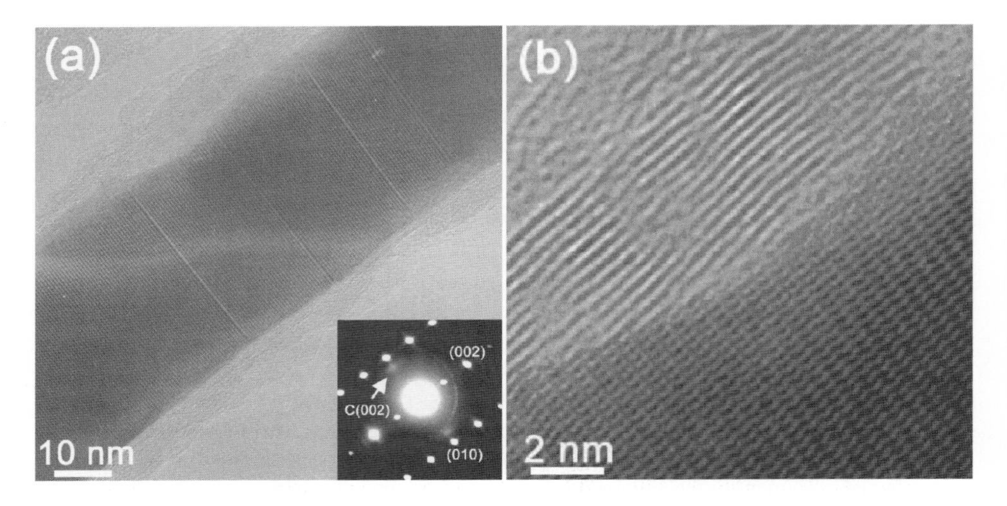

Fig. 5.10. (a) TEM image of the GaN nanowire coated with less than 20 nm thick BCN outerlayers. The SAED pattern shows the [002] direction of the graphitic layers indicated by the arrow (inset). (b) Atomic-resolved image reveals a wurtzite GaN nanowire grown with the [001] direction. The fringes of the crystalline graphitic layers are separated by 0.33–0.34 nm. [69]

For the graphitic outerlayers, the EELS estimates the atomic ratio of B:C:N as about 1:2:1. The XRD and Raman spectroscopy confirm the high crystallinity of the wurtzite GaN crystals and the existence of graphitic BCN layers. It also suggests that the lattice constants of the nanowires would be changed via the strains. The PL exhibits a broad band in the energy range of 2.1–3.6 eV, excited by the 325 nm line of He–Cd laser. The results were possibly explained in terms of the band gap decrease due to the dominant tensile strains. The emission from the BCN outerlayers may contribute in the PL in the lower energy range 2.1–2.9 eV. The GaN nanowires would grow following the vapor–liquid–solid mechanism. They probably act as a template for the formation of the graphitic BCN outerlayers.

Mn-doped GaN nanowires. Diluted magnetic semiconductors (DMS) have attracted considerable research activities because of their great potential as key materials for spintronic devices [70–72]. The demonstration of unique phenomena such as field-effect control of ferromagnetism, efficient spin injection to produce circularly polarized light, and spin-dependent resonant tunneling, opens a rich and various landscape for technological innovation in magnetoelectronics [72]. Using the theory based on bound magnetic polaron model, Curie temperatures (T_C) have been calculated for various 5% Mn-doped p-type III–V and II–VI semiconductors [70(d)]. The (Ga,M)N materials have been extensively studied since they were predicted to be ferromagnetic with a T_C above room temperature. However, many reported experimental values are quite different and even contradictory. A number of groups noticed ferromagnetism near or above room temperature, [73–80], while no ferromagnetism or a very low T_C was also reported [81,82]. These studies have actually focused on the bulk materials, but the integration of DMS materials into electronics will need very low dimensions in order

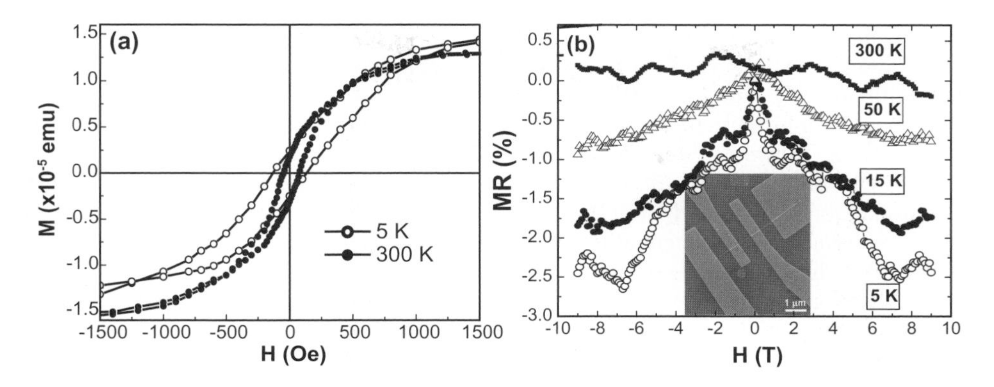

Fig. 5.11. (a) M vs. H at 5 and 300 K, measured by SQUID magnetometry, showing the hysteresis at both temperatures. (b) MR (%) at temperatures 5–300 K, for magnetic field sweeps between −9 T and 9 T. Inset corresponds to SEM image of a nanowire connected to four electrodes. [83]

to make real use of the advantages offered by the spins. Currently, there is a tremendous interest in the one-dimensional (1D) semiconductor nanostructures as well-defined building blocks to fabricate nanoscale electronic and optoelectronic devices. Therefore the 1D DMS nanomaterials would not only be important for the applications in nanoscale spintronic devices, but also provide the fundamental ideas for the role of dimensionality and size in the magnetic-related properties.

We synthesized exclusively high-purity 5% Mn-doped GaN nanowires exhibiting room-temperature ferromagnetism by chemical vapor deposition (CVD) using the reaction of Ga/GaN/$MnCl_2$ with NH_3 [83]. The diameter is uniformly 50 nm and consist of single-crystalline wurtzite GaN structure crystal grown with the [100] direction. The HVEM image detects no other clusters or crystal domains embedded in the GaN crystals. The EELS and line-scanned EDX reveal that Mn dopes homogeneously over the entire nanowire with about 5 at.%. The hysteresis curves measured at 5 and 300 K (Fig. 5.11a), and the temperature-dependent $\Delta M = M_{FC} - M_{ZFC}$ curves with H = 100 Oe provide an evidence for the room temperature ferromagnetism with T_C at around 300 K. The magnetotransport properties of individual nanowires have been measured in the temperature range 5–300 K, showing the negative MR at temperatures below 150 K (Fig. 5.11b). This result is very consistent and is in accordance with the recent work by Choi and coworkers [84]. We conclude that Mn is efficiently incorporating into the GaN nanowires and forming the ferromagnetic semiconductor nanowires.

5.3.2 ZnO nanowires and heterostructures

Zinc oxide (ZnO), a semiconductor with a direct wide band gap (3.37 eV at room temperature) and large exciton binding energy (60 meV), is one of the most promising materials for the fabrication of optoelectronic devices operating in the blue and ultraviolet (UV) region and for gas-sensing applications [85,86]. The synthesis, characterization, and application of various 1D ZnO nanostructures including the rods/wires [87], tubes [88], belts/ribbons [89], tetrapods [90], needles/pins [91,92], columns [93], combs [94], sheets [95], and complex structures [96], are presently the subject of intense research. Moreover, the ZnO heterostructures have been synthesized for nanocables with $Zn(OH)_2$, ZnS, Ga, Al_2O_3, SiO_2, and Zn [88a,b,h,97]; nanorods on CNTs and In_2O_3 nanowires [98]; heterojunction of MgO and Ni [99]. Application of the ZnO nanowire has attracted much attention because of its potential for electronic [99–101] and optoelectronic applications [102,103]. The ZnO nanowire field-effect transistors were implemented as highly sensitive chemical sensors for various gases [104–108]. However, the development of new heterostructures still remains a challenging subject to improve the sensitivity of nanodevices. Herein, we report the CVD synthesis of various ZnO nanostructures with uniformly controlled morphology, enabling to investigate the structural and optical properties.

ZnO nanowires. High-density arrays of ZnO nanowires with a controlled alignment [109,110] were synthesized. Figure 5.12a shows the vertically aligned ZnO nanowires synthesized using Zn powders at 500°C. Figure 5.12b shows the nanowires synthesized using Zn/ZnO powder mixture at 800°C. The vertical alignment is reduced, with a tilted angle of less than 45°. Figure 5.12c displays the randomly tilted nanowires on the substrates. They were synthesized using Zn/ZnO powder mixture at 900°C. As the growth temperature increases to 900°C, the vertical alignment disappears. TEM image shows that the nanowires are all straight with smooth surface and their diameter is uniformly 80 nm (Fig. 5.12d).

The HRTEM images and SAED pattern reveal that the growth direction of ZnO nanowire is dependent on the vertical alignment. All the vertically aligned nanowires have uniform growth direction [001]. In contrast, the randomly tilted nanowires, as shown in Fig. 5.12c, have preferentially the [010] growth direction, but some of them have other growth directions such as [011]. The partially tilted nanowires (as shown in Fig. 5.12b) exhibit mostly the [001] growth direction, but some of them have the [010] growth direction. Therefore we suggest that the vertically aligned ZnO nanowires would prefer to have the [001] growth direction under our growth conditions. With the enhanced growth rate

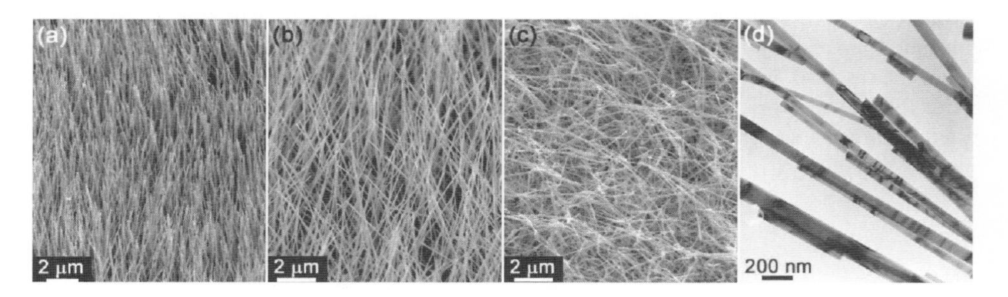

Fig. 5.12. SEM images showing (a) vertically aligned, (b) partially tilted, and (c) all randomly tilted ZnO nanowires. (d) TEM image shows that the average diameter of nanowires is uniformly 80 nm. [110]

at the higher temperature, the vertical alignment cannot maintain and the [010] growth direction would start.

S-doped ZnO nanowires. We synthesized the high-density arrays of vertically aligned S-doped ZnO nanowires [109]. As a first step, pure ZnO nanowires were synthesized on the Au nanoparticles-deposited Si substrates via CVD of Zn at 500°C. The average diameter of ZnO nanowires is 50 nm and the length is about 10 μm. The S-doped ZnO nanowires were synthesized using the CVD of Zn and S powders. The use of the substrate deposited with the vertically aligned ZnO nanowires results in the successful growth of vertically aligned S-doped ZnO nanowires (Fig. 5.13a). The average diameter of S-doped ZnO nanowires is 20 nm and the length is about 10 μm. The HRTEM images and SAED patterns confirm that the ZnO nanowires and the S-doped ZnO nanowires consist of single-crystalline wurtzite ZnO crystal grown with the [001] direction.

EDX and XPS reveal that the ZnO nanowires are doped with 4 at.% S atoms. Elemental mapping of EELS reveals that the S doping takes place mainly at the surface of nanowires with a thickness of a few nm (Fig. 5.13b). The lower angle shift of XRD was observed for the S-doped ZnO nanowires, indicating that the S doping would expand the lattice constants of ZnO. The PL and CL of the S-doped ZnO nanowires show a NBE band at 3.24–3.27 eV (room temperature) and a significantly enhanced green emission compared to that of the ZnO nanowires. The NBE emission of the undoped or less doped inside part would contribute significantly so the peak appears at the same energy as that of the ZnO nanowires. The strong green emission band is probably due to the defects such as oxygen deficiencies at the S doping sites in the surface region of nanowires.

Ga-, In-, and Sn-doped ZnO nanowires. High-density Ga-, In-, and Sn-doped ZnO nanowires were synthesized on the Au nanoparticle-deposited Si substrates via thermal

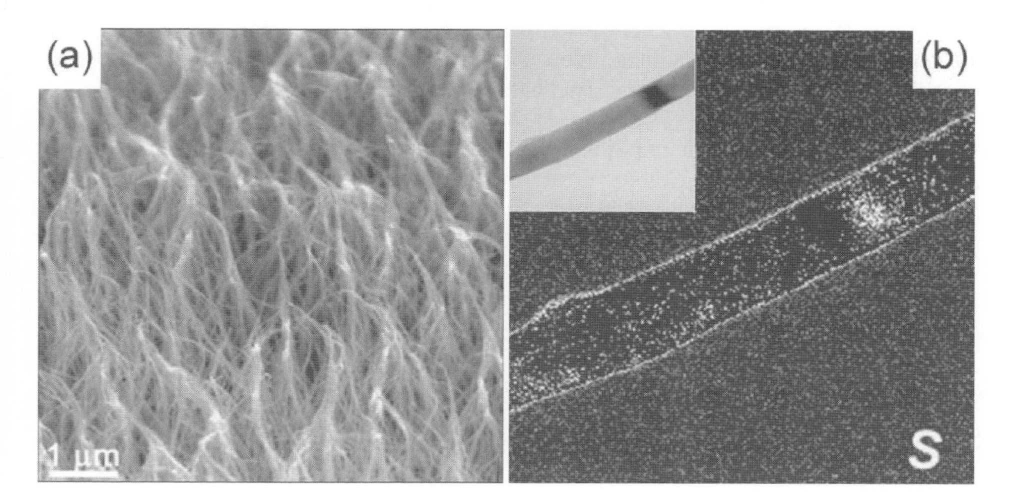

Fig. 5.13. (a) SEM micrograph shows the vertically aligned S-doped ZnO nanowires on a large area of the Si substrate. (b) Elemental maps of sulfur (S) obtained by EELS imaging using inelastic electrons corresponding to the energy loss of the sulfur element. [109]

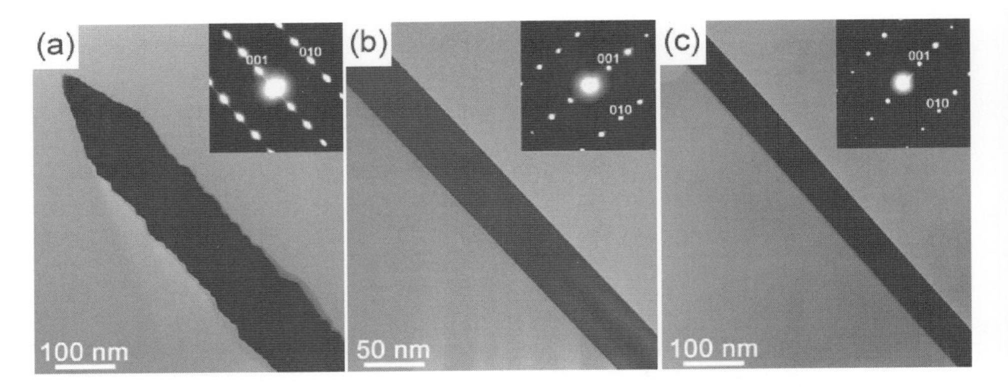

Fig. 5.14. TEM image showing a general morphology of 15% (a) Ga-, (b) In-, (c) Sn-doped ZnO nanowires. The growth direction is [001], [100], and [100] for respective nanowire. [110]

evaporation at 500–1000°C [110]. The average diameter of ZnO nanowires is 80 nm. The HRTEM images and SAED patterns confirm that all ZnO nanowires consist of single-crystalline wurtzite ZnO crystal. The EDX and XPS data reveal that the average content is as high as about 15% for all three dopants. The vertically aligned Ga-doped ZnO nanowires were grown with the [001] direction (Fig. 5.14a). In contrast, the growth direction of randomly tilted In- and Sn-doped ZnO nanowires is [010] (Fig. 5.14b and c). We discuss a correlation between the growth direction and the vertical alignment, using the undoped ZnO nanowires grown with the controlled vertical alignment.

The broader XRD peaks of the doped ZnO nanowires indicate the lattice distortion attributable to the doping. The UV-visible absorption spectrum reveals the E_g decrease that would be originated from the localized band edge states at the doping sites. The PL and CL of the doped ZnO nanowires show a broader NBE band at the lower energy range compared to that of undoped ZnO nanowires, due to the E_g decrease at the doping sites. The Sn doping causes the largest XRD peak broadening, most significant E_g reduction, and strong green emission, which would be due to the largest charge density of Sn. It suggests that the charge density of the doped element would be an important parameter in controlling the optical properties of ZnO nanowires.

We also synthesized 25% In-incorporated ZnO nanowires using the thermal evaporation at 800–1000°C, and compared the structure and optical properties with those of undoped ZnO and 15%-doped ZnO nanowires [111]. All In-containing nanowires exhibit a single-crystalline wurtzite ZnO structure with the identical [010] growth direction. The XRD patterns show that as the In content increases, the peak width becomes broader and the position shifts to a slightly lower angle, indicating that the In incorporation causes the structural defects of ZnO crystals. The XPS data suggests that In withdraws the electrons from Zn and enhances the number of dangling-bond O $2p$ states. Both UV-visible absorption curve and the NBE peak of PL shift to the lower energy and the width becomes broader as the In content increases (Fig. 5.15). The enhancement of the green emission band occurs due to the increased defect sites as predicted from the XRD and XPS data.

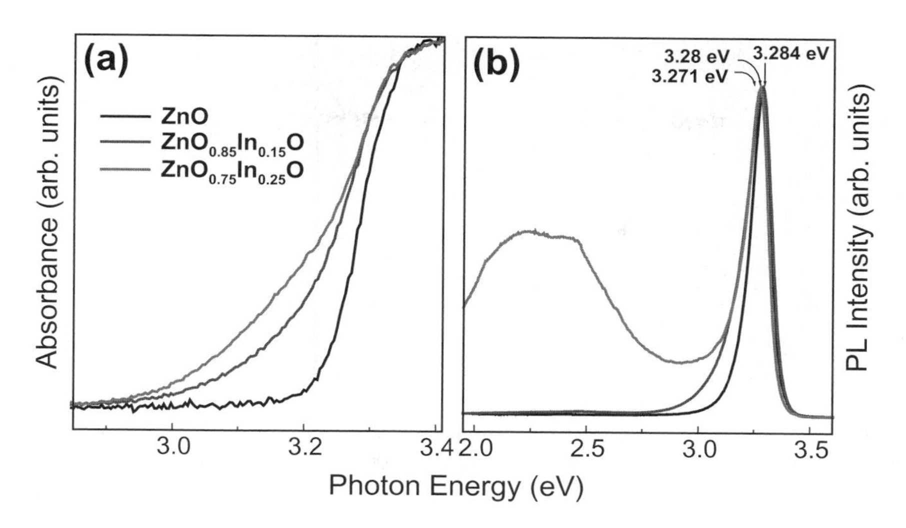

Fig. 5.15. Room-temperature (a) UV-visible absorption and (b) PL spectrum of 15% and 25% In-doped ZnO nanowires, indicating the lower energy shift with the In content. The exciton wavelength is the 325 nm line from a He–Cd laser. [111]

Superlattice structured Sn-doped $In_2O_3(ZnO)_4$ and $In_2O_3(ZnO)_5$ nanowires. Sn-doped $In_2O_3(ZnO)_4$ and $In_2O_3(ZnO)_5$ nanowires were synthesized by thermal evaporation method [112]. As the mixture of ZnO/In/Sn powders was evaporated at 900–1000°C, the nanowires were exclusively grown on the Au nanoparticles-deposited silicon substrates. The diameter of nanowires is periodically modulated in the range of 50–90 nm. The HVEM images and fast-Fourier-transformed (FFT) ED patterns reveal that the Zn–O slabs are composed of wurtzite ZnO crystals grown with the [001] growth direction, and the spacing between the In planes and its nearest Zn planes is approximately 0.3 nm which is significantly larger than the Zn (002) interplanar spacing (Fig. 5.16a–c). The HVEM and EELS imaging show unique longitudinal superlattice structure that one In–O layer and five (or six) Zn–O layer slabs stacked alternately perpendicular to the long axis, with a modulation period of 1.65 (or 1.9) nm. The Sn content is identified to be 6–8%, which is about 1/4 of the In content, by EDX and XPS. The atomic arrangement has been derived from the HVEM images, showing that the two slabs of ZnO on both sides of the In–O layer have opposite polarization (Fig. 5.16d). As the In/Sn content increases, the peak width of XRD becomes broader and new broad peaks appear at the lower angle region, indicative of the structural defects of ZnO crystals. High-resolution XPS data suggests that In/Sn withdraw the electrons from Zn, and enhance the number of dangling-bond O $2p$ states. The valence band absorption of XPS and the UV-visible absorption indicate the E_g reduction due to the incorporation of In/Sn. The PL and CL of superlattice nanowires exhibit the lower energy shift and the broader width of NBE emission as the In/Sn content increases. The enhancement of green emission occurs due to the increased defect sites as predicted from the XRD and XPS data.

Heterostructures of ZnO nanorods with various one-dimensional nanostructures. We reported a large-scale synthesis of various ZnO heterostructures and detailed evaluation of

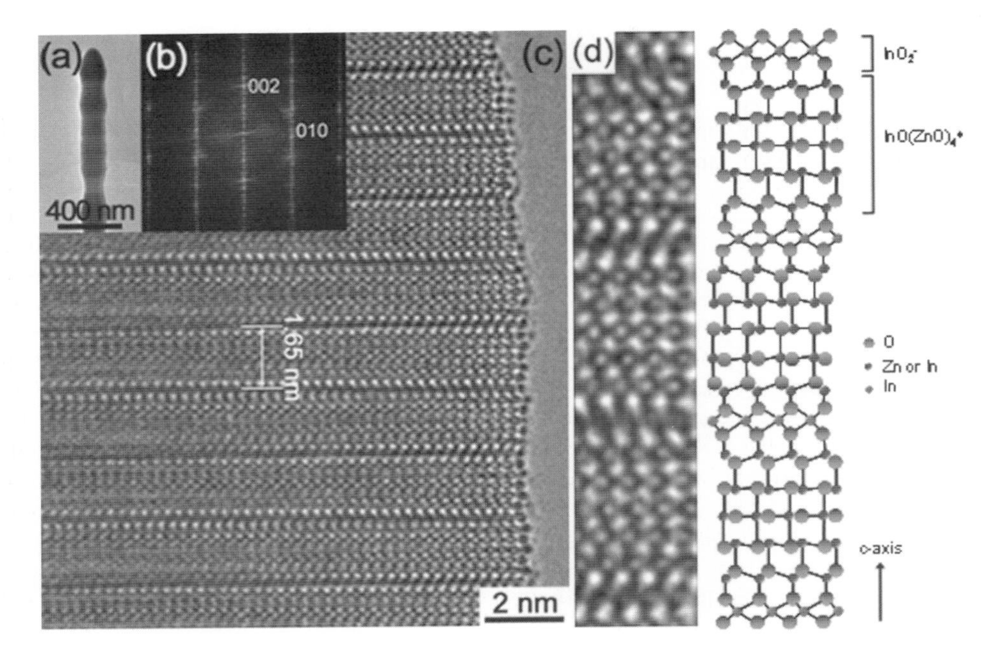

Fig. 5.16. (a) HVEM image of a $In_2O_3(ZnO)_4$ nanowire with corresponding (b) FFT ED pattern and (c) atomic-resolved image. (d) Atomic arrangement of $In_2O_3(ZnO)_4$ nanowires, derived from the HVEM image in the left. [112]

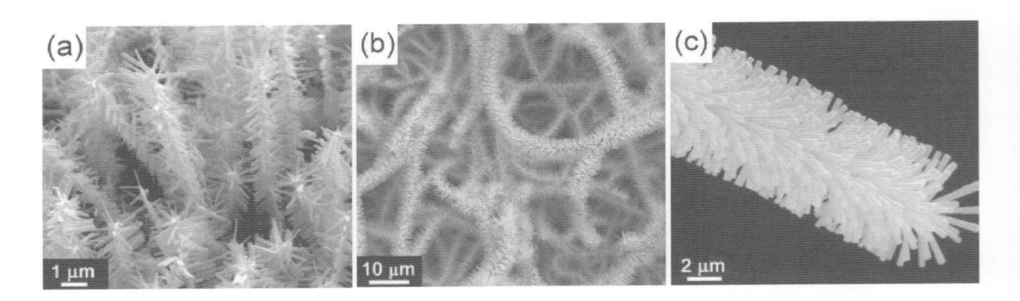

Fig. 5.17. SEM micrographs of ZnO nanorods grown on (a) GaN nanowires, (b) SiC–C nanocables, and (c) its magnified image. [113]

their structural and optical properties [113]. The ZnO nanorods were grown on diverse pre-grown 1D nanostructures via a chemical vapor deposition (CVD) of Zn at a low temperature of 500°C. The pre-grown 1D nanostructures are vertically aligned CNTs, partially vertical-aligned GaN nanowires (Fig. 5.17a), GaP nanowires, SiC nanowires, and SiC core-C shell coaxial nanocables (Fig. 5.17b). The use of Ga or In as a catalyst leads to the high yield of ZnO nanorods (Fig. 5.17c). The size and morphology of the nanostructures were thoroughly examined using the electron microscopy, XRD, and Raman spectroscopy. Then we characterized the optical properties of these ZnO heterostructures by PL and CL.

The results are summarized as follows. (1) The ZnO nanorods align vertically on the wall of these 1D nanostructures. The growth direction of ZnO nanorods is uniform [001]. The symmetric alignment of ZnO nanorods was observed for the GaN–ZnO nanostructures. The diameter of ZnO nanorods is in the range of 80–150 nm and the length is up to 3 μm. (2) The growth of ZnO nanorods was monitored as a function of time, revealing that the deposition of ZnO nanorods starts from the top part of 1D nanostructures. As the deposition time increases, the length and density can increase. The diameter frequently increases with time. We suggest that the growth of ZnO nanorods would follow a vapor–liquid–solid growth mechanism. Zn vapor deposits on the 1D nanostructures and form the nanocable structure that the outer layers encapsulate 1D nanostructure. The ZnO nanorods grow out from the outer layers of nanocables and the diameter is determined by the diameter of nanocables. The triangle facet of the outer layers leads the ZnO nanorods to align in a row on the wall of GaN nanowires. (3) The heterostructures exhibit intense PL at UV region with a peak at 385 nm, corresponding to the NBE peak of ZnO nanorods. For GaN–ZnO nanostructures, the NBE peak shows a blue shift to 380 nm, owing to the emission of GaN nanowires. The CL shows a good consistency with the PL. The relative intensity of green emission band to NBE band increases with the density of ZnO nanorods. This can be due to the higher oxygen vacancies at the increased surface area and more interfaces. We recommend that these heterostructures can be promising candidates for the light emitting diode devices, in respect of the high-density ZnO nanorods forming the 3D hierarchical structures.

Zn alloy nanowires. Zinc gallate ($ZnGa_2O_4$) has attracted much attention in recent years, because it is one of the most promising luminescence oxide materials for blue emission [114–125]. Room temperature resistivity has been reported to an order of 30 mΩcm for polycrystalline $ZnGa_2O_4$ [114]. Therefore it can be used for applications in vacuum fluorescent display (VFD) and field emission display (FED) as a low-voltage CL phosphor. With a bandgap of 4.4–4.7 eV, $ZnGa_2O_4$ is potentially useful as a transparent conducting oxide, particularly when transparency through the violet to near UV region is desired. It can also play an excellent host material for multicolor emitting phosphor layers; manganese-activated $ZnGa_2O_4$ ($ZnGa_2O_4$:Mn) for green emission and $ZnGa_2O_4$:Cr for red emission [116,118]. $ZnGa_2O_4$ material possesses a cubic spinel crystal structure that Zn ions occupy the tetrahedral sites whereas Ga ions occupy the octahedral sites. The green-emitting Mn^{+2} site in $ZnGa_2O_4$: Mn is generally accepted to be tetrahedrally coordinated. Nanocrystalline $ZnGa_2O_4$ was synthesized by hydrothermal reaction [125]. However, this $ZnGa_2O_4$ phosphor material has not yet been reported for the nanowires.

We reported the high-density arrays of vertically aligned $ZnGa_2O_4$ nanowires [126]. They were synthesized on the Au nanoparticles-deposited Si substrates via simple CVD of ZnO–Ga mixture at 1000°C. SEM and TEM images reveal that the average diameter of vertically aligned nanowires is 80 nm and the length is about 5 μm (Fig. 5.18a). The HRTEM images and SAED patterns reveal that the $ZnGa_2O_4$ nanowires consist of single-crystalline cubic structure grown with the [111] direction. The Ga:Zn = 2:1 ratio has been confirmed by XPS. The XRD and Raman data provide another evidence for the cubic spinel structure of $ZnGa_2O_4$ nanowires. The nanowires exhibit strong PL and

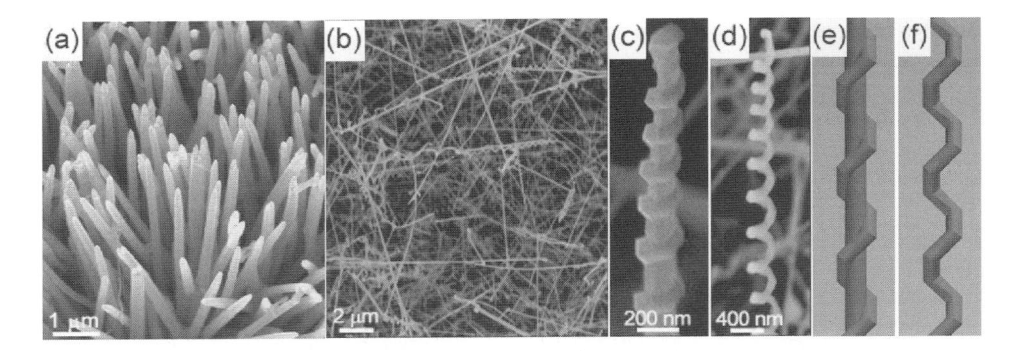

Fig. 5.18. SEM micrographs of (a) vertically aligned straight and (b) spring-like $ZnGa_2O_4$ nanowires. Magnified image reveals that the helical nanowire is (c) rolled around the straight nanowire (nanovines), and (d) self-coiled (nanospring). Schematic models for (e) the nanovines and (f) the nanosprings. [126,127]

CL band centered around 450 nm, which attributes to the Ga–O bonds. The $ZnGa_2O_4$ nanowires can be a promising nanostructure in the applications of high-performance LED devices operating at blue wavelength.

We also synthesized $ZnGa_2O_4$/ZnSe nanovines and $ZnGa_2O_4$ nanosprings by thermal evaporation using the ZnSe nanowires [127]. They were synthesized by a two-step thermal evaporation method. (1) High-purity single-crystalline ZnSe nanowires were synthesized using CoSe/ZnO on an Au nanoparticles-deposited Si substrate at 800°C. (2) The pre-grown ZnSe nanowires were placed near the ZnO/Ga, and the temperature of 600 or 900°C was maintained for 10–60 min, thereby producing the $ZnGa_2O_4$ nanowires. Figure 5.18b shows a SEM image of the spring-like $ZnGa_2O_4$ nanostructures. Figure 5.18c reveals the nanovine structure in which the helical nanowire winds around the straight nanowire support. The magnified image of Fig. 5.19b shows clearly the self-coiled morphology without any support (Fig. 5.18d). Scanning TEM (STEM) and EDX line-scanning reveal that the nanovine is composed of helical $ZnGa_2O_4$ and straight ZnSe nanowires. Both the nanovines and nanosprings have common structure, in which the growth direction of the helical $ZnGa_2O_4$ nanowire changes zigzagged with the same angle. They all consist of single-crystalline cubic $ZnGa_2O_4$ crystals without any dislocation over the entire helical structure, and have four equivalent $\langle 011 \rangle$ growth directions, [011], [101], [011], and [101], with the [001] axial direction. Schematic models for the nanovines and the nanosprings are displayed in Figs. 5.18e and f, respectively. We suggest that the lattice matching with the ZnSe nanowires is an important factor in determining the growth direction of the helical $ZnGa_2O_4$ nanowires.

5.3.3 GaP nanostrucures

Gallium phosphide (GaP) is a popular semiconductor material, due to its large energy bandgap (E_g = 2.26 eV at 300 K) and good thermal stability. The GaP nanowires

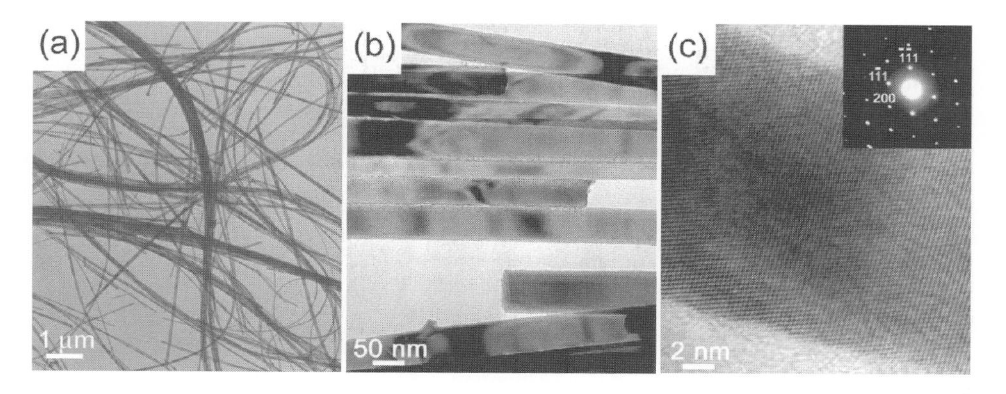

Fig. 5.19. (a) TEM image showing a general morphology of the GaP nanowires. (b) All of them are straight and cylindrical. (c) HRTEM image of one GaP nanowire with its corresponding SAED pattern (inset), showing the zinc blende single-crystalline structure with the identical [111] growth direction. [133]

would be ideally suited to investigate the role of dimensionality and size in the physical properties. The GaP nanorods, nanowires, and nanobelts were synthesized by laser ablation [128,129], carbon nanotube-confined reaction, [130] and thermal decomposition in the presence of surfactants [131]. The fabrication of top-gated GaP nanowire field-effect transistors using core-shell structured GaP nanowires has recently been demonstrated by Kim and coworkers [132]. In this section, we report a successful synthesis of various GaP nanostructures using a catalyst-assisted sublimation method.

GaP nanowires. High-density GaP nanowires were synthesized on the Ni-deposited alumina substrates via the catalyst-assisted sublimation of the ball-milled GaP powders under Ar flow [133]. Figure 5.19a corresponds to an SEM image of the GaP nanowires. The average diameter is 40 nm and the length can be up to 300 μm. All nanowires are straight and consist of single-crystalline zinc blende structure with the identical [111] growth direction (Fig. 5.19b,c). If the millimeter-size pieces are used instead of the ball-milled powders, the curled and polycrystalline nanowires are grown. It shows that the morphology of the nanowires can be controlled by the volatility of the GaP source. The XRD peaks and Raman spectrum confirm the synthesis of the highly crystalline GaP nanowires. The PL spectrum of the nanowires at 8 K shows the broad donor–acceptor-pairs peak and its replicas in the range 2.1–2.2 eV, for the excitation by 488 nm. The typical N isoelectronic exciton peaks appear due to the N-doping on the surface. The increased surface area would become an important factor in determining the optical properties of the nanowires.

N-doped GaP nanowires and nanobelts. The indirect bandgap of GaP usually limits the applications in the optoelectronic fields. But the incorporation of nitrogen (N) leads to a direct bandgap behavior that facilitates its extensive applications in the optoelectronic fields [134–136]. Therefore, the N-doped GaP nanostructures are probably suited to investigate the role of dimensionality and size in the optical properties.

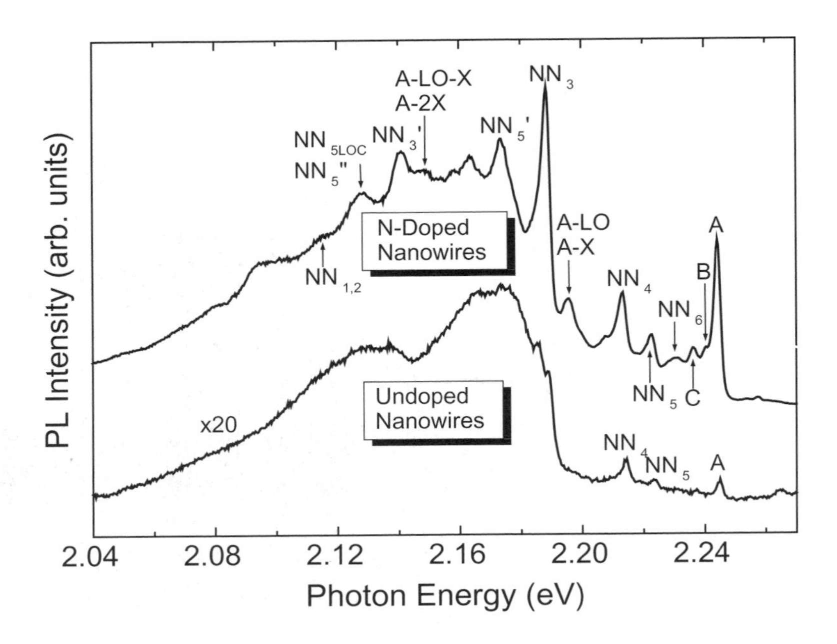

Fig. 5.20. PL spectrum of undoped and N-doped GaP nanowires. The excitation wavelength is the 488 nm line from an argon ion laser. [137]

The N-doped GaP nanowires were directly synthesized via the catalyst-assisted sublimation of ball-milled GaP powders under NH_3 flow [137]. The average diameter is 40 nm and the length is about 500 μm. All nanowires exhibit a single-crystalline zinc blende structure with the identical [111] growth direction. There are few amorphous outerlayers. XRD pattern and Raman spectrum confirm the highly crystalline nature of the nanowires. The PL spectrum shows a series of the isoelectronic bound exciton peaks in the energy range 2.1–2.25 eV, with a greatly enhanced intensity compared to that of the undoped GaP nanowires (Fig. 5.20). We assign the peak origin using the assumption that the peaks shift to the lower energy from those of the GaP crystals. The concentration of the doped N atoms has been estimated to be $\sim 10^{18}$ cm^{-3}.

We also synthesized the N-doped GaP nanobelts where the width is 200–500 nm with an average value of 300 nm and the thickness is about 1/10 of the width [138]. They were grown directly via the sublimation of ball-milled GaP powders under NH_3 flow. The nanobelts consist of single-crystalline zinc blende structured GaP crystal whose [111] direction is parallel to the belt axis. The EELS data reveals that the N doping occurs mainly in the surface region of the nanobelts. The PL spectrum exhibits the typical isoelectronic bound exciton peaks in the range of 2.11–2.25 eV. The concentration of the doped N atoms is estimated to be $\sim 10^{18}$ cm^{-3}. The red shift of the peaks from those of the bulk may be explained by the formation of surface potential. The N doping in the surface region would determine the PL properties of the nanobelts.

Mn-doped GaP nanowires. Another important III–V DMS, (Ga,Mn)P, had been predicted to have $T_C \approx 100$ K [70d]. However, it was reported that the ferromagnetic

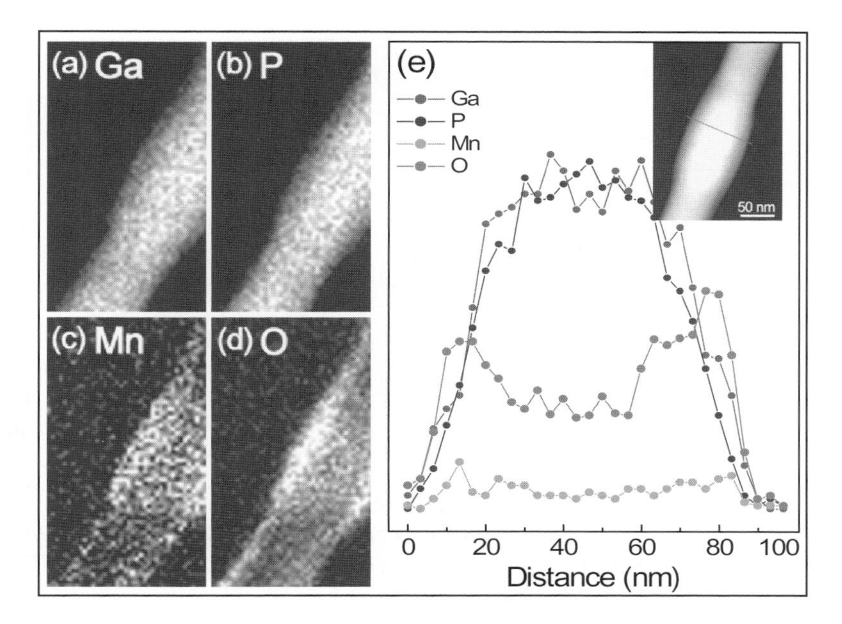

Fig. 5.21. STEM elemental maps of (a) Ga, (b) P, (c) Mn, and (d) O concentrations in a bumpy nanowire, and (e) elemental mapping cross section show that Mn and O concentrates at the outerlayers. The corresponding TEM image is shown in the inset. [141]

behaviors can persist at near room temperature for 3 at.% Mn-doped GaP:C films and T_C can depend on the Mn and hole concentrations [139,140]. Therefore the search for high T_C DMS materials is still a challenging subject with respect to applications as well as theoretical viewpoints.

The Mn-doped GaP nanowires were synthesized by thermal evaporation of ball-milled GaP and Mn powders [141]. The diameter is 40–100 nm. The nanowires are frequently sheathed with the bumpy amorphous outerlayers. The HVEM images reveal a high degree of twin structure in zinc blende GaP crystal grown with the [111] direction. The STEM elemental mapping reveals the inhomogeneous Mn-doping concentrated at the amorphous outerlayers (Fig. 5.21). Average Mn content is estimated to be about 1 at.% for the nanowire parts and about 10 at.% for the outerlayers. The amorphous outerlayers mainly consist of Mn and O. The XRD pattern and Raman spectroscopy confirm that the nanowires have only zinc blende GaP crystals. The hysteresis curves measured at 5 and 300 K, and the temperature-dependent magnetization curves with H = 100 Oe provide an evidence for the ferromagnetism with T_C higher than 330 K (Fig. 5.22a). The magnetotransport properties of individual nanowires have been measured in the temperature range 5–200 K, showing the large negative MR at temperatures over this range (Fig. 5.22b). The absolute value of MR % reaches about 5% at 5 K. We suggest that Mn would efficiently dope into the GaP nanowires and form the DMS nanowires. These Mn-doped GaP nanowires would be encouraging ferromagnetic semiconductor nanowires applicable to the spintronic nanodevices.

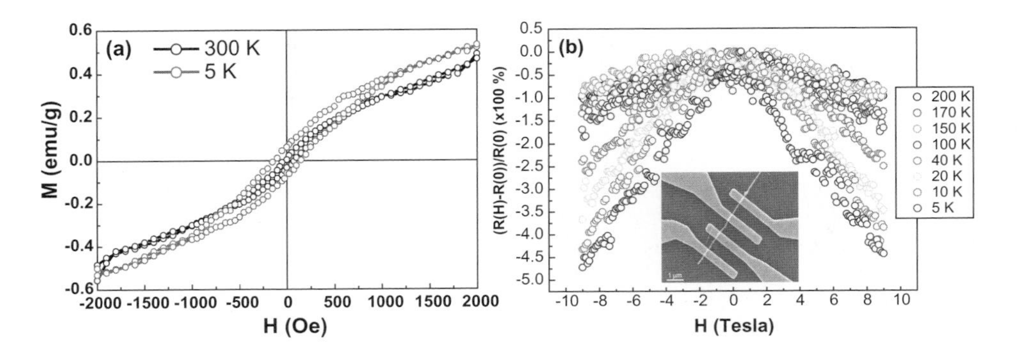

Fig. 5.22. (a) The M vs. H curves at 5 and 300 K, showing a hysteresis curve. (b) MR at various temperatures in the range 5–200 K, for magnetic field sweeps between −9 T and 9 T. Inset corresponds to SEM image of a nanowire connected to four electrodes. [141]

GaP nanocables. Three typed core-shell nanocables, GaP/SiO_x, GaP/C, and GaP/SiO_x/C, were fabricated successfully by controlling the condition of CVD [142]. The key of synthesis methods is summarized as follows. (1) The coaxial GaP–SiO_x nanocables were produced directly on the Au nanoparticles-deposited Si substrates by the CVD of ball-milled GaP powders and Si wafer as Si source. (2) The amorphous C outerlayers uniformly deposit on the GaP nanowires via the CVD of CH_4, following the sublimation of ball-milled GaP powders. (3) The graphitic outerlayers form uniformly on the pre-grown GaP–SiO_x nanocables via the CVD of C_2H_2.

TEM image of GaP–SiO_x–C nanocables shows that the outerlayers sheath the GaP nanowire with uniform thickness (Fig. 5.23a,b). Atomic-resolved image reveals the crystalline GaP crystal, amorphous SiO_x layers, and graphitic outerlayers (Fig. 5.23c). The length of nanocables is up to 500 μm. The GaP nanowire core consists of the zinc

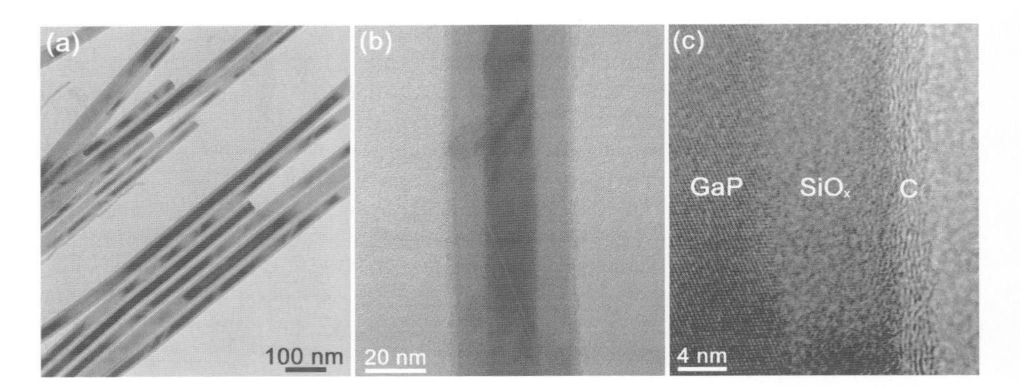

Fig. 5.23. (a) TEM image shows the GaP/SiO_x/C nanocables. (b) The outerlayers sheath the GaP nanowire with uniform thickness. (c) Atomic-resolved image reveals the crystalline GaP crystal, amorphous SiO_x layers, and graphitic outerlayers. [142]

blende structured crystals grown with the [111] direction. The thickness of C outerlayers is controlled by the C deposition time. The degree of crystalline perfection of C outerlayers increases with the deposition temperature. The effect of the outerlayers on the structural properties of GaP nanowires was examined thoroughly using XPS, XRD, Raman spectrum, and PL. Although the crystal defects are frequently produced during the growth process, the outerlayers reduce significantly the surface defects of GaP nanowires by reducing the dangling bonds. We explained the growth process using the vapor–liquid–solid mechanism. The conductivity measurement shows the effect of the insulating SiO_x and metallic C outerlayers on the nanocables. These noble fabrications of the GaP core-shell nanocables with single or double shells in a controlled manner probably facilitate to build a variety of nanodevices.

5.3.4 SiC, Si_3N_4, and BN nanostructures

SiC nanowires and SiC–C nanocables. Silicon carbide (SiC) is a wide-band gap semiconductor material with extreme hardness and high thermal conductivity even at high temperature [143,144]. Lately, one-dimensional SiC nanostructure, SiC nanowires, has attracted much attention because of its potential application for nanostructured composite materials and microelectronic devices. It was reported that the elasticity and strength of zinc blende SiC nanowires (β-SiC or 3C–SiC) are substantially greater than those of the bulk [145]. Various methods, e.g. carbon nanotube-confined reaction, arc discharge, laser ablation, carbothermal reduction, and chemical vapor deposition, have been developed to produce the SiC nanowires [146–151]. Herein we explain a notably simple and efficient method for the large-scale synthesis of highly aligned β-SiC nanowires directly from the silicon (Si) substrates [152].

The n-type Si (100) substrates with an area of 10–20 cm^2 were coated with 0.01 M $FeCl_2{\cdot}4H_2O$ (99%, Aldrich) ethanol solution. The thickness of $FeCl_2$ film was about 300 nm. The substrate was loaded on a quartz boat placed in a quartz tube reactor. About 0.01 g of Ga (99.999%, Aldrich) and GaN powder (99.99+%, Aldrich) with 1:1 volume ratio was placed nearby the substrates. Argon flowed into the tube reactor while raising the temperature. A flow of CH_4 (99.95%) and H_2 (99.999%) mixture was introduced through a mass-flow controller at a rate of 10–40 and 300–500 sccm, respectively. The temperature of the substrates was set at 1100°C and the growth time was 30 min–1 h.

The diameter of aligned SiC NWs is 10–60 nm with an average value of 40 nm and the length is 500 μm (Fig. 5.24a,b). The HRTEM images and SAED patterns reveal the single-crystalline zinc blende structure and the [111] growth direction. If no CH_4 flows, the amorphous Si–NW bundles were grown from the Si substrates under the same growth condition. The dissolution of the Si substrate in molten Ga is crucial for the growth of the SiC and Si nanowires. The XRD, Raman, and IR spectroscopy show the high purity and the stacking faults of the SiC nanowires. We expect the present method to be promising for the mass-production of high purity SiC nanowires.

We also synthesized the aligned SiC–C coaxial nanocables directly from the Si substrates using a novel reaction at 1100°C [153]. After the growth of β-SiCNWs, extra flow of CH_4 deposits the C outerlayers. The diameter of β-SiCNW cores is 5–30 nm with an

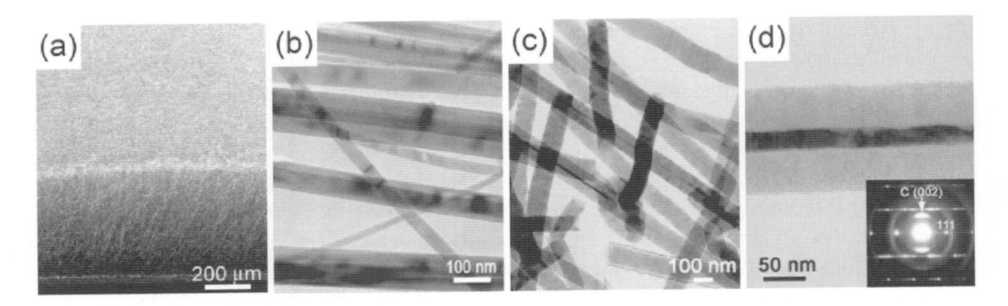

Fig. 5.24. (a) SEM image of aligned SiC nanowire array homogeneously grown on a large area of the Si substrate. The length is 500 μm. TEM images for the general morphology of (b) SiC nanowires and (c) SiC–C nanocables. The average diameter of nanowires and nanocables is 80 and 110 nm, respectively. (d) The thickness of C outerlayer is 50 nm. The SAED pattern shows the graphitic (002) planes parallel to the long axis (inset). [152,153]

average value of 20 nm and the length is up to 150 μm. The thickness of C outerlayers is controllable from 3 to 50 nm using the deposition time variation. Figure 5.24c corresponds to a TEM image showing the general morphology of SiC–C nanocables. The C outerlayer thickness is controlled to be 50 nm (Fig. 5.24d). The SAED pattern shows the graphitic (002) planes parallel to the long axis (inset). All HRTEM images, SAED patterns, and Raman spectroscopy reveal the crystalline zinc blende structure of the SiCNW cores and the crystalline graphitic sheets of the outerlayers. As the thickness of C outerlayers increases, the degree of crystalline perfection increases. The growth of the C outerlayers would take place via layer-by-layer process. The graphitic sheets at the larger diameter would be under lesser stress, forming a more crystalline phase.

Si_3N_4 nanowires. Silicon nitride (Si_3N_4) is a wide band gap (~5.3 eV) semiconductor and ceramic material due to its high strength, lightweight, and good resistance to thermal shock and oxidation [154,155]. The Si_3N_4 nanowires were proven to have a much higher bending strength than the bulk [156]. We synthesized the high-density α-Si_3N_4 nanowires on a large area of the Si substrate using a catalyst-assisted reaction with NH_3 or H_2 at 1200°C, as shown in Fig. 5.25a [157]. The catalyst was Ga, GaN, and the Fe nanoparticles deposited on the Si substrate. The Si substrate and NH_3 (or GaN under H_2 flow) were used as the Si and N sources, respectively. The diameter of the nanowire is 10–80 nm with an average value of 40 nm and the length is about 300 μm (Fig. 5.25b). The HRTEM images and SAED patterns reveal the defect-free single-crystalline α-Si_3N_4 nanowires grown with various growth directions. Figure 5.25c shows the HRTEM image of α-Si_3N_4–NW grown with the [010] growth direction. The insets show the corresponding SAED pattern. The lattice fringes are clearly separated with a distance that is consistent with the lattice constants of the bulk α-Si_3N_4 (JCPDS Card No. 41-0360). No amorphous outerlayers are observed. The XRD pattern and IR spectroscopy confirm pure α-phase Si_3N_4 nanowires. The XRD peaks show a slight shift to higher diffraction angles from those of the bulk, suggesting the strains inside the nanowires. The growth follows a combination of SLS and VLS mechanisms. We look forward to this simple and efficient method application to a mass-production of the Si_3N_4 nanowires.

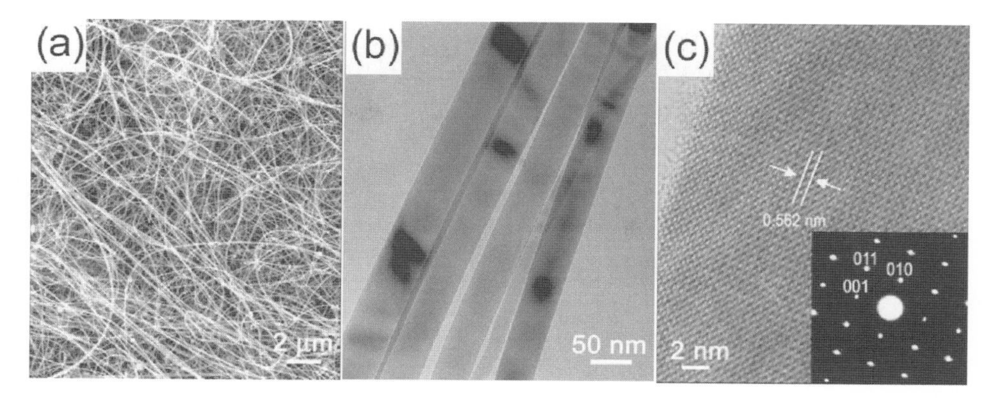

Fig. 5.25. (a) SEM image of high-density α-Si_3N_4 nanowires homogeneously grown on a large area of the Si substrate. (b) TEM image shows the diameter of nanowires in the range of 10–80 nm. (c) HRTEM image of α-Si_3N_4 nanowire grown with the [010] direction. The insets show the corresponding SAED pattern. The lattice fringes are clearly separated with a distance that is consistent with the lattice constants of the bulk α-Si_3N_4 (JCPDS Card No. 41-0360). [157]

BN nanostructures. Boron nitride (BN) nanotubes were first synthesized using arc discharge method in 1995 [158]. The boron nitride nanotubes (BN-NT) has a mechanical strength similar to that of CNT, i.e. elastic modulus is ~1.2 TPa [159]. In contrast to CNTs, the BN nanotubes have semiconducting properties with nearly constant bandgap (5.5 eV) which is independent of the diameter and chirality of the nanotube [160]. Therefore they are considered for many different technological applications exploiting its unique mechanical and electrical properties. The tremendous works show the diverse structure of BN nanotubes depending on the growth method as well as the growth condition. We reported a temperature-dependent growth of multi-walled BN nanotubes in the range 1000–1200°C.

The BN nanotubes were homogeneously grown on the Fe nanoparticles-deposited alumina substrates through the catalyst-assisted reaction of the ball-milled B/BN powder mixture with NH_3 in the temperature range 1000–1200°C [161]. The Fe nanoparticles were formed using $FeCl_2$ethanol solution. The diameter of BN nanotubes is in the range of 40–100 nm and the length is 10–20 μm. At temperatures below 1100°C, the BN nanotubes exhibit exclusively a bamboo-like structure in which the compartment layers appear at a regular distance (Fig. 5.26a,b). As the temperature increases, the growth of the cylindrical BN nanotubes becomes dominant (Fig. 5.26c,d). The BN sheets of the cylindrical BN nanotubes are tilted to the tube axis by an angle of about 25°. The degree of crystalline perfection increases with the temperature. The EELS analysis reveals that the atomic ratio of B and N is ~1 with a trace of O. The Raman scattering spectrum shows that as the crystallinity of the BN sheets decreases, the E_{2g} peak shifts to the lower frequency and its width broadens. The growth would follow a combination of SLS and VLS mechanisms. We explain the growth mechanism of the bamboo-like BN nanotubes by adopting the base-growth mechanism of the vertically aligned CNTs grown using a CVD of acetylene. The compartment layers could be mainly formed by the bulk

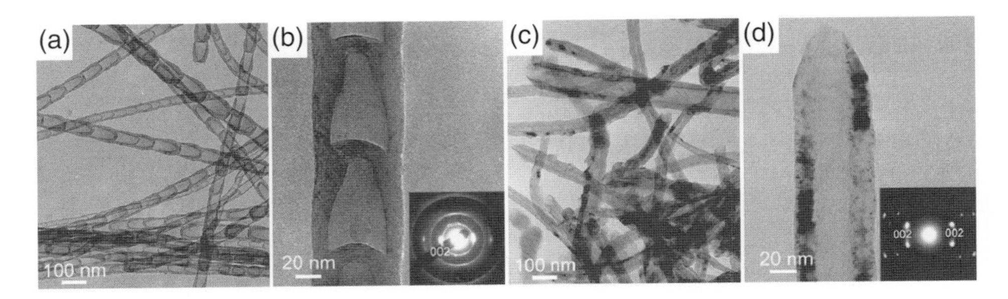

Fig. 5.26. (a) TEM image of the bamboo-like BN nanotubes homogeneously grown at 1050°C. (b) A typical bamboo-like BN nanotube in which the compartment layers appear at a regular distance. (c) TEM images of cylindrical BN nanotubes grown at 1200°C. (d) The cylindrical BN nanotubes have hollow inside and BN sheets tilted to the tube axis by an angle of about 25°. Insets show the corresponding SAED pattern, revealing the [002] direction of hexagonal BN. [161]

diffusion of B/N atoms. The cylindrical structure would result from the strains of highly crystalline BN sheets. We also propose other possibilities for the formation of cylindrical nanotubes. The present synthesis would be a promising method for the synthesis of BN nanotubes at low temperatures and for controlling their structure and crystallinity simply by the growth temperature.

Thorn-like BN nanostructures, i.e. the nanosize hexagonal BN (*h*-BN) layers that are randomly stacked looking like thorns, were synthesized using thermal CVD of B/B_2O_3 under the flow of NH_3 at 1200°C [162]. They can grow self-assembled forming microsize lumps (Fig. 5.27a and b), and also deposit as sheathing layers on the pre-grown SiC nanowires with a controlled thickness in the range 20–100 nm (Fig. 5.27c and d). The spreading of the thorn-like BN layers as the sheathing layers results in a significantly enhanced surface area of 2400 m^2/g.

We investigated the detailed electronic structure of the BN nanotubes and nanothorns using the EELS and angle-resolved XANES [163]. The EELS data reveal the

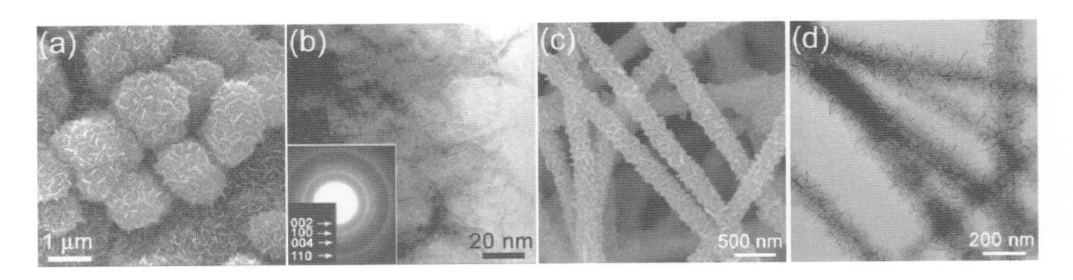

Fig. 5.27. (a) SEM micrograph showing the microsize lumps grown on the substrate. (b) TEM image showing the fibrous morphology with its corresponding SAED (inset). (c) SEM micrograph showing the thorn-like BN layers sheathing the SiC nanowires. (d) TEM image showing thorn-like morphology of BN outerlayer and the single-crystalline SiC nanowire. [162]

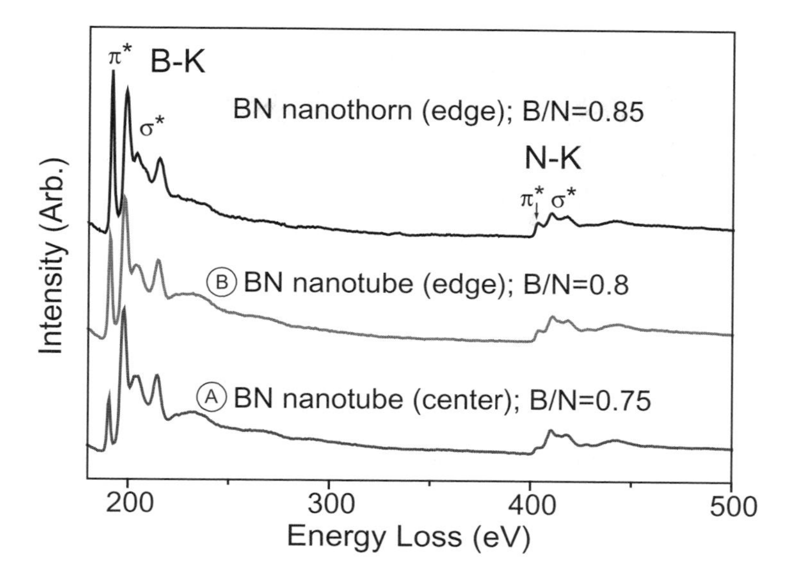

Fig. 5.28. EELS data for the center and edge parts of nanotubes, and the edge part of nanothorns. It shows two distinct absorption features starting at 188 eV and another at 400 eV, corresponding to the known K-shell ionization edges for B and N, respectively. It shows proficient N atom content than the B atom. [163]

N-enrichment with a ratio of B/N = 0.75–0.8 and 0.85 for the nanotubes and the nanothorns, respectively (Fig. 5.28). The B- and N-K edges XANES show the distinct features from that of *h*-BN microcrystals (Fig. 5.29). The nanothorns and nanotubes show the lower energy shift of π^* transition in the N-K edge by about 0.8 and 1.0 eV, respectively, and the enhanced second-order signals of N 1*s* electrons in the B-K edge region. We suggest that the N-rich BN layers would reduce the bandgap of nanostructures relative to that of *h*-BN microcrystals, and enhance the second-order signals of N 1*s*. In both EELS and XANES, the smaller intensity ratio of π^*/σ^* for the nanotubes than for nanothorns was explained by the higher amount of defective N-rich bonding. Raman spectrum of BN nanostructures show the peak shift to the higher frequency region (3 cm^{-1}) and the significant peak broadening compared to that of the *h*-BN microcrystals. The extra broadening of the nanotubes than those of the nanothorns could be due to the increased defects in the more N-rich layers. The present understanding will lead to the understanding of electronic structure of the BN nanostructures, which is a prerequisite to particular applications such as nanoelectronic devices.

5.4 Conclusions

Discovery of novel 1D semiconductor nanostructures, such as GaN, ZnO, GaP, SiC, Si_3N_4, BN, as well as the progress of new experimental techniques, would provide

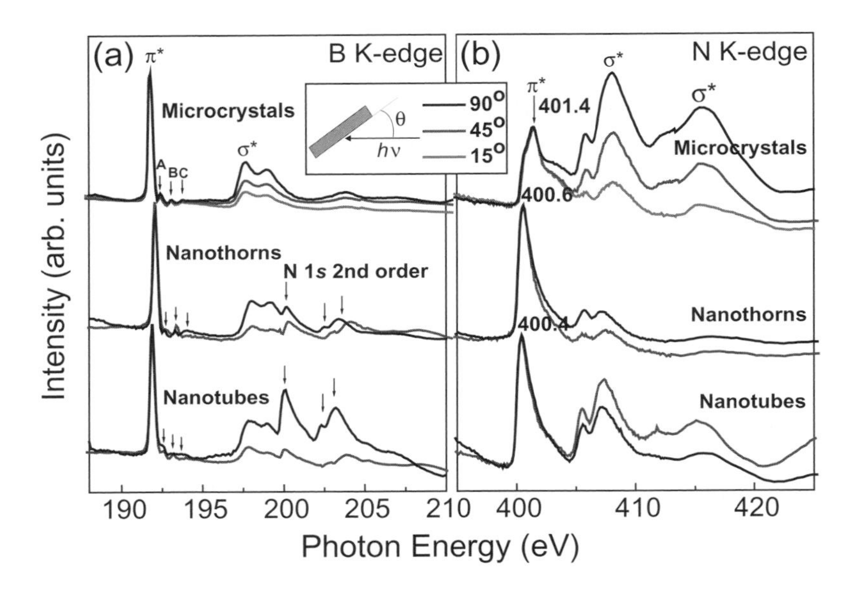

Fig. 5.29. Angle-resolved XANES spectra of (a) B-K edge and (b) N-K edge of *h*-BN microcrystals, nanothorns, and nanotubes. [163]

fresh opportunities for the development of innovative nanosystems and nanostructured materials. The results show that nanowires exhibit many properties that are similar to, and others that are distinctively different from those of their bulk counterparts. The nanowires have the application point of views in that some of the material parameters critical for certain properties can be independently controlled in nanowires but not in the bulk counterparts. Compared to other low-dimensional systems, the nanowires have two quantum-confined with directions, still leaving one unconfined direction for electrical conduction. This allows them to be used in applications which require electrical conduction, rather than tunneling transport. Therefore, the nanowires have shown to provide a promising framework for applying the bottom-up approach for the design of many potential applications. Fabrication of nanoelectronics using individual nanowire allows to build a computer memory to store bits of information, as well as switches, transistors, flat-panel displays, integrated circuits, fast logic gate, nanoscopic lasers, and electrodes in chemical or biosensors, solar cell, and fuel cell.

Acknowledgment

This work is supported by KOSEF (R14-2003-033-01003-0; R02-2004-000-10025-0), KRF (2003-015-C00265), and KIST (2E18740-05-062). SEM, field-emission TEM, HVTEM, and MPMS measurements were performed at Korea Basic Science Institute.

References

[1] (a) X. Duan, Y. Huang, Y. Cui, J. Wang and C.M. Lieber, Nature 409 (2001) 66. (b) M.S. Gudiksen, X. Duan, Y. Cui and C.M. Lieber, Science 293 (2001) 1455. (c) Y. Huang, X. Duan, Y. Cui ,L. J. Lauhon, K.-H. Kim and C.M. Lieber, Science 294 (2001) 1313. (d) X. Duan, Y. Huang, R. Agarwal and C. M. Lieber, Nature 421 (2003) 241.

[2] G.Y. Tseng and J.C. Ellenbogen, Science 294 (2001) 1293.

[3] N.A. Melosh, A. Boukai, F. Diana, B. Gerardot, A. Badolato, P.M. Petroff and J.R. Heath, Science 300 (2002) 112.

[4] Y. Xia, P. Yang, Y. Sun, Y. Wu, B. Mayers, B. Gates, Y. Yin, F. Kim and H. Yan, Adv. Mater. 15 (2003) 353.

[5] C.N.R. Rao, F. L. Deepak, G. Gundiah and A. Govindaraj, Prog. Solid State Chem. 31 (2003) 5.

[6] J. Hu, M. Ouyang, P. Yang and C.M. Lieber, Nature (London) 399 (1999) 48.

[7] Y. Zhang, T. Ichihashi, E. Landree, F. Nihey and S. Iijima, Science 285 (1999) 1719.

[8] (a) W.I. Park, G.C. Yi, M.Y. Kim and S.J. Pennycook, Adv. Mater. 15 (2003) 526. (b) S.W. Jung, W.I. Park, G.C. Yi and M.Y. Kim, Adv. Mater. 15 (2003) 1358.

[9] J. Luo, L. Zhang and J. Zhu, Adv. Mater. 15 (2003) 579.

[10] S.R. Nicewarner-Pena, R.G. Freeman, B.D. Reiss, L. He, D.J. Pena, I.D. Walton, R. Cromer, C.D. Keating and M.J. Natan, Science 294 (2001) 137.

[11] (a) M.S. Gudiksen, L.J. Lauhon, J. Wang, D.C. Smith and C.M. Lieber, Nature (London) 415 (2002) 617. (b) Y. Yu, J. Xiang, C. Yang, W. Lu and C.M. Lieber, Nature (London) 450 (2004) 61.

[12] Y. Wu, R. Fan, P. Yang, Nano Lett. 2 (2002) 83.

[13] R. He, M. Law, R. Fan, F. Kim and P. Yang, Nano Lett. 2 (2002) 1109.

[14] J. Hu, Y. Bando, Z. Liu, T. Sekiguchi, D. Golberg and J. Zhan, J. Am. Chem. Soc. 125 (2003) 11306.

[15] Y. Zhang, K. Suenaga, C. Colliex and S. Iijima, Science 281 (1998) 973.

[16] L.J. Lauhon, M.S. Gudiksen, D. Wang and C.M. Lieber, Nature (London) 420 (2002) 57.

[17] (a) W.-S. Shi, H.-Y. Peng, L. Xu, N. Wang, Y.-H. Tang and S. -T. Lee, Adv. Mater. 12 (2002) 1927. (b) X.-H. Sun, C.-P. Li, W.-K Wong, N.-B. Wong, C.-S. Lee, S.-T. Lee and B.-K. Teo, J. Am. Chem. Soc. 124 (2002) 14464. (c) J.-Q. Hu, X.-M. Meng, Y. Jiang, C.-S. Lee and S.-T. Lee, Adv. Mater. 15 (2003) 70. (d) X.-M. Meng, J.-Q. Hu, Y. Jiang, C.-S. Lee and S.-T. Lee, Appl. Phys. Lett. 83 (2003) 2241. (e) B.K. Teo, C. P. Li, X.H. Sun, N.B. Wong and S.T. Lee, Inorg. Chem. 42(2003) 6723.

[18] (a) Y.-C.Zhu, Y. Bando, D.-F. Xue, F.-F. Xu and D. Golberg, J. Am. Chem. Soc., 125 (2003) 14226. (b) Y. Li, Y. Bando and D. Golberg, Adv. Mater. 16 (2004) 93. (c) C.C.Tang, Y. Bando, T. Sato and K. Kurashima, Adv. Mater. 14 (2002) 1046. (d) Y. Gao, Y. Bando and D. Golberg, Appl. Phys. Lett. 81 (2002) 4133. (e) J. Hu, Y. Bando, and Z. Liu, Adv. Mater. 15 (2003) 1000. (f) C. Tang, Y. Bando, D. Golberg, X. Ding, and S. Qi, J. Phys. Chem. B 107 (2003) 6539. (g) Y.B. Li, Y. Bando, D. Golberg and Z.W. Liu, Appl. Phys.

Lett. 83 (2003) 999. (h) D. Golberg, Y. Bando, K. Fushimi, M. Mitome, L. Bourgeois and C.C. Tang, J. Phys. Chem. B 107 (2003) 8726. (i) Y.-C. Zhu, Y. Bando and R.-Z. Ma, Adv. Mater. 15 (2003) 1377. (j) Y. Gao, Y. Bando, Z. Liu, D. Golberg and H. Nakanishi, Appl. Phys. Lett. 83 (2003) 2913. (k) Y.B. Li, Y. Bando, D. Golberg and Y. Uemura, Appl. Phys. Lett. 83 (2003) 3999. (l) L.W. Yin, Y. Bando, Y.C. Zhu and M.S. Li, Appl. Phys. Lett. 84 (2004) 5314.

[19] (a) X. Wang, P. Gao, J. Li, C.J. Summers and Z.L. Wang, Adv. Mater. 14 (2002) 1732. (b) Y. Ding, X.Y. Kong and Z.L. Wang, J. Appl. Phys. 95 (2004) 306. (c) X.Y. Kong, Y. Ding and Z.L. Wang, J. Phys. Chem. B 108 (2004) 570.

[20] (a) W. Han and A. Zettl, Adv. Mater. 14 (2002) 1560. (b) W. Han and A. Zettl, Appl. Phys. Lett. 81 (2002) 5051.

[21] (a) Y. Wu and P. Yang, Appl. Phys. Lett. 77 (2000) 43. (b) Y. Wu, P. Yang, Adv. Mater. 13 (2001) 520. (c) H.-J. Choi, J.C. Johnson, R. He, S.-K. Lee, F. Kim, P. Pauzauskie, J. Goldberger, R.J. Saykally and P. Yang, J. Phys. Chem. B 107 (2003) 8721.

[22] (a) Y. Yin, Y. Lu, Y. Sun and Y. Xia, Nano Lett. 2 (2002) 427. (b) X. Jiang, B. Mayers, T. Herricks and Y. Xia, Adv. Mater. 15 (2003) 1740.

[23] (a) J.-J. Wu, S.-C. Liu, C.-T. Wu, K.-H. Chen and L.-C. Chen, Appl. Phys. Lett. 81 (2002) 1312. (b) J.-J. Wu, T.-C. Wong and C.-C. Yu, Adv. Mater. 14 (2002) 1643.

[24] (a) H.W. Seo, S.Y. Bae, J. Park, H. Yang and B. Kim, J. Phys. Chem. B 107 (2003) 6739. (b) H.Y. Kim, S.Y. Bae, N.S. Kim and J. Park, Chem. Commum. 20 (2003) 2634.

[25] S.O. Obare, N.R. Jana and C.J. Murphy, Nano Lett. 1 (2001) 601.

[26] L. Manna, E.C. Scher, L.-S. Li and A.P. Alivisatos, J. Am. Chem. Soc. 124 (2002) 7136.

[27] Y. Zhang, N. Wang, S. Gao, R. He, S. Miao, J. Liu, J. Zhu and X. Zhang, Chem. Mater. 14 (2002) 3564.

[28] S. Hofmann, C. Ducati and J. Robertson, Adv. Mater. 14 (2002) 1821.

[29] Y.Q. Zhu, W.K. Hsu, H.W. Kroto and D.R.M. Walton, J. Phys. Chem. B 106 (2002) 7623.

[30] H.-M. Lin, Y.-L. Chen, J. Yang, Y.-C. Liu, K.-M. Yin, J.-J. Kai, F.-R. Chen, L.-C. Chen, Y.-F. Chen and C.-C. Chen, Nano Lett. 3 (2003) 537.

[31] Y.-G. Guo, L.-J. Wan and C.-L. Bai, J. Phys. Chem. B 107 (2003) 5441.

[32] A. Kolmakov, Y. Zhang and M. Moskovits, Nano Lett. 3 (2003) 1125.

[33] N.I. Kovtyukhova, T.E. Mallouk and T.S. Mayer, Adv. Mater. 15 (2003) 780.

[34] Q. Li and C. Wang, J. Am. Chem. Soc. 125 (2003) 9892.

[35] C.-H. Hsia, M.-Y. Yen, C.-C. Lin, H.-T. Chiu and C.-Y. Lee, J. Am. Chem. Soc. 125 (2003) 9940.

[36] Z.-Y. Jiang, Z.-X. Xie, X.-H. Zhang, R.-B. Huang and L.-S. Zheng, Chem. Phys. Lett. 378 (2003) 313.

[37] T. Mokari and U. Banin, Chem. Mater. 15 (2003) 3955.

[38] J. Cao, J.-Z. Sun, J. Hong, H.-Y. Li, H.-Z. Chen and M. Wang, Adv. Mater. 16 (2004) 84.

[39] B. Geng, G. Meng, L. Zhang, G. Wang and X. Peng, Chem. Commun. 20 (2003) 2572.

[40] C. Liang, Y. Shimizu, T. Sasaki, H. Umeharaa and N. Koshizaki, J. Mater. Chem. 14 (2004) 248.

[41] S. Han, C. Li, Z. Liu, B. Lei, D. Zhang, W. Jin, X. Liu, T. Tang, and C. Zhou, Nano Lett. 4 (2004) 1241.

[42] X. Liang, S. Tan, Z. Tang and N.A. Kotov, Langmuir 20 (2004) 1016.

[43] J. Hwang, B. Min, J.S. Lee, K. Keem, K. Cho, M.-Y. Sung, M.S. Lee and S. Kim, Adv. Mater. 16 (2004) 422.

[44] M.K. Lee, H.-J. Shin, Nucl. Instrum. Methods Phys. Rev. A 467 (2001) 508.

[45] G. Fasol, Science 272 (1996) 1751.

[46] S. Nakamura, Science 281 (1998) 956.

[47] F.A. Ponce and D. P. Bour, Nature (London) 386 (1997) 351.

[48] S.N. Mohammad and H. Morkoç, Prog. Quantum Electron. 20 (1996) 361.

[49] R. Calarco, M. Marso, T. Richter, A.I. Aykanat, R. Meijers, A.V.D. Hart, T. Stoica and H. Luth, Nano. Lett. 5 (2005) 981.

[50] F. Qian, Y. Li, S. Gradecak, D. Wang, C.J. Barrelet and C.M. Liber, Nano. Lett. 4 (2004) 1975.

[51] Y. Huang, X. Duan and C.M. Liber, Small 1 (2005) 142.

[52] W. Han, S. Fan, Q. Li, and Y. Hu, Science 277 (1997) 1287.

[53] W. Han, P. Redlich, F. Ernst and M. Rühle, Appl. Phys. Lett. 76 (2000) 652.

[54] X. Duan and C. M. Lieber, J. Am. Chem. Soc. 122 (2000) 188.

[55] J. Y. Li, X. L. Chen, Z.Y. Qiao, Y.G. Cao and Y.C. Lan, J. Crystal Growth 213 (2000) 408.

[56] W. -Q. Han and A. Zettl, Appl. Phys. Lett. 80 (2002) 303.

[57] C.C. Tang, S.S. Fan, H.Y. Dang, P. Li and Y.M. Liu, Appl. Phys. Lett. 77 (2000) 1961.

[58] C.-C. Chen and C.-C. Yeh, Adv. Mater. 12 (2000) 738.

[59] X. Chen, J. Li, Y. Cao, Y. Lan, H. Li, M. He, C. Wang, Z. Zhang and Z. Qiao, Adv. Mater. 12 (2000) 1432.

[60] C.-C. Chen, C.-C. Yeh, C.-H. Chen, M.-Y. Yu, H.-L. Liu, J.-J. Wu, K.-H. Chen, L.-C. Chen, J.-Y. Peng and Y.-F. Chen, J. Am. Chem. Soc. 123 (2001) 2791.

[61] H.-M. Kim, D.S. Kim, Y.S. Park, D.Y. Kim, T.W. Kang and K.S. Chung, Adv. Mater. 14 (2002) 991.

[62] H.W. Seo, S.Y. Bae, J. Park, H. Yang, G.S. Park and S. Kim, J. Chem. Phys. 116 (2002) 9492.

[63] S.Y. Bae, H.W. Seo, J. Park, H. Yang, M. Kang, H. Kim and S. Kim, Appl. Phys. Lett. 82 (2003) 4564.

[64] S.Y. Bae, H.W. Seo, J. Park, H. Yang, J.C. Park and S.Y. Lee, Appl. Phys. Lett. 81 (2002) 126.

[65] S.Y. Bae, H.W. Seo, J. Park, H. Yang and S.A. Song, Chem. Phys. Lett. 365 (2002) 525.

[66] S.Y. Bae, H.W. Seo, J. Park and H. Yang, Chem. Phys. Lett. 373 (2003) 620.

[67] S.Y. Bae, H.W. Seo, J. Park and H. Yang, J. Crystal Growth 258 (2003) 296.

[68] H.W. Seo, J. Park and H. Yang, Chem. Phys. Lett. 376 (2003) 445.

[69] H.W. Seo, S.Y. Bae, J. Park, H. Yang and B. Kim, J. Phys. Chem. B 107 (2003) 6739.

[70] (a) H. Ohno, Science 281 (1998) 951. (b) Y. Ohno, D.K. Young, B. Beschoten, F. Matsukura, H. Ohno, and D.D. Awschalom, Nature (London) 402 (1999) 790. (c) H. Ohno, D. Chiba, F. Matsukura, T. Omiya, E. Abe, T. Dietl, Y. Ohno, and K. Ohtani, Nature (London) 408 (2000) 944. (d) T. Dietl, H. Ohno, F. Matsukura, J. Cibert and D. Ferrand, Science 287 (2000) 1019.

[71] R. Fiederling, M. Keim, G. Reuscher, W. Ossau, G. Schmidt, A. Waag and L.W. Molenkamp, Nature (London) 402 (1999) 787.

[72] S.A. Wolf, D.D. Awschalom, R.A. Buhrman, J.M. Daughton, S. von Molnár, M.L. Roukes, A.Y. Chtchelkanova and D.M. Treger, Science 294 (2001) 1488.

[73] N. Theodoropoulou, A.F. Hebard, M.E. Overberg, C.R. Abernathy, S.J. Pearton, S.N.G. Chu and R.G. Wilson, Appl. Phys. Lett. 78 (2001) 3475.

[74] M.L. Reed, N.A. El-Mastry, H.H. Stadelmaier, M.K. Ritums, M.J. Reed, C.A. Parker, J.C. Roberts and S.M. Bedair, Appl. Phys. Lett. 79 (2001) 3473.

[75] T. Sasaki, S. Sonoda, Y. Yamamoto, K. Suga, S. Shimizu, K. Kindo and H. Hori, J. Appl. Phys. 91 (2002) 7911.

[76] K. Sardar, A.R. Raju, B. Bansal, V. Venkataraman and C.N.R. Rao, Solid State Commun. 125 (2003) 55.

[77] J.M. Lee, K.I. Lee, J.Y. Chang, M.H. Ham, K.S. Huh, J.M. Myung, W.J. Hwang, M.W. Shin, S.H. Han, H.J. Kim and W.Y. Lee, Microelect. Eng. 69 (2003) 283.

[78] I.T. Yoon, C.S. Park, H.J. Kim, Y.G. Kim, T.W. Kang, M.C. Jeong, M.H. Ham and J.M. Myoung. J. Appl. Phys. 95 (2004) 591.

[79] F. Zhang, N. Chen, X. Liu, Z. Liu, S. Yang and C. Chai, J. Cryst. Growth 262 (2004) 287.

[80] R. Giraud, S. Kuroda, S. Marcet, E. Bellet-Amalric, X. Biquard, B. Barbara, D. Frauchart, D. Ferrand, J. Cibert and H. Mariette, Europhys. Lett. 65 (2004) 553.

[81] M. Zajac, R. Doradzinski, J. Gosk, J. Szczytko, M. Lefeld-Sosnowska, M. Kaminska, A. Twardowski, M. Palczewska, E. Grzanka and W. Gebicki, Appl. Phys. Lett. 78 (2001) 1276.

[82] M.E. Overberg, C.R. Abernathy, S.J. Pearton, N.A. Theodoropoulou, K.T. McCarthy and A.F. Hebard, Appl. Phys. Lett. 79 (2001) 1312.

[83] D.S. Han, J. Park, K.W. Rhie, S. Kim and J. Chang, Appl. Phys. Lett. 86 (2005) 132506.

[84] H.J. Choi, H.K. Seong, J.Y. Chang, K.I. Lee, Y.J. Park, J.J. Kim, S.K. Lee, R. He, T. Kuykendall and P. Yang, Adv. Mater. 17 (2005) 1351.

[85] H. Cao, J.Y. Xu, D.Z. Zhang, S.-H. Chang, S.T. Ho, E.W. Seelig, X. Liu and R.P.H. Chang, Phys. Rev. Lett. 84 (2000) 5584.

[86] L.F. Dong, Z.L. Cui and Z.K. Zhang, Nanostructure. Mater. 8 (1997) 815.

[87] (a) M.H. Huang, Y. Wu, H. Feick, N. Tran, E. Weber and P. Yang, Adv. Mater. 13 (2001) 113. (b) M.H. Huang, S. Mao, H. Feick, H. Yan, Y. Wu, H. Kind, E. Weber, R. Russo and

P. Yang, Science 292 (2001) 1897. (c) J.-J. Wu and S.-C. Liu, Adv. Mater. 14 (2002) 215. (d) J.C. Johnson, H. Yan, R.D. Schaller, P.B. Petersen, P. Yang and R.J. Saykally, Nano Lett. 2 (2002) 279. (e) L. Guo, Y.L. Ji, H. Xu, P. Simon and Z. Wu, J. Am. Chem. Soc. 124 (2002) 14864. (f) C. Pacholski, A. Kornowski and H. Weller, Angew. Chem. Int. Ed. 41 (2002) 1188. (g) B. Liu and H.C. Zeng J. Am. Chem. Soc. 125 (2003) 4430. (h) L. Vayssieres, Adv. Mater. 15 (2003) 464. (i) C. Liu, J.A. Zapien, Y. Yao, X. Meng, C.S. Lee, S. Fan, Y. Lifshitz and S.T. Lee, Adv. Mater. 16 (2003) 838. (j) H.T. Ng, J. Li, M.K. Smith, P. Nguyen, A. Cassell, J. Han and M. Meyyappan, Science 300 (2003) 1249. (k) Q.X. Zhao, M. Willander, R.E. Morjan, Q.-H. Hu and E.E.B. Campbell, Appl. Phys. Lett. 83 (2003) 165. (l) J.C Johnson, H. Yan, P. Yang and R.J. Saykally, J. Phys. Chem. B 107 (2003) 8816. (m) S.C. Lyu, Y. Zhang, C.J. Lee, H. Ruh and H. Lee, J. Chem. Mater. 15 (2003) 3294. (n) L.E. Greene, M. Law, J. Goldberger, F. Kim, J.C. Johnson, Y. Zhang, R.J. Saykally and P. Yang, Angew. Chem. Int. Ed. 42 (2003) 3031. (o) J. Zhong, S. Muthukumar, Y. Chen, Y. Lu, H.M. Ng, W. Jiang and E.L. Garfunkel, Appl. Phys. Lett. 83 (2003) 3401. (p) M. Monge, M.L. Kahn, A. Maisonnat and B. Chaudret, Angew. Chem. Int. Ed. 42 (2003) 5321. (q) Y.Q. Chang, D.B. Wang, X.H. Luo, X.Y. Xu, X.H. Chen, L. Li, C.P. Chen, R.M. Wang, J. Xu and D.P. Yu, Appl. Phys. Lett. 83 (2003) 4020. (r) J.-H. Choy, E.-S. Jang, J.-H. Won, J.-H. Chung, D.-J. Jang, Y.-W. Kim, Adv. Mater. 15 (2003) 1911. (s) B. P. Zhang, N.T. Binh, Y. Segawa, Y. Kashiwaba and K. Haga, Appl. Phys. Lett. 84 (2004) 586.

[88] (a) J.-J. Wu, S.-C. Liu, C.-T. Wu, K.-H. Chen and L.-C. Chen, Appl. Phys. Lett. 81 (2002) 1312. (b) J.Q. Hu, Q. Li, X.M. Meng, C.S. Lee and S.T. Lee, Chem. Mater. 15 (2003) 305. (c) X.-H. Zhang, S.-Y. Xie, Z.-Y. Jiang, X. Zhang, Z.-Q. Tian, Z.-X. Xie, R.-B. Huang and L.-S. Zheng J. Phys. Chem. B 107 (2003) 10114. (d) Y.J. Xing, Z.H. Xi, Z.Q. Xue, X.D. Zhang, J.H. Song, R.M. Wang, J. Xu, Y. Song, S.L. Zhang and D.P. Yu, Appl. Phys. Lett. 83 (2003) 1689. (e) R.M. Wang, Y.J. Xing, J. Xu and D.P. Yu, New J. Phys. 5 (2003) 115. (f) H. Kim and W.M. Sigmund, J. Mater. Res. 18 (2002) 2845. (g) X. Kong, X. Sun, X. Li and Y. Li, Mater. Chem. Phys. 82 (2003) 997. (h) X.Y. Kong, Y. Ding and Z.L. Wang, J. Phys. Chem. B 108 (2004) 570.

[89] (a) Z.W. Pan, Z.R. Dai and Z.L. Wang, Science 291 (2001) 1947. (b) Y.B. Li, Y. Bando, T. Sato and K. Kurashima, Appl. Phys. Lett. 81 (2002) 144. (c) Z.R. Dai, Z.W. Pan and Z.L. Wang, Adv. Funct. Mater. 13 (2003) 9. (d) X. Bai, E.G. Wang, P. Gao and Z.L. Wang, Nano Lett. 3 (2003) 1147. (e) X.D. Bai, P.X. Gao, Z.L. Wang and E.G. Wang, Appl. Phys. Lett. 82 (2003) 4806. (f) Y. Yan, P. Liu, J.G. Wen, B. To and M.M. Al-Jassim, J. Phys. Chem. B 107 (2003) 9701. (g) H. Yan, J. Johnson, M. Law, R. He, K. Knutsen, J.R. McKinney, J. Pham, R. Saykally and P. Yang, Adv. Mater. 15 (2003) 1907. (h) X.Y. Kong and Z.L. Wang, Nano Lett. 3 (2003) 1625. (i) W.L. Hughes and Z.L. Wang, Appl. Phys. Lett. 82 (2003) 2886. (j) S.X. Mao, M.H. Zhao and Z.L Wang, Appl. Phys. Lett. 83 (2003) 993. (k) M.S. Arnold, P. Avouris, Z.W. Pan and Z.L. Wang, J. Phys. Chem. B 107 (2003) 659; (l) Y. Li, L. You, R. Duan, P. Shi and G. Qin, Solid State Commun. 129 (2004) 233. (m) C. Ronning, P.X. Gao, Y. Ding, Z.L. Wang and D. Schwen, Appl. Phys. Lett. 84 (2004) 783.

[90] (a) Y. Dai, Y. Zhang, Q.K. Li and C.W. Nan, Chem. Phys. Lett. 358 (2002) 83. (b) H. Yan, R. He, J. Pham and P. Yang, Adv. Mater. 15 (2003) 402. (c) Y. Dai, Y. Zhang and Z.L. Wang, Solid State Commun. 126 (2003) 629. (d) V.A.L. Roy, A.B. Djurisic, W.K. Chan, J. Gao, H.F. Lui and C. Surya, Appl. Phys. Lett. 83 (2003) 141. (e) Y. Zhang, H. Jia, X. Luo, X. Chen, D. Yu and R. Wang, J. Phys. Chem. B 107 (2003) 8289. (f) Q. Wan, K. Yu, T.H. Wang and C.L. Lin, Appl. Phys. Lett. 83 (2003) 2253. (g) V.A.L. Roy, A.B. Djurisic,

H. Liu, X.X. Zhang, Y.H. Leung, M.H. Xie, J. Gao, H.F. Lui, C. Surya, Appl. Phys. Lett. 84 (2004) 756.

[91] (a) W.I. Park, G.-C. Yi, M. Kim and S.J. Pennycook, Adv. Mater. 14 (2002) 1841. (b) Y.W. Zhu, H.Z. Zhang, X.C. Sun, S.Q. Feng, J. Xu, Q. Zhao, B. Xiang, R.M. Wang and D.P. Yu, Appl. Phys. Lett. 83 (2003) 144; (c) Y.-K. Tseng, C.-J. Huang, H.-M. Cheng, I.-N. Lin, K.-S Liu and I.-C. Chen, Adv. Funct. Mater. 13 (2003) 811.

[92] C.X. Xu and X.W. Sun, Appl. Phys. Lett. 83 (2003) 3806.

[93] (a) Z.R. Tian, J.A.Voigt, J. Liu, B. Mckenzie and M.J. Mcdermott, J. Am. Chem. Soc. 124 (2002) 12954. (b) P. Hu, Y. Liu, X. Wang, L. Fu and D. Zhu, Chem. Commun. 11 (2003) 1304.

[94] (a) H. Yan, R. He, J. Johnson, M. Law, R.J. Saykally and P. Yang, J. Am. Chem. Soc. 125 (2003) 4728. (b) Z.L. Wang, X.Y. Kong and J.M. Zuo, Phys. Rev. Lett. 91 (2003) 185502.

[95] (a) S.-H. Yu, J. Yang, Y.-T. Qian and M. Yoshimura, Chem. Phys. Lett. 361 (2002) 362. (b) J.Q. Hu, Y. Bando, J.H. Zhan, Y.B. Li, T. Sekiguchi, Appl. Phys. Lett. 83 (2003) 4414. (c) J.-H. Park, H.-J. Choi, Y.-J. Choi, S.-H. Sohn and J.-G. Park, J. Mater. Chem. 14 (2003) 35.

[96] (a) P. Gao and Z.L. Wang, J. Phys. Chem. B 106 (2002) 12653. (b) J.Y. Lao, J.Y. Huang, D.Z. Wang, Z.F. Ren, Nano Lett. 3 (2003) 235. (c) Z.R. Tian, J.A. Voigt, J. Liu, B. Mckenzie, M.J. Mcdermott, M.A. Rodriguez, H. Konishi and H. Xu, Nat. Mater. 2 (2003) 821. (d) P.-A. Hu, Y.-Q. Liu, L. Fu, X.-B. Wang and D.-B. Zhu, Appl. Phys. A-Mater. 78 (2004) 15.

[97] (a) X. Wang, P. Gao, J. Li, C.J. Summers, Z.L. Wang, Adv. Mater. 14 (2002) 1732. (b) H. Zhou, H. Alves, D.M. Hofmann, W. Kriegseis, B.K. Meyer, G. Kaczmarczyk and A. Hoffmann, Appl. Phys. Lett. 80 (2002) 210. (c) J. Hu, Y. Bando and Z. Liu, Adv. Mater. 15 (2003) 1000. (d) B. Min, J.S. Lee, J.W. Hwang, K.H. Keem, M.I. Kang, K. Cho, M.Y. Sung, S. Kim, M.S. Lee, S.O. Park and J.T. Moon, J. Cryst. Growth 252 (2003) 565. (e) L. Dai, X.L. Chen, X. Zhang, T. Zhou and B. Hu, Appl. Phys. A–Mater. 78 (2004) 557. (f) Y. Ding, X.Y. Kong and Z.L. Wang, J. Appl. Phys. 95 (2004) 306.

[98] (a) H. Kim and W. Sigmund, Appl. Phys. Lett. 81 (2002) 2085. (b) J.Y. Lao, J.G. Wen, Z.F. Ren, Nano Lett. 2 (2002) 1287.

[99] (a) W.I. Park, G.-C. Yi, M. Kim and S.J. Pennycook, Adv. Mater. 15 (2003) 526. (b) S.W. Jung, W.I. Park, G.-C. Yi and M. Kim, Adv. Mater. 15 (2003) 1358.

[100] W.I. Park, J.S. Kim, G.C. Yi and H.J. Lee, Adv. Mater. 17 (2005) 1393.

[101] H.T. Ng, J. Han, T. Yamada, P. Nguyen, Y.P. Chen and M. Meyyappan, Nano. Lett. 4 (2004) 1247.

[102] Y.W. Heo, L.C. Tien, D.P. Norton, S.J. Pearton, B.S. Kang, F. Ren and J.R. Laroche. Appl. Phys. Lett. 85 (2004) 3107.

[103] K. Keem, H. Kim, G.-T. Kim, J.S. Lee, B. Min, K. Cho, M.-Y. Sung and S. Kim, Appl. Phys. Lett. 84 (2004) 4376.

[104] Q. Wan, Q.H. Li, Y.J. Chen, T.H. Wang, Z.L. He, J.P. Li and C.L. Lin, Appl. Phys. Lett. 84 (2004) 3654.

[105] Z. Fan, D. Wang, P.C. Chang, W.Y. Tseng and J.G. Lu, Appl. Phys. Lett. 85 (2004) 5923.

[106] Q.H. Li, Y.X. Liang, Q. Wan and T.H. Wang, Appl. Phys. Lett. 85 (2004) 6389.

[107] Z. Fan and J.G. Lu, Appl. Phys. Lett. 86 (2005) 123510.

[108] H.T. Wang, B.S. Kang, F. Ren, L.C. Tien, P.W. Sadik, D.P. Norton, S.J. Pearton and J. Lin, Appl. Phys. Lett. 86 (2005) 243503.

[109] H.W. Seo, S.Y. Bae and J. Park, J. Phys. Chem. B 108 (2004) 5206.

[110] S.Y. Bae, C.W. Na, J.H. Kang and J. Park, J. Phys. Chem. B 109 (2005) 2526.

[111] S.Y. Bae, H.C. Choi, C.W. Na and J. Park, Appl. Phys. Lett. 86 (2005) 033102.

[112] C.W. Na, S.Y. Bae and J. Park, J. Phys. Chem. B 109 (2005) 12785.

[113] S.Y. Bae, H.W. Seo, H.C. Choi and J. Park, J. Phys. Chem. B 108 (2004) 12318.

[114] T. Omata, N. Ueda, K. Ueda and H. Kawazoe, Appl. Phys. Lett. 64 (1994) 1077.

[115] I.J. Hsieh, K.T. Chu, C.F. Yu and F. S. Feng, J. Appl. Phys. 76 (1994) 3735.

[116] L.E. Shea, R.K. Datta and J.J. Brown Jr. J. Electrochem. Soc. 141 (1994) 2198.

[117] Z. Yan, M. Koike and H. Takei, J. Crystal Growth 165 (1996) 183.

[118] T. Minami, Y. Kuroi, T. Miyata, H. Yamada and S. Takata, J. Lumin. 72 (1997) 997.

[119] I.-K. Jeong, H.L. Park and S.-I. Mho, Solid State Commun. 105 (1998) 179.

[120] Y.E. Lee, D.P. Norton, J.D. Budai and Y. Wei, J. Appl. Phys. 90 (2001) 3863.

[121] S.S. Yi, I.W. Kim, H.L. Park, J.S. Bae, B.K. Moon and J.H. Jeong, J. Crystal Growth 247 (2003) 213.

[122] J.Y. Kim, J.H. Kang, D.C. Lee and D.Y. Jeon, J. Vac. Sci. Technol. A 21 (2003) 532.

[123] S.H. Yang, J. Electrochem. Soc. 150 (2003) H250.

[124] J.S. Kim, H.I. Kang, W.N. Kim, J.I. Kim, J.C. Choi, H.L. Park, G.C. Kim, T.W. Kim, Y.H. Hwang, S.I. Mho, M.-C. Jung and M. Han, Appl. Phys. Lett. 82 (2003) 2029.

[125] Y. Li, X. Duan, H. Liao and Y. Qian, Chem. Mater. 10 (1998) 17.

[126] S.Y. Bae, H.W. Seo, C.W. Na and J. Park, Chem. Commun. 16 (2004) 1834.

[127] S.Y. Bae, J.Y. Lee, H. Jung, J. Park and J.-P. Ahn, J. Am. Chem. Soc. 127 (2005) 10802.

[128] X. Duan and C.M. Lieber, Adv. Mater. 12 (2000) 298.

[129] W.S. Shi, Y.F. Zheng, N. Wang, C.S. Lee and S.T. Lee, J. Vac. Sci. Technol. B 19 (2001) 1115.

[130] C. Tang, S. Fan, M. Lamy de la Chapelle, H. Dang and P. Li, Adv. Mater. 12 (2000) 1346.

[131] Y.-H. Kim, Y.-W. Jun, B.-H. Jun, S.-M. Lee and J. Cheon, J. Am. Chem. Soc. 124 (2002) 13656.

[132] B.K. Kim, J.J. Kim, J.O. Lee, K.J. Kong, H.J. Seo and C.J. Lee, Phys. Rev. B 71 (2005) 153313.

[133] H.W. Seo, S.Y. Bae, J. Park, H. Yang and S. Kim, Chem. Commun. 2564 (2002).

[134] J.I. Pankove, Optical Processes in Semiconductors, Dover Publications, Inc., New York, 1971.

[135] V.K. Bazhenov and V.I. Fistul, Sov. Phys. Semicond. 18 (1984) 843.

[136] B. Gil, J.P. Albert, J. Casmassel, H. Mathieu and C. Benoit à la Guillaume, Phys. Rev. B. 33 (1986) 2701.

[137] H.W. Seo, S.Y. Bae, J. Park, M. Kang and S. Kim, Chem. Phys. Lett. 378 (2003) 420.

[138] H.W. Seo, S.Y. Bae, J. Park, H. Yang, M. Kang, S. Kim, J. C. Park and S.Y. Lee, Appl. Phys. Lett. 82 (2003) 3752.

[139] N. Theodoropoulou, A.F. Hebard, M.E. Overberg, C.R. Abernathy, S.J. Pearton, S.N.G. Chu and R.G. Wilson, Phys. Rev. Lett. 89 (2002) 107203.

[140] M.E. Overberg, B.P. Gila, C.R. Abernathy, S.J. Pearton, N.A. Theodoropoulou, K.T. McCarthy, S.B. Arnason and A.F. Hebard, Appl. Phys. Lett. 79 (2001) 3128.

[141] D.S. Han, S.Y. Bae, H.W. Seo, Y.J. Kang, J. Park, G. Lee and J. P. Ahn, J. Phys. Chem. B 109 (2005) 9311.

[142] S.Y. Bae, H.W. Seo, H.C. Choi, D.S. Han and J. Park, J. Phys. Chem. B 109 (2005) 8496.

[143] V.D. Krstic, J. Am. Ceram. Soc. 75 (1992) 170.

[144] H. Morkoç, S. Strite, G.B. Gao, M.E. Lin, B. Sverdlov and M. Burns, J. Appl. Phys. 76 (1994) 1363.

[145] E.W. Wong, P.E. Sheehan and C.M. Lieber, Science 277 (1997) 1971.

[146] H. Dai, E.W. Wong, Y.Z. Lu, S.S. Fan and C.M. Lieber, Nature 375 (1995) 769.

[147] Z. Pan, H.L. Lai, F.C.K. Au, X. Duan, W. Zhou, W. Shi, N. Wang, C.S. Lee, N.B. Wong, S.T. Lee and S. Xie, Adv. Mater. 12 (2000) 1186.

[148] T. Seeger, P. Kohler-Redlich and M. Rühle, Adv. Mater. 12 (2000) 279.

[149] W. Shi, Y. Zheng, H. Peng, N. Wang, C.S. Lee and S.T. Lee, J. Am. Ceram. Soc. 83 (2000) 3228.

[150] J.Q. Hu, Q.Y. Lu, K.B. Tang, B. Deng, R.R. Jiang, Y.T. Qian, W.C. Yu, G.E. Zhou, X.M. Liu and J.X. Wu, J. Phys. Chem. B 104 (2000) 5251.

[151] Q. Lu, J. Hu, K. Tang, Y. Qian, G. Zhou, X. Liu and J. Zhu, Appl. Phys. Lett. 75 (1999) 507.

[152] H.Y. Kim, J. Park and H. Yang, Chem. Commun. 256 (2003).

[153] H.Y. Kim, S.Y. Bae, N.S. Kim and J. Park, Chem. Commun. 2634 (2003).

[154] G. Ziegler, J. Heinrich and G. Wötting, J. Mater. Sci. 22 (1987) 3041.

[155] D. Muscat, M.D. Pugh, R.A.L. Drew, H. Pickup and D. Steele, J. Am. Ceram. Soc. 75 (1992) 2713.

[156] Y. Zhang, N. Wang, R. He, Q. Zhang, J. Zhu and Y. Yan, J. Mater. Res. 15 (2000) 1048.

[157] H.Y. Kim, J. Park and H. Yang, Chem. Phys. Lett. 372 (2003) 269.

[158] N.G. Chopra, R.J. Luyken, K. Cherrey, V.H. Crespi, M.L. Cohen, S.G. Louie and A. Zettl, Science 269 (1995) 966.

[159] N.G. Chopra and A. Zettl, Solid State Commun. 105 (1995) 297.

[160] A. Rubio, J.L. Corkill and M.L. Cohen, Phys. Rev. B 49 (1994) 5081.

[161] S.Y. Bae, H.W. Seo, J. Park, Y.S. Choi, J.C. Park and S.Y. Lee, Chem. Phys. Lett. 374 (2003) 534.

[162] W.S. Jang, S.Y. Bae, J. Park and J.-P. Ahn, Solid State Commun. 133 (2005) 139.

[163] H.C. Choi, S.Y. Bae, W.-S. Jang, J. Park, H.J. Song and H.-J. Shin, J. Phys. Chem. B 109 (2005) 7007.

CHAPTER 6

Formation, Characterization, and Properties of One-Dimensional Oxide Nanostructures

Jih-Jen Wu, Sai-Chang Liu, and Ko-Wei Chang

Department of Chemical Engineering
National Cheng Kung University
Tainan 701, Taiwan

Metal oxides have long developed technological and industrial interest because of their general characteristics, including high hardness, thermal stability, and chemical resistance. In addition, oxides also possess potential optical, electrical, and magnetic properties for applications to many advanced devices. Recently, research on the subject of one-dimensional (1D) nanostructures such as nanotubes, nanowires, nanorods, and nanobelts has attracted extensive attention due to their contribution to the understanding of fundamental concepts and their potential for future technological applications [1–5]. 1D oxide nanostructures have been successfully synthesized using various bottom–up approaches, including the vapor–liquid–solid process, the template-based method, the thermal oxidation method, the anisotropic growth process, etc. [6]. Several promising properties of the 1D oxide nanostructures have been demonstrated [6]. In this chapter, recent works of 1D ZnO- and Ga_2O_3-based nanostructures are reported. Well-aligned ZnO and $Zn_{1-x}M_xO$ (M = Mg and Co) nanorods with promising optical and ferromagnetic properties were grown on various substrates at low temperatures using metal organic chemical vapor deposition (MOCVD) method. Low-temperature catalytic growth of the well-aligned Ga_2O_3 nanowires has also been achieved using single organometallic precursor. Their 1D heterojunction nanostructures including core–shell nanowires and nanobarcodes have also been formed.

6.1 Low-Temperature Growth of ZnO and $Zn_{1-x}M_xO$ Nanorods

6.1.1 Introduction

ZnO is a versatile material, which has been used considerably for its superior catalytic, electrical, optoelectronic, and photochemical properties [7]. As a gas sensing material, ZnO, one of the earliest discovered and important oxide semiconductor, is sensitive and with satisfactory stability to many sorts of gases [8]. Beside, ZnO has been demonstrated as a promising electron transfer medium for dye-sensitized oxide semiconductor solar cell [9]. ZnO exhibits a hexagonal structure with a direct band gap of 3.37 eV at room temperature, which is very similar to that of GaN. Moreover, it possesses a large exciton binding energy of 60 meV, which is much larger than that of GaN (25 meV) as well as the thermal energy at room temperature ($\sim$26 meV), thus ensuring its efficient exciton emission at room temperature [10,11]. Recently, room temperature UV lasing properties in ZnO epitaxial films [12], microcrystalline thin films [13], nanoclusters [14], and nanowires [15,16] have stimulated intensive interest in the optical properties of ZnO. Alloying ZnO phase with MgO or CdO has been investigated for widening the band gap of the ZnO-based materials [17]. The metastable $Zn_{1-x}Mg_xO$ films with 49 at.% Mg incorporation have been deposited using metalorganic vapor-phase epitaxy (MOVPE) [18,19], although the solid solubility of Mg within hexagonal ZnO phase is limited to only 2 at.% [19]. The optical band gaps of the hexagonal $Zn_{1-x}Mg_xO$ films were tunable from 3.3 to 4.0 eV by adjusting Mg content [19]. In addition to hexagonal $Zn_{1-x}Mg_xO$ films, cubic $Mg_xZn_{1-x}O$ ($1 \leq x \leq 0.5$) films which could possess wider band gap to 6 eV have also been demonstrated [20].

Recently, theoretical prediction of the possibility of room-temperature ferromagnetism in ZnO-based and GaN-based DMSs [21] has stimulated research interest in the field of diluted magnetic semiconductors (DMSs). DMSs are formed by partial replacement of the cations of the nonmagnetic semiconductors by magnetic transition-metal ions [22]. With charge and spin degrees of freedom in a single substance, DMSs have attracted considerable research efforts owing to their great potential for use as spintronic materials [23]. Combining the excellent optical properties with room-temperature ferromagnetism, many practical and versatile functional spintronic devices would be made from the ZnO-based DMSs possibly. Although intensive efforts on growth of transition metal-doped ZnO films have been demonstrated using pulsed laser deposition (PLD) method, their magnetic properties were not yet conclusive [24–26].

It is predicted that the gas sensing ability, photon-to-electron conversion efficiency, and photonic performance would be enhanced by reducing the dimensions of ZnO structures because of the increase in surface area and quantum confinement effect. The synthesis of one-dimensional ZnO nanostructures has also been of growing interest owing to its promising application in nanoscale optoelectronic devices. Growth of 1D single-crystalline ZnO nanostructures have been demonstrated by various high-temperature processes, such as vapor–liquid–solid (VLS) method ($\geq$900°C) [27–28] and thermal evaporation (1400°C) [29]. However, for practical and low-cost application, development of low-temperature synthesis techniques for high-quality and well-ordered 1D ZnO nanostructures with controlled diameters is highly required. In this chapter, we will report our recent works of the low-temperature growth of the ZnO and $Zn_{1-x}M_xO$ (M = Mg and Co) nanorods [30–35]. In comparison with those processes carried out at rather high temperatures, ZnO and $Zn_{1-x}M_xO$ nanorods have been formed through simple thermal reaction of metal acetylacetonate and oxygen at low temperatures. Properties of these ZnO and $Zn_{1-x}M_xO$ nanorods are also demonstrated here.

6.1.2 Experimental section

ZnO nanorods were grown in a quartz tube insert to a two-temperature-zone furnace. The quartz tube was well-sealed to maintain a base pressure of 5.0×10^{-3} torr. Zinc acetylacetonate ($Zn(C_5H_7O_2)_2$), employed as zinc source and placed on a cleaned Pyrex glass container, was loaded into the low-temperature zone of the furnace. The temperature was controlled to be at 130–140°C for vaporizing the solid reactant. The vapor was carried by a N_2/O_2 flow into the higher temperature zone of the furnace where substrates were placed. Fused silica, Si (100), sapphire (001), and sapphire (110) were employed as substrates for ZnO nanorods growth. They were cleaned before being loaded into the quartz tube. It should be noted here that there is no catalytic metal film pre-coated on these substrates. The total pressure of the quartz tube was 200 torr during ZnO nanorods growth.

In the case of growth of the Mg- and Co-doped ZnO nanorods, the nanorods were grown in a quartz tube insert to a three-temperature-zone furnace. Zinc acetylacetonate placed on a cleaned Pyrex glass container was loaded into the low temperature zone where the temperature was controlled in the range of 115–125°C. Magnesium and cobalt metal organic sources, magnesium acetylacetonate ($Mg(C_5H_7O_2)_2$) and cobalt acetylacetonate ($Co(C_5H_7O_2)_2$), which were placed in Pyrex glass containers were loaded into the second zone of the furnace where the temperatures were controlled in the range of 185–215°C and 135–170°C for $Zn_{1-x}Mg_xO$ and $Zn_{1-x}Co_xO$ nanorods formation, respectively. The Pyrex glass containers with diameters of 1.0–1.5 cm were used. The vapor was carried by a N_2/O_2 flow into the high-temperature zone of the furnace in which bare Si(100) substrates were located at 200 torr. Partial pressures of the Zn and the doped metal (Mg or Co) sources in gas phase were tunable through adjusting the vaporizing temperatures of the Zn and the doped metal sources as well as the container diameters of the two sources, resulting in tailoring the doped concentrations of the single phase Mg- or Co-doped ZnO nanorods grown at 500°C in the MOCVD reactor.

6.1.3 Results and discussion

Formation of well-aligned ZnO nanorods. Figures 6.1a–d show the scanning electron microscopy (SEM) images of the ZnO nanorods grown on various substrates including fused silica, Si (100), sapphire (001), and sapphire (110). Among them, sapphire (001) and (110) have been employed as the substrates for epitaxial growth of ZnO films [36,37]. In the case of the ZnO nanorods grown on the fused silica and the Si (100) substrates, as shown in Figs. 6.1a and b, a high density of well-oriented nanorods uniformly grew over the entire substrate. However, the well-aligned characteristic was not observed, as the ZnO nanorods grew on the sapphire (001). In contrast, as revealed in

Fig. 6.1. SEM images of ZnO nanorods grown on (a) fused silica, (b) Si (100), (c) sapphire (001), and (d) sapphire (110) [32].

Fig. 6.1d, the ZnO nanorods grown on the sapphire (110) substrate are not only oriented perpendicularly to the substrate but also formed an orderly array horizontally to the substrate. The diameter and the length of the nanorods were in the ranges of 60–80 nm and 450–500 nm, respectively.

The crystal structure of the nanorods grown on various substrates was examined by X-ray diffraction (XRD) and transmission electron microscopy (TEM). As shown in Fig. 6.2, in addition to the diffraction peaks of substrates, only those of ZnO (002) and (004) appear in the XRD patterns, indicating that the ZnO nanorods grown on fused silica, Si (100), and sapphire (110) are preferentially oriented in the *c*-axis direction. The XRD pattern of the ZnO nanorods grown on sapphire (001) shows that the diffraction peak of ZnO (110) appears in addition to (002) and (004) diffractions. Thus, the epitaxial relationship between ZnO and sapphire (001) substrate does not exist here as SEM observation.

The cross-sectional TEM images of the ZnO nanorods grown on fused silica and Si (100) are shown in Figs. 6.3a and b, respectively. They reveal that most of the nanorods were grown perpendicularly to these two substrates. The selection area electron diffraction (SAED) patterns of the cross-sectional images show discrete ring patterns, indicating that the nanorods are grown on fused silica and Si (100) randomly in the *a*-axis direction

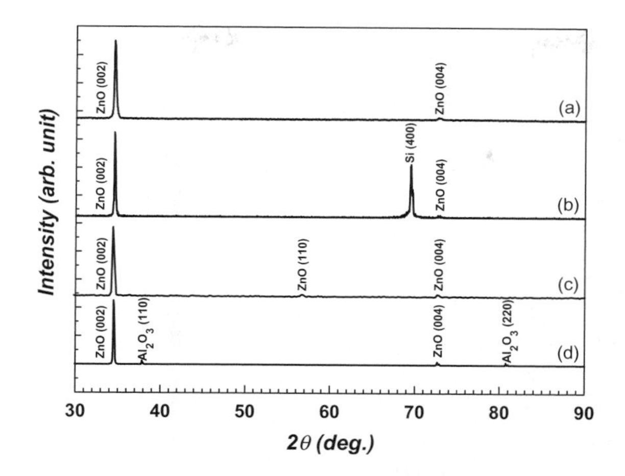

Fig. 6.2. XRD spectra of ZnO nanorods grown on (a) fused silica, (b) Si (100), (c) sapphire (001), and (d) sapphire (110) [32].

although they are preferentially oriented in the c-axis direction. In the case of ZnO nanorods grown on sapphire (110), as shown in Fig. 6.3c, all of the ZnO nanorods grew in a direction perpendicular to the substrate. The SAED pattern taken from the ZnO nanorods portion in Fig. 6.3c is shown in Fig. 6.3d. In contrast to the discrete ring patterns from those grown on fused silica and Si (100) substrates, the single crystal diffraction pattern indicates that the nanorods grown on sapphire (110) are oriented in both the c-axis and the a-axis directions. The SAED pattern taken from the interfacial region of the ZnO nanorods and sapphire (110) is shown in Fig. 6.3e, revealing the existence of an epitaxial relationship.

High-resolution (HR) TEM was employed to study the interfacial regions of the ZnO nanorods and Si (100) as well as sapphire (110) substrates. The HRTEM image of the interfacial region of the ZnO nanorod and Si (100) substrate is shown in Fig. 6.4a. It reveals that the ZnO (001) planes are parallel to Si (100) planes. Moreover, a 3-nm-thick amorphous layer exists between the ZnO nanorod and the Si substrate. Electron energy-loss spectroscopy (EELS) mapping analyses show the amorphous layer is composed of Si and O elements. The HRTEM image of the interfacial region of the ZnO nanorod and sapphire (110) substrate is shown in Fig. 6.4b. The ZnO (001) planes are parallel to the sapphire (110) planes. An abrupt interfacial transition from the sapphire to ZnO is observed, indicating the epitaxial ZnO nanorod grown on sapphire (110) directly without any transition layer in between.

Spectrum I in Fig. 6.5a shows the room-temperature photoluminescence (PL) emission of the well-oriented ZnO nanorods grown on a fused silica substrate. Three emitting bands, including a strong ultraviolet emission at around 386 nm, a very weak blue band (440–480 nm) as well as an almost negligible green band (510–580 nm), are observed in the PL spectrum. The UV emission is contributed by the near band edge emission of the wide band gap ZnO. It has been suggested that the green band emission corresponds to the

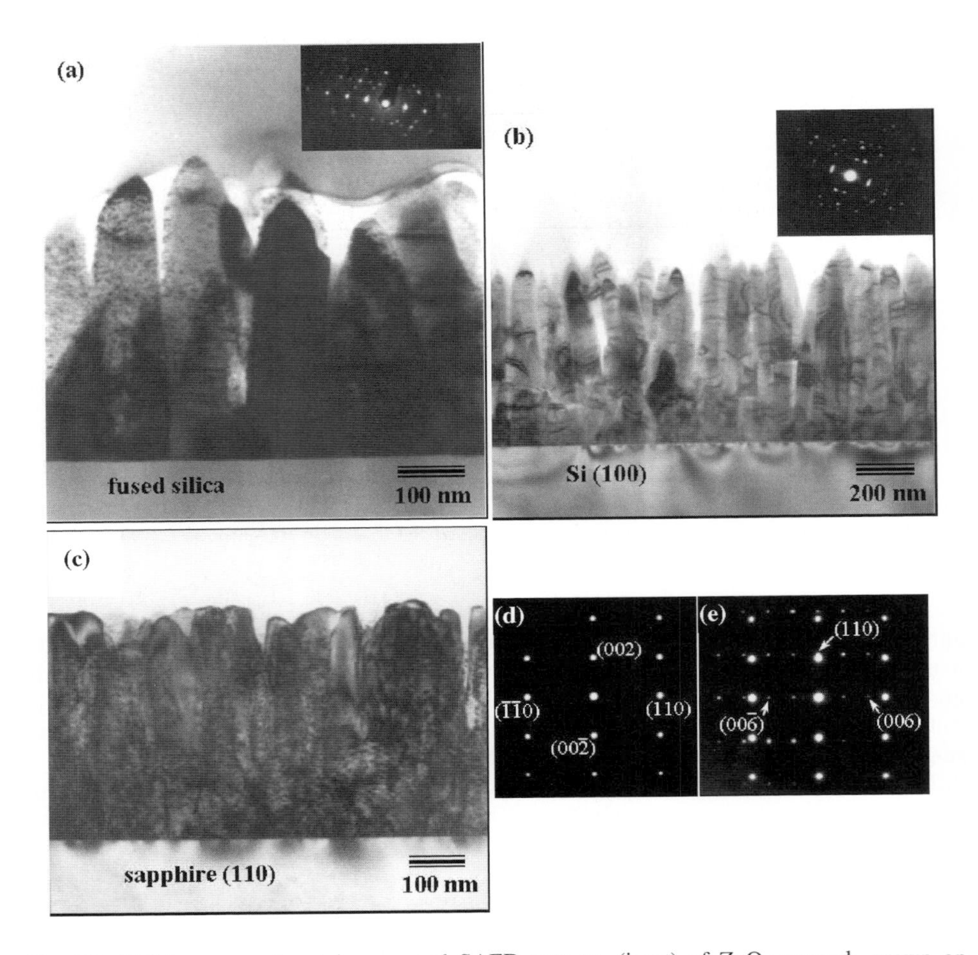

Fig. 6.3. TEM cross-sectional images and SAED patterns (inset) of ZnO nanorods grown on (a) fused silica, (b) Si (100) substrate, and (c) sapphire (110) substrate. SAED patterns of (d) ZnO nanorods only in (c) with zone axis [1$\bar{1}$0] and (e) ZnO nanorods and sapphire (110) substrate in (c) with zone axis [1$\bar{1}$0] [32].

single ionized oxygen vacancy in ZnO [38]. It can be concluded that the almost negligible green band in this figure stands for a very low concentration of oxygen vacancy in the highly oriented ZnO nanorods. The absorption spectrum of the ZnO nanorods obtained at room temperature is shown in spectrum II of Fig. 6.5a. Figure 6.5b presents the dependence of the absorption coefficient as a function of $h\nu$ (E) for the ZnO nanorods. The intercept defining the direct energy gap for the ZnO nanorods is 3.22 eV, which is consistent with the result of PL measurements.

Band gap engineering of well-aligned $Zn_{1-x}Mg_xO$ nanorods. $Zn_{1-x}Mg_xO$ nanorods were grown in a three-temperature-zone furnace. Typical SEM image of the nanorods

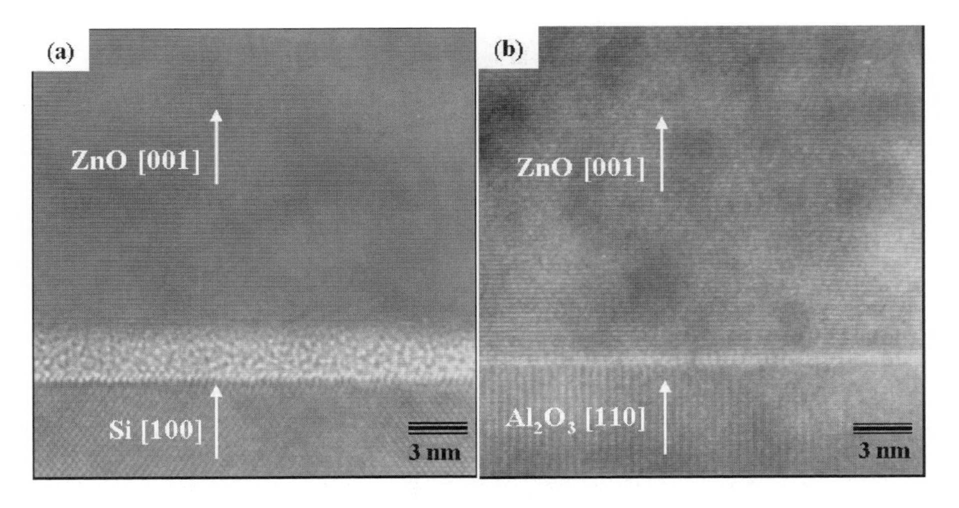

Fig. 6.4. High-resolution (HR) TEM images of the interfacial regions of the ZnO nanorods and (a) Si (100) and (b) sapphire (110) [32].

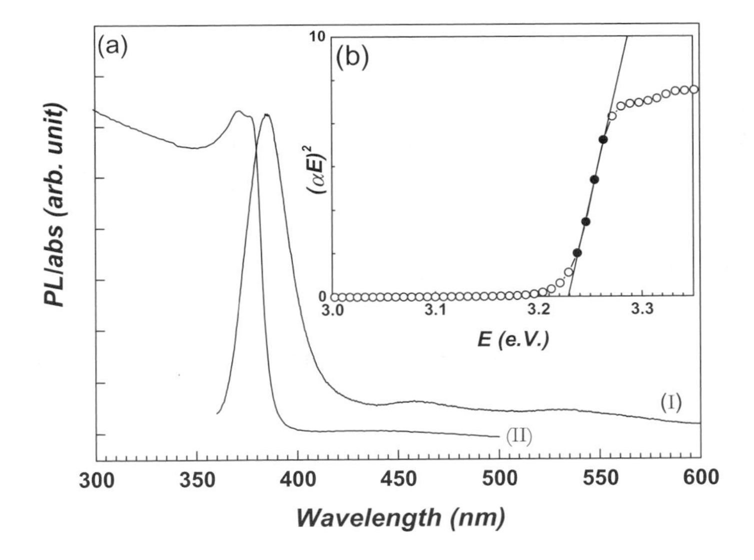

Fig. 6.5. (a) Room-temperature PL (spectrum I) and absorption (spectrum II) spectra of ZnO nanorods on fused silica substrate. (b) Dependence of the absorption coefficient as a function of $h\nu$ for the ZnO nanorods [32].

shown in Fig. 6.6 reveals the formation of well-aligned nanorods of high density. According to Electron Probe X-ray Microanalyzer (EPMA) measurement, Mg content (x) of the $Zn_{1-x}Mg_xO$ nanorods is tunable from 0 to 0.165 through controlling the partial pressures of the Zn and Mg sources in gas phase. In comparison with the $Zn_{1-x}Mg_xO$ films with higher maximum Mg content [18,19], the lower maximum Mg content of the

Fig. 6.6. Typical SEM image of the well-aligned $Zn_{1-x}Mg_xO$ nanorods [34].

$Zn_{1-x}Mg_xO$ nanorods in this study could be due to the restriction for maintaining the nanorod morphology.

Figure 6.7a shows a typical XRD pattern of the well-aligned $Zn_{1-x}Mg_xO$ nanorods. In addition to Si (400) diffraction peak, peaks very close to (00*c*) diffraction peaks of the wurtzite ZnO structure appearing in the pattern shows that the $Zn_{1-x}Mg_xO$ nanorods possess the same structure as that of the ZnO and they preferentially orient in the *c*-axis direction. Moreover, there is no diffraction peak of MgO or Mg crystal presenting in the XRD pattern. The Mg content dependence of the *c*-axis lattice constant (d(002) value) is shown in Fig. 6.7b. It reveals that the d(002) value decreases as the Mg concentration increases. The absence of the diffraction peaks of MgO as well as Mg phases in the XRD pattern and the systematic dependence of the Mg content, with the lattice constant, imply Mg is incorporated within the ZnO nanorods by substituting Zn.

Figure 6.8 shows the room-temperature PL spectra of the hexagonal $Zn_{1-x}Mg_xO$ nanorods with diameters in the range of 50–70 nm. Emission peaks at 377, 365, 355, and 346 nm are observed in the normalized PL spectra of the $Zn_{1-x}Mg_xO$ nanorods with $x = 0$, 0.05, 0.11, and 0.165, respectively. A blue shift of the near-band-edge emission with increasing Mg content is observed. For numerous x values between 0 and 0.165, the Mg content dependence on the near-band-edge emission is illustrated in Fig. 6.8(inset). It reveals that the near-band-edge emission energies of the $Zn_{1-x}Mg_xO$ nanorods measured at room temperature increase monotonically with the Mg content.

Room-temperature ferromagnetism in well-aligned $Zn_{1-x}Co_xO$ nanorods. $Zn_{1-x}Co_xO$ nanorods were also grown in a three-temperature-zone furnace. Figure 6.9a shows a typical SEM image of the $Zn_{1-x}Co_xO$ nanorods grown on Si substrates. Highly dense and well-aligned nanorods with diameters in the range of 60–80 nm are uniformly formed over the entire substrate. The atomic composition ratio of Co/Zn within the nanorods is adjustable from 0 to 0.11 by varying the partial pressures of the Zn and Co sources in gas phase. The typical XRD pattern of the well-aligned $Zn_{1-x}Co_xO$ nanorods is illustrated in Fig. 6.9b. In addition to Si (400) diffraction peak, the peak very close to (002) of the

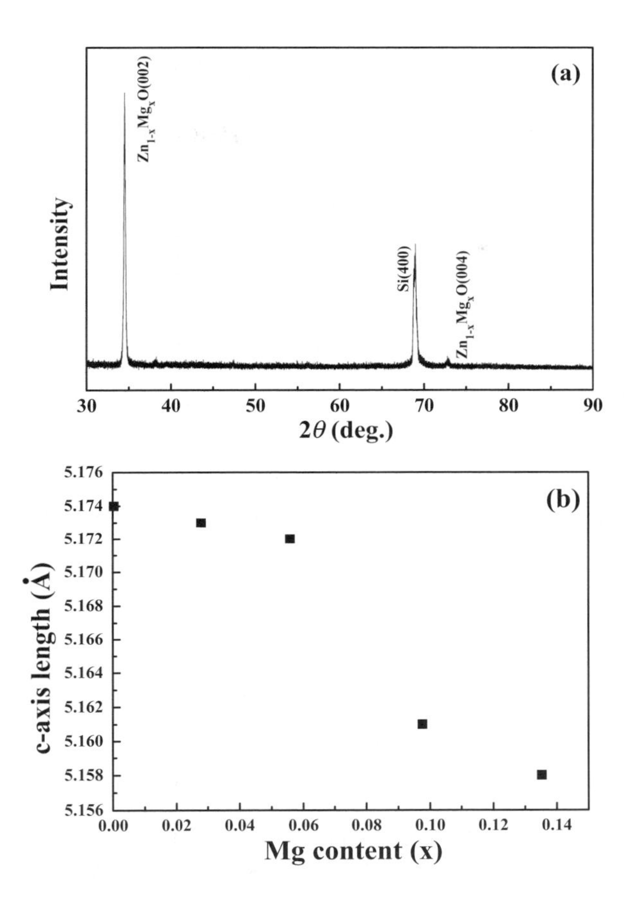

Fig. 6.7. (a) Typical XRD pattern of the $Zn_{1-x}Mg_xO$ nanorods and (b) the Mg content dependence on the c-axis lattice constant [34].

wurtzite ZnO structure shows that the $Zn_{1-x}Co_xO$ nanorods possess the same structure as the ZnO nanorods and are preferentially oriented in the c-axis direction, similar to Mg-doped ZnO nanorods described in the previous section. The absence of the diffraction peaks of CoO and Co structures in the XRD pattern implies to the incorporation of Co within the ZnO nanorods by means of substitution for Zn. The Co content dependence of the c-axis lattice constants (d(002) values) is shown in the inset of Fig. 6.9b. The d(002) values increase with the Co concentrations within the $Zn_{1-x}Co_xO$ nanorods, implying that Co systematically substituted for Zn in the nanorods.

Further structural characterization was performed using HRTEM to examine if tiny impurity phases exist in the $Zn_{1-x}Co_xO$ nanorods. As shown in Fig. 6.4a, there is a SiO_x interfacial layer between the ZnO nanorods and Si substrates using this catalyst-free CVD method, the $Zn_{1-x}Co_xO$ nanorods grown on the fused silica substrate for UV–visible absorption measurement were thus also employed to investigate the structural properties of the nanorods. A cross-sectional TEM image of the $Zn_{1-x}Co_xO$

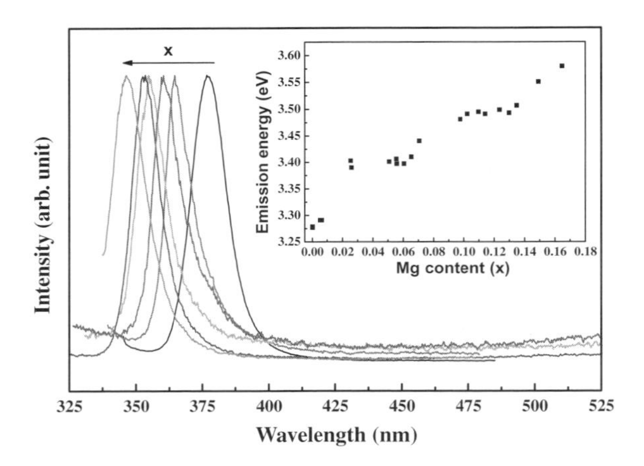

Fig. 6.8. Room-temperature PL spectra of the well-aligned $Zn_{1-x}Mg_xO$ nanorods with various Mg contents and the dependence of the PL peak position as a function of the Mg content (inset) [34].

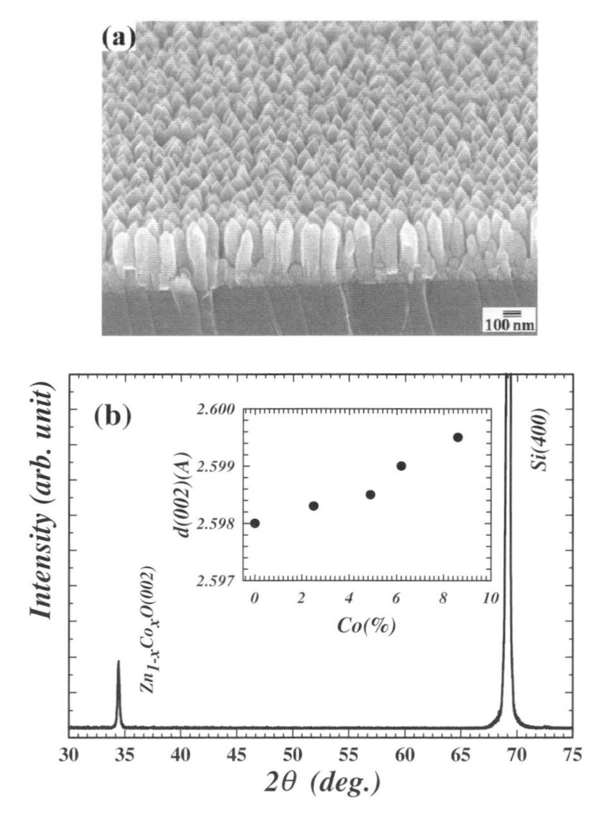

Fig. 6.9. (a) SEM image of the $Zn_{1-x}Co_xO$ nanorods. (b) XRD pattern of the $Zn_{1-x}Co_xO$ nanorods on a Si (100) substrate [35].

($x = 0.037$) nanorods on the fused silica substrate is shown in Fig. 6.10a. It reveals that most nanorods were grown perpendicularly to the substrate. The SAED pattern of the cross-sectional image shows a discrete ring pattern as illustrated in the inset. All the *d*-spacings estimated from the SAED are close to those of the planes of the ZnO structure. It is consistent with the XRD analyses, and is another evidence of non-existence of CoO and Co phases. In addition, the discrete ring pattern indicates that the $Zn_{1-x}Co_xO$ nanorods are grown on the fused silica substrate randomly in the *a*-axis direction though they are preferentially oriented in the *c*-axis direction. Typical bright-field and dark-field images of $Zn_{1-x}Co_xO$ nanorods are illustrated in Figs. 6.10b and c, respectively. The dark-field image indicates that the nanorod possesses the single crystalline structure. Figures 6.10d–f show typical high-resolution TEM images of the bottom, middle, and top regions of an individual nanorod. There is no segregated cluster of impurity phase appearing throughout the nanorod.

The magnetic properties of the nanorods were measured using superconducting quantum interference device magnetometer (SQUID). Figure 6.11a shows the field dependences of magnetization (M–H curves) of the 3.7, 6.3, and 8.7% Co-doped ZnO nanorods grown on the Si substrates measured at 300 K, in which the diamagnetic characteristic of the Si substrates has been subtracted. The magnetic fields were applied perpendicularly to the long axes of the $Zn_{1-x}Co_xO$ nanorods during SQUID measurement. They reveal that hysteresis curves with the coercive field (H_c) of 75 Oe are observed in the $Zn_{1-x}Co_xO$ nanorods, showing their ferromagnetic characteristic at 300 K. Moreover, the saturation magnetization (M_s) per unit volume of the $Zn_{1-x}Co_xO$ nanorods increases with the Co concentration. The M_s of the $Zn_{1-x}Co_xO$ nanorods are estimated to be 0.22 ± 0.01 μ_B/Co site from the M–H curves at 300 K. To investigate the orientation dependence of the ferromagnetic characteristic of the $Zn_{1-x}Co_xO$ nanorods, the M–H curve of the 8.7% Co-doped ZnO nanorods was further measured at 300 K using the magnetic fields applied horizontally to the long axes of nanorods. As shown in Fig. 6.11b, the saturation magnetization is larger in the case of the magnetic fields applied perpendicularly to the long axes of nanorods, indicating the anisotropic ferromagnetism characteristic of the $Zn_{1-x}Co_xO$ nanorods. The temperature dependence of magnetization for the $Zn_{1-x}Co_xO$ nanorods is shown in Fig. 6.11c. A magnetic field of 0.1 T was applied perpendicularly to the long axes of the 3.7, 6.3, and 8.7% Co-doped ZnO nanorods. The ferromagnetic properties are still maintained to a temperature of 350 K and T_c is thus estimated to be higher than 350 K.

The UV–visible absorption spectra of the $Zn_{1-x}Co_xO$ ($x = 0.063$ and 0.091) and ZnO nanorods obtained at room temperature are shown in Fig. 6.12a, indicating the transparency of the $Zn_{1-x}Co_xO$ nanorods in the visible region. Figure 6.12b presents the dependence of the absorption coefficients as a function of $h\nu$ (E) for the Co-doped ZnO nanorods as well as the pure ZnO nanorods. The intercepts defining the direct energy gaps for the 6.3 and 9.1% Co-doped ZnO nanorods are 3.21 and 3.20 eV, respectively, which do not reveal pronounced difference from that of pure ZnO nanorods.

Mechanism of the ZnO and $Zn_{1-x}M_xO$ nanorods growth. There are two well-accepted mechanisms for 1D materials' growth: the VLS mechanism and the template-based

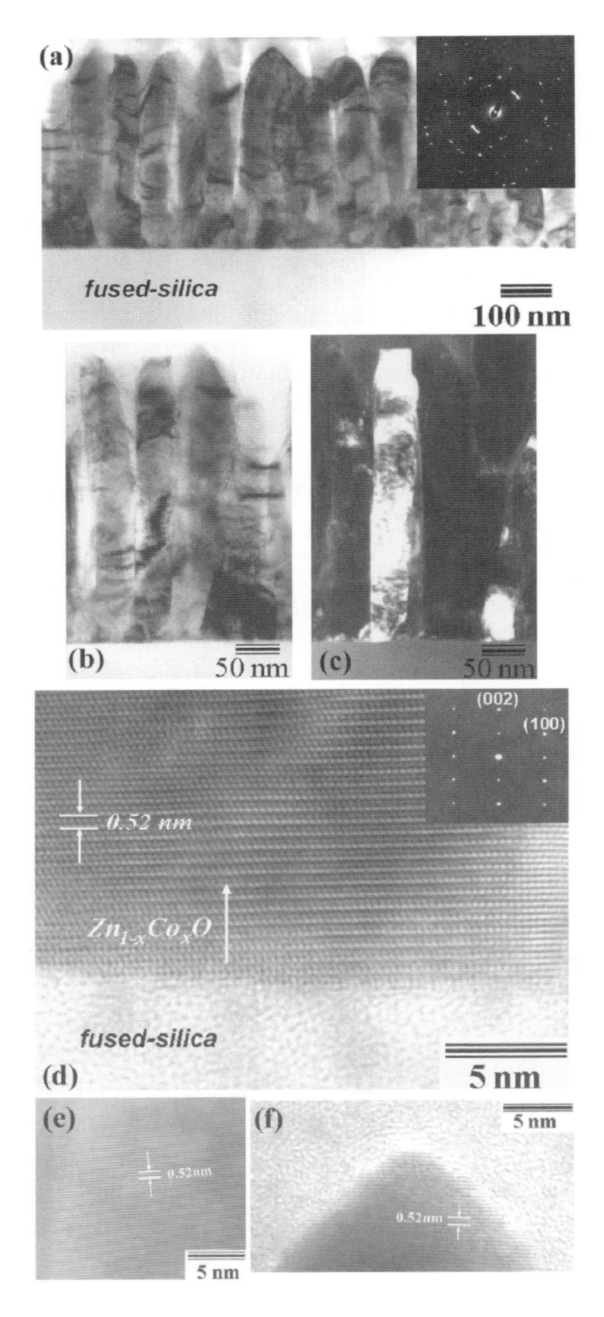

Fig. 6.10. TEM cross-sectional images of $Zn_{1-x}Co_xO$ ($x = 0.037$) nanorods grown on fused silica substrates. (a) Low-magnification image and SAED pattern (inset). (b) and (c) typical bright-field and dark-field images of the $Zn_{1-x}Co_xO$ nanorods, respectively. (d)–(f) High-resolution TEM image of the bottom, middle, and top regions of a single crystalline $Zn_{1-x}Co_xO$ nanorod and the corresponding electron diffraction pattern (inset) [35].

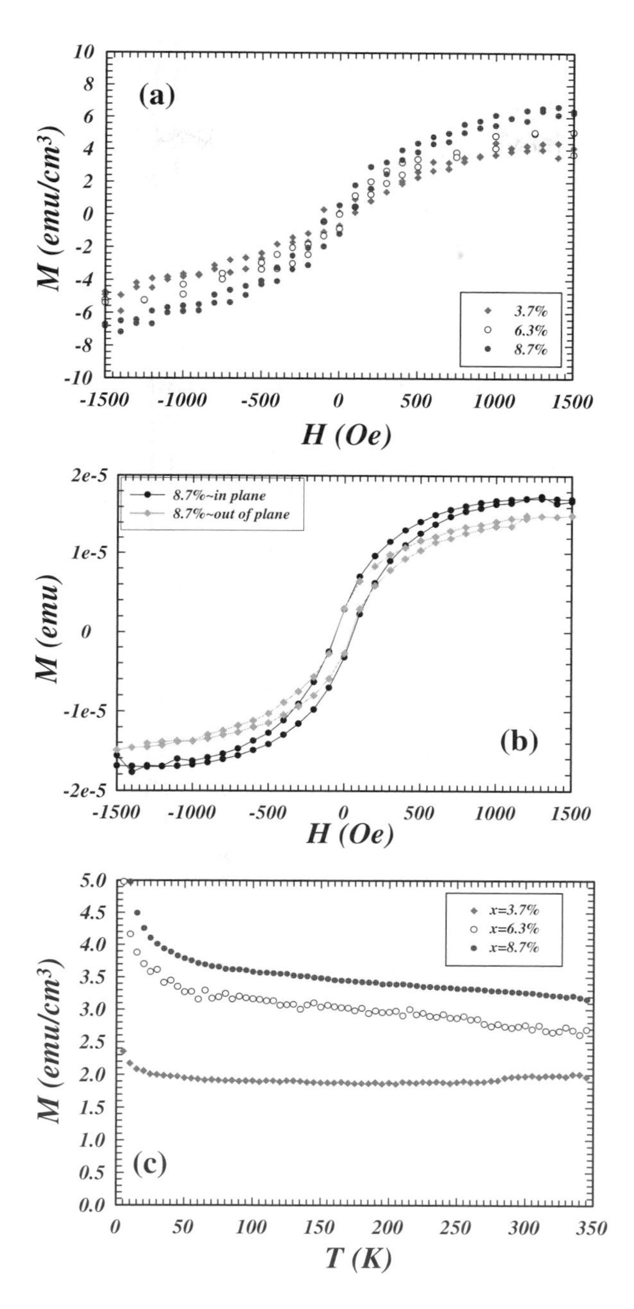

Fig. 6.11. (a) M–H curves of the 3.7, 6.3, and 8.7% Co-doped ZnO nanorods grown on the Si substrate measured at 300 K. (b) M–H curves of the 8.7% Co-doped ZnO nanorods using the magnetic fields applied parallel and perpendicular to the long axes of nanorods. (c) Temperature dependence of magnetization for the $Zn_{1-x}Co_xO$ nanorods [35].

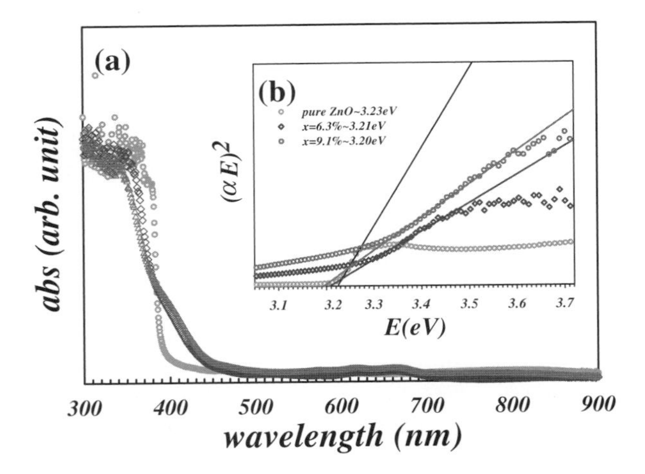

Fig. 6.12. (a) Room-temperature UV–visible absorption spectra and (b) dependence of the absorption coefficients as a function of $h\nu$(E) of the $Zn_{1-x}Co_xO$ ($x = 0.063$ and 0.091) and ZnO nanorods obtained at room temperature [35].

mechanism [39–41]. In the VLS mechanism, metallic particles serving as catalysts for the formation of 1D structures commonly remain at the end of the 1D structures. In this study, no additional metal particle appeared on the top or the bottom of the nanorods, as shown in SEM and TEM images, implying a non-VLS approach to grow the well-aligned ZnO nanorods was achieved. A possible mechanism for the ZnO nanorods growth is proposed as follows: it has been demonstrated that the growth habit of crystals is related to the relative growth rate of various crystal faces bounding the crystal, which is mainly determined by the internal structure of a given crystal as well as is affected by growth conditions [42,43]. The direction of the crystal face with the corner of the coordination polyhedron occurring at the interface possesses the fastest growth rate and the directions of the crystal face with the edge and with the face of the coordination polyhedron occurring at the interface have the second fastest and the slowest growth rates, respectively [42,43]. Figures 6.1, 6.6, and 6.9 suggest that the relationship of the growth rates (R) of the ZnO and $Zn_{1-x}M_xO$ crystal faces is $R_{\langle 001\rangle} > R_{\langle 101\rangle} > R_{\langle 100\rangle}$, resulting in the formation of 1D ZnO and $Zn_{1-x}M_xO$ nanorods with prismatic morphology on the tops.

6.2 Conclusion

Well-aligned ZnO nanorods have been grown on various substrates, such as fused silica, Si (100), and sapphire (110), using a simple catalyst-free MOCVD method at low temperatures. TEM analyses indicate that epitaxial ZnO nanorods have been grown on sapphire (110). In the case of Si (100) substrate, an amorphous SiO_x interfacial layer exists between ZnO nanorods and Si (100). The well-aligned ZnO nanorods on fused silica substrates exhibit a strong UV emission and absorption at around 386 nm under room temperature. Photoluminescence spectrum indicates that there is a very low concentration of oxygen vacancies in the highly oriented ZnO nanorods. In-situ growth of well-aligned $Zn_{1-x}M_xO$ (M = Mg and Co) nanorods on the substrates have also been achieved using MOCVD. Structural analyses indicate that the nanorods grown on substrates are oriented in the *c*-axis direction and the nanorod possesses the single-crystalline hexagonal structure. No phase separation is observed when the Mg and Co content (x) is increased to 0.165 and 0.087, respectively. The *c*-axis constant of the $Zn_{1-x}M_xO$ nanorod decreases with increasing Mg and decreasing Co contents. The PL emission energies of the $Zn_{1-x}Mg_xO$ nanorods measured at room temperature increase monotonically with the Mg content. However, the fundamental absorptions of the $Zn_{1-x}Co_xO$ nanorods estimated from the absorption spectra do not reveal pronounced difference from that of pure ZnO nanorods. Furthermore, the $Zn_{1-x}Co_xO$ nanorods have been determined to possess a Curie temperature higher than 350 K, indicating room-temperature diluted magnetic semiconductor $Zn_{1-x}Co_xO$ nanorods have been synthesized by in-situ doping of Co in ZnO nanorods.

6.3 Ga_2O_3 Nanowires and their 1D Heterojunction Nanostructures

6.3.1 Introduction

Monoclinic gallium oxide (Ga_2O_3), which possesses a wide band gap of 4.9 eV, has potential applications in optoelectronic devices including flat-panel displays, solar energy conversion devices, optical limiter for ultraviolet and high-temperature-stable sensors [44,45]. Ga_2O_3 has also been employed as an active catalyst for dehydrogenation of ethane to ethene in the presence of carbon dioxide [46,47]. Furthermore, Ga_2O_3–TiO_2 catalyst was found to be the most effective Ga_2O_3-loaded catalysts for such reaction [47]. Besides, among a series of titania-based mixed oxides, the Ga_2O_3–TiO_2 mixed oxide exhibits reasonable thermostability up to 1073 K [48]. $ZnGa_2O_4$ has attracted considerable attention for use as a blue phosphor in flat-panel displays in the last decade due to its good cathodeluminescence (CL) characteristic at low voltages and stability in high vacuum in comparison with sulfide-based phosphorus [49–51]. $ZnGa_2O_4$ exhibits a spinel structure, in which Zn^{2+} and Ga^{3+} ions occupy the tetrahedrally and octahedrally coordinated sites, respectively [52]. The origin of a self-activated blue emission band around 430 nm has been reported to correspond to octahedrally coordinated Ga–O group [53,54]. In addition, with a band gap of 4.4 eV, $ZnGa_2O_4$ is also a promising transparent-conducting-oxide material [55–57].

With many promising properties, β-Ga_2O_3 nanowires have been synthesized by various high-temperature methods (800–1200°C), including arc-discharge method [58,59], vapor–liquid–solid (VLS) method [60,61], carbothermal reduction [62,63], thermal oxidation [64,65], etc. In addition to nanowires, Ga_2O_3 nanobelt (nanoribbons) and nanosheets [64,66] have also been demonstrated via a thermal oxidation method. Such high temperature is unavoidable for using gallium sources with high melting points such as Ga_2O_3, GaN, and GaAs in those processes. However, the formation of well-ordered single crystalline nanowires with controlled crystal orientation on the substrates at lower temperatures is needed for application to nanoscale devices. In this section, we will demonstrate our recent works concerning the low-temperature growth of the Ga_2O_3 nanowires [67,68] and the formation of the 1D heterojunction nanostructures using them as templates [69,70].

Epitaxial growth of the Ga_2O_3 nanowires on sapphire (0001) substrate at low temperatures has been achieved via the VLS method using a single precursor of gallium acetylacetonate [67]. The gallium acetylacetonate is the organometallic source with a low decomposition temperature (~196°C) and thus ensures to provide sufficient Ga and O vapors for the Ga_2O_3 nanowires growth in a two-temperature-zone furnace. The concentrations of Ga and O vapors can be controlled by the temperature of the low-temperature zone where the temperature is independent of the Ga_2O_3 nanowires growth in the higher temperature region [67,68]. Ga_2O_3–ZnO and Ga_2O_3–TiO_2 core–shell nanowires were prepared further using a two-step method. The epitaxial Ga_2O_3nanowires were employed as one-dimensional templates for ZnO- and TiO_2-shell formation at a low pressure. Post-annealing of the Ga_2O_3–ZnO and Ga_2O_3–TiO_2 core–shell nanowires were conducted in oxygen ambient to investigate the structural evolutions of the 1D Ga_2O_3-based heterojunction nanostructures. $ZnGa_2O_4$ nanowires and Ga_2O_3–TiO_2 nanobarcodes have been successfully fabricated via annealing of the Ga_2O_3–ZnO and Ga_2O_3–TiO_2 core–shell nanowires at 1000°C, respectively [69,70].

6.3.2 Experimental section

The one-dimensional Ga_2O_3 nanostructures were synthesized on Au-precoated substrates in two-temperature-zone furnace. All substrates after being cleaned were coated with a 1-nm-thick Au layer using an E-beam evaporator. The Ga and O sources, gallium acetylacetonate ($(CH_3COCHCOCH_3)_3Ga$), placed on a cleaned Pyrex glass container was loaded into the low-temperature zone of the furnace, which was controlled to be at 185°C for vaporizing the solid reactant. A N_2 line (so called as N_2(Ga)) separated from the main N_2 stream was connected to the Pyrex glass to carry sufficient Ga and O sources to the high-temperature zone. The pressure for the nanowire growth was set at 100 torr. The Au-precoated substrate was loaded at the center of the high temperature zone where the temperature was set to be at 550°C.

Deposition of the conformal ZnO- or TiO_2-shell onto the surfaces of the as-synthesized Ga_2O_3 nanowires was conducted using metalorganic chemical vapor deposition. The two precursors were titanium acetylacetonate ($Ti(C_{10}H_{14}O_5)$) and zinc acetylacetonate ($(C_5H_7O_2)_2Zn$). The precursor placed in a Pyrex glass container was loaded into the low-temperature zone of the furnace where the temperature was controlled to be at 215°C and 100°C to vaporize the solid reactant for ZnO- and TiO_2-shell growth, respectively. The vapor was carried by a N_2/O_2 flow into the high-temperature zone of the furnace in which substrates were located. In this study, the conformal ZnO and TiO_2 shells were formed on the Ga_2O_3 nanowires under a temperature of 400°C and a pressure of 5 torr. Post-annealing of the Ga_2O_3–ZnO and Ga_2O_3–TiO_2 core–shell nanowires were conducted in oxygen ambient at various temperatures.

6.3.3 Results and discussion

Low-temperature growth of well-aligned β-Ga_2O_3 nanowires using a single-source organometallic precursor. Figures 6.13a–c shows SEM images of the Ga_2O_3 nanowires grown on the Au-precoated sapphire (0001) substrates. Cross-sectional and top-view

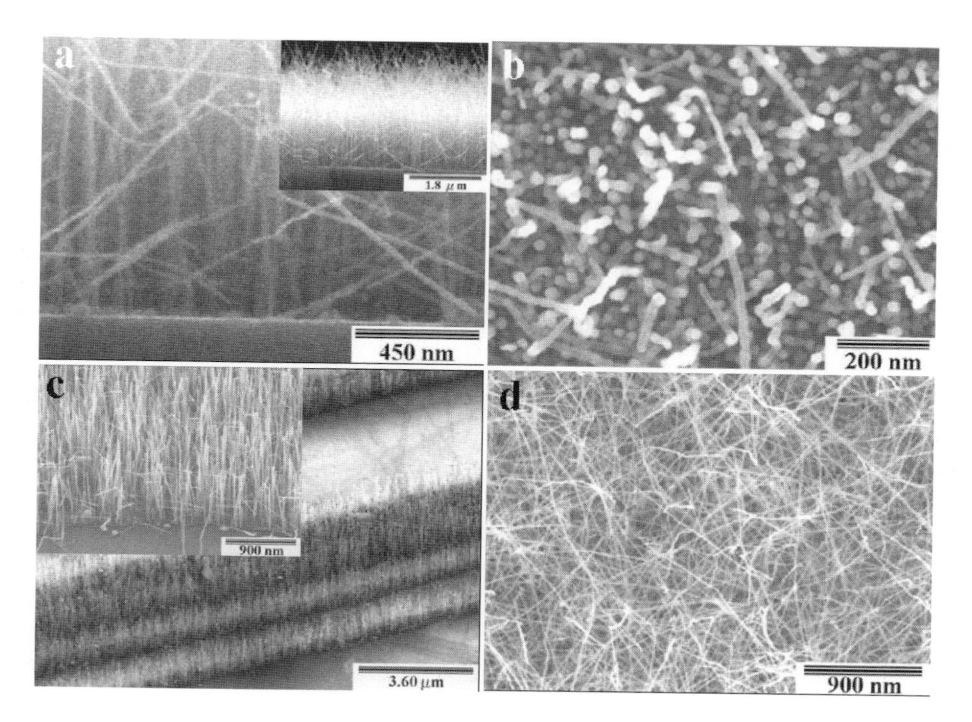

Fig. 6.13. (a) Cross-sectional view, (b) top-view, (c) 45° tilted-view SEM images of well-aligned Ga_2O_3 nanowires on the sapphire (0001) substrates. (d) Ga_2O_3 nanowires on the Si (100) substrate [67].

SEM images of the Ga_2O_3 nanowires, as shown in Fig. 6.13a and b, reveal that the well-aligned nanowires with the average diameter and length of 10 nm and 2.5 μm, respectively, were formed on the sapphire substrates after 30 min of growth. Moreover, selective growth of the well-aligned Ga_2O_3 nanowire arrays was achieved on the Au-patterned sapphire (0001) substrates, as shown in Fig. 6.13c. Nanowires found only on the Au-coated region on the sapphire substrates, show the formation of the nanowires via the VLS mechanism. The well-aligned characteristic of the Ga_2O_3 nanowires is obtained on the sapphire (0001) substrates under the growth conditions of 550–650°C and 100–300 torr. However, as shown in Fig. 6.13d, the aligned Ga_2O_3 nanowires are not observed on the silicon (100) substrate under the same growth conditions. Ga_2O_3 nanowires were randomly grown on the silicon substrate as those have been reported [58–65].

Typical XRD pattern of the well-aligned Ga_2O_3 nanowires is shown in Fig. 6.14. Apart from the (0001) diffraction peak of the sapphire substrate, the rest three peaks appearing in the pattern are indexed as the high order diffraction peaks of $(\bar{2}01)$ of the monoclinic structure of β-Ga_2O_3 ($a = 12.23$ Å, $b = 3.04$ Å, $c = 5.80$ Å, $\alpha = 90°$, $\beta = 103.7°$, and $\gamma = 90°$) according to JCPDS, powder diffraction file No. 41-1103, indicating the β-Ga_2O_3 nanowires are preferentially oriented in the $[\bar{2}01]$ direction on the sapphire (0001) substrate. Fig. 6.15a illustrates the cross-sectional TEM image of the

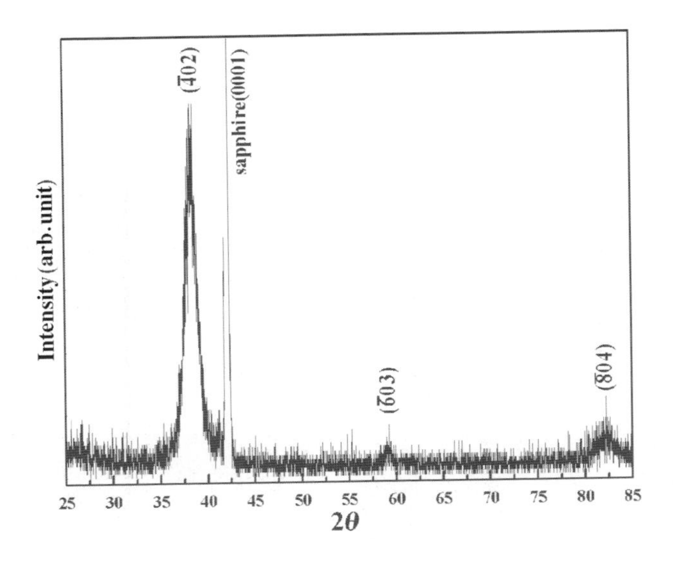

Fig. 6.14. XRD pattern of β-Ga_2O_3nanowires grown on sapphire (0001) substrate [67].

β-Ga_2O_3 nanowires grown for 20 min on sapphire (0001). It reveals that all nanowires capped with nanoparticles grew in the direction perpendicular to the substrate surface. The presence of the nanoparticles on top of the nanowires further proves the VLS mechanism of the Ga_2O_3 nanowire growth in this study. Moreover, the corresponding dark-field TEM image shown in Fig. 6.15b indicates the single structure of the β-Ga_2O_3 nanowires. A high-resolution TEM image of the Ga_2O_3 nanowires on the sapphire (0001) is shown in Fig. 6.15c. The lattice spacing of 0.47 nm corresponds to the d-spacing of ($\bar{2}$01) planes of β-Ga_2O_3 structure, confirming the XRD analysis that the β-Ga_2O_3 are preferentially oriented in the [$\bar{2}$01] direction on the sapphire (0001). In addition, the [2$\bar{1}$1] direction of the Ga_2O_3 nanowires is also parallel to the sapphire [11$\bar{2}$0] direction. The diffraction patterns of the center β-Ga_2O_3 nanowire, the interfacial region of the Ga_2O_3 nanowire on the sapphire, and the sapphire substrate in Fig. 6.15c taken from Fast Fourier Transfer (FFT) are shown in Figs. 6.15d–f, respectively. Figure 6.15d reveals that the β-Ga_2O_3 nanowire possesses a single-crystal structure. Furthermore, Fig. 6.15e indicates that the β-Ga_2O_3 nanowire exhibits an epitaxial relationship with the sapphire (0001) substrate.

In addition to the selection of a proper catalyst for VLS mechanism to proceed at low temperatures, we suggest that having sufficient precursors is another crucial point for low-temperature growth of nanowires via VLS method. In the case of VLS-Ga_2O_3 nanowires synthesized using Ga metal, Ga source was heated together with catalyst-pretreated substrates at temperatures higher than 780°C and nanowires were formed on the substrates at a position near the source. The product with desired morphology as well as diameter was obtained only in a limited region [71]. It implies that the reactant depletion problem could be a concern in the growth of Ga_2O_3 nanowires using Ga metal because of the low vapor pressure of Ga metal ($\sim 10^{-4}$ torr at 900°C [72]). In contrast to Ga metal, the low decomposition temperature (~196°C) of the organometallic source of gallium acetylacetonate can provide more sufficient Ga vapor than can Ga metal.

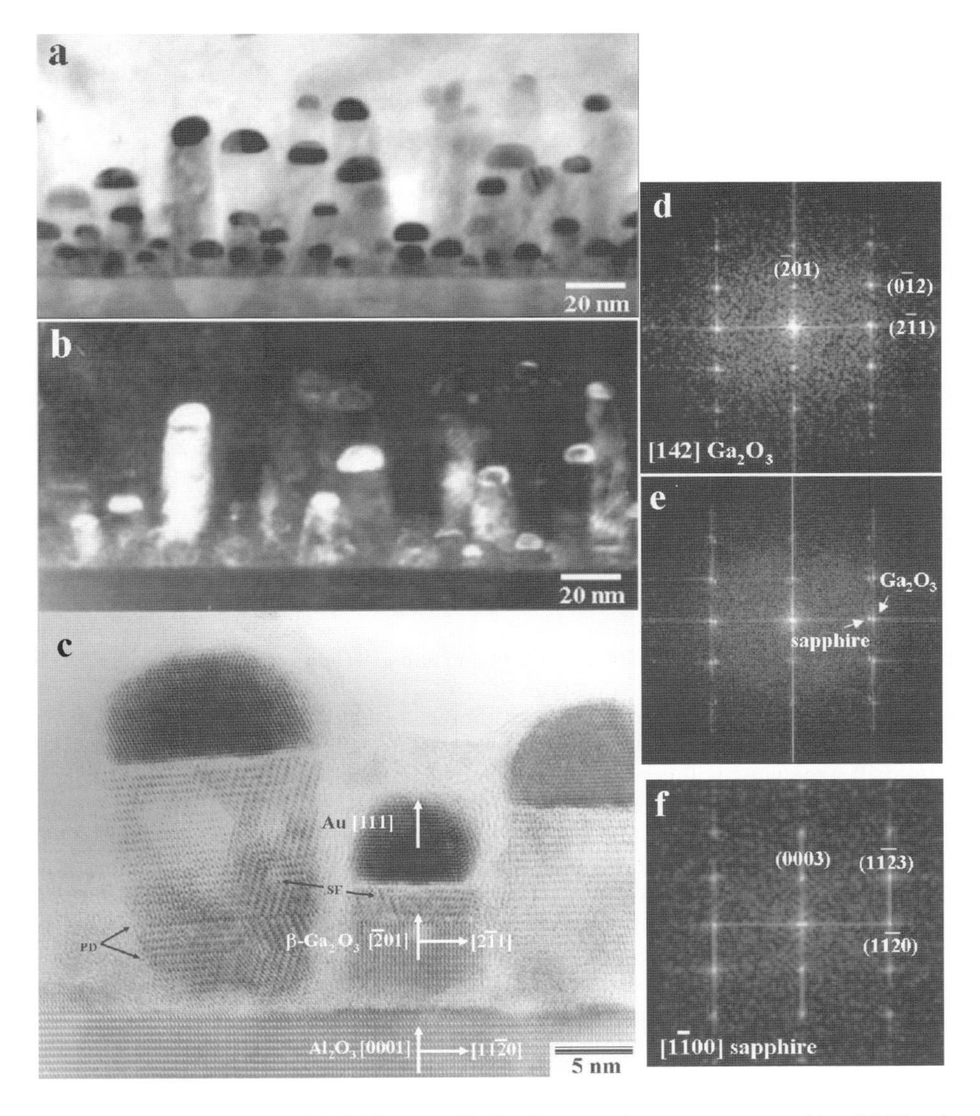

Fig. 6.15. (a) Cross-sectional TEM image of β-Ga_2O_3 nanowires grown on sapphire (0001) substrate and (b) the corresponding dark-field image. (c) Cross-sectional HRTEM image of the interfacial region of the β-Ga_2O_3 nanowires on the sapphire (0001) substrate and the corresponding diffraction patterns taken from Fast Fourier Transform (FFT) of (d) Ga_2O_3, (e) interfacial, and (f) sapphire (0001) regions [67].

In addition, gallium acetylacetonate also provides enough oxygen vapors for the growth of Ga_2O_3 nanowires. By employing a two-temperature-zone furnace, the concentrations of Ga and O vapors can be controlled by the temperature of the low-temperature zone where the temperature is independent of the Ga_2O_3 nanowires formation region in the second zone. Thus, the catalyst forming a miscible liquid phase with Ga and O remains the only restriction for low-temperature growth of Ga_2O_3 nanowires. In this work, it has been achieved that the Ga_2O_3 nanowires can be synthesized at a temperature as low as 550°C using Au as the catalyst under the sufficient supplement of precursors.

Formation of well-aligned $ZnGa_2O_4$ nanowires from Ga_2O_3–ZnO core–shell nanowires. Ga_2O_3–ZnO core–shell nanowires were synthesized via a two-step process. Catalytic growth of well-aligned β-Ga_2O_3 nanowires was first performed on Au-precoated sapphire (0001) substrates using gallium acetylacetonate, as described in the previous section. The ZnO shells were then deposited on the well-aligned β-Ga_2O_3 nanowires using zinc acetylacetonate under O_2 flow at a temperature of 400°C. Figure 6.16a shows a typical SEM image of the Ga_2O_3–ZnO core–shell nanowires. It reveals that the well-aligned feature is maintained after forming the core–shell structure. Figures 6.5b and c show the TEM analyses of the Ga_2O_3–ZnO core–shell nanowires in which the ZnO shell was deposited for 15 min. In Fig. 6.16b, a low-magnification TEM image of an individual nanowire reveals that a conformal sheath is formed on the surface of the nanowire. A HRTEM image (Fig. 6.16c), shows that the core possesses the single-crystal structure and the lattice spacing of around 0.28 nm corresponds to the d-spacing of (002) crystal planes of monoclinic Ga_2O_3. Furthermore, it illustrates that nanocrystals embed in the amorphous matrix of the shell portion. The average grain size of the ZnO nanocrystals is around 2–10 nm. The corresponding SAED pattern is illustrated in the inset of Fig. 6.16c. It shows a set of sharp diffraction spots and several discrete diffraction rings that are ascribed to the single-crystalline Ga_2O_3 core and nanocrystalline ZnO shell, respectively.

The Ga_2O_3–ZnO core–shell nanowires were further annealed for the $ZnGa_2O_4$ nanowires formation. In order to obtain the stoichiometric $ZnGa_2O_4$ nanowire, the Ga_2O_3–ZnO core–shell nanowires with various shell thickness prepared through ZnO-shell deposition time of 10, 15, and 20 min, were further annealed at a temperature of 1000°C for 1 h. Figure 6.17 shows the glancing-angle-mode XRD patterns of the annealed nanowires. In the cases of the 10- and 20-min-grown ZnO shells the diffraction peaks of $ZnGa_2O_4$ as well as Ga_2O_3 and ZnO structures are present in the XRD patterns of the annealed nanowires as illustrated in pattern I and III, respectively, resulting from inappropriate ZnO–Ga_2O_3 ratio in the original core–shell nanowires. On the other hand, only diffraction peaks of the $ZnGa_2O_4$ phase present in pattern II indicate the successful formation of the single phase $ZnGa_2O_4$ after annealing of the Ga_2O_3–ZnO core–shell nanowires with 15-min-grown ZnO shells. Figure 6.18a shows the SEM image of the single phase $ZnGa_2O_4$ nanowires formed by annealing the Ga_2O_3–ZnO core–shell nanowires. It illustrates that the nanowires still possess the well-aligned feature after 1000°C annealing. A power-diffraction-mode XRD pattern of the $ZnGa_2O_4$ nanowires, as shown in Fig. 6.18b, was further taken for examining the well-aligned characteristic of the $ZnGa_2O_4$ nanowires. Apart from the diffraction peaks attributed to the sapphire substrate and the sample holder for XRD analyses, other peaks appearing here correspond to the diffraction peaks of the $ZnGa_2O_4$ crystal. In comparison with the $ZnGa_2O_4$ power diffraction file, JCPDS No. 86-0413, XRD pattern in Fig. 6.18b indeed reveals the (111) preferential orientation of the $ZnGa_2O_4$ nanowires.

TEM analyses of the annealed nanowires shown in Fig. 6.19 are further carried out to investigate the mechanism of formation of the $ZnGa_2O_4$ nanowires from 1D core–shell nanostructures. Figures 6.19a–d shows the TEM analyses of the nanowire

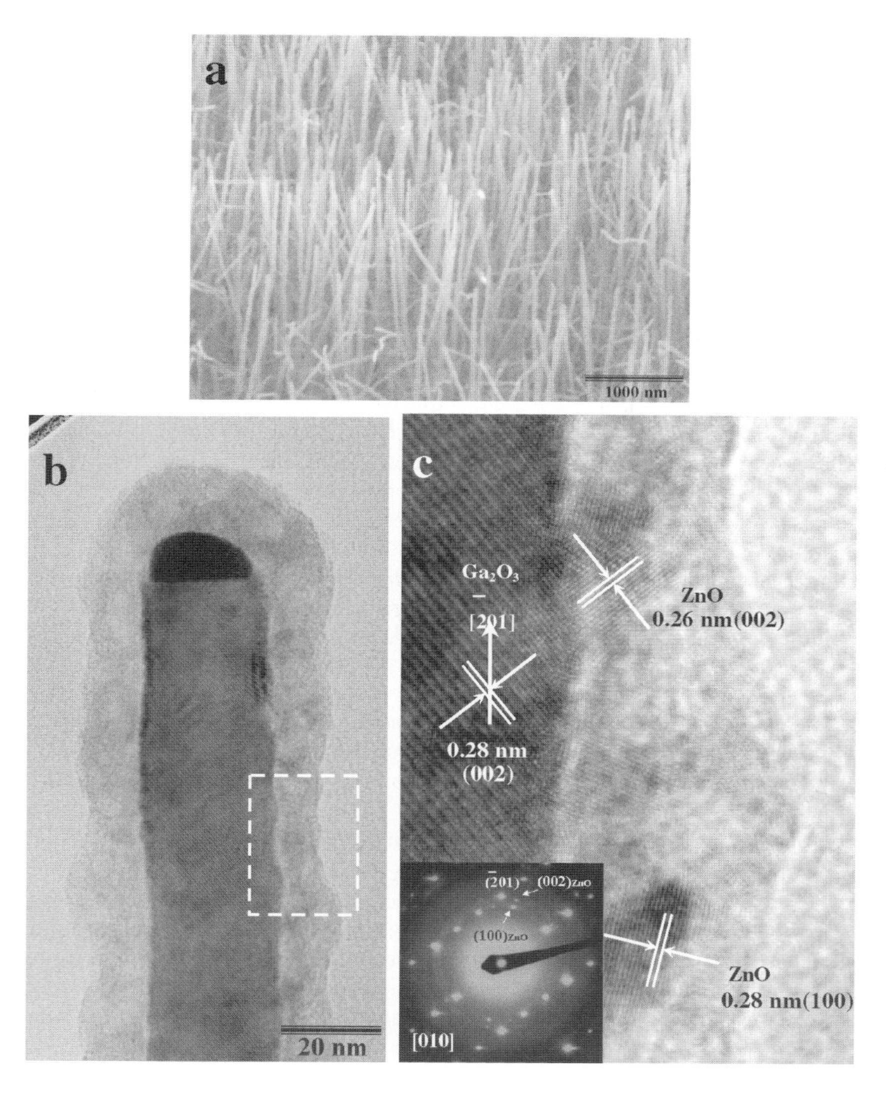

Fig. 6.16. (a) SEM image of the Ga_2O_3–ZnO core–shell nanowires. (b) Low-magnification TEM image of an individual core–shell nanowire. (c) HRTEM image and the corresponding electron diffraction pattern (inset) of the interfacial region of the core–shell nanowire [69].

formed via annealing the Ga_2O_3 core/10-min-grown ZnO shell nanowire at 1000°C for 1 h. A typical bright-field image of an individual nanowire reveals that the core–shell 1D nanostructure is preserved, as shown in Fig. 6.19a. Nevertheless, the corresponding dark-field image illustrated in Fig. 6.19b indicates that instead of nanocrystals embedded in the amorphous matrix before annealing, a single-crystal structure is formed in the shell portion of the annealed nanowire. Figure 6.19c shows the HRTEM image of the

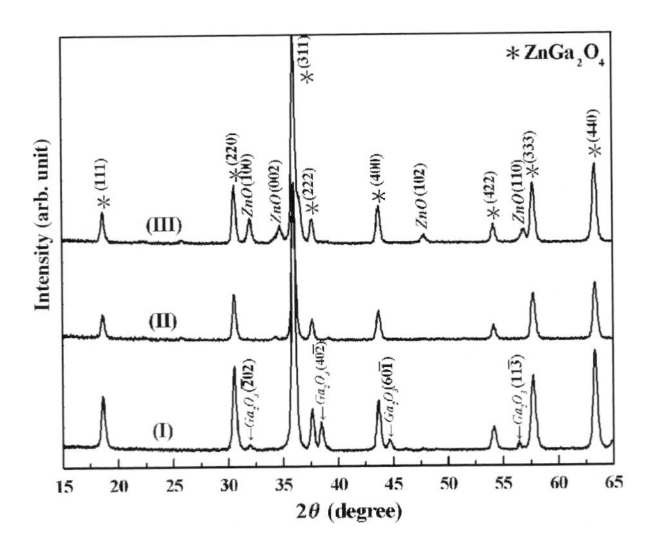

Fig. 6.17. Glancing-angle-mode XRD patterns of the nanowires with various shell thickness annealed at 1000°C. The original nanowires possess various ZnO-shell thicknesses prepared through shell deposition times of 10 (pattern I), 15 (II), and 20 min (III) [69].

interfacial region of the core–shell 1D nanostructure, as denoted in Fig. 6.19a. Well-crystalline structures of both the shell and the core are demonstrated in this HR image. The parallel fringes with the spacing of 0.24 nm along the longitudinal axis direction of the nanowire are observed in both regions. The corresponding diffraction pattern, as shown in Fig. 6.19d, indicates two sets of single-crystalline diffraction dots with an epitaxial relationship.

Figures 6.20a–c illustrate the TEM images of the nanowire formed via annealing the Ga_2O_3 core/20-min-grown ZnO shell nanowire at 1000°C for 1 h. A low-magnification TEM image reveals that the nanowire became twisted and inlaid with nanocrystals on the surface, as shown in Fig. 6.20a. The HRTEM image of the core portion of the nanowire (region A in Fig. 6.20a) is shown in Fig. 6.20b. It reveals that the core possesses the single-crystal structure and the lattice spacing of around 0.48 nm corresponds to the d-spacing of (111) crystal planes of cubic $ZnGa_2O_4$. The corresponding SAED pattern is illustrated in the inset of Fig. 6.20b. In the surface region (B in Fig. 6.20a), as shown in Fig. 6.20c, the HRTEM image illustrates that the laid nanocrystal possesses the ZnO structure, confirming the XRD analysis that the co-existence of the $ZnGa_2O_4$ and ZnO phases of the annealed nanowires, as shown in pattern III of Fig. 6.17.

In the case of the Ga_2O_3 core/15-min-grown ZnO shell nanowire, the slightly twisted nanowires with clear surfaces were found after 1000°C annealing for 1 h, as shown in Fig. 6.21a. There is no nanocrystal inlaid on the surfaces of the annealed nanowires. The HRTEM image of the nanowire (denoted in Fig. 6.21a) is shown in Fig. 6.21b. It reveals that the nanowire possesses the single-crystal structure and the lattice spacing of around 0.48 nm along the longitudinal axis direction corresponds to the d-spacing

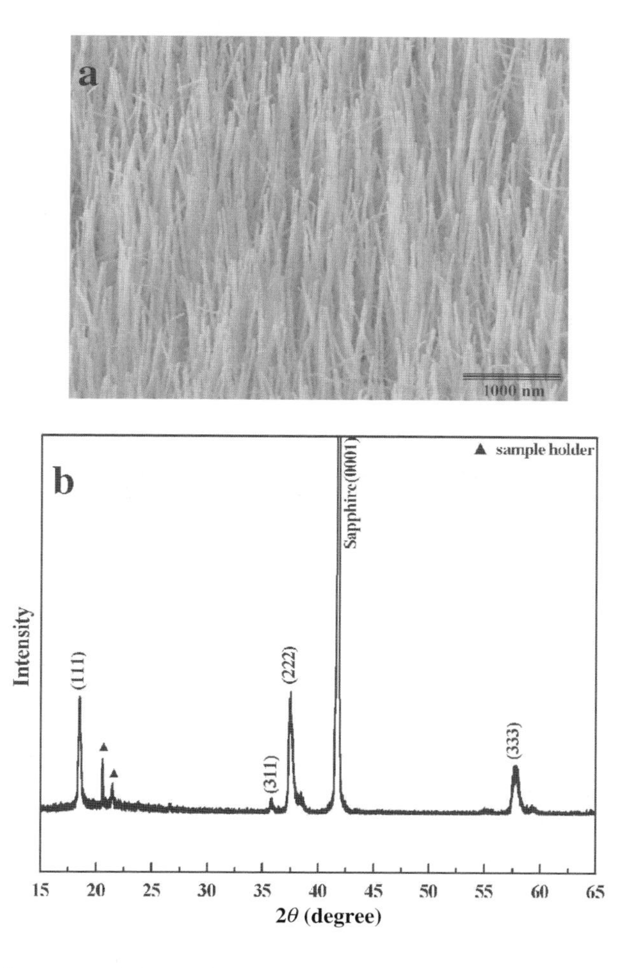

Fig. 6.18. (a) SEM image of the single phase $ZnGa_2O_4$ nanowires. (b) Power-diffraction-mode XRD pattern of the $ZnGa_2O_4$ nanowires [69].

of (111) crystal planes of cubic $ZnGa_2O_4$, further confirming the XRD analysis of the formation of the (111) preferentially oriented $ZnGa_2O_4$ nanowires. The corresponding SAED pattern is illustrated in the inset of Fig. 6.21b.

The phase diagram of the bulk Ga_2O_3–ZnO system shows that $ZnGa_2O_4$ is formed above 400°C when the molar ratio of Ga_2O_3–ZnO is unity [73]. A solid-state reaction of Ga_2O_3 and ZnO at high temperatures has been employed to prepare $ZnGa_2O_4$ for a long time [53–55]. Based on the XRD and the TEM analyses in this study, the Ga_2O_3–$ZnGa_2O_4$ core–shell nanowires, the single-crystalline $ZnGa_2O_4$ nanowires, and the $ZnGa_2O_4$ nanowires inlaid with ZnO nanocrystals were obtained via annealing the Ga_2O_3–ZnO core–shell nanowires, with various ZnO–shell thicknesses. It demonstrates that formation of the single-crystalline $ZnGa_2O_4$ nanowires has been achieved through

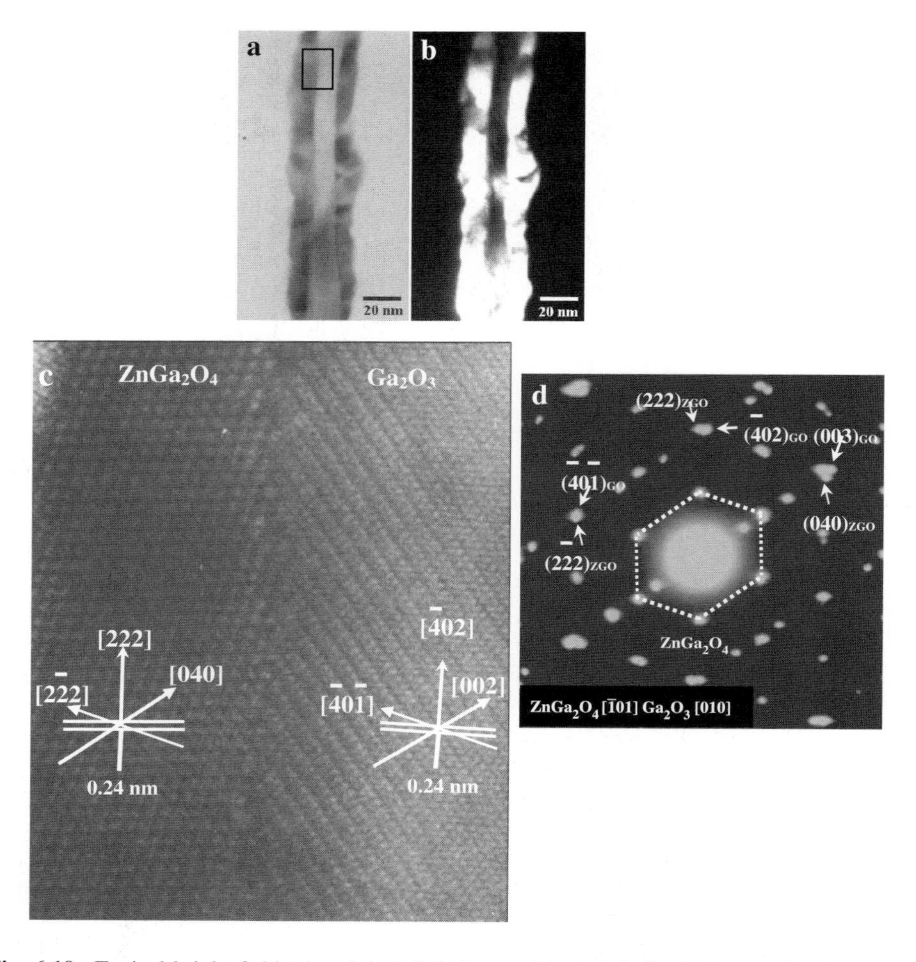

Fig. 6.19. Typical bright-field (a) and dark-field image (b) of an individual nanowire formed via annealing the Ga_2O_3 core/10-min-grown ZnO shell nanowire at 1000°C for 1 h. (c) HRTEM image of the interfacial region of the annealed core–shell 1D nanostructure. (d) The corresponding diffraction pattern of the annealed core–shell nanowire [69].

the solid-state reaction of the Ga_2O_3–ZnO core–shell nanowires along the radial direction. The thickness of the original ZnO shell and the thermal budget of the annealing process play crucial roles for preparing the single-crystalline $ZnGa_2O_4$ nanowires. Moreover, as shown in Fig. 6.19, TEM analyses reveal that $ZnGa_2O_4$ and Ga_2O_3 phases exhibit an epitaxial relationship during the solid-state reaction.

Figure 6.22 shows the room-temperature CL spectra of the single-crystalline $ZnGa_2O_4$ nanowires. Unlike the absence of any emission peak in the spectra of the as-grown Ga_2O_3–ZnO core–shell nanowires, a strong emission band centered at 360 nm and a small tail centered at 680 nm appear in the CL spectrum of the single-crystalline $ZnGa_2O_4$

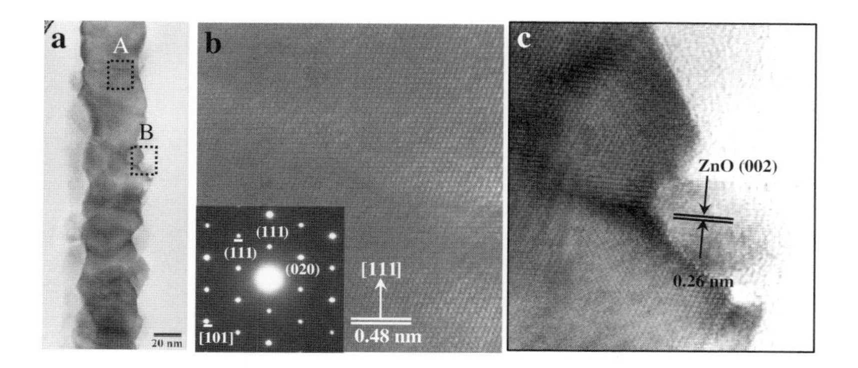

Fig. 6.20. (a) Low-magnification TEM image of the nanowire formed via annealing the Ga_2O_3 core/20-min-grown ZnO shell nanowire at 1000°C for 1 h. (b) HRTEM image and the corresponding SAED pattern of the core portion of the nanowire (region A in (a)). (c) HRTEM image of the surface region of the nanowire (region B in (a)) [69].

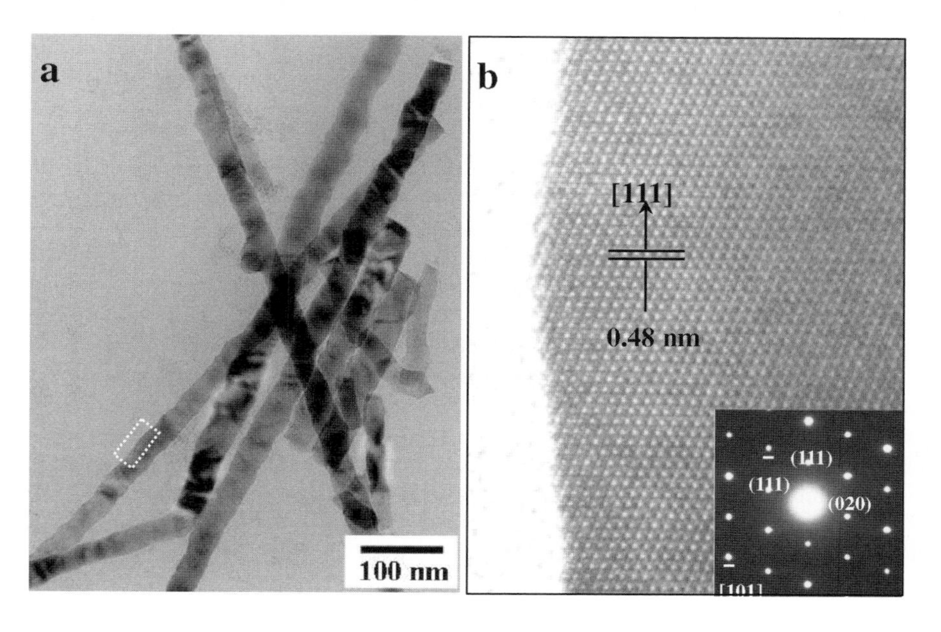

Fig. 6.21. (a) Low-magnification TEM image of the nanowire formed via annealing the Ga_2O_3 core/15-min-grown ZnO shell nanowire at 1000°C for 1 h. (b) HRTEM image and the corresponding SAED pattern of the nanowire (denoted in (a)) [69].

nanowires. The emission peaks at 360 and 680 nm have been observed from the $ZnGa_2O_4$ after being annealed in a reducing atmosphere [74]. Formation of the single oxygen vacancies after annealing in a reducing atmosphere has been demonstrated to relate to these emissions [74]. In this study, although the single-crystalline $ZnGa_2O_4$ nanowires are formed via annealing of the Ga_2O_3–ZnO core–shell nanowires in an oxidizing ambient,

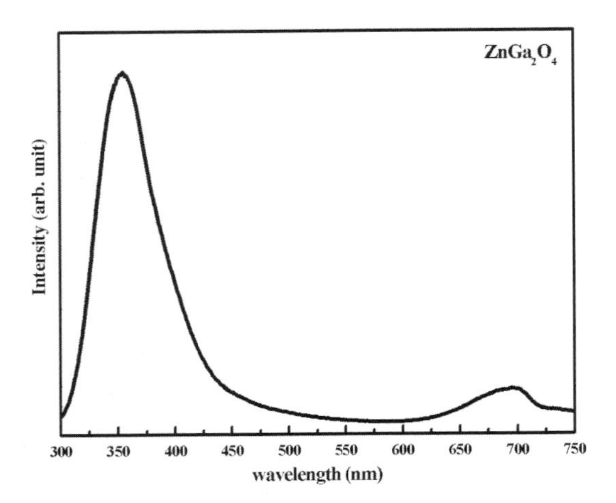

Fig. 6.22. Room-temperature CL spectra of the single-crystalline $ZnGa_2O_4$ nanowires [69].

the appearances of the peaks at 360 and 680 nm in the CL spectrum suggest the existence of the oxygen vacancies within the $ZnGa_2O_4$ nanowires.

Formation of well-aligned Ga_2O_3–TiO_2 nanobarcodes from Ga_2O_3–TiO_2 core–shell nanowires. For the synthesis of the Ga_2O_3–TiO_2 core–shell nanowires, a two-stage process was employed. Well-aligned β-Ga_2O_3 nanowires were first grown on the Au-precoated sapphire (0001) substrates using gallium acetylacetonate, as described in the previous section. The Ga_2O_3 nanowires were then used as one-dimensional templates for the TiO_2 shell deposition. The TiO_2 shells were formed on the Ga_2O_3 nanowires using titanium acetylacetonate under a N_2/O_2 flow at a temperature of 400°C. Figure 6.23a shows the Ga_2O_3–TiO_2 core–shell nanowires on the sapphire substrates, revealing that the well-aligned morphology of the nanowires has still been maintained after the formation of the TiO_2 shells. Further structural analyses of the as-synthesized Ga_2O_3–TiO_2 core–shell nanowires were performed by TEM. As shown in Fig. 6.23b, a low magnification TEM image of an individual nanowire reveals that the 1D nanostructure possesses a coaxial structure, i.e. a thin and conformal sheath with lighter contrast is formed outside the surface of the nanowire structure of dark contrast. A HRTEM image demonstrates the single-crystal structure core, as shown in Fig. 6.23c. The lattice spacings of the crystalline core around 0.297 nm are consistent with the d-spacing of ($\bar{4}00$) crystal planes of monoclinic Ga_2O_3. In addition, the HRTEM image reveals that the light-contrast shell is composed of both nanocrystalline and amorphous structures. The lattice spacings of the nanocrystal around 0.35 nm, as shown in Fig. 6.23c, are consistent with the d-spacing of (101) crystal planes of anatase TiO_2. The corresponding SAED pattern is also illustrated in the inset of Fig. 6.23c. It shows a set of sharp diffraction spots, which can be identified as monoclinic Ga_2O_3 with zone axis [010].

TiO_2 shells were also deposited on the well-aligned Ga_2O_3 nanowires at 600°C. In Fig. 6.24a, a typical low-magnification TEM image of an individual nanowire reveals that a conformal sheath composed of nanocrystals is formed on the surface of the

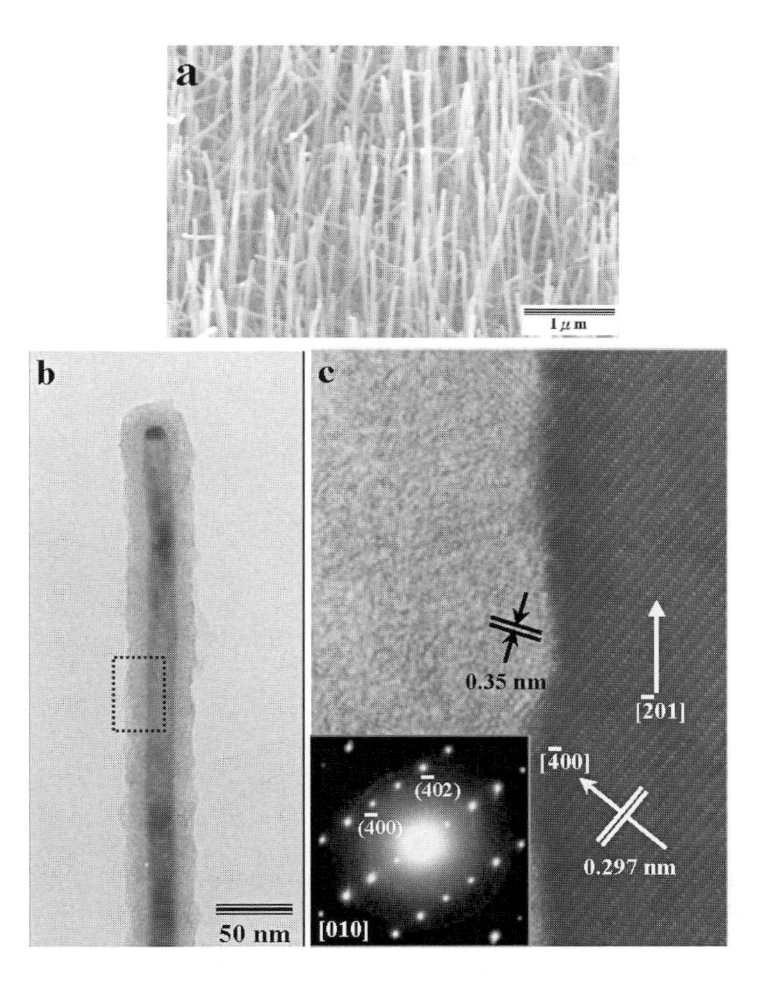

Fig. 6.23. Well-aligned Ga_2O_3–TiO_2 core–shell nanowires with 400°C-grown TiO_2 shell on sapphire (0001) substrates. (a) SEM image, (b) low-magnification TEM image of an individual Ga_2O_3–TiO_2core–shell nanowire, and (c) high-resolution (HR) TEM image of the TiO_2 shell and Ga_2O_3 core with the corresponding SAED pattern (inset) [70].

nanowire. Figure 6.24b, a HRTEM image denoted in (a), shows that the lattice spacing of the nanocrystal of around 0.35 nm is consistent with the d-spacing of (101) crystal planes of anatase TiO_2. The average grain size of the TiO_2 nanocrystals is around 5–8 nm. The corresponding SAED pattern is illustrated in the inset of Fig. 6.24b. A set of sharp diffraction spots with another ring pattern are ascribed to the single-crystalline Ga_2O_3 core and nanocrystalline anatase TiO_2 shell, respectively.

Deposition of TiO_2 on the Ga_2O_3 nanowires at a higher temperature of 1000°C was further performed. As shown in Fig. 6.25a, the core–shell characteristic is not observed in the TEM image of an individual nanowire. The HRTEM image of the nanowire

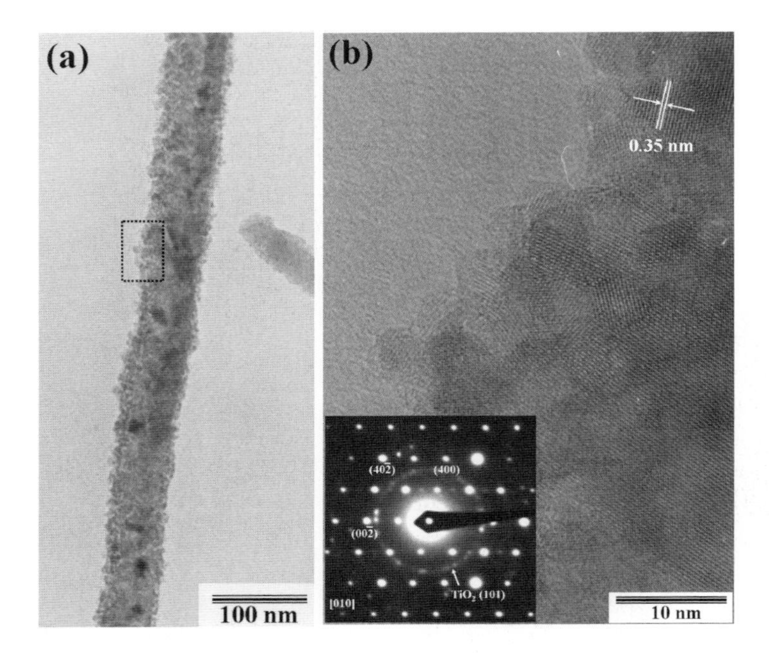

Fig. 6.24. (a) Low-magnification TEM image and (b) HRTEM image denoted in (a) with the corresponding SAED pattern (inset), of an individual Ga_2O_3–TiO_2 core–shell nanowire with 600°C-grown TiO_2 shells.

(denoted in Fig. 6.25a) is shown in Fig. 6.25b. It reveals that the nanowire possesses the single-crystal structure and the lattice spacing of around 0.47 nm along the longitudinal axis direction. The lattice spacing corresponds to the d-spacing of ($\bar{2}$01) crystal planes of monoclinic Ga_2O_3. The corresponding SAED pattern in Fig. 6.25c shows a set of sharp diffraction spots, which can be attributed from the monoclinic Ga_2O_3 with zone axis [142]. EDS analysis shows that the nanowire is mainly composed of Ga and O elements, as illustrated in Fig. 6.25d. Copper and carbon signals appearing in this spectrum should be ascribed to the copper grids coated with porous carbon film for supporting sample. It is suggested that the sticking coefficient of the Ti atom on the surface of the Ga_2O_3 nanowires is rather low at 1000°C, resulting in the absence of the TiO_2 shells on the Ga_2O_3 nanowires.

Post-annealing of the Ga_2O_3–TiO_2 core–shell nanowires, in which the TiO_2 shells were formed at 400°C, was further conducted in oxygen ambient at 1000°C for 1 h. Figure 6.26a is the SEM image of the post-treated nanowires. The aligned morphology of the 1000°C-annealed nanowires is still maintained. The diameters of the annealed nanowires are in the range of 30–70 nm. Figure 6.26b demonstrates the interlacements of the dark and light contrasts along the nanowires, suggesting that the nanowires possess the 1D multiple junction nanostructures after a 1000°C-annealing process. Figures 6.27a–d show that the EELS mapping images of Ga, Ti, and O elements and the corresponding TEM image of an individual multiple junction nanowire. They reveal that the nanowire

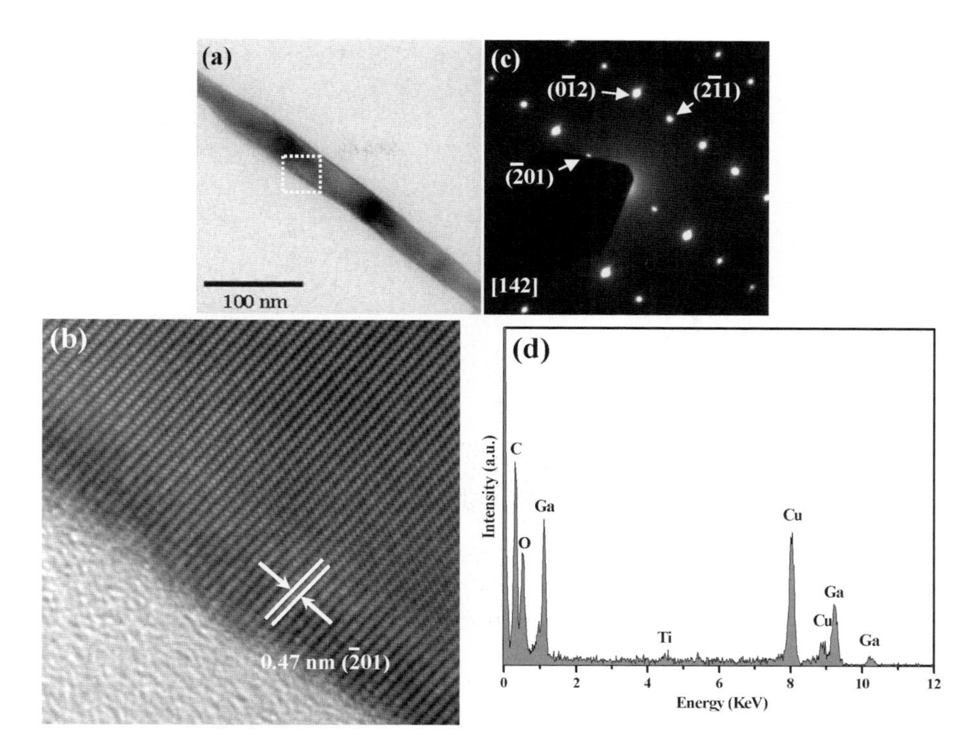

Fig. 6.25. Ga_2O_3 nanowire after 1000°C deposition process for the TiO_2 shells. (a) Low-magnification TEM image, (b) HRTEM image, and (c) the corresponding SAED pattern of the nanowire (denoted in a). (d) EDS spectra of the nanowire.

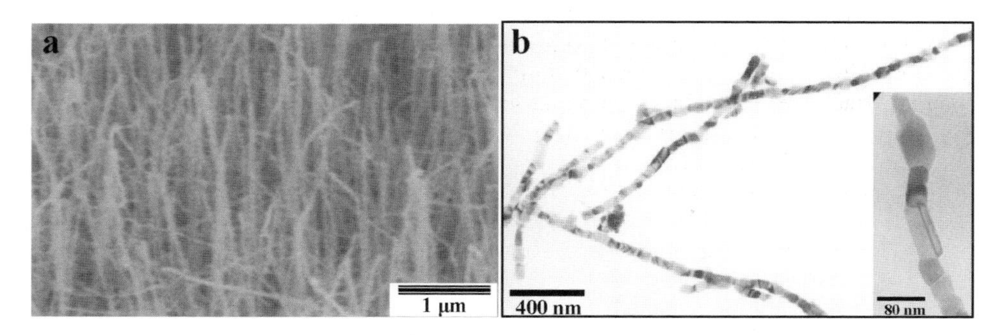

Fig. 6.26. Ga_2O_3/400°C-TiO_2 core–shell nanowires annealed at 1000°C. (a) SEM image and (b) low-magnification TEM images of the nanowires [70].

is composed of Ga/O and Ti/O elements interlacing along the longitudinal direction, resulting in the barcode-like nanostructure. Spatially resolved EDS analyses with an electron beam probe size of ca. 10 nm is illustrated in Fig. 6.27e. It also demonstrates the interlacement of the Ga_2O_3 and TiO_2 nanocrystals along the nanobarcode indexed in Fig. 6.27d. Moreover, each Ga_2O_3 segment possesses a diameter similar to that of

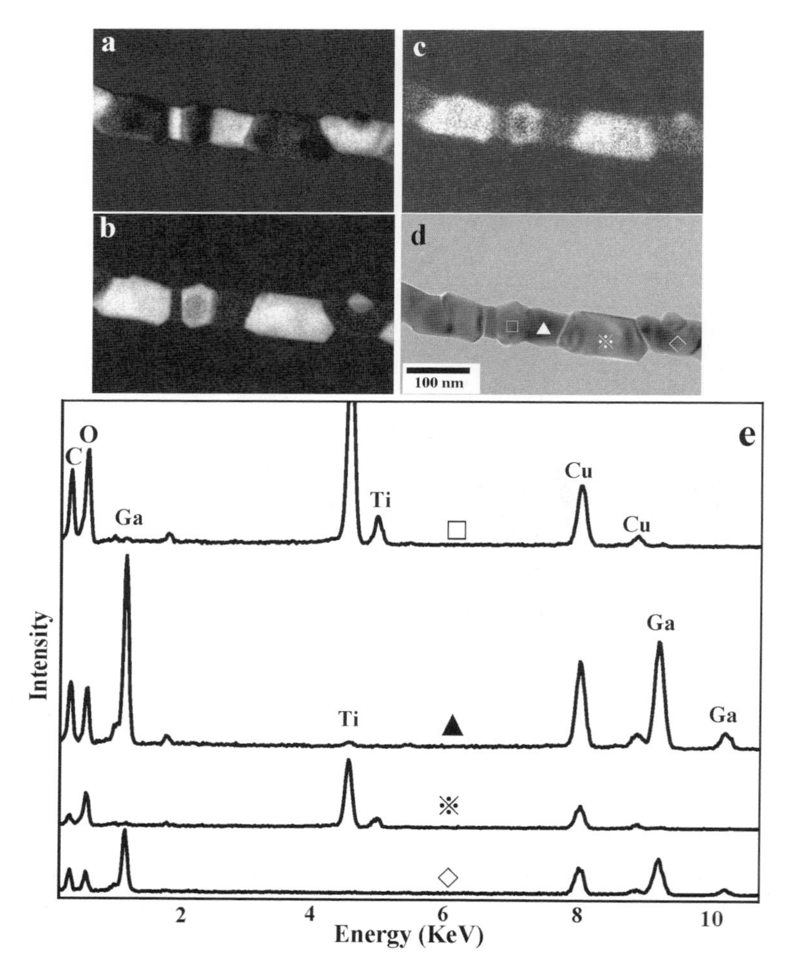

Fig. 6.27. EELS mapping images of the (a) Ga, (b) Ti, and (c) O elements with (d) the corresponding TEM image of an individual Ga_2O_3–TiO_2 multiple-junction nanowire. (e) Spatially resolved EDS analysis with an electron-beam probe size of ca. 10 nm; indexed as in (d) [70].

the as-grown Ga_2O_3 nanowire whereas the polygonal TiO_2 segment has a larger average diameter, resulting in the variation of the diameter along the nanowire.

HRTEM was further used to study the detailed structures of the Ga_2O_3–TiO_2 nano-barcodes. Compositional analysis by EDS indicates that the nanowire shown in the low-resolution TEM image of Fig. 6.28a, was composed of Ga_2O_3–TiO_2–Ga_2O_3–TiO_2–Ga_2O_3 segments from left to right. HRTEM images of different sections of the Ga_2O_3 and TiO_2 segments (Section b–d, as labeled in Fig. 6.28a) and their corresponding SAED patterns are shown in Fig. 6.28b–e. The HRTEM image of section b (from a Ga_2O_3 segment) reveals two groups of parallel fringes with spacings of 0.23 and 0.28 nm, corresponding

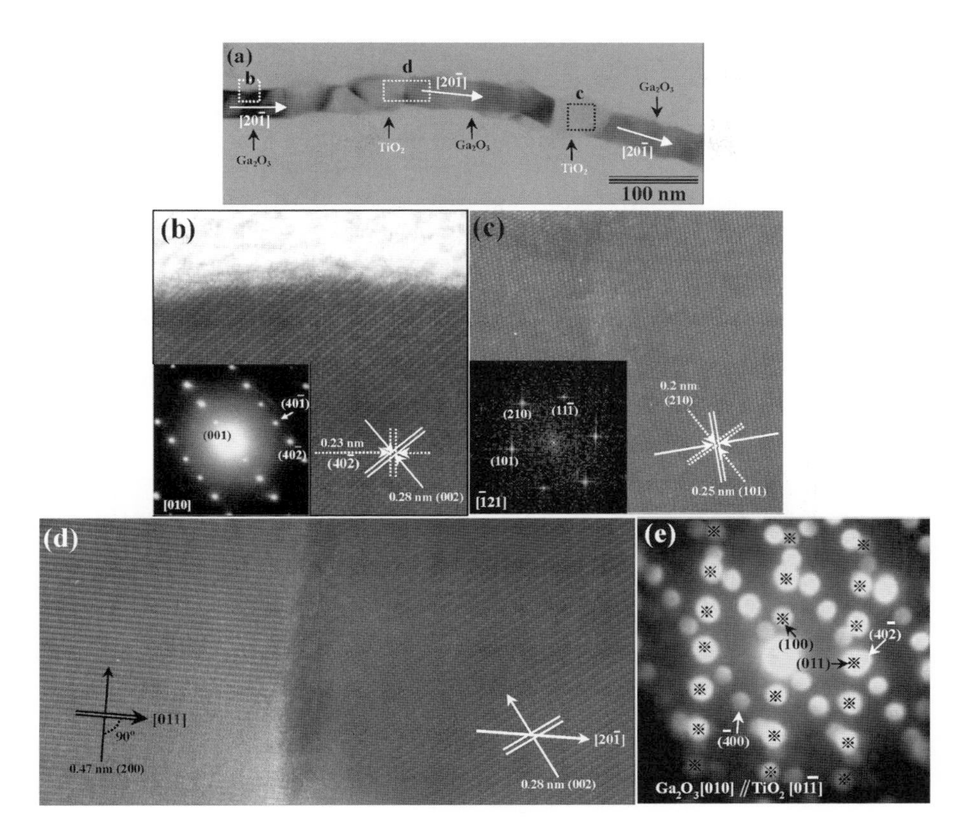

Fig. 6.28. (a) Low-magnification TEM image of Ga_2O_3–TiO_2–Ga_2O_3–TiO_2–Ga_2O_3 (from left to right) segments of nanobarcodes. HRTEM images of the (b) Ga_2O_3, (c) TiO_2, and (d) TiO_2–Ga_2O_3 interfacial region denoted in (a). The insets and (e) show the corresponding SAED patterns.

to the (40$\bar{2}$) and (002) planes of monoclinic Ga_2O_3, respectively. It also indicates that the longitudinal axis of the Ga_2O_3 segment lies in the [20$\bar{1}$] direction, which is consistent with the direction of the longitudinal axis of the Ga_2O_3 core of the as-grown Ga_2O_3–TiO_2 core–shell nanowires, shown in Fig. 6.23c. The corresponding SAED of the [010] zone axis is illustrated in the inset of Fig. 6.28b. Figure 6.28c and its inset show the HRTEM image of section c (a TiO_2 segment) and the corresponding SAED rutile TiO_2 pattern of the [$\bar{1}$21] zone axis obtained by Fast Fourier Transform (FFT), respectively. The lattice fringes, with interlayer distances about 0.2 and 0.25 nm, are quite consistent with the d-spacings of (210) and (101) crystal planes of rutile TiO_2, respectively. The HRTEM image of section e (a Ga_2O_3–TiO_2 interfacial region) is shown in Fig. 6.28d. It shows the existence of a perfect, crystalline interface without obvious defects or amorphous phase. The corresponding diffraction pattern is illustrated in Fig. 6.28e. In addition to the one attributed to the Ga_2O_3 segment, the other is indexed to be the diffraction pattern of a rutile TiO_2 structure, thus confirming a substantial lattice mismatch at the interface between Ga_2O_3 and TiO_2.

To investigate the effect of annealing temperatures on the formation of the Ga_2O_3–TiO_2 nanobarcodes, post-annealing of the Ga_2O_3–TiO_2 core–shell nanowires was further performed at temperatures of 750 and 850°C. TEM images of the 700°C- and 850°C-annealed nanowires are illustrated in Fig. 6.29. Figures 6.29a and b show a typical bright-field and the corresponding dark-field images of an individual nanowire formed via annealing the Ga_2O_3–TiO_2 core–shell nanowire at 700°C for 1 h, respectively, revealing that the core–shell 1D nanostructure is preserved. Instead of nanocrystals being embedded in the amorphous matrix before annealing shown in Fig. 6.23, anatase TiO_2 nanocrystals with larger sizes were formed on the Ga_2O_3 nanowire surface by recrystallization of the original TiO_2 shell. The corresponding SAED pattern is illustrated in the inset of Fig. 6.29a, which is similar to that shown in the inset of Fig. 6.24b. Figure 6.29c shows the HRTEM image of the interfacial region of

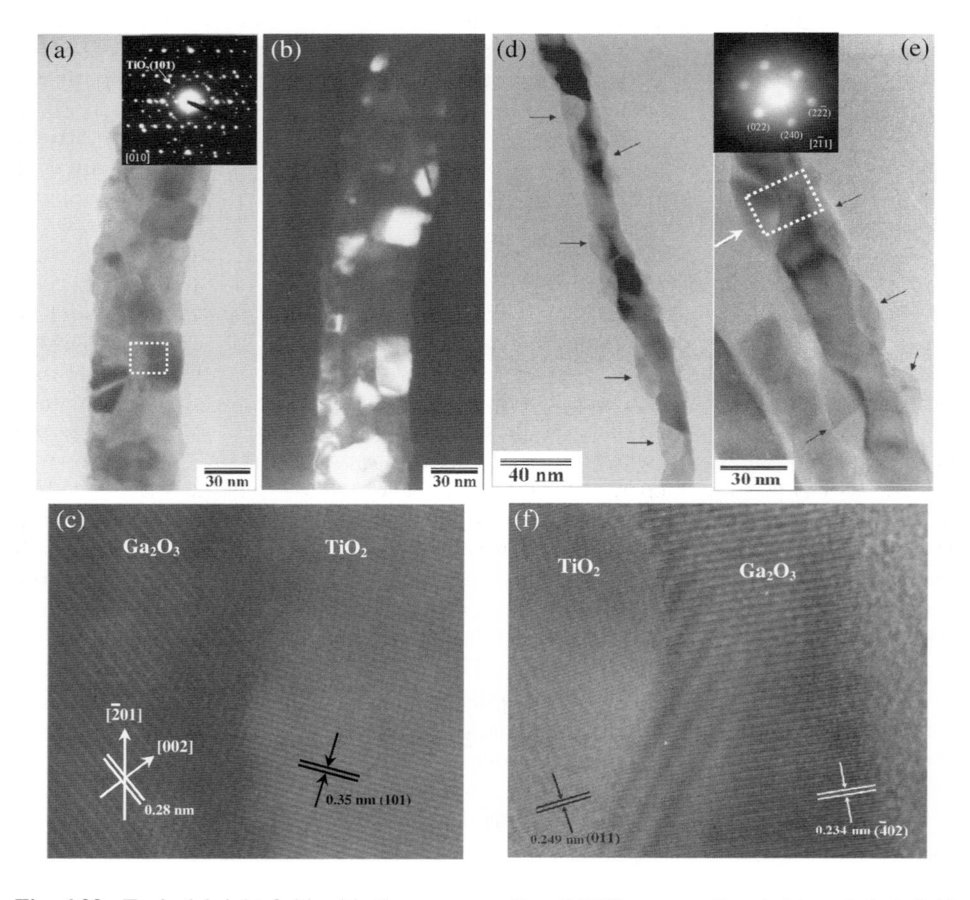

Fig. 6.29. Typical bright-field with the corresponding SAED pattern (inset) (a) and dark-field image (b) of individual nanowire annealed at 700°C for 1 h. (c) HRTEM image of Ga_2O_3–TiO_2 interfacial region denoted in (a). (d) and (e) Low-magnification TEM images of the nanowires annealed at 850°C. (f) HRTEM image of Ga_2O_3–TiO_2 interfacial region denoted in (e).

the core–shell 1D nanostructure, as denoted in Fig. 6.29a. Well-crystalline structures of both the core and shell are demonstrated in this HR image. Figures 6.29d and e reveal that the nanowires became inlaid with nanocrystals (denoted by arrows), which were formed after annealing at 850°C. The SAED pattern illustrated in the inset of Fig. 6.29e shows the rutile TiO_2 phase of the inlaid nanocrystals. Figure 6.29f shows the HRTEM image of the interfacial region between the inlaid-nanocrystal, rutile TiO_2 and Ga_2O_3, as marked in Fig. 6.29e. As shown inFig. 6.29, Ga_2O_3–TiO_2 nanobarcodes are not available by annealing of the Ga_2O_3–TiO_2 core–shell nanowire for 1 h at temperatures lower than 1000°C.

Fabrication of the core–shell nanowires by conformal growth of the shell layers on the surface of the core nanowires using the CVD method has been demonstrated in many systems [75–77]. In the case of the Ga_2O_3–TiO_2 system, both the nanocrystalline and amorphous structures of the shell could be attributed to a substantial lattice mismatch at the interface between $Ga_2O_3$and TiO_2. Based on TEM analysis of the nanowires annealed at various temperatures, one possible mechanism for the formation of the Ga_2O_3–TiO_2 nanobarcodes is proposed, as shown in Fig. 6.30. The phase diagram of the bulk Ga_2O_3–TiO_2 system shows that a solid solution of the $Ga_4Ti_{m-4}O_{2m-2}$ formed at temperatures higher than 1190°C, whereas the separated rutile TiO_2 and β-Ga_2O_3 phases are formed below this temperature [78,79]. Since the surface area of the material in nanostructure form is larger than that in the bulk form, the temperature for the formation of the solid solution (or for phase separation) in the nanostructure form would be lower than that in the bulk. After annealing at 700°C, the sizes of the anatase TiO_2 nanocrystals are increased and the Ga_2O_3–TiO_2 core–shell structure is maintained, suggesting that the solid solution of $Ga_4Ti_{m-4}O_{2m-2}$ is not formed at the annealing temperature. In the case of 850°C, the TiO_2 shell of the as-grown core–shell nanowires, which is composed of amorphous and nanocrystalline TiO_2, preferentially diffuses along the surface of the Ga_2O_3 core to form rutile TiO_2 nanocrystals during the annealing process. Simultaneously, the $Ga_4Ti_{m-4}O_{2m-2}$ solid solutions are formed at the interfacial regions of Ga_2O_3 and TiO_2 due to diffusion of the solid state in the radial direction. The twisted Ga_2O_3 nanowires interlaid with rutile TiO_2 nanocrystals form as the temperature was lowered slowly to room temperature. This would have resulted from the incomplete formation of $Ga_4Ti_{m-4}O_{2m-2}$ solid solutions along the radial direction during the 1-h long annealing process at 850°C. In the case of annealing at 1000°C, temperature was high enough for $Ga_4Ti_{m-4}O_{2m-2}$ solid solutions to form along the entire radial direction,

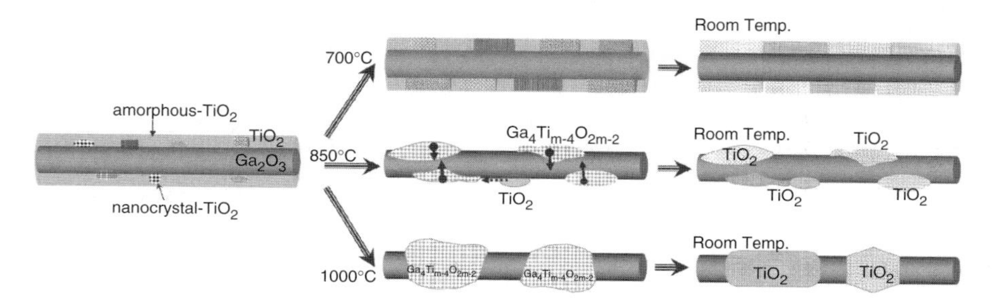

Fig. 6.30. Schematic of the proposed mechanism for the formation of Ga_2O_3–TiO_2 nanobarcodes.

formation of the multiple-junction nanostructures of Ga_2O_3 and $Ga_4Ti_{m-4}O_{2m-2}$ would take place, as shown in Fig. 6.30. While the temperature was slowly decreased, Ga_2O_3 phase would precipitate in the interfacial region of Ga_2O_3 and $Ga_4Ti_{m-4}O_{2m-2}$. Phase separation of TiO_2 and Ga_2O_3, form their respective nanocrystals along the longitudinal directions of the nanowires that occurred when the temperature was slowly brought back to room temperature. The nanocrystals had rather faceted morphologies, which was probably due to the minimum total energy state of the nanobarcodes configuration for this system at room temperature. The Ga_2O_3–TiO_2 nanobarcodes were thus formed after annealing the Ga_2O_3/400°C-grown TiO_2 core–shell nanowires at 1000°C for 1 h.

6.3.4 Conclusion

Well-aligned β-Ga_2O_3 nanowires have been synthesized on the sapphire (0001) at low temperatures using a single precursor of gallium acetylacetonate through the vapor–liquid–solid (VLS) mechanism. Structural analyses reveal that the well-aligned Ga_2O_3 nanowires are epitaxially grown on the sapphire (0001). In addition, the well-aligned Ga_2O_3–TiO_2 and Ga_2O_3–ZnO core–shell nanowires were synthesized using a two-stage MOCVD method, i.e. the well-aligned Ga_2O_3 nanowires are employed to be templates for a conformal deposition of TiO_2 and ZnO shell on their surfaces. Aligned Ga_2O_3–TiO_2 nanobarcodes have further been formed through thermal annealing of the Ga_2O_3–TiO_2 core–shell nanowires. Formation of the well-aligned and single-crystalline $ZnGa_2O_4$ nanowires on sapphire (0001) substrates has also been achieved via annealing of the Ga_2O_3–ZnO core–shell nanowires. Structural analyses of the annealed nanowires reveal the existence of an epitaxial relationship between $ZnGa_2O_4$ and Ga_2O_3 phases during the solid-state reaction.

Acknowledgment

The financial support of this work, by the National Science Council in Taiwan under Contract No. NSC 93-2214-E-006-022, is gratefully acknowledged.

References

[1] J. Hu, M. Ouyang, P. Yang and C.M. Leiber, Nature 399 (1999) 48.

[2] L.J. Lauhon, M.S. Gudiksen, D. Wang and C.M. Leiber, Nature 420 (2002) 57.

[3] M.S. Gudilsen, L.J. Lauhon, J. Wang, D.C. Smith and C.M. Leiber, Nature 415 (2002) 617.

[4] R. He, M. Law, R. Fan, F. Kim and P. Yang, Nano. Lett. 2 (2002) 1109.

[5] M.T. Bjork, B.J. Ohlsson, T. Sass, A.I. Persson, C. Thelander, M.H. Magnusson, K. Depperk, L.R. Wallenberg and L. Samuelson, Appl. Phys. Lett. 80 (2002) 1058.

[6] Y. Xia, P. Yang, Y. Sun, Y. Wu, B. Mayers, B. Gates, Y. Yin, F. Kim and H. Yan, Adv. Mater. 15 (2003) 353.

[7] L. Vayssieres, K. Keis, A. Hagfeldt and S.E. Lindquist, Chem. Mater. 13 (2001) 4386.

[8] T. Seiyama and A. Kato, Anal. Chem. 34 (1962) 1502.

[9] K. Hara, T. Horiguchi, T. Kinoshita, K. Sayama, H. Sugihara and H. Arakawa, Sol. Energy Mater. Sol. Cells 64 (2000) 115.

[10] Y. Chen, D.M. Bagnall, H. Koh, K. Park, K. Hiraga, Z. Zhu and T. Yao, J. Appl. Phys. 84 (1998) 3912.

[11] A. Ohtomo, M. Kawasaki, Y. Sakurai, I. Ohkubo, R. Shiroki, Y. Yoshida, T. Yasuda, Y. Segawa and H. Koinuma, Mater. Sci. Eng. B 56 (1998) 263.

[12] D.M. Bagnall, Y.F. Chen, Z. Zhu, T. Yao, S. Koyama, M.Y. Shen and T. Goto, Appl. Phys. Lett. 70 (1997) 2230.

[13] P. Zu, Z.K. Tang, G.K.L. Wong, M. Kawasaki, A. Ohtomo, H. Koinuma and Y. Segawa, Solid State Commun. 103 (1997) 459.

[14] H. Co, J.Y. Xu, E.W. Seelig and R.P.H. Chang, Appl. Phys. Lett. 76 (2000) 2997.

[15] M.H. Huang, S. Mao, H. Feick, H. Yan, Y. Wu, H. Kind, E. Weber, R. Russo and P. Yang, Science 292 (2001) 1897.

[16] J.C. Johnson, H. Yan, R. Schaller, L.H. Haber, R.J. Saykally and P. Yang, J. Phys. Chem. 105 (2001) 11387.

[17] T. Makno, Y. Sagawa, M. Kawasaki, A. Ohtomo, R. Shiroki, K. Tamura, T. Yasuda and H. Koinuma, Appl. Phys. Lett. 78 (2001) 1237.

[18] A.K. Sharma, J. Narayan, J.F. Muth, C.W. Teng, C. Jin, A. Kvit, R.M. Kolbas and O.W. Holland, Appl. Phys. Lett. 75 (1999) 3327.

[19] W.I. Park, G.C. Yi and H.M. Jang, Appl. Phys. Lett. 79 (2001) 2022.

[20] W. Yang, S.S. Hullavarad, B. Nagaraj, I. Takeuchi, R.P. Sharma, T. Venkatesan, R.D. Vispute and H. Shen, Appl. Phys. Lett. 82 (2003) 3424.

[21] T. Dietl, H. Ohno, F. Matsukura, J. Cibert and D. Ferrand, Science 287 (2000) 1019.

[22] H. Ohno, Science 281 (1998) 951.

[23] S.J. Pearton, C.R. Abernathy, D.P. Norton, A.F. Hebard, Y.D. Park, L.A. Boatner and J.D. Budai, Mater. Sci. Eng. R 40 (2003) 137.

[24] K. Ueda, H. Tabata and T. Kawai, Appl. Phys. Lett. 79 (2001) 988.

[25] T. Fukumura, Z. Jin, M. Kawasaki, T. Shono, T. Hasegawa, S. Koshihara and H. Koinuma, Appl. Phys. Lett. 78 (2001) 958.

[26] J.H. Kim, H. Kim, D. Kim, Y.E. Ihm and W.K. Choo, J. Appl. Phys. 92 (2002) 6066.

[27] Y.C. Kong, D.P. Yu, B. Zhang, W. Fang and S.Q. Feng, Appl. Phys. Lett. 78 (2001) 407.

[28] M.H. Huang, Y. Wu, H. Feick, N. Tran, E. Weber and P. Yang, Adv. Mater. 13 (2001) 113.

[29] Z.W. Pan, Z.R. Dai and Z.L. Wang, Science 291 (2001) 1947.

[30] J.J. Wu and S.C. Liu, Adv. Mater. 14 (2002) 215.

[31] S.C. Liu and J.J. Wu, J. Mater. Chem. 12 (2002) 3125.

[32] J.J. Wu and S.C. Liu, J. Phys. Chem. B 106 (2002) 9546.

[33] J.J. Wu, H.I. Wen, C.H. Tseng and S.C. Liu, Adv. Funct. Mater. 14 (2004) 806.

[34] C.H. Ku, H.H. Chiang and J.J. Wu, Chem. Phys. Lett. 404 (2005) 132.

[35] J.J. Wu, S.C. Liu and M.H. Yang, Appl. Phys. Lett. 85 (2004) 1027.

[36] Y. Chen, D.M. Bagnall, H. Koh, K. Park, K. Hiraga, Z. Zhu and T. Yao, J. Appl. Phys. 84 (1998) 3912.

[37] P. Fons, K. Iwata, A. Yamada, K. Matsubara, S. Niki, K. Nakahara, T. Tanabe and H. Takasu, Appl. Phys. Letts. 77 (2000) 1801.

[38] K. Vanheusden, W.L. Warren, C.H. Seager, D.R. Tallant, J.A. Voigt and B.E. Gnade, J. Appl. Phys. 79 (1996) 7983.

[39] L. Wang, H. Wada and L.F. Allard, J. Mater. Res. 7 (1992) 148.

[40] G.W. Sear, Acta Metall. 1 (1953) 457.

[41] R.S. Wagner and W.C. Ellis, Appl. Phys. Lett. 4 (1964) 89.

[42] R.A. Laudise and A.A. Ballman, J. Phys. Chem. 64 (1960) 688.

[43] W.J. Li, E.W. Shi, W.Z. Zhong and Z.W. Yin, J. Cryst. Growth (1999) 186.

[44] M. Ogita, N. Saika, Y. Nakanishi and Y. Hatanaka, Appl. Surf. Sci. 142 (1999) 188.

[45] D.D. Edwards, T.O. Mason, F. Goutenoir and K.R. Poeppelmeier, Appl. Phys. Lett. 70 (1997) 1706.

[46] K. Nakagawa, M. Okamura, N. Ikenaga, T. Suzuki and T. Kobayashi, Chem. Commun. (1998) 1025.

[47] K. Nakagawa, C. Kajita, K. Okumura, N. Ikenaga, M. Nishitani-Gamo, T. Ando, T. Kobayashi and T. Suzuki, J. Catal. 203 (2001) 87.

[48] B.M. Reddy, I. Ganesh, E.P. Reddy, A. Fernandez and P.G. Smirniotis, J. Phys. Chem. B 105 (2001) 6227.

[49] I.J. Hsieh, K.T. Chu, C.F. Yu and M.S. Feng, J. Appl. Phys. 76 (1994) 3735.

[50] I.K. Jeong, H.L. Park and S.I. Mho, Solid State Commun. 108 (1998) 823.

[51] Y.E. Lee, D.P. Norta, C. Park and C.M. Roulean, J. Appl. Phys. 89 (2001) 1653.

[52] A.F. Wells, Structural Inorganic Chemistry, Oxford University Press: London, 1975, p. 489.

[53] L.E. Shea, R.K. Datta and J.J.Jr. Brown, J. Electrochem. Soc. 141 (1994) 1950.

[54] I.K. Jeong, H.L. Park and S.I. Mho, Solid State Commun. 105 (1998) 179.

[55] T. Omata, N. Ueda and K. Ueda, Appl. Phys. Lett. 64 (1994) 1077.

[56] Z. Yan, H. Takei and H. Kawazoe, J. Am. Ceram. Soc. 81 (1998) 180.

[57] H. Kawazoe and K. Ueda, J. Am. Ceram. Soc. 82 (1999) 3330.

[58] Y.C. Choi, W.S. Kim, Y.S. Park, S.M. Lee, D.J. Bae, Y.H. Lee, G.S. Park, W.B. Choi, N.S. Lee and J.M. Kim, Adv. Mater. 12 (2000) 746.

[59] W.Q. Han, P. Kohler-Redlich, F. Ernst and M. Ruhle, Solid State Commun. 115 (2000) 527.

[60] C.H. Liang, G.W. Meng, G.Z. Wang, Y.W. Wang and L.D. Zhang, Appl. Phys. Lett. 78 (2001) 3202.

[61] H.J. Chun, Y.S. Choi, S.Y. Bae, H.W. Seo, S.J. Hong, J. Park and H. Yang, J. Phys. Chem. B 107 (2003) 9042.

[62] X.C. Wu, W.H. Song, W.D. Huang, M.H. Pu, B. Zhao, Y.P. Sun and J.J. Du, Chem. Phys. Lett. 328 (2000) 5.

[63] G. Gundiah, A. Govindaraj and C.N.R. Rao, Chem. Phys. Lett. 351 (2002) 89.

[64] Z.R. Dai, Z.W. Pan and Z.L. Wang, J. Phys. Chem. B 106 (2002) 902.

[65] B.C. Kim, K.T. Sun, K.S. Park, K.J. Im, T. Noh, M.Y. Sung, S. Kim, S. Nahm, Y.N. Choi and S.S. Park, Appl. Phys. Lett. 80 (2002) 479.

[66] Z.L. Wang, Adv. Mater. 15 (2003) 432.

[67] K.W. Chang and J.J. Wu, Adv. Mater. 16 (2004) 545.

[68] K.W. Chang and J.J. Wu, J. Phys. Chem. B 108 (2004) 1834.

[69] K.W. Chang and J.J. Wu, J. Phys. Chem. B 109 (2005) 13572.

[70] K.W. Chang and J.J. Wu, Adv. Mater. 17 (2005) 241.

[71] J.Y. Li, Z.Y. Qiao, X.L. Chen, L. Chen, Y.G. Cao, M. He, H. Li, Z.M. Cao and Z. Zhang, J. Alloys Compd. 306 (2000) 300.

[72] R.E. Honig and D.A. Kramer, RCA Rev. 30 (1969) 285.

[73] C.W.W. Hoffman and J.J. Brown, J. Inorg. Nucl. Chem. 30 (1968) 63.

[74] J.S. Kim, H.I. Kang, W.N. Kim, J.I. Kim, J.C.Choi, H.L.Park, G.C.Kim, T.W. Kim, Y.H. Hwang, S.I. Mho, M.C. Jung and M. Han, Appl. Phys. Lett. 82 (2003) 2029.

[75] L.J. Lauhon, M.S. Gudiksen, D. Wang and C.M. Leiber, Nature 420 (2002) 57.

[76] J. Cao, J.Z. Sun, J. Hong, H.Y. Li, H.Z. Chen and M. Wang, Adv. Mater. 16 (2004) 84.

[77] H.M. Lin, Y.L Chen, J. Yang, Y.C. Liu, K.M. Yin, J.J. Kai, F.R. Chen, L.C. Chen, Y.F. Chen and C.C. Chen, Nano. Lett. 3 (2003) 537.

[78] R.M. Gibb and J.S. Anderson, J. Solid State Chem. 5 (1972) 212.

[79] S. Kamiya and R.J.D. Tilley, J. Solid State Chem. 22 (1977) 205.

CHAPTER 7

Synthesis of Hyperbranched Conjugative Polymers and Their Applications as Photoresists and Precursors for Magnetic Nanoceramics

Jacky Wing Yip Lam[a], Matthias Häußler[a], Hongchen Dong[a], Anjun Qin[a], and Ben Zhong Tang[a,b,*]

[a]*Department of Chemistry*
The Hong Kong University of Science & Technology
Clear Water Bay, Kowloon
Hong Kong, China

[b]*Department of Polymer Science & Engineering*
Zhejiang University
Hangzhou, 310027, China

7.1 Introduction

Synthesis of hyperbranched polymers is a hot research topic in recent years because of the expectation that their unique molecular architectures will impart novel properties [1,2]. Especially, those containing metal elements have aroused much interest owing to their potential as precursors for the fabrication of nanostructured materials and advanced ceramics. Compared to dendrimers, hyperbranched polymers can be readily prepared by economic one-pot, single-step polymerization procedures, accessible through different synthetic strategies [3–13]. The most commonly used approach is self-condensation polymerization of AB_x-type multifunctional monomers with $x \geq 2$, dating back to the theoretical work of Flory in 1952 [3]. This procedure is normally carried out in concurrent mode and thus leads to uncontrolled polymerization propagations with rather broad polydispersities. Slow addition of monomer or the presence of a core molecule may overcome this detrimental problem, which can control the polymer growth and the resulting polymer architecture [14–19]. Due to the limited commercial supply of, and difficult synthetic

*Corresponding author. Ben Zhong Tang. E-mail: tangbenz@ust.hk.

access to, multifunctional monomers equipped with two or more different functional groups, alternative approaches such as co-polymerization of A_2 monomers with B_n co-monomers ($n \geq 3$) have been successfully employed [20–25]. The stoichiometric requirements, however, between the pairs of the functional comonomers are practically difficult to meet, which often limit the growth of propagating species and the formation of high molecular weight products. Recently, other methods including self-condensing vinyl polymerizations, initiated by carbocationic [26] or radical species [27,28], as well as ring-opening multibranching polymerizations [29–33] have been utilized to synthesize hyperbranched macromolecules.

Attracted by the academic and application prospects, our group has worked on the development of new methodologies for the construction of hyperbranched polymers. Through elaborate efforts, we succeeded in exploring a facile approach to conjugated hyperbranched poly(alkenephenylene)s [34–40] and polyarylenes [41–50] via diyne polycyclotrimerizations catalyzed by transition metal complexes. The resulting polymers show excellent thermal stability, high photoluminescence efficiency, and strong optical nonlinearity [39–43]. The transition metal-mediated reaction is, however, intolerant of functionality and cannot be used for the syntheses of polymers bearing functional groups [43]. The reaction produces random mixtures of 1,2,4- and 1,3,5-trisubstituted benzenes, making the structures of the polymers irregular [39,40]. Recently, we succeeded in exploring a new route to hyperbranched polymers that is functionality-tolerant and regioselective. Using base-catalyzed alkyne cyclization [51–53], polymers with carbonyl functionalities, namely poly(aroylarylene)s, are synthesized in perfect 1,3,5-regioregular fashion [54–55]. In another strategy, we have used polycoupling to build hyperbranched polyynes and poly(ferrocenylene)s. In this chapter, we report our works on the synthesis of these polymers and present their applications as photoresists and precursors for magnetic nanoceramics.

7.2 Results and Discussion

7.2.1 Hyperbranched poly(aroylarylene)s

We have succeeded in synthesizing hyperbranched poly(alkenephenylene)s and polyarylenes by transition metal-catalyzed polycyclotrimerization of alkynes. All the polymers, however, possess regioirregular structures consisting of 1,3,5/1,2,4-substituted triylphenyl isomers. The structural irregularity may be an advantage in terms of the polymer's solubility and processability. This, on the other hand, makes it a challenging job to completely characterize their molecular structures by spectroscopic analyses. Furthermore, most of the transition metal catalysts are highly moisture-sensitive and intolerant of polar functional groups. Such limitations thus intrigue us to pursue new synthetic routes to construct hyperbranched polymers. In 1980, Balasubramanian *et al.* report the trimerization of aryl ethynyl ketones in refluxing *N,N*-dimethylformamide (DMF), giving 1,3,5-triaroylbenzenes in good yields [51]. Later, Matsuda and coworkers found that addition of a trace amount of diethylamine as a catalyst can further improve the results [57]. Because this reaction is simple and strictly regioselective due to the ionic mechanism involved, it has been widely used for the preparation of crystalline inclusion hosts and cyclophane ring system [52,58]. Moreover, using the cyclotrimerization of 3,5-dibenzylphenyl ethynyl ketone as a key step, partially dendritic molecules are prepared as precursors of high-spin hexacarbenes [59]. Because of the unique advantages of base-catalyzed alkyne polycyclization, we thus explored the possibility of utilizing this reaction to synthesize hyperbranched polymers from aroyldiynes bearing functional groups.

Monomer synthesis and polymerization. Different types of organic and organometallic alkyne monomers with carbonyl functionalities imbedded between the acetylenic and aromatic moieties are designed and synthesized (Scheme 1). Most of the aroylacetylene

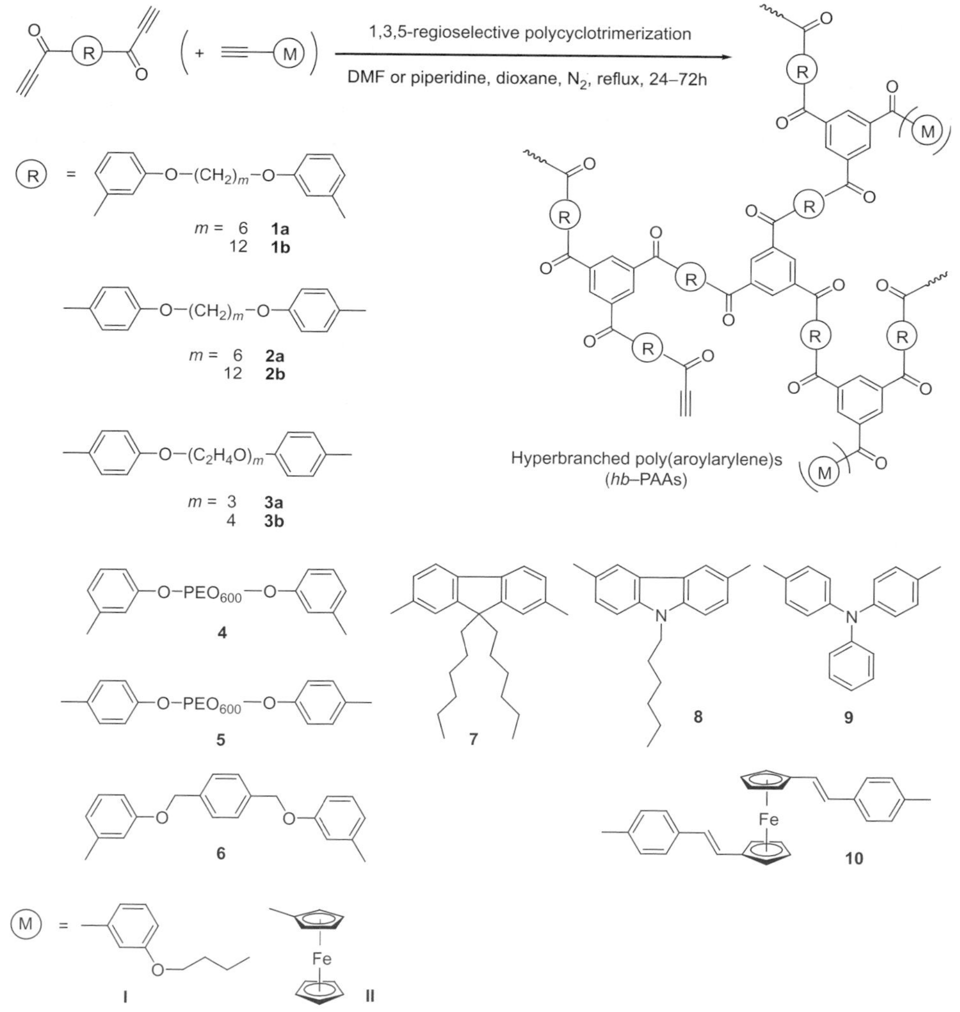

1,3,5-regioselective polycyclotrimerization

DMF or piperidine, dioxane, N_2, reflux, 24–72h

R = —O—$(CH_2)_m$—O—

m = 6 **1a**; 12 **1b**

—O—$(CH_2)_m$—O—

m = 6 **2a**; 12 **2b**

—O—$(C_2H_4O)_m$—

m = 3 **3a**; 4 **3b**

—O—PEO_{600}—O— **4**

—O—PEO_{600}—O— **5**

6 **7** **8** **9** **10**

Hyperbranched poly(aroylarylene)s (*hb*–PAAs)

M = **I** **II**

Scheme 1. Base-catalyzed 1,3,5-regioselective polycyclotrimerization of bis(aryl ethynyl ketone)s.

monomers are accessible through etherification of *m*- or *p*-hydroxybenzaldehyde with alkyl or aryl dibromide, followed by Grignard addition and alcohol oxidation [54–56]. All the monomers are fully characterized spectroscopically, from which satisfactory analysis data corresponding to their molecular structures are obtained.

Since aroylacetylenes readily cyclotrimerize when refluxed with DMF or in the presence of diethylamine [51–53], we thus first tried to polymerize our monomers in DMF or its mixtures. The catalyst for the reaction comes from a trace amount of dimethylamine formed by the decomposition of DMF at high temperatures [53]. The polymerization is normally performed at high temperatures for a long period of time (up to 72 h) to ensure complete consumption of monomers and encourage the formation of high molecular weight polymers. While the reaction carried out in the presence of diethylamine produces polymers in low yields (7–21%) [55], good to excellent yields are obtained when the solvent is changed to DMF/tetralin (Table 7.1, nos. 1–6). The polymerizations of **2b** and **6** in pure DMF, however, give polymers in lower yields. Attempts to shorten the reaction time and increase the polymer yield by employing diphenylamine as a catalyst have

Table 7.1. Polycyclotrimerization of bis(aryl ethynyl ketone)s.

No.	Polymer	Yield (%)	M_w^a	M_w/M_n^a
Polymerization in the presence of in-situ generated catalyst from DMF[b]				
1	**P1a**(D)	71	12 700	3.5
2	**P2a**(D)	73	10 800	3.8
3	**P2b**(D)	45	7 200	2.0
4	**P6**(D)	67	5 400	2.7
5	**P7**(D)	84	46 100	5.1
6	**P8**(D)	96	10 100	2.3
Polymerization with piperidine as the externally added catalyst[c]				
7	**P1a**	99	14 500	5.3
8[d]	**P1a/II-1**	67	22 900	4.3
9[d]	**P1a/II-2**	54	15 000	3.3
10[d]	**P1a/II-3**	55	11 400	3.0
11	**P2a**	60	7 900	4.4
12	**P2b**	95	24 300	6.0
13	**P3a**	90	11 500	3.6
14	**P3b**	99	15 300	4.0
15	**P5**	65	6 100	1.7
16	**P6**	85	15 000	4.0
17	**P6/II**	64	7 600	3.2
18	**P7**	66	10 800	2.6
19	**P8**	78	8 500	2.2
20	**P9**	86	11 300	2.8
21	**P10**	52	9 100	2.8
22[d]	**P10/I**	43	10 400	3.4

[a]Determined by SEC in THF on the basis of a polystyrene calibration.
[b]Reflux under nitrogen in tetralin/DMF (1:1 v/v) for 72 h. $[M_0] = 0.14$ M.
[c]Carried out in chlorobenzene (no. 7), dioxane (nos. 8–12 and 15–22) or dichloroethane (nos. 13 and 14) for 24–48 h under nitrogen. $[M_0] = 0.14–0.18$ M. [piperidine] = 0.3 mole equivalent.
[d]Copolymerization with monomer ratio of 1:1 (no. 8), 1:2 (no. 9), 1:3 (no. 10) or 1:0.75 (no. 22).

failed. Delightfully, when piperidine, a secondary amine, is used, polymers are produced in much shorter reaction times and higher yields (Table 7.1, nos. 7–22). For example, **P1a** and **P3b** are formed in nearly quantitative yields in only 24 h.

Molecular structure. The structures of the polymers are characterized by spectroscopic methods with satisfactory results. An example of the ^{1}H NMR spectrum of **P1a** is given in Fig. 7.1a. The spectrum corresponds well to its expected molecular structure, with no peaks unassignable. The acetylene proton of its monomer **1a** absorbs at δ 3.50 (a), which completely disappears after polymerization. Meanwhile, new peaks emerge at δ ~7.9–8.5, corresponding to the proton absorptions of the triaroylbenzene groups formed by the alkyne polycyclotrimerization. Compare the spectrum of **P1a** with that of a model compound, the peak at δ ~8.33 (i) is assigned to the protons of the new benzene rings formed in the polymerization. The peaks at δ 8.43 (k) and 7.91 (j) arise from the resonances of tribenzoylphenyl protons of the cyclophane formed by the end-capping of a triple bond in a polymer branch by two triple bonds in a monomer [52]. It should be noted that the end-capping reaction sometimes does not go to completion, thus leaving a

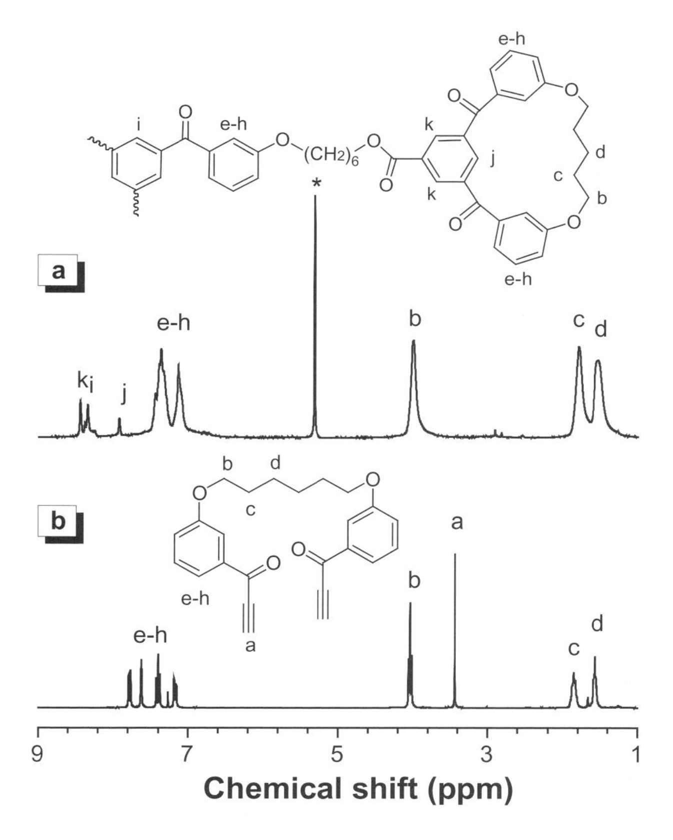

Fig. 7.1. ^{1}H NMR spectra of (a) hyperbranched poly(aroylarylene) **P1a** and (b) its monomer **1a** in dichloromethane-d_2.

few unreacted alkyne terminal groups in the polymers. Obviously, the regioregular structures of the polymers formed by the polycyclotrimerization of bis(aryl ethynyl ketone)s make the spectroscopic analyses much easier and simpler.

Photo-cross-linking. Benzophenone has been introduced into natural (e.g. protein) and synthetic polymers (e.g. polyimide) to serve as a photo-cross-linker [60,61]. The hyperbranched poly(aroylarylene)s contain many triaroylbenzene units and may thus show high photo-cross-linking efficiencies. This is indeed the case: for example, a thin film of **P6** on a glass plate can be readily cross-linked by irradiation from a hand-held UV lamp at room temperature. The cross-linking may have proceeded via the well-established radical mechanism [60–62]: one carbonyl moiety abstracts hydrogen from another benzyl unit, creating a stable benzyl radical. Subsequently, coupling of different radicals leads to cross-linking and hence, the gel formation.

Figure 7.2a depicts the dose effect on the gel formation of **P1a** and **P6** films after they have been exposed to a weak UV light ($\sim$1 mW/cm^2). Though the photo-cross-linking conditions have not been optimized, both the hyperbranched poly(aroylarylene)s already exhibit much higher sensitivities ($D_{0.5}$ = 50–180 mJ/cm^2) than commercial poly(amic ester) photoresists (650–700 mJ/cm^2) [63]. Compared with **P1a**, **P6** exhibits lower $D_{0.5}$ and higher $\gamma_{0.5}$ (contrast) [64,65]. This is possibly due to the faster reduction of the ketone group in **P6**, thanks to the ease in abstracting hydrogen from its benzyl unit. This hypothesis is supported by the fact that the intensity of the *K* band of the triaroylbenzene units of **P6** drops much faster than that of **P1a** [55].

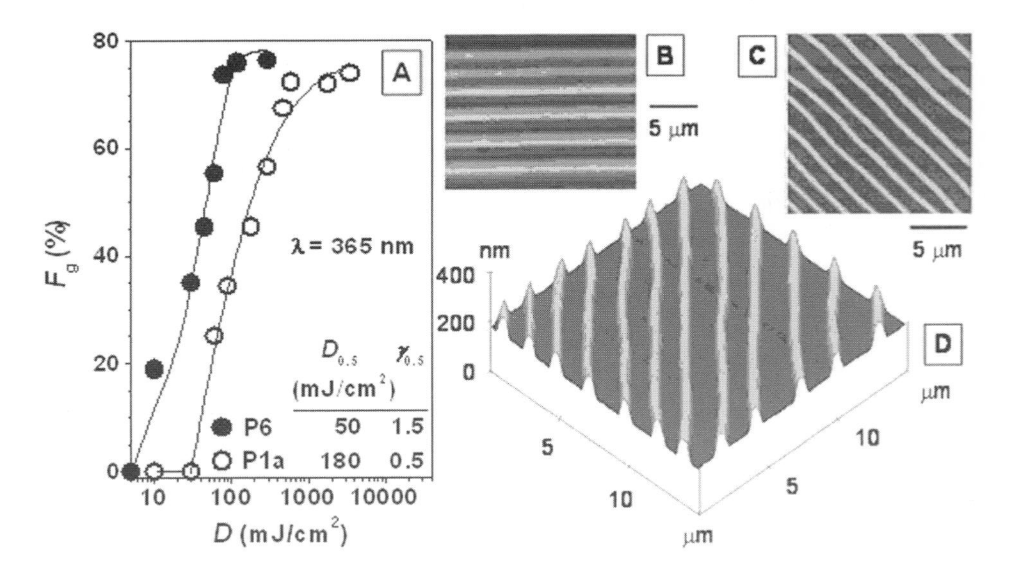

Fig. 7.2. (a) Plots of gel fractions (F_g) of **P1a** and **P6** films vs. exposure doses (D) and AFM images of (b) micro- and (c and d) nano-scale patterns obtained from the **P6** films exposed to 1 J/cm^2 of UV irradiation. Samples taken from Table 7.1, no. 7 and 16.

Well-resolved patterns with line widths of 1.0 and 1.5 μm are readily obtained when a film of **P6** has been exposed to a UV dose of 1 J/cm^2 (Fig. 7.2b). Patterns with submicron resolutions (line width down to 500 nm) are also achievable, as demonstrated by the examples given in Fig. 7.2 (panels C and D). Clearly, **P6** is an excellent photoresist material.

Thermal stability and pyrolytic ceramization. The thermal properties of the polymers are investigated by thermogravimetric analysis (TGA) under nitrogen. Figure 7.3 shows the TGA thermograms of a few polymers. Thanks to their rigid structures, all the polymers are thermally very stable and lose only ~5% of their weights when heated to temperatures as high as 440°C. The amount of residue after pyrolysis at 800°C is high for all polymers, suggesting that they are promising precursor candidates for ceramics (Table 7.2).

We have previously found that hyperbranched organometallic polymers, in comparison to their linear counterparts, are better precursors of magnetic ceramics in terms of ceramization yield and magnetic susceptibility, because the three-dimensional cages of hyperbranched polymers enable better retention of pyrolyzed species and steadier growth of magnetic crystallite [66,67]. Because **P1a/II-1**, **P1a/II-2**, **P1a/II-3**, **P6a/II**, and **P10** contain ferrocene moieties, it is envisioned that their ceramics would show intriguing magnetic properties. Pyrolizing the polymers in a tube furnace at 1000°C for 1 h under a stream of nitrogen gives black ceramics **C1**–**C5** in 30–58% yields, which are in good agreement with their TGA results (Scheme 2).

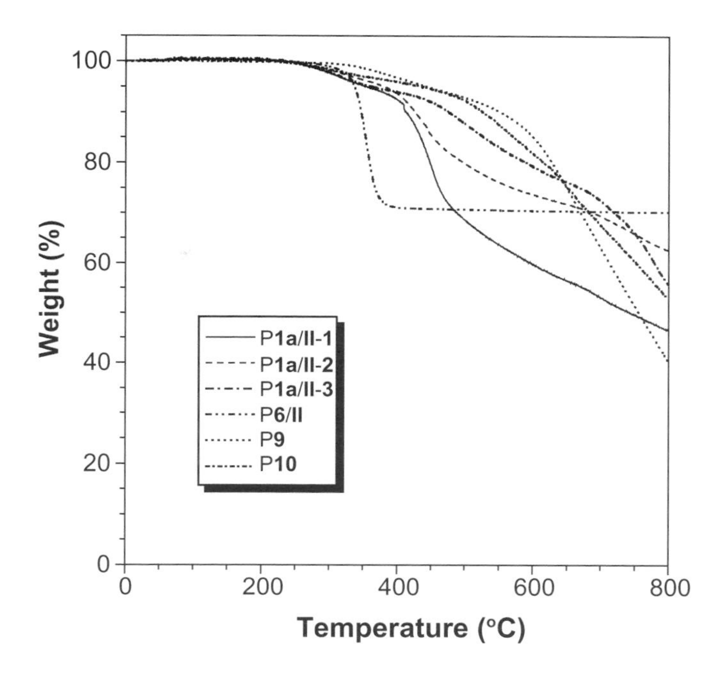

Fig. 7.3. TGA thermograms of **P1a/II–1**, **P1a/II–2**, **P1a/II–3**, **P6a/II**, **P9**, and **P10** recorded under nitrogen at a heating rate of 20°C/min. Samples taken from Table 7.1, no. 8, 9, 10, 17, 20, and 21.

Table 7.2. Thermal properties of hyperbranched poly(aroylarylene)s (*hb*-PAAs) and magnetic properties of their ceramics[a].

hb-PAA	T_d^b (°C)	W_r^c (%)	Ceramic	Yield (wt %)	M_s^d (emu/g)	H_c^e (kOe)
P1a/II-1	355	43.5	**C1**	29.5	11.9	0.07
P1a/II-2	381	58.9	**C2**	32.8	13.2	0.15
P1a/II-3	336	70.1	**C3**	57.1	20.0	0.33
P6/II	363	46.0	**C4**	57.9	10.3	0.12
P10	434	52.9	**C5**	58.1	22.2	0.13

[a] Fabricated by pyrolysis at a high temperature (1000°C) in nitrogen atmosphere.
[b] Degradation temperature (T_d) for 5% weight loss.
[c] Weight residue (W_r) at 800°C.
[d] M_s = saturation magnetization in an external field of 10 kOe.
[e] H_c = coercivity at zero magnetization.

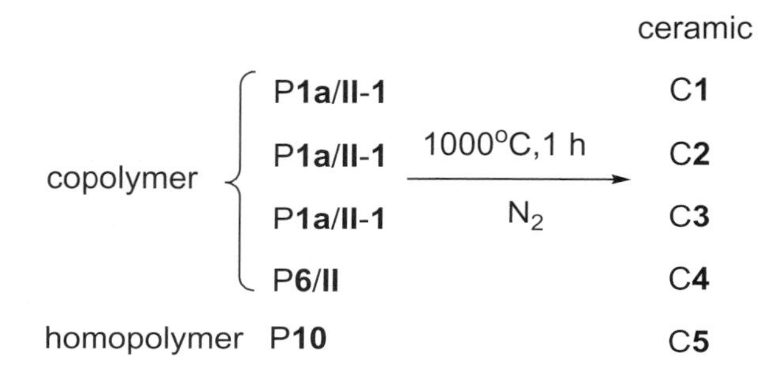

Scheme 2. Ceramization of hyperbranched poly(aroylarylene)s to magnetic ceramics.

To get a first impression of the resulting metalloceramics, we used scanning electron microscopy (SEM) to examine their surface morphologies. The SEM measurements can be carried out without gold coating. The clear images of the SEM photomicrographs indicate that the ceramics are electrically conductive and contain conductive graphite carbon and metallic iron species. As shown in Fig. 7.4, the ceramics produced at 1000°C under nitrogen are mesoporous, revealing that the pyrolytic decomposition first strips off some of the organic moieties from the skeletons of the hyperbranched polymer spheres and the following fast evaporation of the volatile fragments at high temperatures leaves behind the mesoscopic pores. In the meantime, the naked reactive inorganic residues agglomerate into bulky clusters [67]. Ceramics **C2** and **C3** show relatively smooth surfaces (Fig. 7.4b and c), whereas surfaces decorated with nano- and microcrystallizes are observed in **C1** and **C4** (Fig. 7.4a and d) [68].

Compositions. X-ray photoelectron spectroscopy (XPS) is a powerful tool for the estimation of surface compositions. XPS analyses reveal that the ceramics obtained from the pyrolysis of hyperbranched poly(aroylarylene)s contain the expected atomic compositions of carbon, iron, and oxygen elements (Table 7.3). The surfaces of the ceramics are mainly comprised of carbon (as high as ~95.6 atom%) with a small amount of iron.

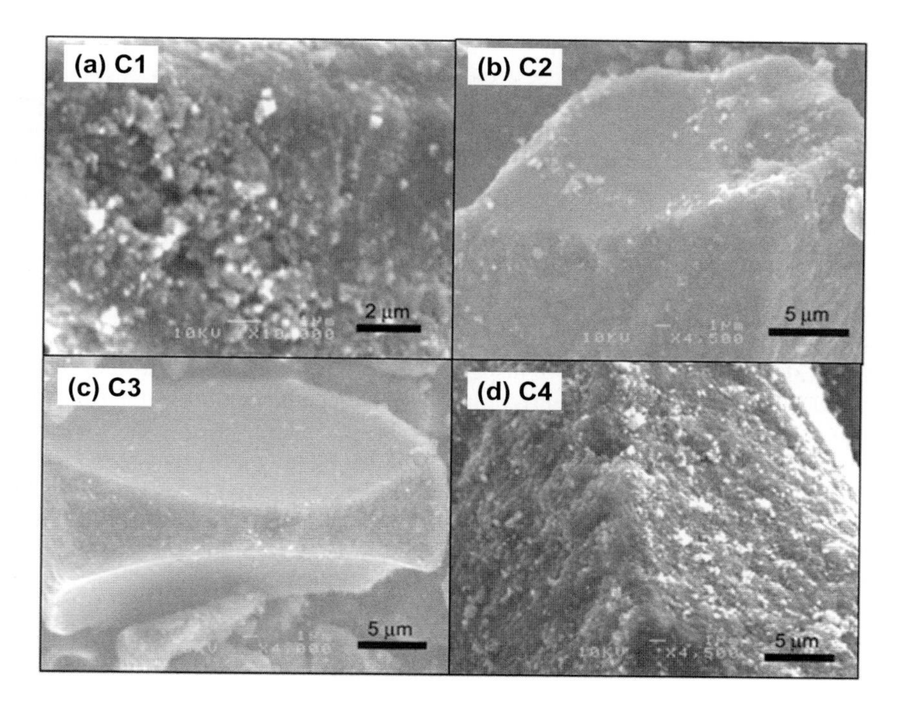

Fig. 7.4. SEM images of ceramics **C1**–**C4**.

Table 7.3. Atomic composition of ceramics **C1**–**C5** estimated by XPS and EDX analyses.

	Atomic composition (%)		
Ceramic	C	Fe	O
C1 (XPS)	94.52	1.41	3.97
C1 (EDX)	82.73	4.01	13.26
C2 (XPS)	95.30	1.10	3.48
C2 (EDX)	90.41	6.53	3.06
C3 (XPS)	95.47	1.06	3.36
C3 (EDX)	87.47	9.81	2.72
C4 (XPS)	95.66	0.84	3.28
C4 (EDX)	91.65	4.10	4.25
C5 (XPS)	94.52	1.41	3.97
C5 (EDX)	85.02	11.64	3.34

A considerable amount of oxygen (up to ~37 atom%) is detected on the surface. Oxygenic species have often been found in the ceramic products prepared by the polymer precursor routes, although the pyrolyses have been conducted in sealed quartz tubes under inert-gas atmosphere or in tube furnaces under nitrogen steam [69,70]. Since the polymer precursors contain carbonyl functionalities, it is understandable why oxygenic species are found in the ceramics. Bulk compositions of the ceramics are analyzed by energy-dispersion X-ray (EDX) spectroscopy. As can be seen from the data listed in Table 7.3, compared to the values obtained from the XPS surface analyses, the content of C in the bulk generally decreases, whereas the amount of Fe generally increases. This gradient distribution from the bulk to the surface reveals that the ceramization process starts from the pyrolysis-induced formation and aggregation of the iron species inside the organic matrix. The ceramics may thus be imagined as iron inner cores covered by carbonic and oxide outer mantles.

The valence shell electrons are known to contribute to the net force experienced by the core-level electrons, and the core-level binding energies thus change with the variation in the chemical environments [71]. We inspected the Fe2p core level photoelectron spectra of **C1** and **C4** in order to gain an insight into the chemical structures of the iron species on the surfaces of the ceramics. Two major peaks at ~724.6 and 711.0 eV associated with the $Fe2p_{1/2}$ and $Fe2p_{3/2}$ core-level binding energies of Fe_2O_3, respectively, are observed in the spectra (Fig. 7.5) [72,73]. The spectra of **C1** and **C4** also show a small peak at 707.4 eV associated with the absorption of Fe. The Fe2p photoelectron peaks are rather broad, probably due to the unresolved multiple splitting of the core electron peaks caused by the exchange interaction of the unpaired valence electrons [68].

To gain more information on the compositional species in the ceramic materials, X-ray diffraction (XRD) patterns of the ceramics are measured. Although the hyperbranched

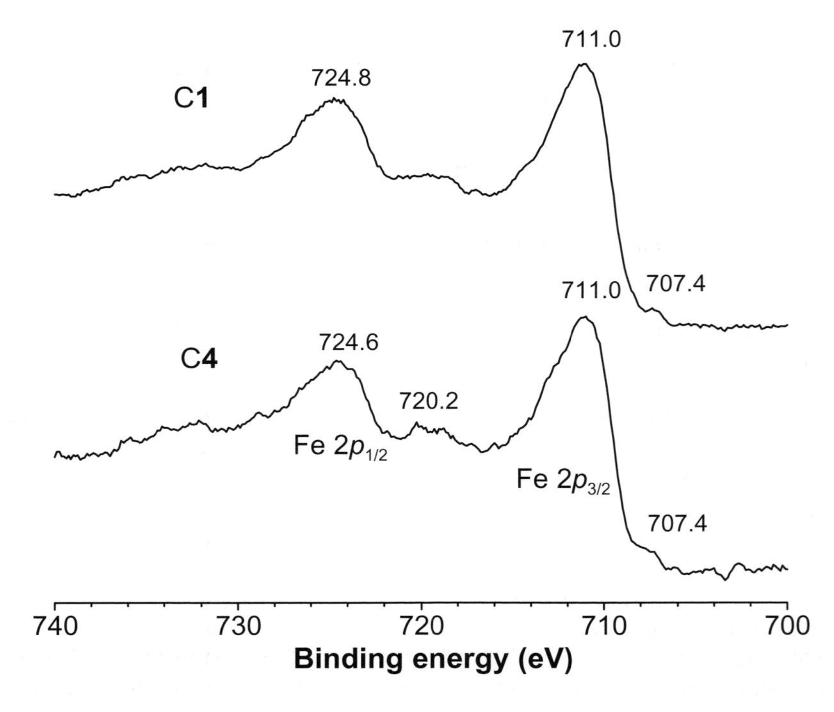

Fig. 7.5. Fe2p photoelectron spectra of ceramics **C1** and **C4**.

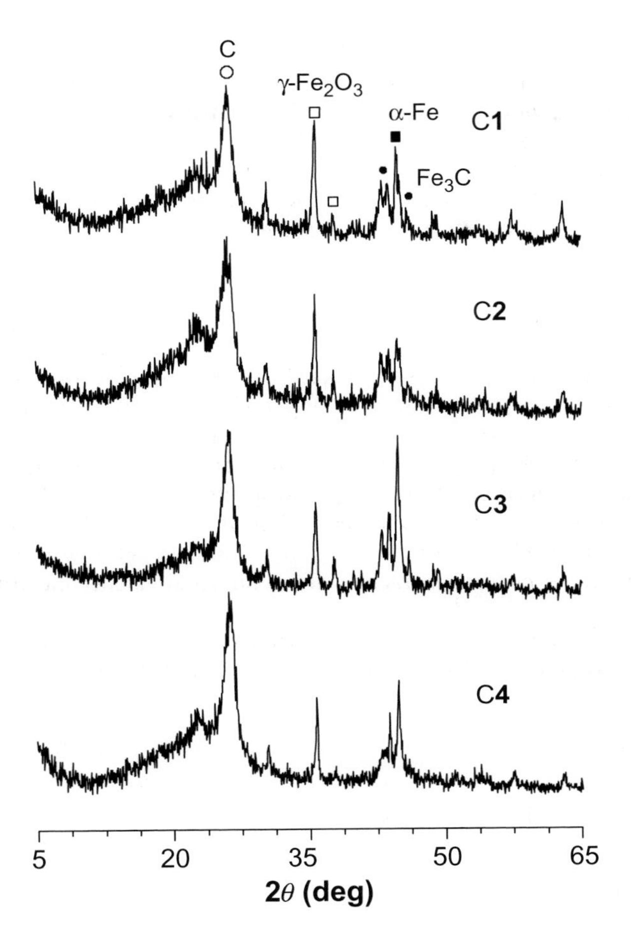

Fig. 7.6. XRD diffractograms of ceramics **C1**–**C4**.

polymer precursors are amorphous in nature, their ceramic products show diffraction patterns with numerous Bragg reflections (Fig. 7.6). We used the data files in the databases of the Joint Committee on Powder Diffraction Standards of the International Center for Diffraction Data (JCPDS-ICDD) to identify the reflections. The results of ceramic **C3** are summarized in Table 7.4 as an example. All the ceramics display characteristic diffraction patterns of γ-Fe_2O_3 and iron carbide Fe_3C. Using the full widths at half-maximums of the refection peaks, the sizes of γ-Fe_2O_3 particles calculated by Scherrer equation in **C1**–**C4** are ~22–37 nm [74]. The Fe_3C crystals should be small in size, as evidenced by their broad lines in the XRD diagrams. The iron carbide cannot grow in large size because it is an intermediate formed during the graphitization of amorphous carbon by iron catalyst at high temperatures [74]. The existence of Fe_3C suggests that the process of graphitization has not completed. Actually, we have found an intense peak associated with the (002) reflection of graphite at $2\theta = 26.4°$ ($d = 3.38$ Å) in the XRD diffractograms of all ceramics. The diffraction diagrams of **C3** and **C4** show an intense reflection peak

Table 7.4. Nanocrystals in ceramic **C3** identified by XRD analysis.

No.	Crystal	2θ (deg)/d spacing (Å)[a]	ICDD file
1	C	26.25/3.39 (3.35)	26–1080
2	Fe_2O_3	30.45/2.93 (2.95)	39–1346
		35.70/2.51 (2.52)	
3	Fe_3C	37.80/2.38 (2.38)	34–001
		43.75/2.07 (2.07)	
4	Fe	44.75/2.02 (2.03)	06–0696

[a]The value given in the parentheses are taken from the powder diffraction files of the database of the International Center for Diffraction Data (ICDD).

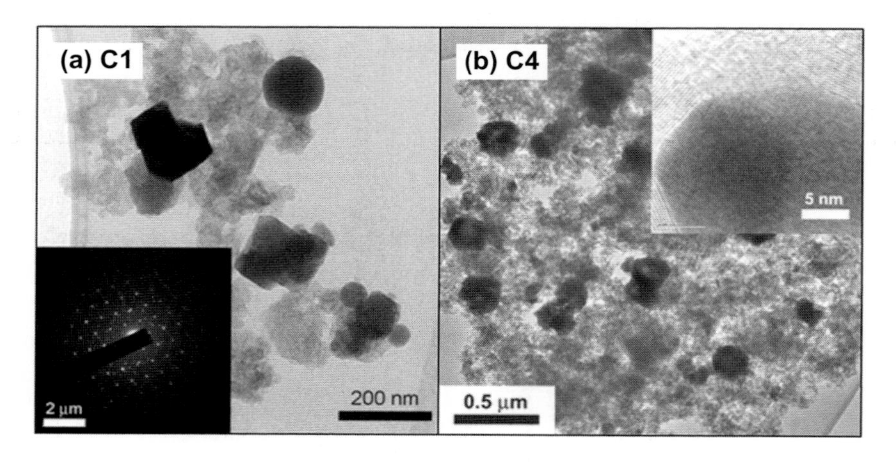

Fig. 7.7. TEM images of ceramics (a) **C1** and (b) **C4**. Inset: ED spectrum of **C1** and enlarged portion of TEM image of **C4**.

of α-Fe crystal at 44.70° (2.0 Å). The secondary order reflection, however, cannot be detected, probably due to imperfect packing structure of the nanocrystals. The crystal sizes of α-Fe in **C3** and **C4** are estimated to be 23 and 17 nm, respectively. It is difficult to detect the presence of α-Fe in the spectra of **C1** and **C2** because of its low concentration and small ceramic size.

The structures of the ceramics are studied by high-resolution transmission electron microscopy (HRTEM). Figure 7.7 shows typical TEM images of ceramics **C1** and **C4** and their corresponding electron diffraction (ED) patterns. As we can see, the iron-rich nanocrystallites (dark grains) are well embedded in the abundant graphite matrix (light areas). Interestingly, the sizes of most iron particles are ~100–200 nm, much larger than those calculated from the XRD diffractograms. This suggests that the nanocrystallites are easily clustered together during the ceramization process. The ED pattern of one iron particle (Fig. 7.7a) reveals that it contains more than one nanocrystallite with perfect packing structures. From the enlarged portion of Fig. 7.7b, the graphite is identified with a lattice spacing of 3.4 Å. The graphite ribbon appears to encapsulate the Fe particles as a shell, which may act as protective coating toward oxidation of the nanoparticles.

Magnetism. Because the ceramics contain nanoscopic iron species, it is expected that they may show novel magnetic properties. Indeed, all the ceramic products can be readily attracted to a bar magnet at room temperature; that is, they are readily magnetizable. The magnetization behaviors of the ceramics are studied using superconducting quantum interference device (SQUID) techniques. As shown in Fig. 7.8, **C1** is swiftly magnetized at 300 K, as evidenced by an immediate raise in magnetization upon its exposure to a magnetic field. Its magnetization rapidly increases with an increase in the strength of the applied field and becomes saturated at ~12 emu/g when the external field reaches ~7 kOe. While γ-Fe_2O_3, α-Fe, and Fe_3C are all well-known magnetic materials, if comparison is made, taking into consideration the iron content present in the samples, the relatively low M_s values of the ceramics are understandable [75,76]. Actually, the magnetization does increase when the Fe atom% in the ceramic becomes higher. Among the ceramics, **C3** with up to 9.8 atom% of iron has the highest M_s value (~20 emu/g).

The hysteresis loops of the magnetoceramics are small. From the enlarged *H*–*M* plots shown in the inset of Fig. 7.8, the coercivities (H_c) of **C1** and **C2** are found to be

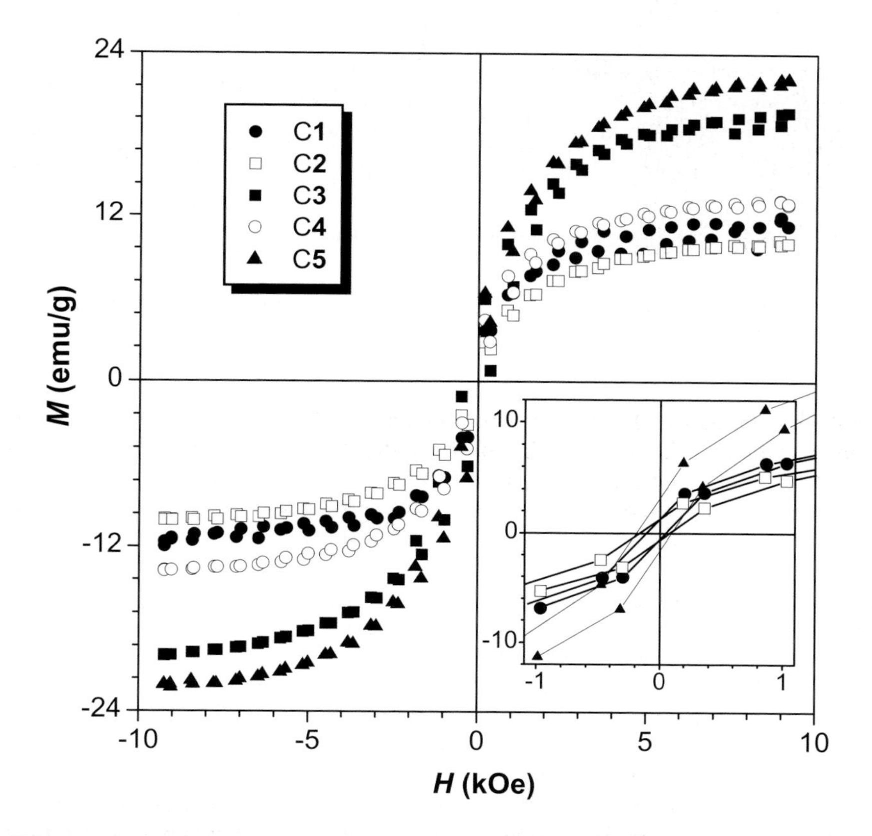

Fig. 7.8. Plots of magnetization (*M*) vs. applied magnetic field (*H*) at 300 K for magnetoceramics **C1**–**C5**. Inset: enlarged portions of the *M*–*H* plots in the low strength region of the applied magnetic field.

Fig. 7.9. Optical micrographs of photopatterns of **P10** fabricated by UV photolithography before (left) and after (right) pyrolysis (1000°C, 1 h) under nitrogen.

0.075 and 0.120 kOe, respectively. Low coercivity is a general property of this new family of magnetic ceramics. Since a ferromagnetic material with coercivity smaller than 0.126 kOe (or 10^4 A/m) is termed a soft magnet, most of our ceramics are thus good soft ferromagnetic materials, which may find high-tech applications in various electromagnetic systems [77–80].

Ceramic pattern. The previous section shows that **P1a** and **P6** can be readily crosslinked by UV irradiation to give patterns with nanometer resolutions. Is it possible for the ferrocene-containing copolymers to be used as resists for UV photolithography? The answer is a firm yes. Negative-tone patterns with micron resolutions (line width less than 10 μm) are readily achieved when a thin film of **P10** is exposed to UV irradiation, as demonstrated by the example given in Fig. 7.9 (left panel). Pyrolysis the pattern at 1000°C under nitrogen for 1 h fabricates magnetic ceramic lines with good shape retention (Fig. 7.9; right panel). The rough surfaces of the lines indicate a transformation of the surface topologies of the patterned area from a uniform thin film to a collection of small ceramic islands. The composition of these islands is probably comparable with that of ceramic product **C4** in the bulk, which contains iron-rich nanocrystallites well embedded in the abundant graphite matrix. It shows a good example of combining conventional lithographic techniques with easily processable organometallic polymer resists [81].

7.2.2 Hyperbranched poly(ferrocenylene)s

We have used desalt coupling for the synthesis of hyperbranched organometallic poly(ferrocenylenesilyne)s [66,67] and found that the ceramics prepared from pyrolysis of the polymers show outstanding soft ferromagnetic behaviors with practically no hysteresis loss. Encouraged by these results, we have extended our research efforts on the integration of other groups of 14 and 15 elements including germanium, phosphorus, and antimony into the poly(ferrocenylene) framework.

Polymer synthesis. All the polymers are prepared by reaction of dilithioferrocene ($FcLi_2$) with various metalloid chlorides under vigorously dried and strictly controlled polymerization conditions (Scheme 3). $FcLi_2$ readily reacts with $PhSiCl_3$, furnishing poly[1,1′-ferrocenylene-(phenyl)silyne] [**11**(Ph)] in a high yield (~62%, Table 7.5, no. 1). The resulting polymer is, however, only partially soluble, with a molecular weight of the THF-soluble fraction reaching an M_w value of ~3100. It is known that the size-exclusion chromatograph (SEC), when calibrated with standards of linear polymers (in this case polystyrene), often underestimates the molecular weights of hyperbranched polymers. The determination of the absolute molecular weight of a previously prepared but structurally similar hyperbranched poly[1,1′-ferrocenylene(*n*-octadecenyl)silyne] by an SEC system equipped with light-scattering and viscometer detectors gives an M_w value of 5×10^5, which is much higher than the relative M_w value estimated by the SEC system calibrated with the linear polystyrene standards (M_w value of ~12000) [82]. Thus, the molecular weights of the hyperbranched polymers shown in Table 7.5 might be dramatically underestimated. Other monomers also show high polymerizability and hyperbranched poly(ferrocenylene)s containing other metalloid species of Si, Ge, P, and Sb are successfully prepared, with the reaction of $FcLi_2$ with PCl_3 furnishing a polymer

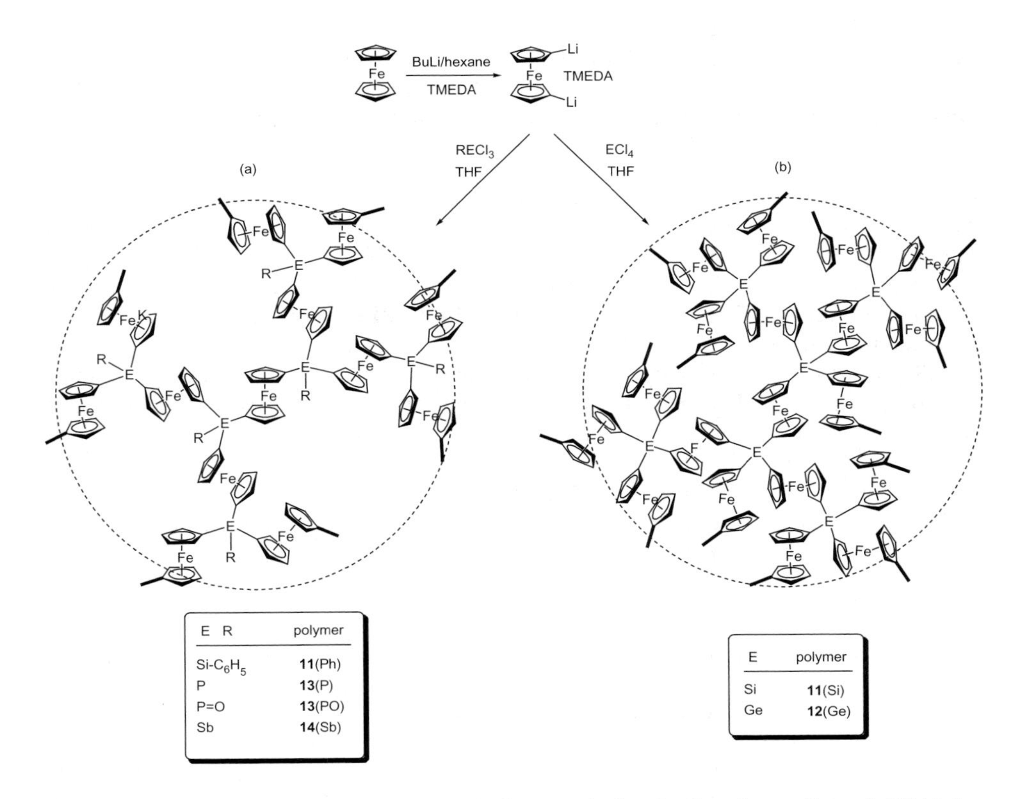

Scheme 3. Synthesis of hyperbranched poly(ferrocene)s by desalt polycoupling of dilithioferrocene with (a) trichlorides of silicon, phosphorus and antimony and with (b) tetrachlorides of silicon and germanium.

Table 7.5. Synthesis of hyperbranched poly(ferrocenylene)s containing groups 14 and 15 elements[a].

No.	Polymer	Yield (wt %)	Solubility[b]	M_w^c	M_w/M_n^c	Appearance
1	**11**(Ph)	62.0	Δ	3 100[d]	3.8[d]	brown powder
2	**11**(Si)	44.5	Δ	870[d]	1.7[d]	brown powder
3	**12**(Ge)	76.5	Δ	1 150[d]	1.9[d]	amber powder
4	**13**(P)	81.7	×			dark brown powder
5	**13**(PO)	33.0	×			dark brown powder
6	**14**(Sb)	73.4	×			golden yellow powder

[a]Carried out under nitrogen in THF at room temperature for 16 h.
[b]Tested in common solvents such as dichloromethane, chloroform, toluene, dioxane, acetone, THF, DMF, and DMSO; symbols: × = insoluble, Δ = partially soluble, and √ = completely soluble.
[c]Estimated by SEC in THF on the basis of a polystyrene calibration.
[d]For the THF-soluble fraction.

with the highest yield of ~82%. Unfortunately, all the polymers are only partly soluble or even completely insoluble.

Structural characterization. Since some of the hyperbranched poly(ferrocenylene)s are partially or completely insoluble, we can only characterize their molecular structures by IR spectroscopy. The overtone and combination bands of the ferrocene moieties (Fc) in the range of 1750–1600 cm^{-1} as well as their vibration and deformation bands with peak maximums at ~1036 and ~831 cm^{-1} are readily observed in the spectra of the polymers (Fig. 7.10) [83,84]. Additionally, the metalloid-ferrocene (Efc) bands appear at 1421 and 1166 cm^{-1}, 1431 and 1151 cm^{-1}, 1420 and 1183 cm^{-1}, and 1411 and 1136 cm^{-1} for the Si-, Ge-, P-, and Sb-containing polymers, respectively. These spectroscopic assignments confirm that the structures of the polymers consist of ferrocene and metalloid moieties. Furthermore, no absorption bands associated with such structures as silane (Si–H) or siloxane (Si–O–Si) are observed at ~2200 and ~1100 cm^{-1} [83].

Ceramization. The linear poly(ferrocenylenesilene)s have been utilized as polymer precursors to ceramics [85,86]. Will the hyperbranched poly(ferrocenylene)s play the same role? Investigations of thermolysis behaviors by thermogravimetric analysis (TGA) reveal that most of the hyperbranched polymers are thermally stable, losing little of their weights when heated to ~350–400°C. The polymers undergo a rapid thermolytic degradation in the temperature region of ~400–550°C, after which, the TGA curves are almost leveled off. Little further weight losses are recorded when the polymers are further heated to ~680°C, indicating that the polymers have been ceramized by the pyrolysis processes. Indeed, heating the polymers to high temperatures and then isothermally sintering them in an inert-gas atmosphere afford ceramic products in up to ~69 wt% yields (Table 7.6). The high char yields can be attributed to the outstanding thermal stability of the hyperbranched polymer frameworks and to the better retention of the pyrolysis-generated species inside the three-dimensional macromolecular structures.

Ceramic compositions. We first analyzed the ceramic compositions by XPS. XPS analyses reveal that the ceramics obtained from the pyrolysis of hyperbranched poly(ferrocenylene)s contain the expected carbon and iron atoms as well as the respective elements of groups 14 or 15 (Table 7.7). The surfaces of the ceramics are mainly

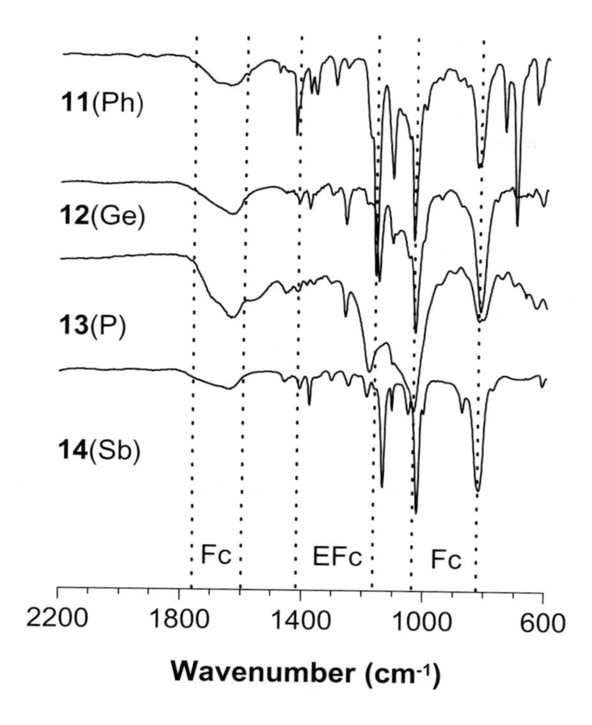

Fig. 7.10. IR spectra of hyperbranched poly(ferrocenylene)s containing groups 14 and 15 elements; E = element; Fc = 1,1′-ferrocenylene.

Table 7.6. Pyrolytic ceramization of hyperbranched poly(ferrocenylene)s

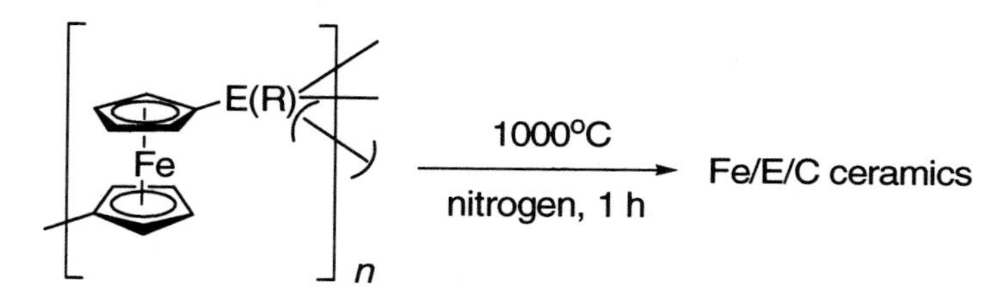

E	Ceramic	Yield (wt%)
SiC_6H_5	C**6**(Ph)	57.4
Si	C**6**(Si)	60.4
Ge	C**7**(Ge)	40.6
P	C**8**(P)	68.8
PO	C**8**(PO)	62.5
Sb	C**9**(Sb)	25.2

Table 7.7. Ceramic compositions[a].

No.	Ceramic[a]	XPS (atom %)					EDX (atom %)			
		C	N	O	E	Fe	C	O	E	Fe
1	**C6**(Ph)	81.2	2.6	10.5	4.8	0.9	77.3	10.3	4.3	8.0
2	**C6** (Si)	45.0	0.6	36.7	16.1	1.6	45.2	38.2	12.7	3.9
3	**C7**(Ge)	78.7	0	16.7	1.5	3.1	69.0	15.4	10.2	5.4
4	**C8**(P)	86.4	0	10.1	2.6	0.9	77.4	14.4	3.4	4.8
5	**C8**(PO)	83.5	0	13.1	2.6	0.6	76.9	12.4	4.4	6.3
6	**C9**(Sb)	59.3	0.1	27.3	3.5	9.8	68.3	17.4	8.1	6.2

[a]All ceramics were fabricated by pyrolysis at 1000°C under an atmosphere of nitrogen.

comprised of carbon (up to ~87 atom %) with small amounts of iron (0.6–9.8 atom %) and groups 14 or 15 elements (0.6–16.1 atom %). A considerable amount of oxygen (up to ~37 atom %) is detected on the surface. Since the polymer precursors, except for **C8**(PO), contain no oxygen atoms, the oxygenic species may thus be introduced by moisture absorbed by the polymer samples prior to pyrolysis and/or by postoxidation of the ceramics during the handling and storage processes. The amount of iron as well as groups 14 and 15 elements in the bulk detected by EDX are much higher than those on the surface, while the contents of C and O become lower. This suggests that the ceramization process starts from the formation of iron nanocluster inner core, onto which other ceramic species are deposited along with the pyrolytic processes.

The Fe2p core level photoelectron spectra reveal two sets of $Fe2p_{1/2}$ and $Fe2p_{3/2}$ splitted peaks (Fig. 7.11), which can be assigned to Fe (720 and 707 eV) and Fe_2O_3 species (725.3 and 711.4 eV) [72,73]. The ratio of the species varies in different ceramics. The Si- and Sb-containing ceramics contain almost exclusively Fe_2O_3 species, whereas the Ge- and P-containing ceramics contain both the species. The peaks associated with the Fe species are noticeably higher in the spectra of ceramics **C8**(P) and **C8**(PO). Additionally, for most of the Si-containing ceramics, small amounts of nitrogenic species (up to 2.6 atom %) have been detected. It has been reported that amorphous Si_3N_4, Fe_2N, and Fe_4N species have been formed in the ceramization processes of the linear poly(ferrocenylenesilene)s due to the incorporation of nitrogen from the pyrolysis atmosphere into the ceramic products [87].

To gain more information on the compositions of the ceramic materials, we carried out the XRD analysis. Although the hyperbranched polymer precursors are amorphous in nature, their ceramic products exhibit well-defined XRD patterns. The Bragg reflection peaks are identified according to the database files of the Joint Committee on Powder Diffraction Standards of the International Center for Diffraction Data (JCPDS-ICDD). All of the Si-containing ceramics display diffraction patterns characteristic of α−Fe and carbon steel (Fe,C; Fig. 7.12). The presence of the latter is rather unusual, as its *hkl* values suggest a face-centered-cubic (fcc) phase of γ−Fe at room temperature. While α−Fe is stable and crystallizes in a body-centered-cubic (bcc) phase, γ−Fe is normally unstable and is hardly found in ceramics prepared from the pyrolysis of linear poly(ferrocenylenesilene)s [87]. It is possible that the carbonic nanostructures have helped to induce the crystallization of the γ−Fe phase inside the ceramics. More detailed studies are underway in our laboratories to clarify the appearance of this unusual phase in the ceramic products.

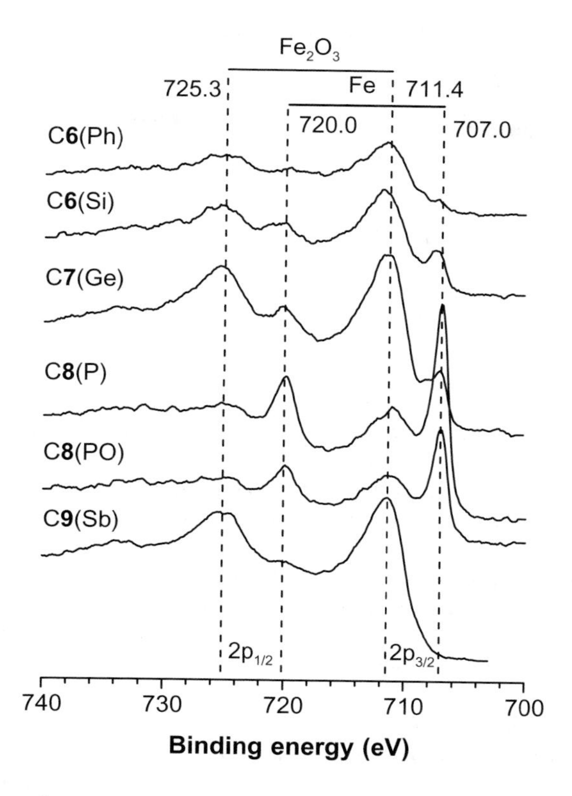

Fig. 7.11. Fe2p photoelectron spectra of ceramics prepared by pyrolyses of poly(ferrocenylene)s at 1000°C under nitrogen. Spectra calibrated to the position of C1s peak at 284.8 eV.

No peaks associated with Fe_2O_3 can be identified, revealing that Fe_2O_3 is present only as a thin surface layer but not in the bulk of these ceramics. The sizes of the α−Fe and Fe,C crystallites calculated by the Scherrer equation are in the range of 15–25 nm. Other ceramics show no peaks of elemental iron but ironic alloys ($\beta-Fe_{1.6}Ge$ and ε−SbFe) or iron phosphides (FeP and Fe_2P). The crystal sizes of the Ge– and Sb–Fe alloys are 22 and 33 nm, respectively, whereas the sizes of the iron phosphides are as small as ~0.8 nm. It is noteworthy that the broad peaks at $2\theta \sim 26°$, identifiable as graphite, are observed only in the XRD spectra of the ceramics containing elemental iron, indicating that only Fe can efficiently catalyze graphite crystallization.

The structures of the nanocrystallites in the ceramics are studied by high-resolution transmission electron microscopy (HRTEM), examples of which are given in Fig. 7.13. The shapes and sizes of the inorganic nanoparticles vary in different ceramics. Most Fe particles in C**6**(Ph) take irregular round shapes (Fig. 7.13, image A1). The nanoparticles in C**7**(Ge) look like Chinese dumplings (image B1), with the darker core of metal alloy intimately surrounded by the lighter shell of carbonaceous species, as clearly shown by image B2. For ceramic C**9**(Sb), it is difficult to identify isolated particles because most of them are clustered together in the amorphous environment (images C1 and C2). These findings confirm that Ge and Sb readily form alloys with Fe.

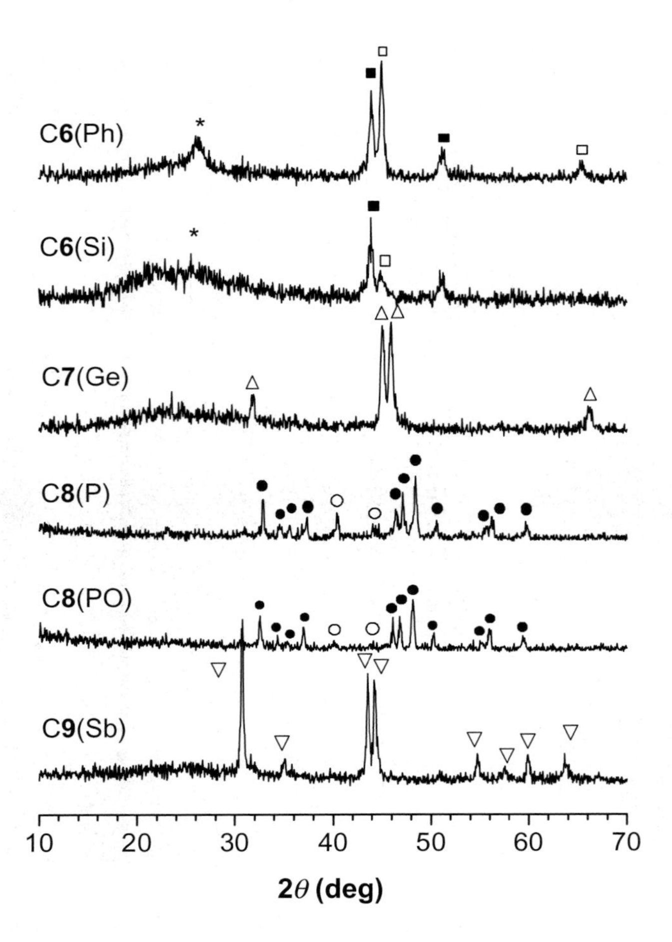

Fig. 7.12. XRD diffractograms of ceramics containing crystalline species of (□) α–Fe (JCPDS-ICDD file #06-0696) (■) Fe,C (#31-0619), (△) β–$Fe_{1.6}Ge$ (#18-0556), (○) FeP (#39-0809), (•) Fe_2P (#33-0670), (▽) ε–SbFe (#34-1053), and (*) graphite (#26-1080).

Fe is known to catalyze the crystallization of graphite and the formation of nanotubes. From the HRTEM images of **C6**(Ph), graphite can be easily identified with a lattice spacing of 3.4 Å. The Fe nanocrystals are covered by the graphite sheets (Fig. 7.13, image A2), which may have acted as protective coating toward oxidation of the nanoparticles. In addition, small, empty nanotube-like structures are clearly seen in the HRTEM images and their diffraction patterns of selected area (image A3). Different from ceramic **C6**(Ph), ceramics **C7**(Ge) and **C9**(Sb) exhibit no images and diffraction patterns related to the graphite layers, in accordance with the powder XRD observations discussed above (cf., Fig. 7.12).

Magnetic susceptibilities. In our previous studies, the ceramics prepared by pyrolysis at 1200°C under an atmosphere of argon are readily magnetized upon their exposure to a magnetic field and become saturated at ~49 emu/g when the external field

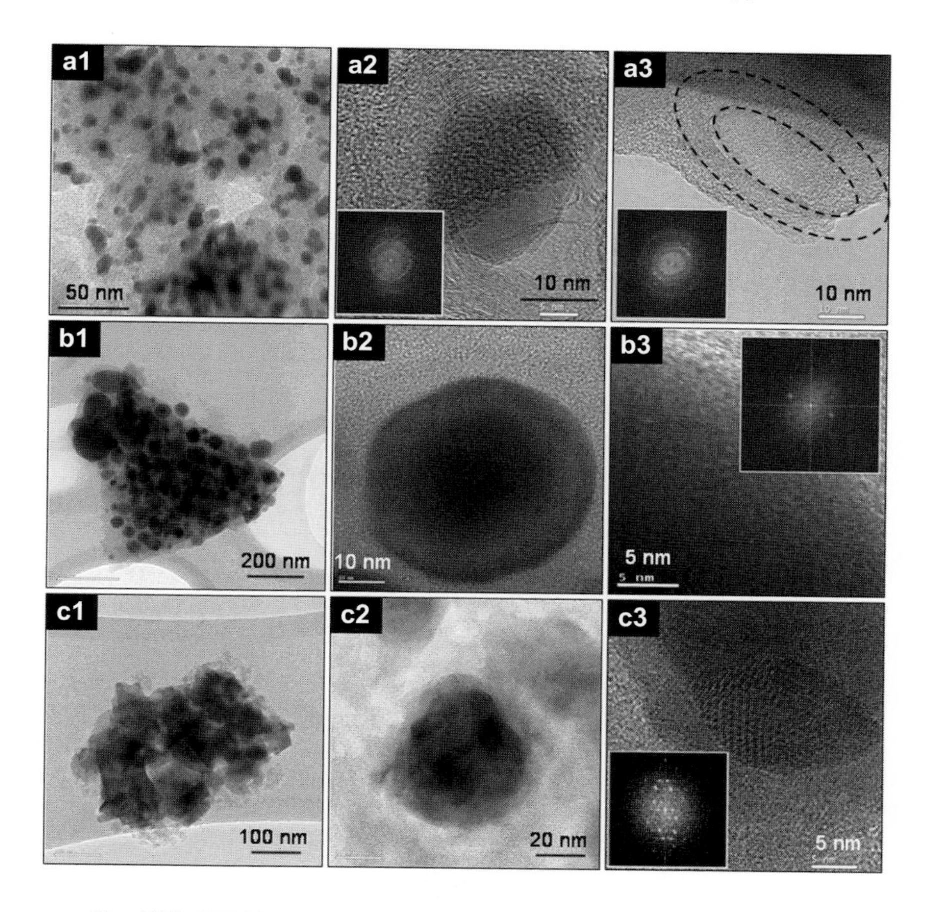

Fig. 7.13. TEM images of ceramics (a) C**6**(Ph), (b) C**7**(Ge), and (c) C**9**(Sb).

reaches ~5 kOe [66,67]. The magnetization plots furthermore show no hysteresis loops, that is, the magnetic remanence (M_r) and coercivity (H_c) are practically zero. The ceramics prepared from the hyperbranched poly(ferrocene) precursors at 1000°C under nitrogen show different magnetization behaviors. All the magnetization plots contain hysteresis loops, as can be seen from the inset of Fig. 7.14. Among the four metalloid species, the Ge-containing ceramic C**7**(Ge) shows the best performance: its saturation magnetization (M_s) value is high (21 emu/g) and its hysteresis loop is narrow (with a H_c value as small as ~0.07 kOe; Table 7.8, no. 3). The Si- and Sb-containing ceramics exhibit M_s values of 11–19 and 9 emu/g, respectively. The magnetic susceptibilities of the P-containing ceramics are very weak. The reason for this may be due to the ionic iron phosphide species and/or their small crystallite sizes (~0.8 nm) as revealed by the XRD analyses (cf., Fig. 7.12).

It should be stressed that the majority of the magnetoceramics prepared from the pyrolysis of the hyperbranched poly(ferrocene) precursors display very low H_c values: most of them are lower than 0.1 kOe and some of them are as low as 0.06 kOe. A ferromagnetic

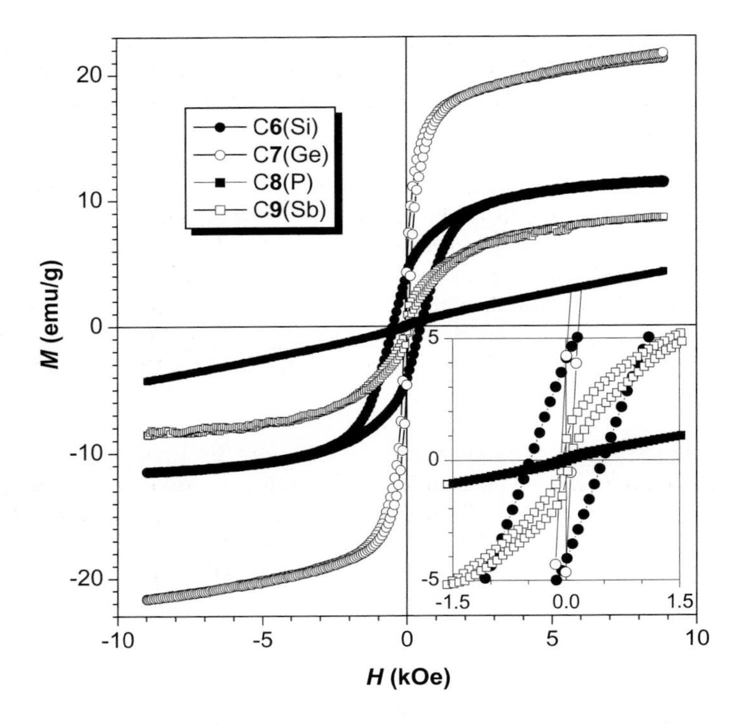

Fig. 7.14. Plots of magnetization (M) vs. applied magnetic field (H) at 300 K for ceramics **C6**(Si), **C7**(Ge), **C8**(P), and **C9**(Sb). Inset (lower right panel): enlarged portion of the plots in the low magnetic field region.

Table 7.8. Magnetic properties of the ceramics[a].

No.	Ceramic	M_s (emu/g)	M_r (emu/g)	H_c (kOe)
1	**C6**(Ph)	19	4.0	0.39
2	**C6**(Si)	11	4.2	0.44
3	**C7**(Ge)	21	4.2	0.07
4	**C8**(P)	5	0.1	0.09
5	**C8**(PO)	<1	~ 0	0.08
6	**C9**(Sb)	9	0.8	0.06

[a]Determined by VSM. Abbreviations: M_s = saturation magnetization (in an external field of ~10 kOe), M_r = magnetic remanence (at zero external field), and H_c = coercivity (at zero magnetization).

material with a coercivity smaller than 0.126 kOe (or 10^4 A/m) is termed a soft magnet [77–80]. Most of the magnetoceramics prepared are thus soft magnetic materials.

7.2.3 Hyperbranched polyynes

In addition to constructing hyperbranched polymers using polycyclotrimerization of diyne monomers, we also explored the possibility of using triynes as monomer units to assemble hyperbranched polyynes. Such carbon-rich materials exhibit an array of interesting thermal, optical, and electronic properties. Furthermore, they are rich in internal and external acetylenic triple bonds and hence function as macroligands for forming organometallic macromolecular complexes.

Polymer Preparation. We have synthesized different aromatic triynes with polar functional groups of ether, amine, and phosphorous oxide (Scheme 4). Homopolycoupling of triethynylbenzene (**15**) in *o*-dichlorobenzene by CuCl in air proceeds rapidly and produces gels quickly [88]. Introduction of a flexible alkoxy group results in the formation of completely soluble, high-molecular weight hyperbranched polyyne (**P16**) in a high yield (76%) (Table 7.9). The three nonplanar aryltriynes (**17–19**) also show high polymerizability. Delightfully, two of their polymers, i.e. **P18** and **P19**, are soluble, although **17** undergoes cross-linking reaction.

Copolymerization with comonomers and introduction of long alkoxy spacers are useful to suppress cross-linking reactions and to impart solubility to polymers. These approaches also work in our system: almost all the copolycoupling reactions of triynes **15–19** with monoyne **III** afford soluble copolyynes. While homopolycouplings of **15** and **17** easily run out of control, their copolycouplings with monoyne **III** proceed smoothly, yielding high molecular weight copolyynes with high solubility in common organic solvents.

Molecular structure. All of the soluble hyperbranched polyynes are fully characterized spectroscopically and give satisfactory analysis data corresponding to their expected molecular structures. IR and $^{1}H/^{13}C$ NMR spectra reveal clearly the depletion of the free acetylene moieties and the formation of the new diacetylene bonds [88]. Figure 7.15a shows the ^{1}H NMR spectrum of **P18**, a homopolyyne, as an example. All the peaks are readily assignable: the resonance signals at δ 7.4, 7.0 (a) and 3.1 (b) are due to the absorptions of the aromatic and acetylenic protons, respectively.

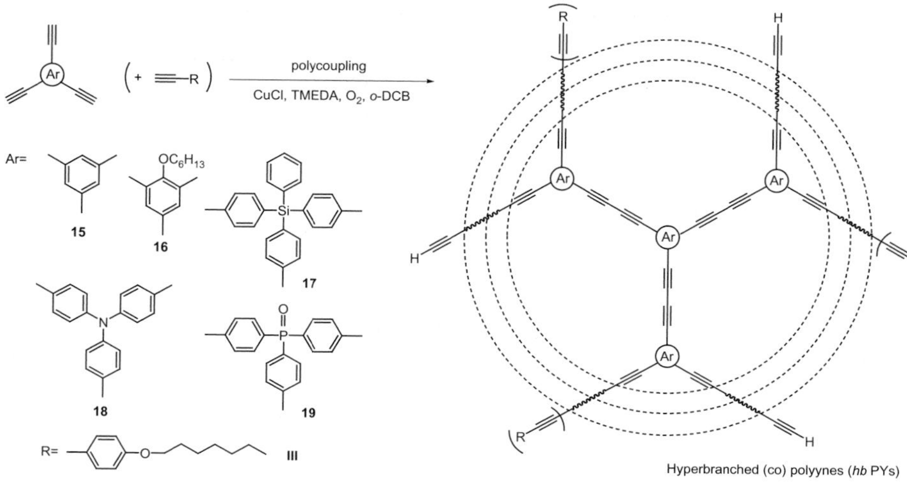

polycoupling

CuCl, TMEDA, O_2, *o*-DCB

Hyperbranched (co) polyynes (*hb* PYs)

Scheme 4. Homopolycouplings of triynes and their copolycouplings with monoyne.

Table 7.9. Syntheses[a] and properties of hyperbranched polyynes (*hb*–PYs).

No.	*hb*-PY	Yield (wt %)	M_w^b	M_w/M_n^b	T_d^c (°C)	W_r^d (wt %)
1	**P16**	76.1	30 700	3.6	377	50.4
2	**P18**	51.7	24 100	1.6	516	78.0
3	**P19**	37.1	5 100	1.4	549	84.0[e]
4	**P15/III**	46.9	17 900	4.7	412	60.3
5	**P17/III**	67.4	13 000	7.2	411	59.2
6	**P18/III**	57.2	18 200	5.3	456	73.3
7	**P19/III**	19.1	7 500	1.4	430	65.9

[a]Carried out at 50°C with a flow of air; [aryltriynes] = 80 mM; [CuCl] = 4 mM, [TMEDA] = 13.8 mM. For copolycoupling, [aryltriynes]/[III] = 1:1.5 (by mol). Reaction time: 8–180 min.
[b]Determined by SEC in THF on the basis of a polystyrene calibration.
[c]Temperature for 5% weight loss.
[d]Weight residue at 900°C.
[e]At 850°C.

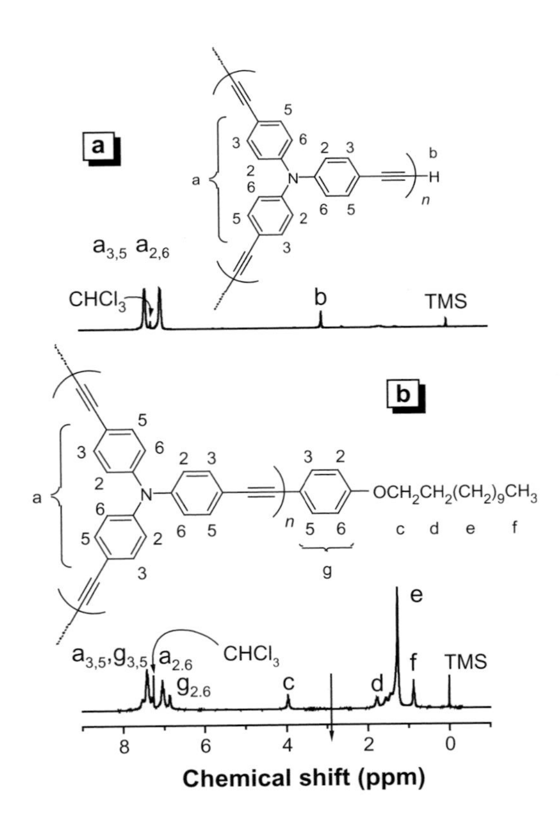

Fig. 7.15. ^{1}H NMR spectra of (a) **P18** and (b) its end-capped product **P18/III** in chloroform-*d*.

Thermal and optical properties. Similar to hyperbranched poly(aroylarylene)s and poly(ferrocenylene)s, all the hyperbranched polyynes are thermally very stable and carbonized in high yields at high temperatures [88]. The temperature for 5% weight loss and the weight residue at 900°C of the polyynes are high, being 377–549°C and 50.4–78.0%, respectively. Among the polyynes, **P19** shows the highest thermal stability, which loses little of its weight when heated to ~550°C and carbonizes with 84% yield when pyrolyzed at 850°C.

Theoretical predictions show that polymers consisting of groups with high polarizabilities and small volumes can exhibit high refractivities [89]. Thanks to its polarizable aromatic rings and slender triple-bond rods, a thin film of **P18** shows refractive indexes (n) of 1.861–1.770 in the spectral region of 600–1700 nm [88,90] (Fig. 7.16), which are much higher than those of the well-known commercial 'organic glasses' such as poly(methyl methacrylate) ($n = 1.497$–1.489), polycarbonates ($n = 1.593$–1.576), and polystyrene ($n = 1.602$–1.589) [91,92]. The polyyne film is transparent and shows high transmittance in the long wavelength region.

Metal complexation. Carbon–carbon triple bonds are versatile ligands in organometallic chemistry [93–95]. Examples of acetylene-metal reactions include facile complexations of one triple bond with $Co_2(CO)_8$ [96–98] and of two triple bonds with $CpCo(CO)_2$ [99,100]. The hyperbranched polyynes contain many triple bonds and thus should be easily metallified through their complexations with the cobalt carbonyls. When a mixture of **P18** and octacarbonyldicobalt with a $[Co_2(CO)_8]/[C \equiv C]$ ratio of 1:1 is stirred

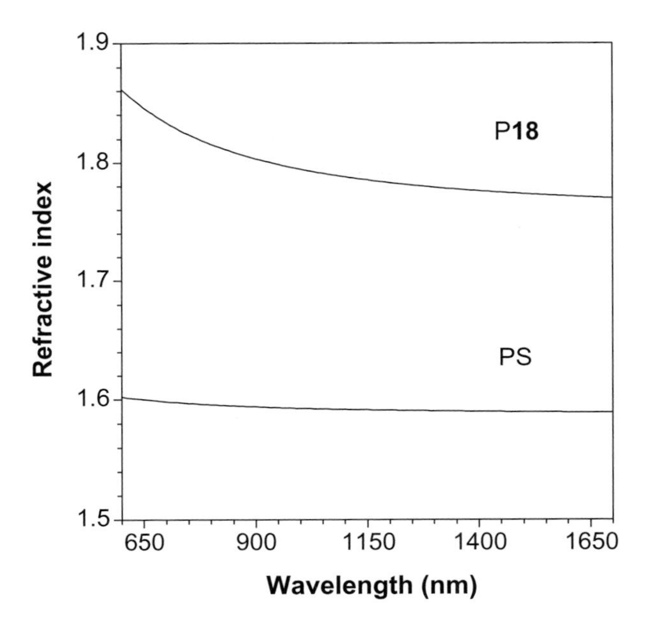

Fig. 7.16. Wavelength dependence of the refractive index of a thin film of **P18**. Data for a thin film of polystyrene (PS) is shown for comparison.

P18 — $Co_2(CO)_8$, THF, rt, N_2, 1 h → P18–{$[Co(CO)_3]_2$}$_m$ **20**

P18 — $CpCo(CO)_2$, THF, rt, N_2, 1 h → P18–[$Co(Cp)_1]_2$}$_m$ **21**

20, **21** — Δ, N_2 → C_xCo_y (**C10**, **C11**)

Scheme 5. Complexation with Cobalt Carbonyls and Ceramization to Magnetic Ceramics.

in THF at room temperature, the solution color changes from yellow to brown accompanying CO gas evolution (Scheme 5). The mixture remains homogenous towards the end of reaction, and the organometallic hyperbranched polymer **P20** is obtained by pouring the THF solution into hexane with 88.6% yield (Table 7.10). Although **P20** is completely soluble in the reaction mixture, it becomes insoluble after purification, possibly due to the formation of supramolecular aggregates during the precipitation and drying processes. An analogous result is obtained when a higher $[Co_2(CO)_8]/[C \equiv C]$ feed ratio (1.5:1) is used: the reaction solution is homogeneous but the purified product (**P20b**) is insoluble. The yield of **P20b** is higher, suggesting a more complete metal/triple bond complexation and thus a higher cobalt loading. Complexes **P20a** and **P20b** are stable in air, whose IR spectra show strong vibration bands typical of cobalt carbonyl absorptions at 2090, 2055, and 2025 cm^{-1} [101], verifying the integration of the metallic species into the polyyne structure at the molecular level. Alternatively, **P18** is cobalt-metallized through intra- and/or intermolecular complexations of its triple bonds with $CpCo(CO)_2$, giving a hyperbranched polyyne complex carrying cyclopentadienylcyclobutadienylcobalt (CpCbCo) moieties (**P21**), however, in lower yields (59.5%).

Elemental analysis is a useful tool to investigate the atomic compositions of materials. While organometallic hyperbranched polymer **20a** exhibits a cobalt content of 27.8 wt% (Table 7.10), its congener **20b** displays a much higher value (36.7 wt%). The cobalt content of **21** is similar to that of **20b**. This seems to be odd at first sight because one

Table 7.10. Synthesis of cobalt-containing polyynes and ceramics[a].

No.	Co complex	$[Co]:[C \equiv C]$[b]	Yield[c] (%)	Yield[d] (%)	Elemental analysis (wt%)			
					C	H	N	Co[f]
1	$Co_2(CO)_8$	1.0:1.0	88.6 (**20a**)	49.9 (**C10a**)	43.8	3.1	2.5	27.8[g]
2	$Co_2(CO)_8$	1.5:1.0	93.8 (**20b**)	52.6[e] (**C10b**)	29.6	2.0	1.7	36.7[g]
3	$CpCo(CO)_2$	1.5:1.0	59.5 (**21**)	64.8 (**C11**)	58.1	4.2	2.7	35.0

[a]Carried out under nitrogen; [**P18**] = 6 mg/mL.
[b]Ratio of cobalt complex to acetylene bond in the feed mixture.
[c]Calculation based on 100% substitution.
[d]Weight of the residue left after pyrolysis at 1000°C for 1 h under a steam of nitrogen (flow rate: ~0.2 L/min).
[e]The weight of the residue left after pyrolysis at 1200°C for 1 h under a steam of nitrogen (flow rate: ~0.2 L/min) was 42.4%.
[f]Calculated cobalt content by subtraction of the %C, %H, and %N from 100%.
[g]Assuming that Co is bonded as $Co_2(CO)_6$ complex with a [Co]:[O] ratio of 1:3.

Co-atom of [$CpCo(CO)_2$] needs two triple bonds for a covalent binding vs. two Co-atoms of [$Co_2(CO)_8$] needing just one. It, however, becomes easily understandable if someone takes into account that simultaneous incorporation of the two cobalt moieties six CO-ligands are introduced, understating the apparent Co content of **20b** compared to polymer **3**.

Homogenous brown-colored films of the metallified polymers **20** can be readily obtained from its freshly prepared complexation solution by spin-coating. Afterwards, the dried films are irradiated through a Cu-negative mask for 30 min. Interestingly, the color of the illuminated parts is bleached, leaving behind the pattern of the photomask (Fig. 7.17). The enlarged micrograph clearly reveals the sharp edges of the patterns.

Inspired by the UV light-induced color change of **20**, we investigated its optical properties in more detail. Figure 7.18 shows the results of the wavelength-dependent refractive indexes of its unexposed and exposed thin films. Similar to its parent **P18**, the metallified polymer shows very high refractive indexes (n = 1.813–1.714) in the spectral region of 600–1600 nm. Surprisingly, the refractive index drops after UV irradiation significantly (n = 1.777–1.667). A polymer with such a high refractive index change is promising for photonic applications: for example, it may function as photorefractive material in holographic devices [102] or work as high refractive index optical coatings [103].

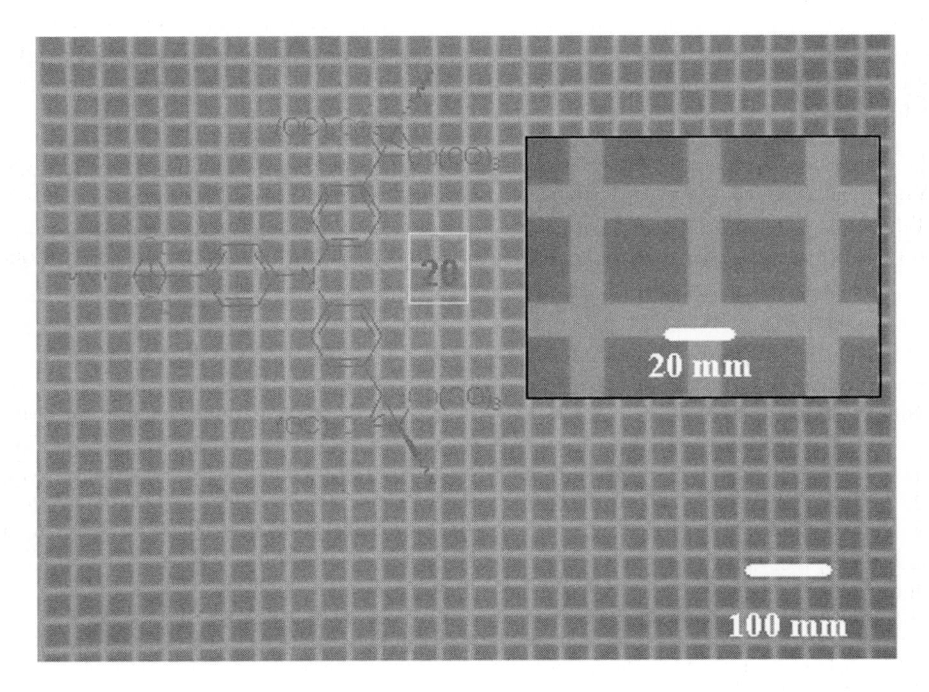

Fig. 7.17. Optical micrograph of the photopattern generated by UV photolysis of **20** through a Cu-negative mask. Insets: molecular structure of **20** and enlarged fraction of the optical micrograph.

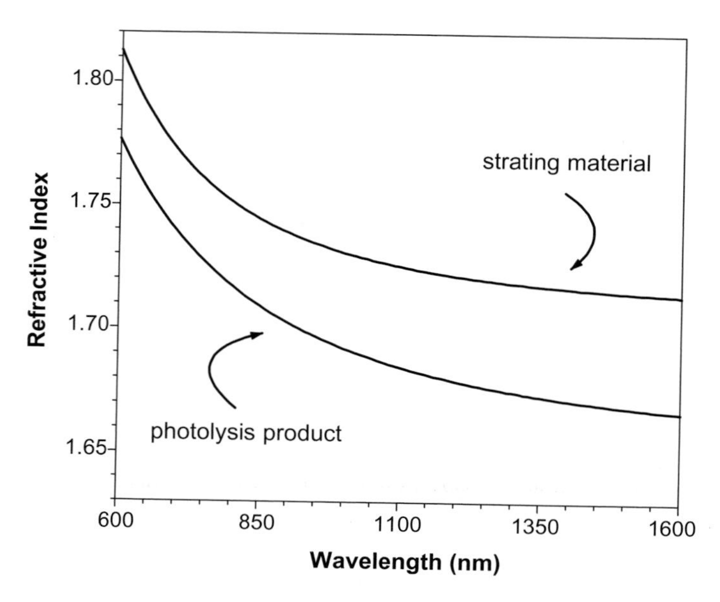

Fig. 7.18. Wavelength dependences of the refractive indexes of a thin film of **20** and its photolysis product after UV irradiation for 30 min.

Ceramization and magnetic properties. Since the hyperbranched polymers can be easily metallified by complexations with cobalt carbonyls and possess high thermal stability, we thus tried to use **P20** and **P21** as precursors for the preparation of metalloceramics. Ceramization of the complexes in a tube furnace at 1000 or 1200°C for 1 h under a steam of nitrogen gives ceramic products **C10a**, **C10b**, and **C11** (Scheme 5 and Table 7.10) in ~42–65% yields. All the ceramics are magnetizable and can be readily attracted to a bar magnet. Analyses by XPS and EDX show the presence of C, N, O, and Co elements inside the ceramics. In order to gain more insights into the chemical structure of the bulk, we investigated the ceramics by powder XRD. Figure 7.19 shows the XRD pattern of the organometallic polymer **P20** and the ceramics **C10b** and **C11**. Whereas **P20** shows no Bragg reflections and is thus amorphous in nature, ceramics **C10b** and **C11** exhibit three peaks located at 2θ angles of 44.2°, 51.5°, and 75.9°. The reflection peaks can be identified according to the database of JCPDS-International Centre for Diffraction Data (ICDD) as reflections for the cobalt metal (ICDD-data file 15-0806). No peaks associated with the presence of Co_xO_y have been found, suggesting only native amorphous cobalt oxide layers on the surfaces as detected by XPS. Comparing the XRD patterns, the reflection peaks of **C11** are broader, suggesting that the crystallites are less perfect and smaller in size. Using the full widths at half-maximums of the peak reflections, a crystallite size of ~33.8 and ~26.3 nm for **C10** and **C11**, respectively, can be calculated from the Scherrer equation.

The magnetization curves of the ceramics are shown in Fig. 7.20. With an increase in the field strength of the external magnetic field, the magnetization of ceramic **C10b** swiftly increases and eventually levels off at a saturation magnetization (M_S) of ~118 emu/g, which is much higher than the M_S value of the commonly known magnet or maghemite

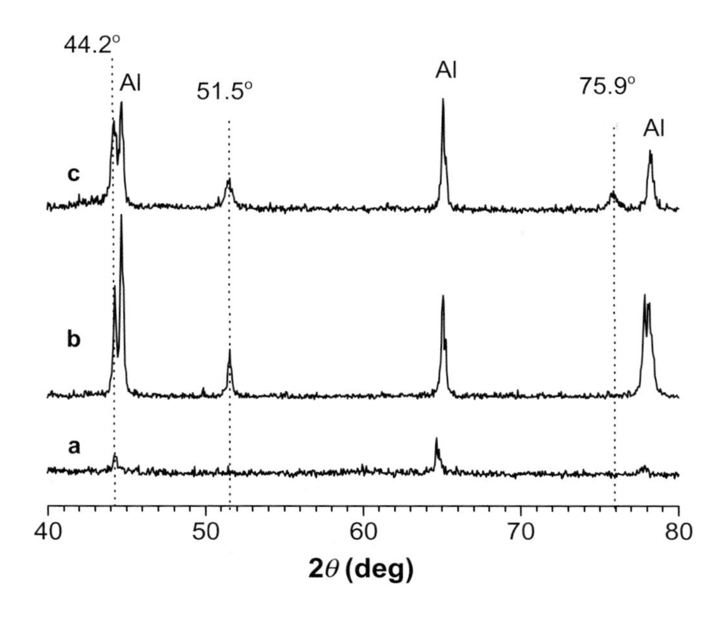

Fig. 7.19. XRD diffractograms of (a) **P20b**, (b) **C10b**, and (c) **C11**. The refraction peaks labeled with Al are caused by the aluminum sample holder.

(γ-Fe_2O_3; M_s = 74 emu/g) [104]. The M_s value of ceramic **C11** is somewhat lower (~26 emu/g), which is understandable, because the cobalt content of its precursor complex (**21**) is lower. The hysteresis loops of our magnetoceramics are very small. From the enlarged *H*–*M* plots shown in the inset of Fig. 7.20, the coercivities (H_c) of **C10b** and **C11** are found to be 0.058 and 0.142 kOe, respectively. The high magnetizability ($M_s \sim 118$ emu/g) and low coercivity ($H_c \sim 0.06$ kOe) of **C10b** make it an excellent soft ferromagnetic material, which may find an array of high-technology applications in various electromagnetic systems.

7.3 Conclusions

In this work, a group of new hyperbranched polyarylenes, polyynes, and poly(ferrocenylene)s with high molecular weights are prepared in high yields by polycyclotrimerization and polycoupling of acetylenic monomers, and polycondensation reaction of dilithioferrocene with tri- or tetrachlorides of group 14 and 15 elements. All the polymers possess high thermal stability, losing little of their weights when heated to >400°C. Hyperbranched poly(aroylarylene)s and polyynes show high photosensitivities and can be readily photocrosslinked to give photoresist patterns with nanometer resolutions. Thin films of hyperbranched polyynes exhibit very high refractive indexes (*n* up to 1.86). Ceramizations of ferrocene-containing poly(aroylarylene)s, cobalt-polyyne complexes, and poly(ferrocenylene)s at high temperatures under nitrogen afford soft nanoceramics with magnetic susceptibilities up to ~118 emu/g and near-zero magnetic losses.

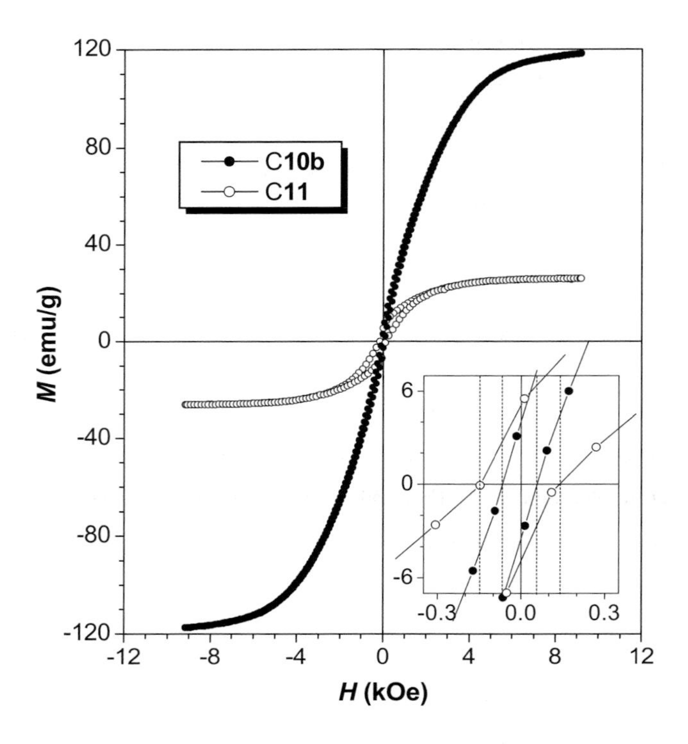

Fig. 7.20. Plots of magnetization (M) vs. applied magnetic field (H) at 300 K for magnetoceramics **C10b** and **C11**. Inset: enlarged portion of the M–H plots in the low strength region of the applied magnetic field.

Acknowledgment

This work partially supported by the Hong Kong Research Grants Council (Project Nos.: 603505, 603304, 604903, N_HKUST606_03, HKUST6085/02P, 6121/01P and 6187/99P), the University Grants Committee of Hong Kong through an Area of Excellence (AoE) Scheme (AoE/P-10/01-1-A), and the National Science Foundation of China. B.Z.T. thanks the support from the Cao Guangbiao Foundation of Zhejiang University.

References

[1] A. Hult, M. Johansson and E. Malmström, Adv. Polym. Sci. 143 (1999) 1.

[2] S.M. Grayson and J.M.J Frechet, Chem. Rev.101 (2001) 3819.

[3] P.J. Flory, J. Am. Chem. Soc. 74 (1952) 2718.

[4] P.A. Gunatillake, G. Odian and D.A. Tomalia, Macromolecules 21 (1988) 1556.

[5] Y.H. Kim and O.W. Webster, J. Am. Chem. Soc. 112 (1990) 4592.

[6] D.T. Tomalia and J.M.J. Frechet, J. Polym. Sci. Polym. Chem. 40 (2002) 2719.

[7] B. Voit, J. Polym. Sci. Polym. Chem. 38 (2000) 2505.

[8] K. Inoue, Prog. Polym. Sci. 25 (2000) 453.

[9] C.J. Hawker, Curr. Opin. Coll. Interf. Sci. 4 (1999) 117.

[10] T.M. Miller, T.X. Neenan, R. Zayas and H.E. Bair, J. Am. Chem. Soc. 114 (1992) 1018.

[11] B. Voigt, C. R. Chimie 6 (2003) 821.

[12] Y.H. Kim, J. Polym. Sci. Polym. Chem. 36 (1998) 1685.

[13] C.J. Hawker, R. Lee and J.M.J. Frechet, J. Am. Chem. Soc. 113 (1991) 4583.

[14] I. Sendijarevic, A.J. McHugh, L.J. Markoski and J.S. Moore, Macromolecules 34 (2001) 8811.

[15] W. Radke, G. Litvinenko and A.H.E. Müller, Macromolecules 31 (1998) 239.

[16] R. Hanselmann, D. Hölter and H. Frey, Macromolecules 31 (1998) 3790.

[17] A. Sunder, R. Hanselmann, H. Frey and R. Mühlhaupt, Macromolecules 32 (1999) 4240.

[18] P. Bharati and J.S. Moore, J. Am. Chem. Soc. 119 (1997) 3391.

[19] P. Bharati and J.S. Moore, Macromolecules 33 (2000) 3212.

[20] F. Wang, M.S. Wilson, R.D. Rauh, P. Schottland and J.R. Reynolds, Macromolecules 33 (1999) 4272.

[21] C. Gao and D. Yan, Macromolecules 34 (2001) 156.

[22] T. Emrick, H.T. Chang and J.M.J. Frechet, Macromolecules 32 (1999) 6380.

[23] M. Jikei, S.H. Chon, M. Kakimoto, S. Kawauchi, T. Imase and J. Watanabe, Macromolecules 32 (1999) 2061.

[24] S. Russo, A. Boulares, A. da Rin, A. Mariani and M.E. Cosulich, Macromol. Symp. 143 (1999) 309.

[25] S.M. Ahoni, Polym. Adv. Technol. 6 (1995) 373.

[26] J.M.J. Frechet, M. Henmi, I. Gitsov, S. Aoshima, M. Leduc and R.B. Grubbs, Science 269 (1995) 1080.

[27] C.J. Hawker, J.M.J. Frechet, R.B. Grubbs and J. Dao, J. Am. Chem. Soc. 117 (1995) 10763.

[28] S.G. Gaynor, S.Z. Edelman and K. Matyjaszewski, Macromolecules 29 (1996) 1079.

[29] A. Dworak, W. Walach and B. Trzebicka, Macromol. Chem. Phys. 196 (1995) 1963.

[30] M. Suzuki, S. Yoshida, K. Shiraga and T. Saegusa, Macromolecules 31 (1998) 1716.

[31] H. Magnusson, E. Malmström and A. Hult, Macromol. Rapid Commun. 20 (1999) 453.

[32] M. Bednarek, T. Biedron, J. Helinski, K. Kaluzynski, P. Kubisa and S. Penczek, Macromol. Rapid. Commun. 20 (1999) 369.

[33] A. Sunder, R. Hanselmann, H. Frey and R. Mülhaupt, Macromolecules 32 (1999) 4240.

[34] K. Xu and B.Z. Tang, Chinese J. Polym. Sci. 17 (1999) 397.

[35] J.W.Y. Lam, J. Luo, H. Peng, Z. Xie, K. Xu, Y. Dong, L. Cheng, C. Qiu, H.S. Kwok and B.Z. Tang, Chinese J. Polym. Sci. 19 (2000) 585.

[36] B.Z. Tang, K. Xu, Q. Sun, P.P.S. Lee, H. Peng, F. Salhi and Y. Dong, ACS Symp. Ser. 760 (2000) 146.

[37] Y. Mi and B.Z. Tang, Polym. News 26 (2001) 170.

[38] B.Z. Tang, K. Xu, H. Peng, Q. Sun and J. Luo, US Patent Appln No. 10/109 (2002) 316.

[39] K. Xu, H. Peng, Q. Sun, Y. Dong, F. Salhi, J. Luo, J. Chen, Y. Huang, D. Zhang, Z. Xu and B.Z. Tang, Macromolecules 35 (2002) 5821.

[40] R.H. Zheng, H.C. Dong, H. Peng, J.W.Y. Lam and B.Z. Tang, Macromolecules 37 (2004) 5196.

[41] H. Peng, J.W.Y. Lam and B.Z. Tang, Macromole. Rapid Commun. 26 (2005) 673.

[42] H. Peng, J.W.Y. Lam and B.Z. Tang, Polymer 46 (2005), 46, 5746.

[43] H. Peng, L. Cheng, J.D. Luo, K.T. Xu, Q.H. Sun, Y.P. Dong, F. Salhi, P.P.S. Lee, J.W. Chen and B.Z. Tang, Macromolecules 35 (2002) 5349.

[44] Z. Xie, J.W.Y. Lam, Y. Dong, C. Qiu, H.S. Kwok and B.Z. Tang, Opt. Mater. 21 (2002) 231.

[45] H. Peng, J. Luo, L. Cheng, J.W.Y. Lam, K. Xu, Y. Dong, D. Zhang, Y. Huang, Z. Xu and B.Z. Tang, Opt. Mater. 21 (2002) 315.

[46] Z.L. Xie, H. Peng, J.W.Y. Lam and B.Z. Tang, B. Z. Macromol. Symp. 195 (2003) 179.

[47] J.W.Y. Lam, J.W. Chen, C.C.W. Law and B.Z. Tang, Macromol. Symp. 196 (2003) 289.

[48] J.W. Chen, H. Peng, C.C.W. Law, Y.P. Dong, J.W.Y .Lam, I.D. Williams and B.Z. Tang, Macromolecules 36 (2003) 4319.

[49] M. Häußler, J.W.Y. Lam, R. Zheng, H. Peng, J. Luo, J. Chen, C.C.W. Law and B. Z. Tang, C. R. Chimie 6 (2003) 833.

[50] C.C.W. Law, J.W. Chen, J.W.Y. Lam, H. Peng and B.Z. Tang, J. Inorg. Organomet. Polym. 14 (2004) 39.

[51] K. Balasubramanian, S. Selvaraj and P.S. Venkataramani, Synthesis 29 (1980) 29.

[52] F.C. Pigge, F. Ghasedi and N.J. Rath, Org. Chem. 67 (2002) 4547.

[53] S. Saito and Y. Yamamoto, Chem. Rev. 10 (2000) 2901.

[54] H.C. Dong, R.H. Zheng, J.W.Y. Lam, M. Häußler and B.Z. Tang, Polym. Prepr. 45 (2004) 825.

[55] H. Dong, R. Zheng, J.W.Y. Lam, M. Häußler and B.Z. Tang, Macromolecules, 38 (2005) 6382.

[56] H.C. Dong, Mhil. Dissertation, The Hong Kong University of Science & Technology, 2005

[57] K. Matusda, N. Nakamura and H. Iwamura, Chem. Lett. 9 (1994) 1765.

[58] V.S.S. Kumar, F.C. Pigge, N.P. Rath, Cryst. Growth. Des. 4 (2004) 651.

[59] K. Matsuda, N. Nakamura, K. Takahashi, K. Inoue, N. Koga, H. Iwamura, J. Am. Chem. Soc. 117 (1995) 5550.

[60] Y. Luo, J. Leszyk, Y.D. Qian, J. Gergely and T. Tao, Biochemistry 38 (999) 678.

[61] M. Hasegawa and K. Horie, Prog. Polym. Sci. 26 (2001) 529.

[62] N.J. Turro, Modern Molecular Photochemistry, Benjamin-Cummings: Menlo Park, CA, 1978.

[63] K.H. Kim, S. Jang and F.W. Harris, Macromolecules 34 (2001) 8925.

[64] L.F. Thompson, C.G. Wilson and J.M.J. Fréchet, Materials for Microlithography, ACS, Washington, D.C, 1984.

[65] E. Reichmanis and L.F. Thompson, Chem. Rev. 89 (1989) 1273.

[66] Q. Sun, K. Xu, H. Peng, R. Zheng, M. Häußler and B.Z. Tang, Macromolecules, 36 (2003) 2309.

[67] Q. Sun, J.W.Y. Lam, K. Xu, H. Xu, J.A.P. Cha, P.C.L. Wong, G. Wen, X. Zhang, X. Jing, F. Wang and B.Z. Tang, Chem. Mater, 12 (2000) 2617.

[68] C.R. Bruncle, T.J. Chung, K. Wandelt, Surf. Sci. 68 (1977) 459.

[69] R.T. Paine and C.K. Narula, Chem. Rev, 90 (1990) 73.

[70] D.R. Messier and W.J. Croft, Preparation and Properties of Solid State Materials, W.R. Wilcox, Ed., Dekker, New York, Vol. 7, Chapter 2: 1982.

[71] O. Mills, J.L. Sullivan, J. Phys. D: Appli. Phys. 14 (1983) 723.

[72] J.F. Moulder, W.F. Stickle, P.E. Sobol and K.D. Bomben, Handbook of X-ray Photoelectron Spectroscopy: a Reference Book of Standard Spectra for Identification and Interpretation of XPS Data, J. Chastain, Ed., Physical Electronics Division, Perkin-Elmer Corp., Eden Prairie, MN, 1992.

[73] C.S. Fradley, Electron Spectroscopy, D.A. Shirley, Ed., North-Holland, Amsterdam, 1972.

[74] H.P. Klug and L.E. Alexander, Powder X-ray Diffraction Techniques, Wiley: New York, 1974.

[75] M. Yudasakaa and R. Kikuchi, Graphitization of Carbonaceous Materials by Ni, Co and Fe, Supercarbon, Springer, 1998.

[76] C. Pham-Huu, C. Estournes, B. Heinrich and M.J. Ledoux, J. Chem. Soc., Faraday Trans. 94 (1998) 435.

[77] X.Q. Zhao, Y. Liang, Z.Q. Hu, B.X. Liu, J. Appl. Phys. 80 (1996) 5857.

[78] R.C. O'Handley, Modern Magnetic Materials: Principles and Applications, Wiley: New York, 2000.

[79] A. Goldmann, Handbook of Ferromagnetic Materials, Kluwer, Boston, MA, 1999.

[80] Concise Encyclopedia of Magnetic & Superconducting Materials, J.E. Evetts, Ed., Pergamon, New York, 1992.

[81] D.R. Askeland, The Science and Engineering of Materials, 3rd ed., PWS, Boston, MA, 1994.

[82] A.Y. Cheng, S.B. Clendenning, G. Yang and I. Manners, Chem. Commun. 7 (2004) 780.

[83] T. Imai, T. Satoh, H. Kaga, N. Kaneko and T. Kakuchi, Macromolecules 37 (2004) 3113.

[84] R.M. Silverstein and F.X. Webster, Spectrometric Identification of Organic Compounds, Wiley, Chapter 3: 1998.

[85] R.T. Bailey and E.R. Lippincott, Spectrochim. Acta 21 (1965) 389.

[86] B.Z. Tang, R. Petersen, D.A. Foucher, A. Lough, N. Coombs, R. Sodhi and I. Manners, J. Chem. Soc. Chem. Commun. (1993), 523.

[87] R. Petersen, D.A. Foucher, B.Z. Tang, A. Lough, N.P. Raju, J.E. Greedan and I. Manners, Chem. Mater. 7 (1995) 2045.

[88] M. Ginzburg, M.J. MacLachlan, S.M. Yang, N. Coombs, T.W. Coyle, N.P. Raju, J.E. Greedan, R.H. Herber, G.A. Ozin and I. Manners, J. Am. Chem. Soc. 124 (2002) 2625.

[89] M. Häußler, R.H. Zheng, J.W.Y. Lam, H. Tong, H.C. Dong and B.Z. Tang, J. Phys. Chem. B 108 (2004) 10645.

[90] C. Badarau and Z.Y. Wang, Macromolecules 37 (2004) 147.

[91] M. Häußler, J.W.Y. Lam and B.Z. Tang, Polym. Prepr. 45 (1) (2004) 448.

[92] J.C. Seferis, Polymer Handbook, 3rd ed., Wiley: New York, 1989, VI/451.

[93] N.J. Mills, Concise Encyclopedia of Polymer Science & Engineering, Wiley: New York, 1990, 683.

[94] F. Babudri, G.M. Farinola and F.J. Naso, J. Mater. Chem. 14 (2004) 11.

[95] N.J. Long and C.K. Williams, Angew. Chem. Int. Ed. 42 (2003) 2586.

[96] U.H.F. Bunz, J. Organomet. Chem. 683 (2003) 269.

[97] G.R. Newkome, E.F. He and C.N. Moorefield, Chem. Rev. 99 (1999) 1689.

[98] B.P.S. Chauhan, R.J.P. Corriu, G.F. Lanneau, C. Priou, N. Auner, H. Handwerker and E. Herdtweck, Organometallics 14 (1995) 1657.

[99] W.Y. Chan, A. Berenbaum, S.B. Clendenning, A.J. Lough and I. Manners, Organometallics 22 (2003) 3796.

[100] H. Nishihara, M. Kurashina and M. Murata, Macromol. Symp. 196 (2003) 27.

[101] M. Altmann and U.H.F. Bunz, Angew. Chem. Int. Ed. 34 (1995) 569.

[102] W.Y. Chan, A. Berenbaum, S.B. Clendenning, A.J. Lough and I. Manners, Organometallics 22 (2003) 3796.

[103] E. Hendrickx, C. Engels, M. Schaerlaekens, D. Van Steenwinckel, C. Samyn and A.J. Persoons, J. Phys. Chem. B. 106 (2002) 4588.

[104] C. Lu, C. Guan, Y. Liu, Y. Cheng and B. Yang, Chem. Mater. 17 (2005) 2448.

[105] B.Z. Tang, Y. Geng, J.W.Y. Lam, B. Li, X. Jing, X. Wang, F. Wang, A.B. Pakhomov and X.X. Zhang, Chem. Mater. 11 (1999) 1581.

CHAPTER 8

Deformation Characteristics of Nanocrystalline Metals

S.C. Tjong

Department of Physics and Materials Science
City University of Hong Kong
Tat Chee Avenue, Kowloon, Hong Kong

8.1 Introduction

Recently, increasing attention has been paid to the study of deformation behavior of nanocrystalline metals by materials scientists and physicists. Nanocrystalline metals are attractive for use in structural applications in which enhanced mechanical and physical properties are needed. Nanocrystalline metals have been found to exhibit very different microstructures and mechanical behaviors compared to their conventional coarse-grained polycrystalline counterparts. Nanocrystalline metals possess a relatively large number of grain boundaries and can be expected to influence the mechanical properties dramatically. The grain-boundary structure, boundary angle, and boundary sliding are important factors that determine the mechanical behavior of nanocrystalline metals. Moreover, a decrease in grain size significantly affects the yield strength, hardness, and ductility. In this respect, nanocrystalline metals exhibit very high yield strength and hardness but limited tensile plasticity at room temperature. Limited tensile ductility results from a low strain-hardening capacity or plastic instability of nanocrystalline metals during tension and from porosity in the specimens. This is the main problem of a nanocrystalline structure that hinders the structural use of nanocrystalline metals to their full potential. Efforts to improve ductility of nanocrystalline metals by modification of processing methods or microstructural control have shown limited progress. At higher temperatures, grain refinements to a nanoscale regime produces a considerable decrease in temperatures for superplasticity as the grain-boundary-mediated deformation mechanisms prevail when the grain size decreases. Thus, nanoscale refinement opens up the possibilities of superplastic forming of metals at low temperatures with high strain rates to achieve substantial reductions in the processing time and energy cost.

The yield strength of conventional polycrystalline metals is known to increase with decreasing grain size. The increase in strength results from the grain boundaries acting as barriers to dislocation motion. Such an increase can be predicted by the Hall–Petch (H–P) effect, which relates the yield stress to the inverse square root of the grain size. Thus, dislocation slip is the dominant deformation for coarse-grained metals. When the grain size reaches a critical value (ca. 20 nm), the strength appears to decrease with further grain refinement owing to the fact that the dislocation source inside the grains ceases to operate. The dislocation slip would be replaced by the grain-boundary-mediated deformation mechanism. At this stage, inverse or negative Hall–Petch effect, i.e. softening is observed. The softening behavior of nanocrystalline metals still remains controversial. Several mechanisms have been proposed to explain the anomalous deformation behavior of nanocrystalline metals with the grain size below the critical value. These include grain-boundary sliding, grain-boundary diffusion creep, triple junction effect, presence of nanopores, and impurities, etc.

The microstructure of nanocrystalline metals is generally known to be strongly dependent on the synthesis techniques [1]. The relationship between the structure of the material and the conditions of its preparation must be established. Bulk nanocrystalline metals for the mechanical tests are commonly prepared via consolidation of inert gas condensed (IGC) nanoparticles and electrodeposition. Consolidated IGC samples are generally known to contain internal impurities and pores that could lead to softening of the mechanical strength and hardness in the nanometer regime. Pore-free and dense electrodeposited samples also contain impurities inherited from electroplating bath solutions and organic additives. Moreover, the composition and microstructure of electrodeposited samples vary from batch to batch. This raises the issue of whether negative H–P behavior is inherited from the intrinsic effect of nanograin size or has resulted from extrinsic defects introduced into the samples during fabrication.

The search for comprehensive understanding of the basic deformation mechanisms is essential in the development of novel nanomaterials with unique properties for structural engineering applications. Unfortunately, dense nanocrystalline metals with grain size smaller than 20 nm and free from grain-boundary impurities are difficult to prepare. In this respect, molecular dynamics (MD) modeling has been used to simulate the mechanical deformation mechanism of nanocrystalline materials. The simulation specimens are considered to exhibit idealized model microstructures having internal grains separated by high angle boundaries and containing no porosities and impurities. Microstructures can be synthesized either by simulation of a supercooled melt into which small crystalline seeds of random orientations are inserted or by filling the space according to the Voronoi construction with randomly nucleated seeds and crystallographic orientations. MD simulations for fcc nanometals demonstrated that partial dislocation activity takes over in nanocrystalline metals at grain sizes above ~8–18 nm, below which grain-boundary sliding occurs via atomic shuffling. In the former case, stacking faults are produced associated with the gliding of partial dislocations generated and absorbed in opposite grain boundaries. The formation of stacking faults and deformation twins in nanocrystalline metals have been observed experimentally. On the other hand, MD simulations performed at temperatures above 0.7 T_m imply that a Coble creep mechanics prevails, i.e. grain-boundary sliding is governed by grain-boundary diffusion. It is noted that all MD simulations are carried out on idealized model microstructures under high stresses (~1–3 GPa) and high strain rates ($\sim 10^8$ s^{-1}) to meet the capacity of computation. It seems that the results of MD simulation do not reflect the practical deformation behavior of bulk nanocrystalline metals containing nanopores and impurities subjected to lower applied stresses and strain rates.

8.2 Inverse Hall–Petch Effect

The strengthening of conventional polycrystalline metals and alloys with grain refinement can be described by the classic Hall–Petch (H–P) equation:

$$\sigma_y = \sigma_o + k\,d^{-1/2} \tag{8.1}$$

where σ_y is the yield stress, d is grain size, σ_o is friction stress resisting the motion of gliding dislocation, and k is the H–P slope, which is associated with a measure of the resistance of the grain boundary to slip transfer. A similar behavior is observed in the plot of hardness vs. grain size curves. The H–P relationship is almost obeyed in ductile metals with grain sizes of ~1 μm down to 100 nm. The strengthening is attributed to the pile-up of dislocations at grain boundaries acting as efficient barriers to dislocations motion. By decreasing the grain size of metals from 100 nm to a critical value (ca. ~20 nm), the H–P slope becomes smaller or even approaches zero. With further refinement below a critical grain size, the H–P slope is negative. This behavior is referred to as the inverse H–P effect. The interpretation of this inverse H–P effect still remains controversial. Several factors, such as grain-boundary sliding, creep diffusion, triple junctions, pores, and impurities could contribute to inverse H–P relation in metals and alloys [2–6]. Figures 8.1a–c show the grain-size dependence of the hardness of Cu, Pd, and Ni metals. For Cu and Pd metals, the plots were reconstructed by Takeuchi [7] on the basis of the hardness-grain size results of several research groups including Chokshi *et al.* [3], Nieman *et al.* [8], Fougere *et al.* [9], and Sanders *et al.* [10]. Apparently, inverse H–P behavior can be observed in these metals when $d < 20$ nm.

Bulk nanocrystalline metals of nanograin sizes (<100 nm) for mechanical tests are usually prepared via in-situ compaction of nanoparticles prepared from inert gas condensation (IGC), consolidation of ball-milled nanopowders, and electrodeposition [11–20]. Inert gas condensation proceeds via evaporation of a metal in ultrahigh vacuum using a variety of hot sources. The vaporized species then migrate into a cooler gas, typically helium by a combination of convective flows and diffusion. A high degree of material vapor supersaturation is needed to condense the nanoparticles. The particles collected at cold finger are consolidated in situ in the same facility. Consolidated IGC samples generally contain internal impurities and pores and insufficient interparticle bonding [21]. The cleanliness of the vacuum environment, gas purity, and outgassing rate of the vacuum system determine the presence of impurities in the material obtained [22]. The intrinsic mechanical deformation behavior of consolidated IGC samples is often overshadowed by responses from the internal pores.The pores and other defects developed in the consolidated IGC powders are detrimental to the mechanical properties such as a reduction of the modulus [2,10]. Other major limitations of IGC include high cost of the equipment and a relatively low yield of nanoparticles. In the case of ball-milling, raw powder particles placed inside a container undergo a repetitive cold-welding and fracturing mechanism

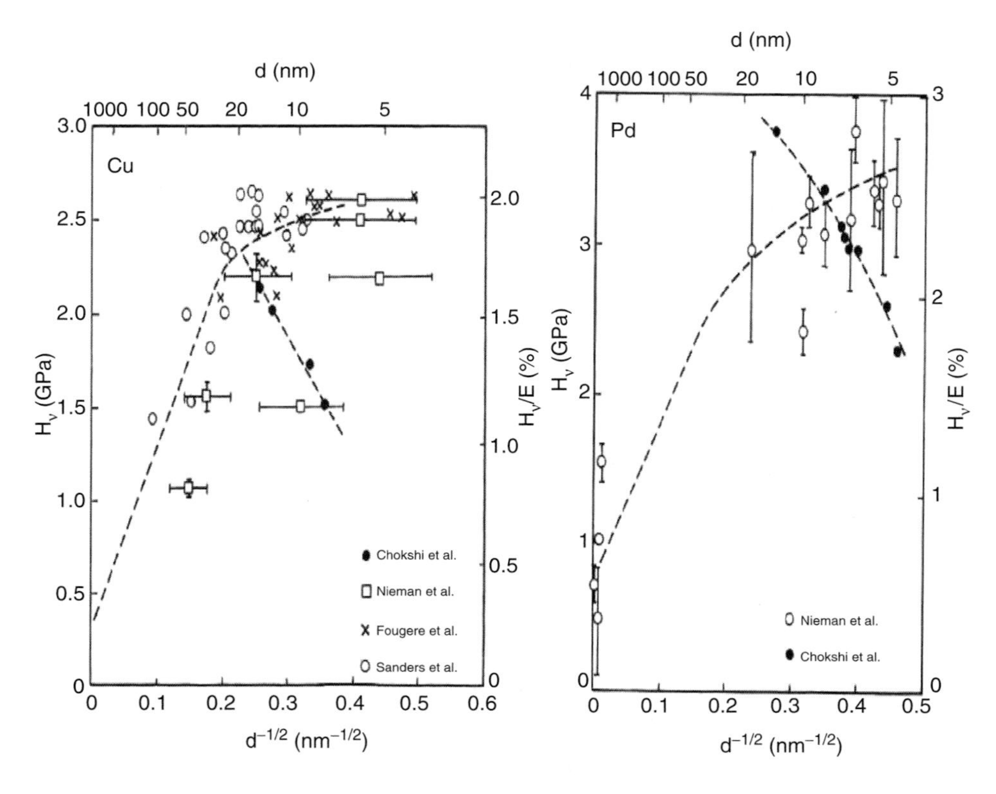

Fig. 8.1. Variation of hardness as a function of grain size for (a) Cu, (b) Pd, and (c) Ni. Data for polycrystalline Ni with grain size from nano-, ultrafine-, and microcrystalline regimes are indicated by the blue-colored points. The red points with scatter ranges denote data extracted from instrumented nanoindentation experiments on thin Ni foils produced by pulsed laser deposition (reprinted from [7a and 7b] with permission from Elsevier).

subject to the colliding action of stainless steel balls under an appropriate ball/powder ratio. The disadvantage of ball-milling for making nanocrystalline powders is the contamination of products from the milling media (balls and vial) and atmosphere. Moreover, recrystallization and grain growth occur readily during high-temperature consolidation of ball-milled powders in forming bulk samples. Compared with mechanical consolidation, the advantages of electrodeposition for forming dense nanocrystalline materials are: (a) low cost and industrial applicability, as it involves little modification of existing electroplating technologies, (b) simple operation, as the electrodeposition parameters can be easily tailored to meet the required crystal grain size, microstructure, and chemistry of products, (c) versatility, as the process can produce a wide variety of pore-free materials and coatings, and (d) high production rates. Nanocrystalline deposits are formed on a cathode surface during plating by properly controlling the electrodeposition parameters, e.g. bath composition, temperature, pH, etc. It should be noted that the contamination from the components in the electroplating bath is the major problem. Electrodeposited samples may contain sulfur, hydrogen, and carbon impurities that originate from the

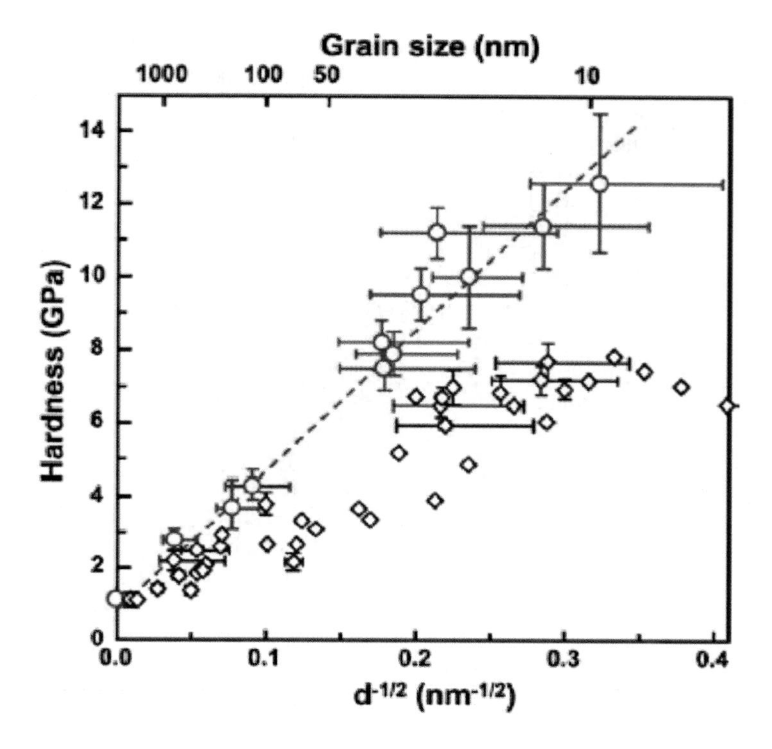

Fig. 8.1. *cont'd*

chemicals and organic additives used. The microstructure may also be inhomogeneous due to the formation of textured grains [23]. These impurities could affect mechanical performances of the electrodeposits, and their effects remain poorly understood, except for sulfur [24]. Sulfur in Ni has a very high diffusivity and is expected to segregate to the grain boundaries, thereby causing embrittlement [25,26]. Therefore, the fabrication of bulk nanocrystalline metals that are free from porosities and free of contaminants is still a big challenge for materials scientists.

Though all the data results presented in Figs. 8.1a and b are derived from the IGC samples, but the impurity chemistries and contents of IGC samples differ from one research group to another. Consequently, the production and consolidation of IGC powders will greatly affect the mechanical properties of nanocrystalline metals. For pore-free and dense electroplated Ni samples, El-Sherik *et al.* also observed a deviation from H–P relation [27]. As electrodeposited samples may contain impurities and exhibit inhomogeneous microstructure, it is unclear whether inverted H–P behavior is inherited from intrinsic effect of the nanograins or resulted from extrinsic flaws introduced into the samples during processing.

From this, it appears that conventional dislocation mechanisms cease to operate when the grain sizes are below a critical value. This is because the size of a Frank–Read source cannot exceed the grain size. As nanocrystalline metals contain a high fraction of atoms at interfaces that contain free volumes, there is a widespread speculation that deformation occurs primarily at the grain-boundaries. A number of possible mechanisms have been proposed to explain the grain-boundary deformation mechanisms for nanocrystalline metals, e.g. grain-boundary diffusion creep (Coble creep), grain-boundary sliding, grain-boundary rotation, triple junction effect, etc. In particular, grain-boundary diffusion creep and/or grain-boundary sliding are considered as the main deformation mechanisms in nanocrystalline metals. However, there are no conclusive experimental evidences to support such mechanisms.

Creep deformation normally occurs at elevated temperatures for conventional coarse-grained metals and alloys. In contrast, nanocrystalline metals exhibit ambient temperature creep. In the diffusion creep of coarse-grained metals, atoms are transported from grain boundaries experiencing compressive stresses to those subjected to tensile stresses. Therefore, such grains would elongate in the tensile direction under diffusion transport of atoms. As a result, individual grains tend to elongate, producing macroscopic strains. Such a change in the grain shape must be accommodated by grain-boundary sliding to prevent the formation of internal voids or cracks during diffusional creep. For the mechanism involving thermally activated processes, the observed creep rate ($d\varepsilon/dt$) due to Coble creep for a given applied stress σ is given by [28]:

$$\dot{\varepsilon} = A\frac{\sigma\Omega}{kT}\frac{\delta D_{\mathrm{GB}}}{d^3} \tag{8.2}$$

where δ is the effective diffusional width of the grain boundary (GB), Ω is the activation or atomic volume, $\sigma\Omega$ is the work performed by the stress during an elementary diffusion jump in the grain boundary, D_{GB} is the grain-boundary diffusion coefficient, and A is a constant. Gifkins indicated that the grain-boundary sliding mechanism is either controlled by lattice or grain-boundary diffusion by taking the Nabarro–Herring and Coble creep mechanisms into consideration [29]. For the lattice diffusion-controlled grain-boundary sliding mechanism, the constitutive equation for the creep rate is given by [30]:

$$\dot{\varepsilon} = BD_{\mathrm{L}}\frac{Gb}{kT}\left(\frac{b}{d}\right)^2\left(\frac{\sigma}{G}\right)^2 \tag{8.3}$$

where b is the Burgers vector, G is the shear modulus, D_{L} is the coefficient of lattice diffusion, B is a constant and the other parameters are specified in Eq. 8.2. For the grain-boundary diffusion-controlled grain-boundary sliding, creep rate is

$$\dot{\varepsilon} = B'D_{\mathrm{GB}}\frac{Gb}{kT}\left(\frac{b}{d}\right)^3\left(\frac{\sigma}{G}\right)^2 \tag{8.4}$$

where B' is a constant. From these, it appears that the grain-boundary sliding models give rise to the parabolic relation between the creep rate and the applied stress, i.e. a stress exponent of 2.

Chokshi *et al.* attributed the inverse H–P effect in nanocrystalline Cu to rapid diffusion creep (Coble creep) at room temperature [3]. However, the inverse cube dependence of the strain rate on the grain size could not be verified experimentally at room temperature. This is because the creep rate of Cu at low to intermediate temperatures is two to four orders of magnitude slower than the values predicted by the equation for Coble creep. The samples show logarithmic creep typical of the behavior of conventional materials at room temperature [8]. Recently, Lu and coworkers have performed tensile creep tests on electrodeposited nanograined Cu (30 nm) in the temperature range 293–323 K [31]. The creep rates of nanograined Cu are of the same order of magnitude as those calculated from Coble creep. The activation energy for creep is determined to be 0.72 eV, which is close to that of grain-boundary diffusion in nanocrystalline Cu. Such creep behavior is termed 'interface-controlled diffusional creep.' The discrepancy in the creep results of Lu's group with those of Nieman *et al.* and Sanders *et al.* is attributed to the difference in the processing conditions of the samples. The samples of Lu's group were prepared by the electrodeposition process whereas consolidated IGC Cu samples containing internal nanopores were used by Nieman *et al.* and Sanders *et al.* Such internal pores would affect the creep behavior of Cu.

Yin *et al.* [32] studied the creep deformation of electrodeposited Ni with 30 nm grains at room temperature (290 K) and at 373 K. The stress exponent values for nanocrystalline Ni at room temperature and 373 K are determined to be 1.1 and 6.5, respectively. The rate of diffusional Coble creep exhibits a linear dependence on true applied stress. Thus the deformation of Ni at room temperature is dominated by a Coble creep mechanism, in which enhanced diffusion along grain boundaries, triple junctions, and quadruple nodes control the mass transport process. At 373 K, lattice dislocation gliding and grain-boundary sliding may play an important role in the deformation of nanocrystalline Ni. Very recently, Yin *et al.* [24] further investigated the creep deformation behavior of electrodeposited Ni (an average grain size of 26 nm) containing impurity species (0.03 to 0.122 at.% S; 0.085 at.% B) in the temperature range 290–473 K. They determined the activation energy for creep to be ~90 kJ/mol. This value is close to the activation energy for grain-boundary diffusion (107 kJ/mol) in coarse-grained Ni. This suggests that grain-boundary diffusion or grain-boundary sliding may dominate the creep deformation of electrodeposited Ni. The stress exponent values for impure Ni at room temperature and 373 K are determined to be 2 and 5, respectively. Because the stress exponent is 2 at room temperature and not unity as it should be in Coble creep, grain-boundary sliding is expected to be the dominant deformation mechanism [24]. As mentioned above, the chemical compositions and microstructure of electrodeposited samples differ from batch to batch. This is because impurities derived from the bath solutions and additives can incorporate into the samples. These impurities would influence the mechanical characteristics of nanocrystalline metals. As a result, many of the literature results are inconclusive in showing the intrinsic deformation behavior of electrodeposited samples.

8.3 Superplasticity

Superplasticity is a flow process in which microcrystalline metals exhibit high elongations prior to final failure. Superplasticity in metallic alloys with micrograins is attained at low strain rates from 10^{-5} to 10^{-3} s^{-1} and at temperatures higher than 0.5 T_m. Moreover, superplasticity occurs at low flow stresses without pronounced strengthening. The precipitate particle or second phase of microcrystalline alloys can inhibit the growth of micrograins significantly, thereby promoting superplasticity. However, such a strain rate range is rather low for industrial forming of structural materials. Superplasticity of nanocrystalline metals and alloys follows the general trend of the constitutive relations (Eqs. 8.3 and 8.4) but with important differences in the level of stress and strain hardening rates. From Eqs. 8.3 and 8.4, the grain-size exponent (p) has a value between 2 and 3, depending upon whether lattice diffusion or grain-boundary diffusion is the controlling mode of deformation. These equations reveal that a reduction in the grain size can lead to an increase in the superplastic strain rate at constant temperature, or a reduction in the superplastic temperature at constant strain rate [33,34]. High strain rate superplasticity in nanocrystalline metals/alloys is often characterized by very high flow stresses or pronounced strengthening. Grain-boundary sliding is considered to be the dominant deformation mode for superplasticity in nanocrystalline materials [33]. Both the increase of superplastic strain rate and reduction of superplastic temperature in nanocrystalline metals are of technological significance as they would facilitate the advent of superplasticity-forming capability at lower production time and cost.

McFadden *et al.* [35,36] reported that electrodeposited nanocrystalline Ni (35 nm) exhibits superplastic deformation behavior at 350–420°C ($\sim$0.36 T_m) at a strain rate of 10^{-3} s^{-1}. An elongation of 295% was achieved when samples were deformed at 350°C. A maximum elongation of 875% was achieved when samples were deformed at 420°C. Such a temperature range is considerably lower than that observed during the normal superplastic deformation of microcrystalline metals. They referred this as the low temperature high strain-rate superplasticity. McFadden and Mukherjee further demonstrated that sulfur impurity is needed to achieve low-temperature superplasticity in electrodeposited Ni. [37]. The segregation of sulfur at grain boundaries at temperatures $\geq$350°C could reduce the grain-boundary mobility and promote grain boundary sliding. Superplasticity was not achieved in nanocrystalline Ni electrodeposited from a sulfamate bath with very low sulfur content of less than 0.002 wt%. They attributed this to rapid grain growth of Ni without sulfur at 350°C (Fig. 8.2). More recently, Dalla Torre *et al.* [38] also investigated the superplastic behavior of electrodeposited Ni containing sulfur (746 ppm). They indicated that the formation of nickel sulfide phase associated with the segregation of sulfur to the grain-boundaries at 330°C plays a key role for superplasticity in Ni. They suggested that the nickel sulfide phase acts as a lubricant owing to its low melting point. Figure 8.3a shows the tensile stress–strain curves for electrodeposited Ni from room temperature to 330°C under a strain rate of 2×10^{-4} s^{-1}. It is apparent the mechanical behavior changes substantially between room temperature and 330°C. At room temperature, the tensile deformation behavior is characterized by a high tensile strength of 1800 MPa but limited tensile strain of $\sim$5.3%. This is the general characteristic of nanocrystalline metals. Poor tensile ductility at room temperature results from low strain hardening capacity. At 210°C, the tensile strength reduces to 800 MPa but the tensile strain increases to 9.2%. At 330°C, the yield strength drops to 105 MPa and the maximum elongation reaches 82.6%. This implies that superplasticity occurs at 330°C ($\sim$0.2 T_m). Morerover, a relatively high strain rate sensitivity value of 0.3–0.4 was determined at 330°C, compared to the 0.02 at room temperature [38]. The tensile strain rates also exhibit a dramatic influence on the tensile deformation behavior of electrodeposited Ni (Fig. 8.3b). From these, it appears that superplasticity does not

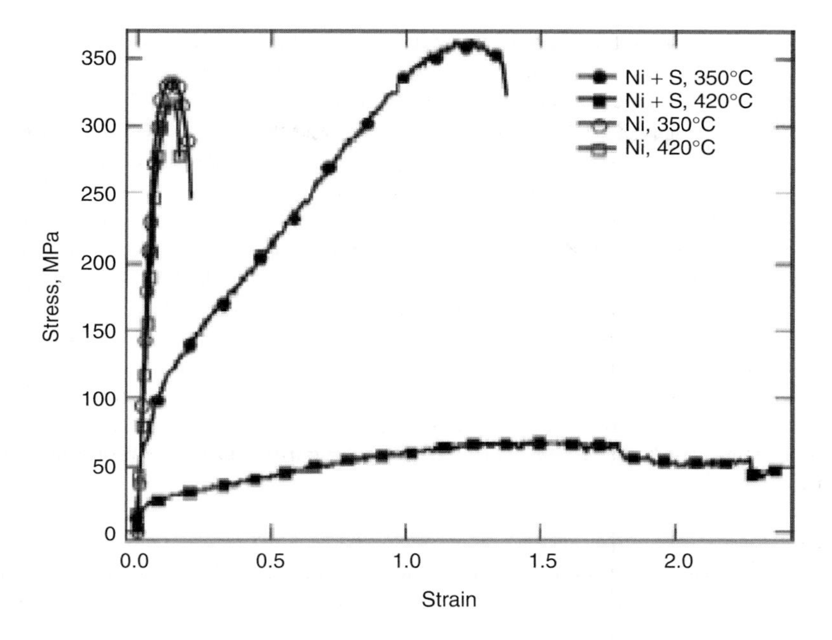

Fig. 8.2. Stress–strain curves for electrodeposited Ni with and without sulfur (reprinted from [37] with permission from Elsevier).

take place at ambient temperature in contrast to the occurrence of room temperature creep in nanocrystalline metals.

8.4 In-situ TEM Observation

Basic understanding of structure–property relationships is essential to elucidate the mechanisms responsible for the deformation of nanocrystalline metals. The technique of in-situ straining of TEM foil specimens is effective to explore the deformation and grain-boundary structures of nanocrystalline materials [39–41]. TEM observation is focused within localized small area of the samples, hence the results may not be representative of the bulk behavior. Ke *et al.* conducted in-situ deformation on nanocrystalline gold films with grain diameters of ~10 nm [39]. They found, by observing relative fringe rotation in different grains, that deformation occurred by grain-boundary sliding and grain rotation. No evidence of dislocation activity was detected during straining. Youngdahl *et al.* [41] strained nanocrystalline IGC Cu foils in a TEM, and videotaped the experiments. The grain size distribution of the Cu sample was very broad, ranging from 2 to 500 nm, with the majority lying between 50 and 80 nm. Dislocation activity-induced deformation was observed in grains with diameters of ~30–100 nm. There was no direct evidence for grain-boundary sliding or rotation. The presence of dislocation activity and absence of the grain-boundary sliding or rotation in strained Cu foils are derived from their larger grain sizes. This result is in sharp contrast to the observation

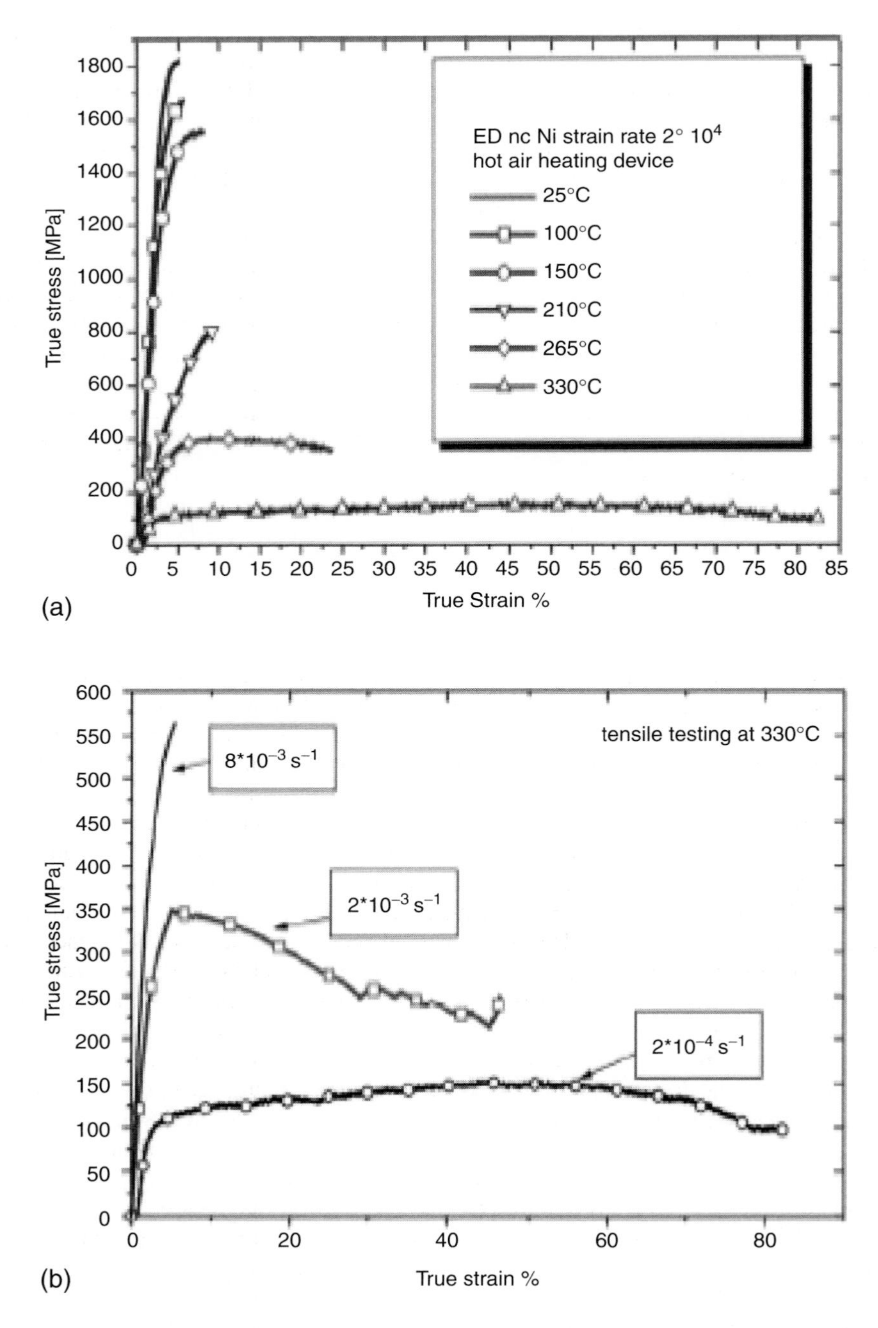

Fig. 8.3. (a) Tensile true stress vs. true strain curves for electrodeposited nickel at various temperatures under a strain rate of 2×10^{-4} s^{-1}. (b) Stress–strain curves at 330°C under different strain rates (reprinted from [38] with permission from Elsevier).

of Ke *et al.* mentioned above. Legros *et al.* [42] designed a microsample tensile testing device to strain dog bone-shaped microsample tensile specimens (3 mm in overall length with a 200 × 200 μm cross-sectional gauge). Subsequent TEM examination found no dislocation activity in nanocrystalline Ni with an average grain size of 28 nm. However, SEM observation revealed extensive dislocation-based plasticity in aluminum (Al) with a large grain size, i.e. 250 nm. In a recent in-situ-strained TEM study, Kumar *et al.* [40] observed copious dislocation activity in the region around the crack tip of electrodeposited Ni with nanograins of ~30 nm. Voids were formed at grain boundaries and triple junctions in the regions ahead of the crack tip. Thus, nanocracks generated at triple junctions acted as sites for nucleation of dimples. Ex-situ SEM examination confirmed the formation of dimpled failure in nanocrystalline Ni. The size of dimples was six to ten times larger than the average grain size of Ni. It is worth noting that the grain size of their electrodeposited Ni (~30 nm) is larger than a critical grain size (12 nm) predicted by the molecular dynamic simulations as discussed in the next section. The copius dislocation activity observed could have resulted from the emission of Schockley partials from the grain boundaries.

8.5 Theoretical Modeling

8.5.1 Molecular dynamic simulation

As contamination-free and dense nanocrystalline metals with grain size smaller than 20 nm are difficult to acquire at present, large-scale molecular dynamic (MD) modeling proves to be a powerful tool to simulate the interfacial structure and mechanical deformation mechanism of nanocrystalline materials [43–54]. Simulations can provide the atomic level details and deformation structures that are not accessible by experimental routes. The simulations are strongly dependent on the appropriate selection of empirical models for atomic interactions and systems with realistic grain structures. Most simulations are performed on idealized model microstructures in which fully dense metals (e.g. Ni, Cu, and Pd) have their adjacent grains separated by high angle boundaries, and are free from impurities. The mathematical model that most ideally describes the metallic microstructure consists of a construction of randomly selected Voronoi cells. The very construction of a Voronoi space shows a strong resemblance to the evolution of a real microstructure during which more or less randomly distributed nuclei grow to form grains. The approach has shown to produce nanocrystals with a log-normal grain size distribution similar to that of experimental samples. According to the literature, three-dimensional (3D) nanocrystalline grains are either created by filling the space according to the Voronoi construction with randomly nucleated seeds and crystallographic orientations [43–45], or by crystallization from a melt [54–56]. Van Swygenhoven and coworkers reported that the grain-boundary region of nanocrystalline Cu and Ni constructed from a Voronoi model has a high degree of structural order that is not fundamentally different from the grain boundaries in polycrystals. In the simulation, a tight binding potential in the Parrinello–Rahman approach was employed. Periodic boundary conditions and fixed orthorhombic angles were imposed on the computational cell. On the other hand, Keblinski *et al.* proposed the formation of highly disordered or amorphous ‘glue-like’ grain-boundary structures. Molecular dynamic simulations using the Stillinger–Weber three-body potential were used to synthesize fully dense nanocrystalline silicon with a grain size up to 7.3 nm by crystallization from the melt [54]. A nanocrystalline structure was generated in modeling where a liquid was solidified in the presence of randomly oriented crystalline seeds. The system was then quenched, and the liquid crystallized on the seeds, thus yielding a nanocrystalline structure [54]. In this case, high atomic mobility takes place easily at less-perfect atomic packing at nanograin boundaries.

The MD simulations are particularly useful to predict the nanocrystalline plasticity in the grains with diameters below 10 nm. Such simulations are commonly performed under high stresses of about 1–3 GPa and high strain rates ($\sim 10^8$ s^{-1}). Schiotz *et al.* determined the yield stress from MD simulations as a function of the grain size, whereas Van Swygenhoven *et al.* concentrated on the strain vs. time behavior from which the strain rate and activation energy can be evaluated. These two groups recognized the grain-boundary sliding as the main deformation mechanism [46–51]. From the earlier MD simulations of the deformation of nanocrystalline Cu samples with grain sizes range from 3.3 to 6.6 nm at 300 K, Schiotz *et al.* [50] reported softening of the yield stress with decreasing grain size. The grains were grown according to the Voronoi construction. The nanocrystal contained approximately 10^5 atoms arranged in 16 grains. An interaction potential approach based on the effective-medium theory was adopted to calculate the forces acting between the atoms in the simulations. Uniaxial deformation was applied by stretching the nanocrystal in one direction. The movement of individual atoms was calculated, and their positions were recorded. The average stress in the sample as a function of the amount of deformation was calculated. Figure 8.4a shows typical MD stress–strain curves for nanocrystalline Cu with various grain sizes $\leq$6.56 nm at 300 K. The variations of flow (stress at the flat regime of stress–strain curves) and yield stresses with grain sizes at 300 K are also shown in Figs. 8.4b and 8.4c, respectively. Simulations

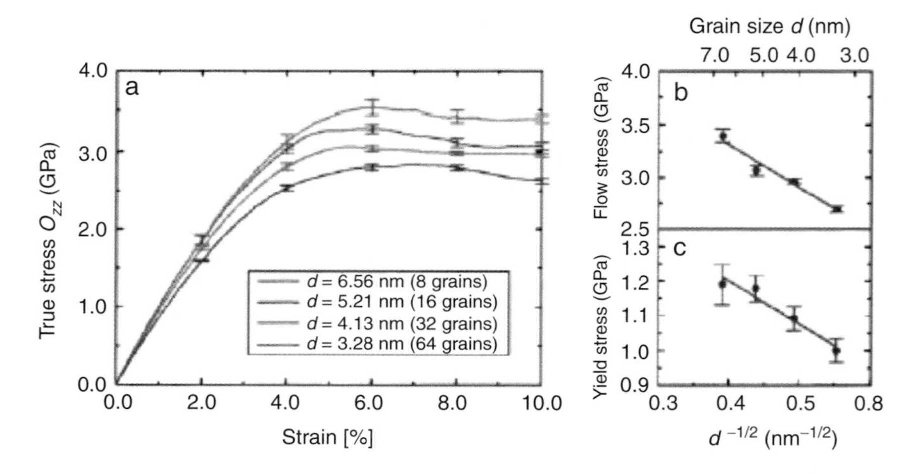

Fig. 8.4. (a) Atomic-scale simulation of the stress–strain curves for nanocrystalline Cu at 300 K for varying grain sizes. Grain-size dependence of (b) yield stress and (c) flow stress (reprinted from [50] with permission from Nature).

yield a negative H–P relation for both flow and yield stresses at 300 K for pore-free Cu samples. They attributed this to the sliding of the grain boundaries. It is noted that a crossover from the 'inverse' behavior from nanograin sizes to the normal Hall–Petch regime at larger grain sizes is not attained due the the limitation and cost of computation processes. They further simulated the deformation of such nanocrystalline Cu at 100 K for grain sizes of up to 13.2 nm. Similarly, grain-boundary sliding can occur without any thermally activated process at 0 K (athermal), and driven by high stress only [51]. Very recently, Schiotz *et al.* [52] further simulated plastic deformation of nanocrystalline Cu with system sizes up to 100 millions. Their new model incorporates a large number of grains and a wider distribution in grain size, i.e. the size varies from 5 to 50 nm. A crossover from normal H–P to inverse H–P occurs at ~14 nm in the plot of flow stress vs. grain size (Fig. 8.5). Such crossover is caused by a shift in the microscopic deformation mechanism from dislocation-mediated plasticity to grain-boundary sliding. The dislocation-mediated process in nanocrystalline Cu with grain sizes above a critical value is associated with the nucleation of many single Schockley partials, producing a large number of stacking faults (Fig. 8.6). In the case of high stacking fault energy metal like Al, Yamakov *et al.* reported that the crossover (critical) grain size is 18 nm on the basis of MD simulations [55].

In recent simulations on the tensile deformation of nanocrystalline Ni and Cu with grain sizes of 3–12 nm, Van Swygenhoven *et al.* indicated that grain-boundary sliding is the dominant deformation mechanism for Ni with grain sizes below a critical value (d_c), i.e. <12 nm. The sliding is triggered by atomic shuffling and to some extent by stress-assisted free volume migration. Above 12 nm, intergrain deformation or Schockley partial dislocation emission from the grain boundaries occurs. Such partials move through the grains and trap into the opposite grain boundaries, leaving trails of stacking faults. In Cu, the stacking faults are observed at smaller grain sizes than in Ni, i.e. 8 nm since Cu has

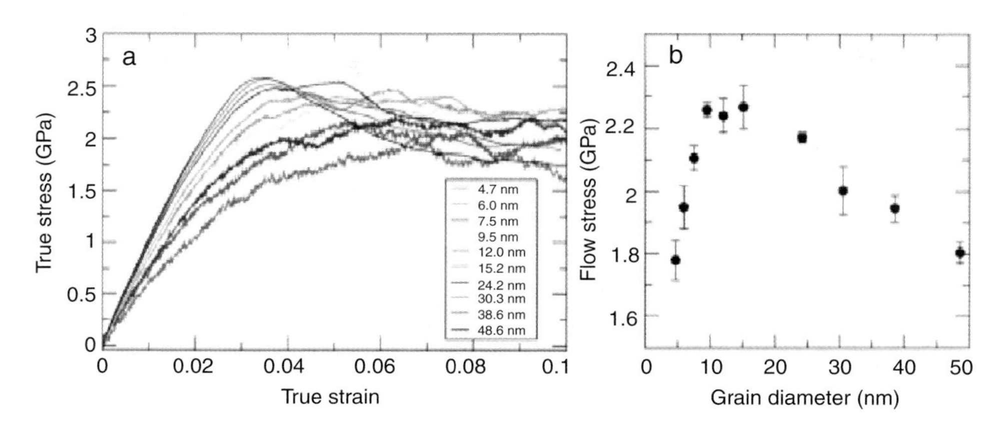

Fig. 8.5. MD simulation in nanocrystalline Cu showing (a) stress–strain curves for varying grain sizes and (b) grain-size dependence of flow stress. A crossover from normal H–P to inverse H–P relation occurs at ~14 nm (reprinted from [52] with permission from the American Association for the Advancement of Science).

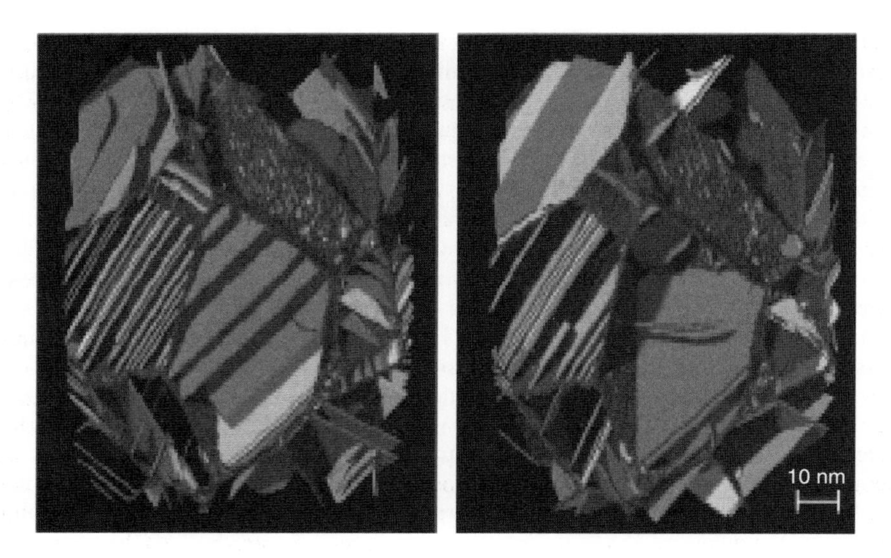

Fig. 8.6. MD simulations showing the formation of planar faults in nanocrystalline Cu with an average grain size of 39 nm, deformed at 5×10^8 s^{-1} (left) and 5×10^7 s^{-1} (right). Stacking faults are dark and twin boundaries are light. All atoms in fcc lattice have been removed for the purpose of clarity. In the right panel, a dislocation that splits into two partials separated by a stacking fault ribbon, is moving through the large central grain (reprinted from [53] with permission from Elsevier).

lower stacking fault energy [44,46,48]. From the MD simulations of Swygenhoven *et al.* and Shiotz *et al.* it appears that no dislocation activity exists below d_c. In this case, plastic deformation takes place entirely by means of grain-boundary sliding. Sliding of the grain boundaries in porosity-free nanocrystalline metals seems to be accommodated by stress-induced atom shuffling that resembles grain-boundary diffusion. However, there is no direct experimental evidence for the grain-boundary sliding in nanocrystalline metals with grain sizes below ~8–12 nm.

As mentioned above, MD simulations of Swygenhoven *et al.* and Schiotz *et al.* predict the emissions of dislocations grain boundaries, because of the lack of dislocation Frank–Read sources within nanograins with sizes above a critical value, e.g. 12 nm for Ni. Intrinsic stacking faults are produced by the motion of Schockley partial dislocations generated and absorbed in opposite grain boundaries. Thus, the grain boundaries are considered to be acting as sources and sinks for the dislocation activity. Yamakov *et al.* carried out MD simulations of nanocrystalline Al (d = 45 nm) under relatively high stresses of ~2.5 GPa and large plastic strains of ~12%. Both heterogeneous and homogeneous nucleation of deformation twins is observed. The heterogeneous mechanism involves the successive emission of Schockley partials from the grain boundaries onto neighboring slip planes [57]. The nucleation of deformed twins seems to be characteristic of MD simulations with nanograins and high strain rate. Deformation twinning is generally known to occur in microcrystalline fcc metals, such as Cu and Ni at low deformation temperatures or high strain rates. However, deformation twins are not observed in polycrystalline Al. More recently, Chen *et al.* observed the deformation twins in nanocrystalline Al films with grain sizes of ~10–20 nm using high-resolution TEM. The Al films were deformed by microindentation. Their results underscore a transition from deformation controlled by normal slip to those controlled by partial dislocation activity when grain size decreases to tens of nanometers [59]. From these, it is apparent that partial dislocation activity takes over in nanocrystalline metals at grain sizes around 10–20 nm, below which grain-boundary sliding plays a dominant role in the plastic deformation on the basis of MD simulations. It is noted that most MD simulations are focused mainly on fcc metals, little information is available on the MD simulations of mechanical behavior of nanocrystalline bcc and hcp metals. The bcc metals generally exhibit a rapid increase of yield strength at low temperatures and deform in brittle mode accordingly. More recently, Schiotz and coworkers demonstrated that the plastic deformation in nanocrystalline molybdenum with MD simulations involves grain-boundary sliding, dislocation migration and twin formation. Molybdenum is easily susceptible to interganular fracture beyond 6% of deformation [60a]. For hcp nanocrystalline metals, Zheng *et al.* indicated that both partial dislocation and dissociated dislocation activities are prevalent in nanocrystalline cobalt (10 nm) with low stacking fault energy. Moreover, deformation twinning is rare in spite of high applied stress (2.8 GPa) during the MD simulation [60b].

It should be noted that MD simulations of the deformation of nanocrystalline materials are performed at high loads, and under extremely high strain rates, e.g. on the order of $\sim 10^8\ s^{-1}$ [51]. Such high strain rates can cause serious plastic deformation in materials during practical tensile tests. The MD simulations also have certain limitations. The time and size scales that can be studied with the MD method are restricted. These impose constraints on the number or the size of the grains to be simulated. The small number of grains compared with real samples may induce anomalies in the measured properties. Further, the deformation can only be initiated at early stages, thus extremely high strain rates are needed to get any reasonable deformation within the available time [45]. In this respect, simulations related to time-dependent deformation, such as creep become difficult.

Gleiter and coworkers and Yamakov *et al.* conducted MD simulations to study the creep deformation in silicon and palladium at high temperatures [56,58]. The creep behavior is controlled by grain-boundary diffusion and described quantitatively by Coble creep (Eq. 8.2). In this regard, there is no doubt that grain-boundary diffusion is a thermally activated process associated with high-temperature simulations. For simulations of nanocrystalline Pd at 900–1300 K, the activation energy for grain-boundary diffusion creep (0.61 eV) is the same as that for grain-boundary self-diffusion (Fig. 8.7). Although extrapolation of Coble creep deformation to room temperature has been considered, it is unclear whether the results of simulation conducted at high temperatures truly reflect the creep deformation of nanocrystalline metals at room or ambient temperatures. Such low temperature deformation is considered to be associated with athermal process [31,32].

When the grain-boundary width, δ, is non-negligible compared with the grain size (d) in the nanocrystalline metals, the grain-size dependence of the strain rate changes from d^{-3} to d^{-2}. Taking grain-boundary width into consideration, the Coble creep rate

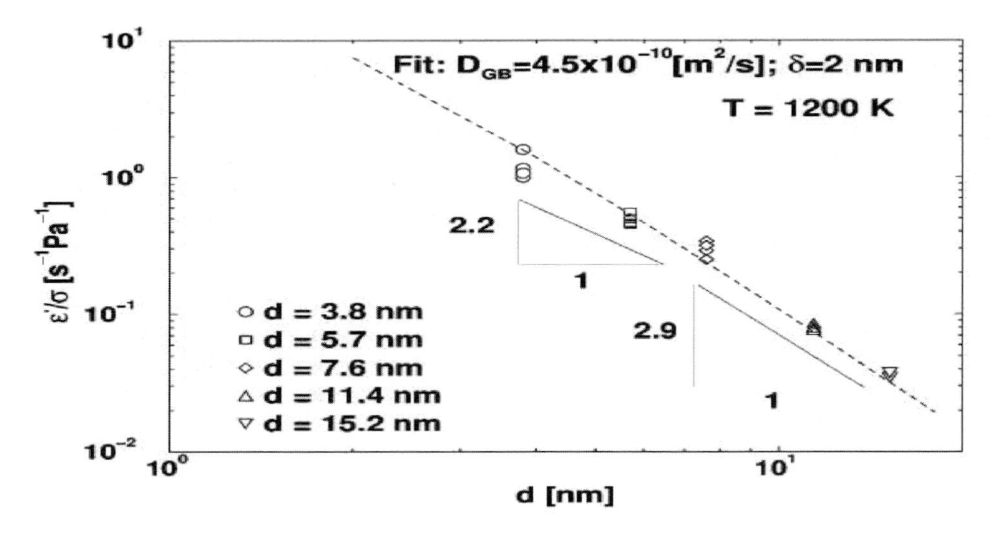

Fig. 8.7. MD simulation in nanocrystalline Pd showing log–log plot of $\dot{\varepsilon}/\sigma$ with at 1200 K. The dashed line shows a fit with Eq. 8.5 with estimated values of the fitted parameters $D_{\mathrm{GB}} = 4.5 \times 10^{-10}$ m^2/s and $\delta = 2$ nm (reprinted from [58] with permission from Elsevier).

equation is modified to:

$$\dot{\varepsilon} = 4\mathrm{K}\frac{\sigma \Omega_D D_{\mathrm{GB}}}{kT}\left(\frac{2\delta}{d^3} - \frac{\delta^2}{d^4}\right) \tag{8.5}$$

where K is a material constant. In the limit of large grain size, i.e. $d \geq \delta$, the second term inside the bracket of Eq. 8.5 is negligible compared to the first one, yielding the Coble creep as expressed in Eq. 8.2. In the limit of small grain size, $d \approx 2\delta$, the second term is comparable to the first one, and the resulting creep rate becomes

$$\dot{\varepsilon} = 3\mathrm{K}\frac{\sigma \Omega_D D_{\mathrm{GB}}}{kT}\frac{1}{d^2} \tag{8.6}$$

It is evident that the grain-size dependence of grain-boundary diffusion creep becomes identical to the Nabarro-Herring creep model ($\dot{\varepsilon} \propto \frac{1}{d^2}$) when the grain-boundary width becomes comparable to the grain size. Therefore, Eq. 8.5 extends the validity of the Coble creep equation from coarse-grained polycrystals to nanocrystalline materials, provided that the creep mechanism remains under the control of grain-boundary diffusion process. Moreover, grain-boundary sliding is needed to accommodate the homogeneous grain elongation induced by Coble creep, with the entire deformation process being controlled by grain-boundary diffusion [58]. In other words, grain-boundary diffusion requires grain-boundary sliding as an accommodation process in order to avoid the formation of microcracks. Throughout the simulations, Gleiter and coworkers used the marker lines to detect the grain boundary sliding. After ~4% elongation in the tensile direction, the marker lines experienced shifts across all the grain boundaries. Their results disagree with the MD prediction of Swygenhoven and Caro [47] who conducted MD simulations for creep deformation of nanocrystalline Ni samples under 3 GPa and 10^8 s^{-1} at temperatures below 120 K. They reported that the mechanisms responsible for creep deformation at low temperature are grain-boundary sliding, grain rotation, and grain-boundary motion. In other words, grain-boundary sliding, without diffusion accommodation, is proposed to be the principal room-temperature deformation mechanisms in fcc nanocrystalline Ni. It is considered that the diffusion of atoms does not take place at low temperatures in contrast to the simulations of Gleiter *et al.* at high temperatures in which diffusion of atoms does occur. Nevertheless, MD simulations clearly demonstrate that grain-boundary sliding tends to occur in nanocrystalline metals during creep tests despite the fact that the relevant accommodation process for sliding differs, i.e. one is grain-boundary sliding accommodated by stress-induced atom shuffling whilst the other is grain-boundary diffusion accommodated by grain-boundary sliding.

8.5.2 Analytical modeling

Apart from MD simulations, several analytical models have been developed to describe the grain-size dependency of the yield stress in nanocrystalline metals [61–67]. Among these approaches, the composite concept is a simple and effective method for modeling. Additional factors relating to the structure and elastic–plastic deformation of materials are incorporated into the formulations by individual researchers according to the simulation needs. For example, Kim and coworkers treated nanocrystalline material as a composite comprising of the two major phases, i.e. interior grains (GIs) and grain boundaries (GBs) [61,62]. They treated GIs collectively as one phase, while the GBs are considered as a separate second phase. The rule of mixture can then be applied to characterize some macroscopic physical quantities, such as yield strength and Young's modulus of nanocrystalline metals. According to the rule of mixture, the yield stress under the iso-strain condition is given by:

$$\sigma_y = f\sigma_{GB} + (1 - f)\sigma_{GI} \tag{8.7}$$

where f is the volume fraction of the grain-boundary phase, σ_{GB} and σ_{GI} are the stresses acting in the grain boundaries and grain interiors, respectively. The deformation mechanism for the grain-boundary phase is modeled as a diffusional flow of matter through the grain-boundary diffusion, i.e.

$$\dot{\varepsilon}_{GB} = A' \frac{\sigma\Omega}{kT} \frac{D_{GB}}{d^2} \tag{8.8}$$

where A' is a constant.

The plastic flow in GIs is governed by dislocation glide and diffusion mechanisms. The diffusion mechanism is a combination of the grain-boundary diffusion (Coble creep; Eq. 8.2) and the lattice diffusion (N–H creep). In other words, the total strain rate of a crystallite is calculated by the summation of the contribution of dislocation, boundary diffusion, and lattice diffusion mechanisms. In this respect, the plastic strain rate for the GI phase is:

$$\dot{\varepsilon}_{\mathrm{GI}} = \dot{\varepsilon}_{\mathrm{disl}} + \dot{\varepsilon}_{\mathrm{diff}} \tag{8.9}$$

where

$$\dot{\varepsilon}_{\mathrm{diff}} = \dot{\varepsilon}_{\mathrm{Coble}} + \dot{\varepsilon}_{\mathrm{N-H}} \tag{8.10}$$

$$\dot{\varepsilon}_{\mathrm{diff}} = 14\pi \frac{\Omega \sigma_{\mathrm{GI}}}{kT} \cdot \frac{\delta}{d} \cdot \frac{D_{\mathrm{GB}}}{d^2} + 14\pi \frac{\Omega_{\mathrm{GI}}}{kT} \frac{D_{\mathrm{L}}}{d^2} \tag{8.11}$$

in which D_{L} is the lattice diffusion coefficient and δ is the grain-boundary width. Taking the other parameter relating to dislocation density (ρ) into consideration, it is possible to evaluate the yield strength of nanocrystalline metals, strain rate sensitivity, and strain hardening exponent ($\partial\sigma/\partial\ln\varepsilon$) as a function of the grain size by solving the above-mentioned constitutive equations [61,62].

Figure 8.8 shows the calculated grain size dependence of yield strength for nanocrystalline Cu at various strain rates. For the purposes of comparison, the experimental results from several researchers are also included in the plot [8,68–71]. Generally, good agreement is obtained between the model and the experimental data points, although the

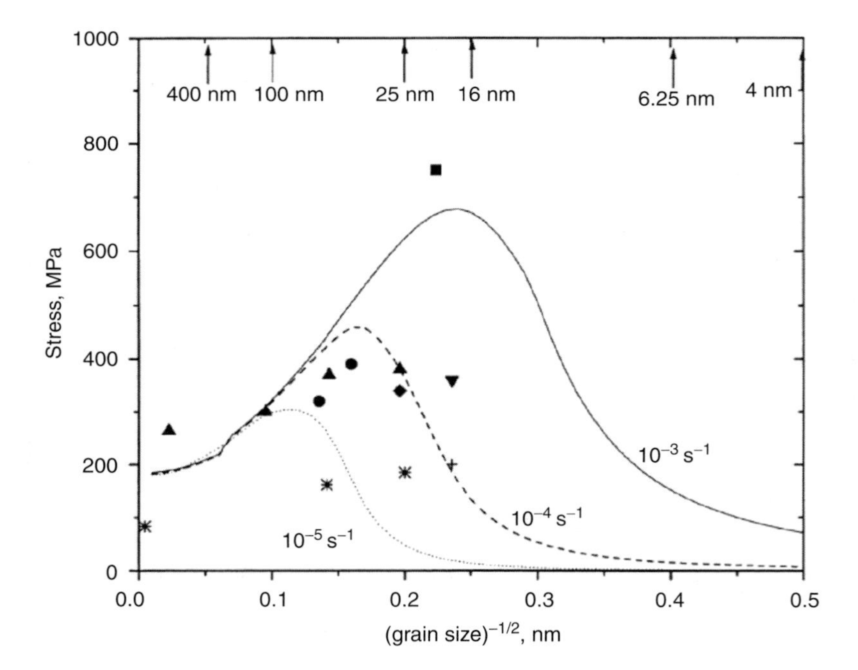

Fig. 8.8. Calculated grain-size dependence of yield stress for nanocrystalline Cu for strain rates of 10^{-3}, 10^{-4}, and 10^{-5} s^{-1} (reprinted from [61] with permission from Elsevier).

model gives overestimated values. Kim and coworkers attributed this to the experimental uncertainties, residual porosities, and grain-size distributions of nanocrystalline samples. Both normal and inverse H–P relations are predicted by the composite model. It can be seen that the point of reversal or critical grain size (d_c) moves to larger grain sizes with decreasing imposed strain rates. Moreover, the magnitude of the peak stress decreases with a decreasing strain rate as expected [61]. The calculated strain rate sensitivity (m) and strain-hardening exponent vs. grain size at various strain rates for Ni are shown in Figs. 8.9 and 8.10, respectively. For strain rates higher than 10^{-1} s^{-1}, the strain rate sensitivity is very low because the prevalent deformation mechanism is dislocation glide. The strain rate sensitivity is found to increase with decreasing grain size at slow strain rates $\leq 10^{-3}$ s^{-1} and reaches almost unity for $d \leq 20$ nm due to a growing role of the diffusion-controlled mechanisms with grain refinement. Similarly, the model predicts the elevated m values for nanocrystalline metals as observed in practice but overestimates them. The elevated strain rate sensitivity of nanocrystaline metals will be discussed in the next section. From Fig. 8.10, the strain-hardening exponent decreases when the strain rate and grain size are reduced. It vanishes at a strain rate of 10^{-3} s^{-1} for $d \leq 10$ nm, and at 10^{-5} s^{-1} for $d \leq 40$ nm. Such low strain-hardening capacity of metals with grain sizes in nanometer regime results in poor tensile ductility as reported in the practical cases.

Meyers and coworkers [72,73] also treated the nanocrystalline material as a composite, comprised of the grain interior, with flow stress σ_{fG} and grain-boundary work-hardened

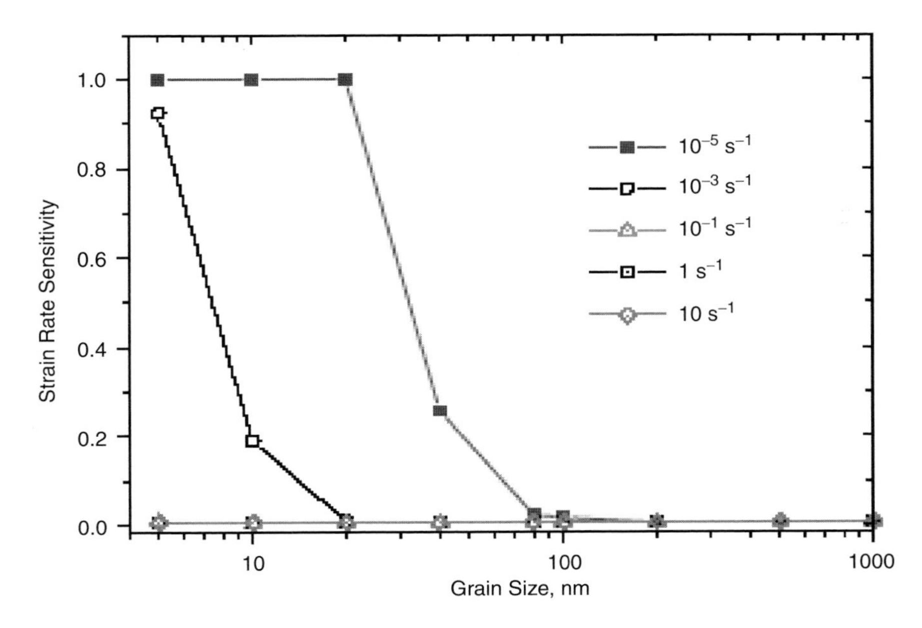

Fig. 8.9. Strain rate sensitivity at 5% strain as a function of the grain size calculated for various strain rates (reprinted from [62] with permission from Elsevier).

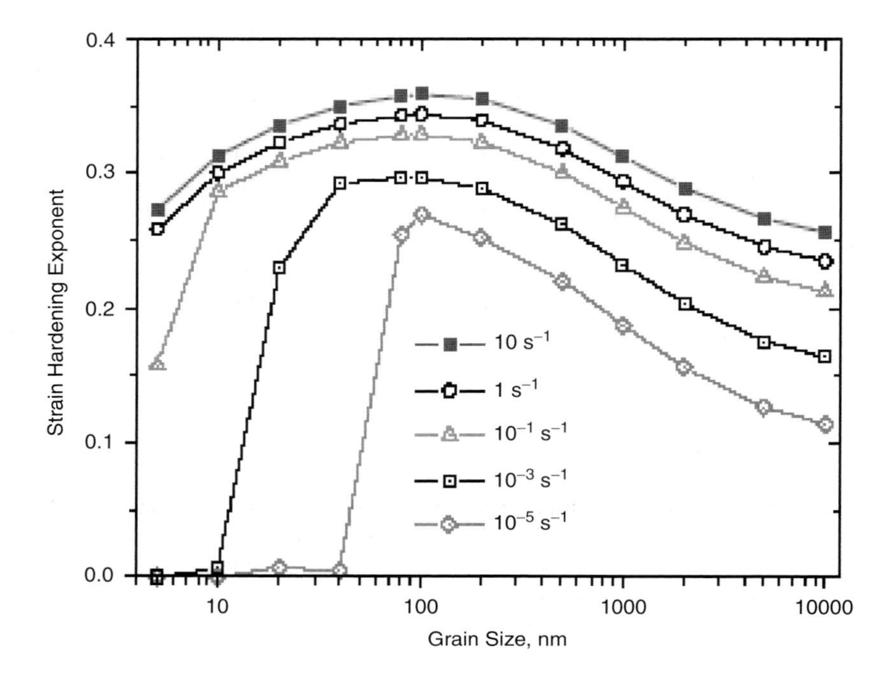

Fig. 8.10. Strain-hardening exponent (n) as a function of the grain size calculated for various strain rates (reprinted from [62] with permission from Elsevier).

layer, with flow stress $\sigma_{f\mathrm{GB}}$. Four principal factors contribute to grain-boundary strengthening. These include: (a) the grain boundaries act as barriers to plastic flow, (b) the grain boundaries act as dislocation sources, (c) elastic anisotropy causes additional stresses in the grain-boundary surroundings, and (d) multislip is activated in the grain-boundary regions, whereas grain interiors are initially dominated by single slip. From these, they derived the following equation:

$$\sigma_y = \sigma_{f\mathrm{G}} + 8\kappa(\sigma_{f\mathrm{GB}} - \sigma_{f\mathrm{G}})d^{-1/2} - 16\kappa^2(\sigma_{f\mathrm{GB}} - \sigma_{f\mathrm{G}})d^{-1} \tag{8.12}$$

For large grain size in the micrometer range, the $d^{-1/2}$ term dominates and a Hall–Petch relationship is obtained. The Hall–Petch slope, k_{HP}, is equal to:

$$k_{\mathrm{HP}} = 8\kappa(\sigma_{f\mathrm{GB}} - \sigma_{f\mathrm{G}}) \tag{8.13}$$

As the grain size is decreased, the d^{-1} term becomes progressively dominant; the σ_{y} vs. $d^{-1/2}$ goes through a maximum. This occurs at $d_{\mathrm{c}} = (4\kappa)^2$. For values of $d < d_{\mathrm{c}}$ it is assumed that the flow stress reaches a plateau. Thus, the decrease of the H–P slope in the nanocrystalline domain is explained and corresponds to a grain size for which the thickness of the work hardened layer is equal to one half of the grain diameter. It is noted that their computational approach use time-independent plasticity. Therefore, time-dependent phenomena, such as thermal activation, grain-boundary sliding, and diffusion mechanisms are not taken into account. Computational predictions are compared with experimental results for Cu and Fe, reported by Mallow and Koch, Weertman *et al.* Feltham and Meakin, Andrade *et al.* and Abrahamson II [74–78] as shown in Figs. 8.11a and b. Apparently, the H–P slope for both Fe and Cu decreases and asymptotically approaches a plateau when the grain size is progressively reduced. As time-dependent plasticity is not taken into consideration, it appears that Fu *et al.* model can be used to explain the experimental hardness vs. grain size curve for Cu in which the H–P slope approaches the plateau regime when the grain size decreases (Fig. 8.1a). In the case in which time-dependent plasticity is considered, the model based on Kim and coworkers is more appropriate to explain the softening H–P behavior of Cu as shown in Fig. 8.1a.

8.6 Strain Rate Sensitivity

Despite the fact that nanocrystalline metals exhibit high strength and hardness, they are very brittle with little tensile ductility compared to their coarser-grained counterparts [42]. Poor tensile ductility of nanocrystalline metals is associated with their low strain hardening capability. Ductility is of utmost importance in many forming operations and in the prevention of catastrophic failure during load-bearing applications. The tensile elongation of consolidated IGC Cu and Pd metals with grain sizes below 50 nm is quite low, i.e. in the range of 1.6–4%. In contrast, Cu with a grain size of 110 nm exhibits an elongation to failure >8% [70]. Compressional ductility in nanocrystaline Cu samples is found to be higher than the tensile ductility, with strains of 12–18% prior to failure, but still much less than the coarse grain form. The low ductility of these nanocrystalline metals resulted from inferior sample quality due to the presence of microvoids and impurities [70].

For nanocrystalline Ni, Dalla Torre *et al.* [23] reported that the plastic strain of commercial electrodeposited Ni samples with a mean grain size of ~21 nm was very low and less than 4%. They attributed this to the presence of impurities and porosities in the electrodeposited nanocrystalline Ni. Two batches of electroplated Ni having the same mean grain size but different microstructures and impurities contents were used. Metallic impurities and sulfur contaminants originating from the saccharin inhibitor were detected in these two batches. Nanopores could possibly originate from hydrogen being incorporated during electrodeposition. Nearly full density of 99.5 ±0.5% was achieved despite the presence of hydrogen-filled nanopores. Mechanical properties of nanocrystalline materials are also found to be strain-rate-dependent. Dalla Torre *et al.* demonstrated that the average tensile strength of Ni samples is ~1388 ±61 MPa from 5.5×10^{-5} to 5.5×10^{-2}, but it increased dramatically to 2500 MPa at a strain rate of 10^3 s^{-1}. The tensile ductility appears to decrease dramatically with increasing strain rates from strain rates of 5.5×10^{-5} to 5.5×10^{-2} s^{-1} [23]. More recently, Schwaiger *et al.* employed both tensile and depth-sensing indentation techniques to examine the fracture behavior and damage evolution of electrodeposited nanocrystalline Ni (20 nm), ultrafine grained Ni (100–1000 nm), and microcrystalline Ni (>1 μm) specimens [79]. From the results of both measurements, they indicated that the flow stress of nanocrystalline Ni increases with increasing strain rate. This effect was not observed in ultrafine grained

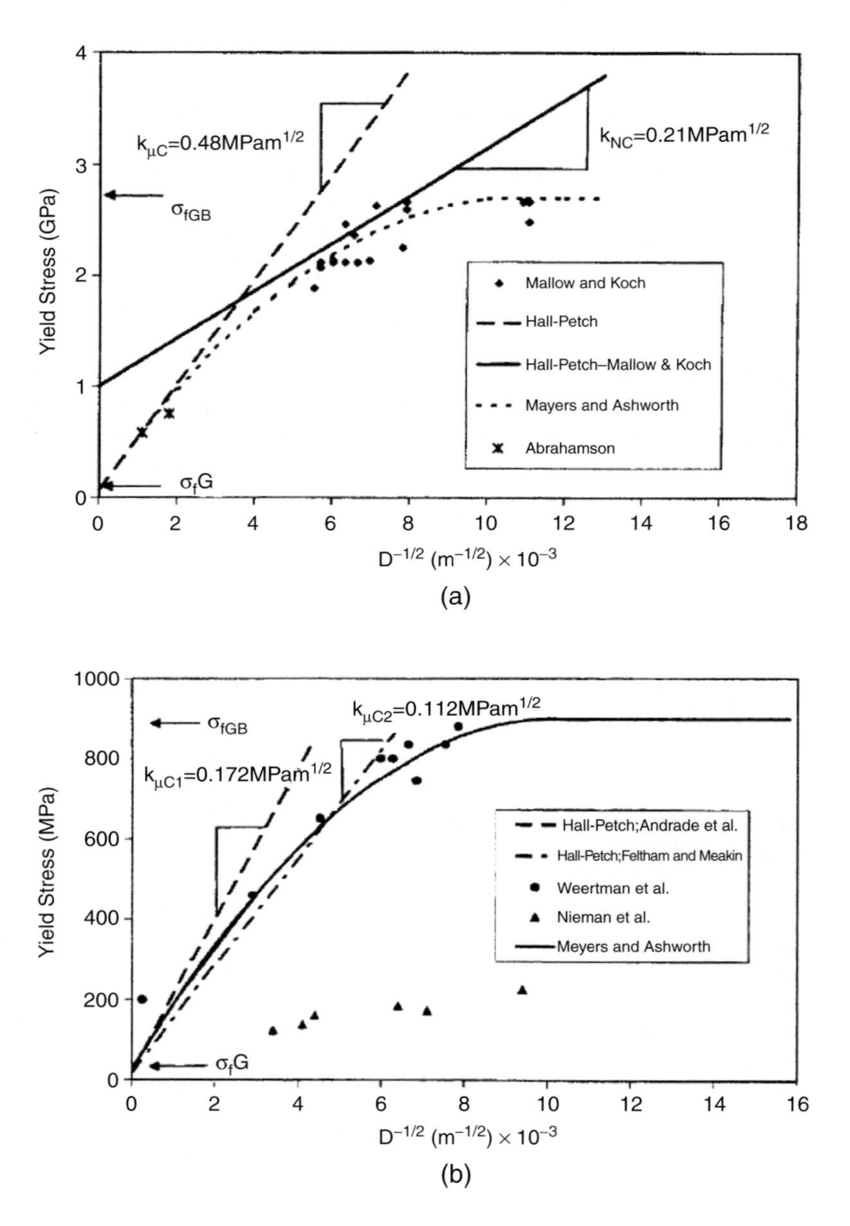

Fig. 8.11. Comparison of Meyers and coworkers' prediction with experimental data for (a) iron and (b) copper (reprinted from [72] with permission from Elsevier).

and microcrystalline Ni specimens. The yield strength and flow stress of microcrystalline Ni are known to be rate-insensitive.The positive strain rate sensitivity in the flow stress of nanocrystalline Ni was attributed to the localized plastic deformation in grain boundaries and the nearby regions.

In general, consolidated nanocrystalline samples are too small to be used in conventional, large-scale mechanical tests. In this regard, indentation method appears to be an effective route to study the creep deformation of nanocrystalline materials. Indentation creep, in which the extent of deformation is measured during both loading and unloading, has been studied in nanocrystalline materials by monitoring the changes in the indentation size occurring beneath the indenter as a function of time. In the process, the indenter is pushed into the surface at a fixed rate of indentation until a predetermined load or penetration depth is reached, then the load is held fixed while the indenter continues to creep into the material. Li and coworkers [80,81] demonstrated the use of indentation and impression creep in investigating the creep deformation mechanism of materials at high homologous temperatures. In other words, indentation creep technique can be used to determine the power law stress exponent and activation energy. The results obtained from indentation creep translate readily into those determined from tensile loading. In this regard, the activation volume (V^*) is determined preferentially during nanoindentation measurements [82–85]. V^* is defined as a measure of the work done by the external stress during the activation process. This corresponds to Burger's vector (b) times the area swept out by dislocation during the process of thermal activation, the so-called 'activation area' (A^*), i.e. $V^* = bA^*$ [86]. Elmustafa and Stone demonstrated that the activation volume data determined from nanoindentation creep for high purity aluminum and alpha brass samples scales with the activation volume measured using conventional uniaxial loading [84].

Considering an applied uniaxial stress (σ) induces a shear stress (τ) that acting on dislocations and sufficient to cause plastic strain, γ. As the plastic deformation is a thermally activated process, the rate equation can be described by an Arrhenius-type equation,

$$\dot{\gamma} = \dot{\gamma}_o \exp\left(\frac{-\Delta G(\tau^*)}{\kappa T}\right) \tag{8.14}$$

where $\dot{\gamma}$ is shear strain rate, ΔG is the activation Gibb's free energy required for the dislocation to overcome the obstacles; $\dot{\gamma}_0 = b\nu_0 l^2 N$ where ν_0 the frequency, l the distance between the obstacles and N the number of glide dislocations per unit volume; τ^* is the effective stress defined by $\tau = \tau^* + \tau_i$ (τ_i being the internal stress arising from obstacles). Differentiating Eq. 8.14 for τ^* yields the activation volume V^* [87] at a fixed temperature,

$$V^* = -\left(\frac{\partial \Delta G}{\partial \tau^*}\right)_T = kT\left(\frac{\partial \ln(\frac{\dot{\gamma}}{\dot{\gamma}_0})}{\partial \tau^*}\right)_T \tag{8.15}$$

In practice, activation enthalpy ΔH rather than ΔG is used ($\Delta H = \Delta G - T\Delta S$). Further simplification of Eq. 8.15 leads to

$$V* = \kappa T \left(\frac{\partial \ln \dot{\gamma}}{\partial \tau} \right)_T \quad (8.16)$$

and the strain rate sensitivity, m is related to $V*$ as:

$$m = \frac{\kappa T}{\tau V*} \quad (8.17)$$

Therefore, thermally activated m and $V*$ parameters provide quantitative measures of the sensitivity of flow stress to loading rate and show insights on the controlling plastic deformation mechanisms. For fcc metals, $V*$ is given as:

$$V* = b \cdot \xi \cdot l* \quad (8.18)$$

where ξ is the distance swept out by mobile dislocation during one activation event, and $l*$ is the length of dislocation segment involved in the thermal activation [88]. Combining Eqs. 8.16 and 8.17, we have

$$m = \frac{\kappa T}{\tau \cdot l* \cdot \xi \cdot b} \quad (8.19)$$

For nanocrystalline materials, Wei *et al.* [89] demonstrated that m takes the form,

$$m = \frac{\kappa T}{\xi b} \cdot \frac{1}{\chi[\alpha(\rho)^{1/2}d + \beta(d)^{1/2}]} \quad (8.20)$$

where ρ is the density of dislocation, α, β, and χ are constants. Equation 8.20 implies that m should increase with decreasing d. The m values of nanocrystalline fcc metals are much higher, whereas those of bcc nanocrystalline metals are much reduced compared to their microcrystalline counterparts [23,79,89,90]. The reduced m in nanocrystalline bcc metals would induce plastic instability more easily, leading to shear localization during deformation. According to the literature, typical m values for microcrystalline Cu and Ni are ~0.006 [91] and 0.004 [23,92], respectively. Therefore, fcc metals are characterized by low strain rate sensitivity and deform by dislocation slip mechanism. The m value of electrodeposited Ni increases to ~0.01–0.03 as the grain sizes reduces to nanoscale level [23,79]. The activaton volumes of fcc polycrystalline Cu and Ni are $\sim 1000b^3$ and $800\text{–}2000b^3$, respectively [93,94]. With decreasing grain size, the dislocation gliding is suppressed and grain-boundary-mediated mechanisms (e.g. grain-boundary sliding, atomic shuffling) increases. The activation volume of nanocrystalline Ni is about two orders of magnitude smaller, i.e. $10\text{–}20b^3$ [89,95]. The larger m but smaller $V*$ values in nanocrystalline metals indicate a strong temperature dependence of yield or flow stress. In this case, the grain-boundary-mediated deformation mechanisms would dominate when the grain size is reduced to nanoscale regime.

The incorporation of nanoscale twins during the processing of metals with ultrafine grains is also known to increase the loading rate sensitivity by almost an order of magnitude and decrease the activation volume by two orders of magnitude as compared to the values observed in microcrystalline metals. Lu *et al.* reported that the introduction of nanoscale twins within ultrafine crystalline metals, leads to significant increases in flow stress and hardness [96,97]. The extent of such strengthening is comparable to that achievable by nanocrystalline grain refinement [7b]. A typical plot showing the dependence of activation volume on grain size for pure Cu and Ni is shown in Fig. 8.12. The activation volume for the ultrafine grained Cu specimens with nanotwins is also indicated in this figure, with the twin width replacing the grain size as the characteristic structural length scale. It can be seen that a 100-fold increase in the activation volume as the spacing of the internal interface is varied about 20–100 nm [7b].

In defect-free nanocrystalline metals, the ability to strain harden is an important factor for stabilizing uniform tensile deformation. In tensile tests, the onset of non-uniform deformation or necking can be predicted by the Considere criterion [98], i.e.

$$\left(\frac{d\sigma}{d\varepsilon}\right) \leq \sigma \tag{8.21}$$

where σ and ε are true stress and true strain, respectively. For nanocrystalline materials, its strain hardening rate $(d\sigma/d\varepsilon)$ is low and strength (σ) is high. This results in the

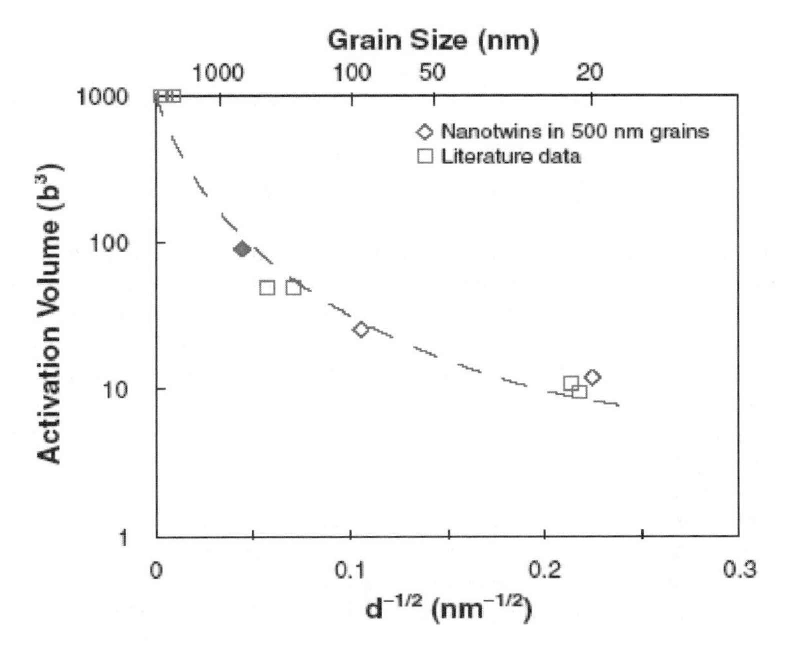

Fig. 8.12. A plot of the effect of grain size on the activation volume, measured in units of b^3, for pure Cu and Ni from available information in the literature (reprinted from [7b] with permission from Elsevier).

onset of plastic instability or necking occurs in nanocrystalline metals at very small strains during tensile deformation. Early plastic instability in nanocrystalline metals is manifested in the form of shear banding that concentrates large strains and stresses, leading to rapid catastrophic failure due to excessive localized deformation. Evidence of shear bands has been seen in nanocrystalline metals like fcc Cu, Pd, Ni, and bcc Fe under tension [6,8,23] or bcc metals under compression [99–101]. For fcc metals, there appear to be no reports for shear localization during compression. Compression test is particularly attractive for evaluation of the strain-hardening response of nanocrystalline metals because the compressive behavior is not affected by internal flaws, not subjected to necking instabilities, and less likely to suffer from immediate catastrophic failure even when shear bands form [102]. Figure 8.13a shows typical shear bands developed in bcc Fe with grain sizes $\leq$268 nm at a plastic strain of 3.7% under uniaxial compression. With increasing the plastic strain to 7.8%, broadening of existing shear bands (Bands I, II, and III) and propagation of bands (Band III) into several branches are observed. New shear bands (IV) are also nucleated at this stage. The development of a network of shear bands at about $\pm 45°$ to the loading axis at 7.8% plastic strain is summarized in Fig. 8.13b. TEM observations reveal that grains in the band are elongated along the shear direction. In contrast, equiaxed grains are observed outside the band.

From the above discussion, it is clear that the elimination of impurities and mechanical instability can lead to improved tensile strength and ductility of nanocrystalline metals.

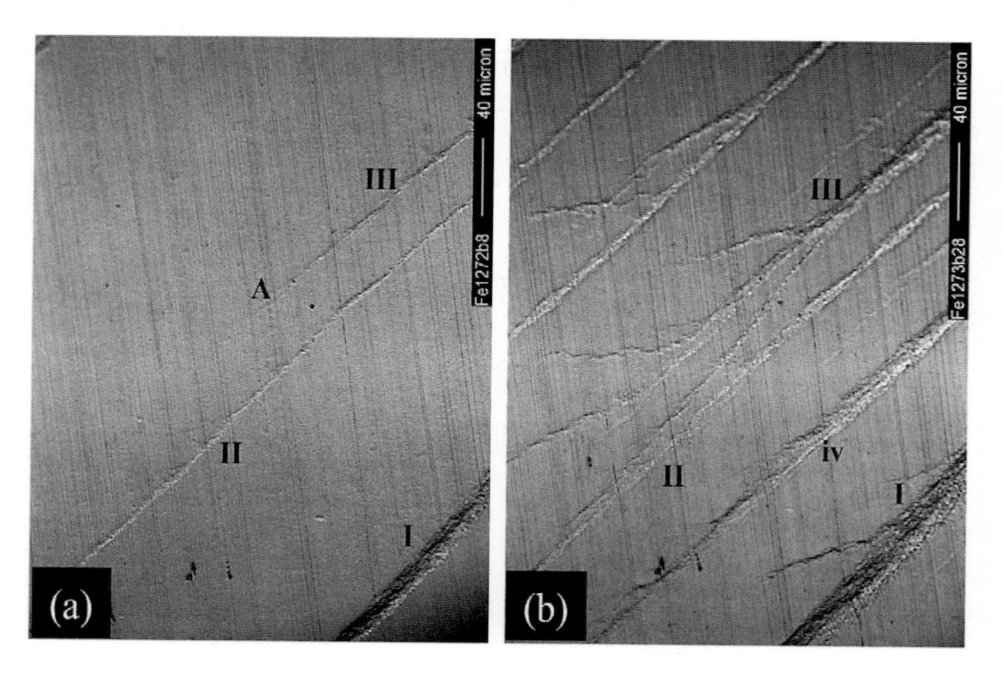

Fig. 8.13. Formation and development of shear bands in 268 nm Fe after uniaxial compression to plastic strain levels of (a) 3.7% and (b) 7.8% at 1.4×10^{-4} s^{-1}. Loading axis is vertical (reprinted from [101] with permission from Elsevier).

For electrodeposited metals, contamination from the ingredients in the electroplating baths is a long-standing issue. In this aspect, low-temperature annealing at temperatures below 150°C can improve the tensile properties of electroplated nanocrystalline Ni. Annealing at 200°C can be used to tailor the grain-size distribution for improved ductility in combination with good strength [103]. Youssef *et al.* prepared artifact-free nanocrystalline Cu disk with an average grain size of 23 nm via combinations of ball-milling at liquid nitrogen temperature and room temperature. Miniaturized disk bend test (MSBT) was used for mechanical characterization [104]. The nanocrystalline Cu exhibits an extraordinarily high yield strength (770 MPa) and good ductility. Such Cu disk exhibited high strain hardening, thereby preventing plastic instability during the membrane-stretching regime. These researchers later used a similar technique but modified the procedures to prepare tensile samples of naocrystalline Cu (average grain size of 54 nm) [105]. Figure 8.14 shows the stress–strain curves of nanocrystalline Cu at various strain rates. At a strain rate of 10^{-4} s^{-1}, tensile elongation of ~12% is observed. Apparently, tensile plasticity of cryomilled Cu improves considerably when compared to other nanocrystalline Cu prepared by other routes. The strain rate sensitivity (m) of the flow stress at a fixed strain of 2.5% is determined to be 0.0272; this is a fourfold increase over that of conventional coarse-grained Cu of 0.006 [91]. A larger value of m is generally associated with a greater resistance to neck development, thereby delaying tensile failure. For superplastic materials, m value is on the order of 0.3–0.8. From this, it is obvious that the m values of cryomilled Cu are too small for superplastic deformation to occur. In order to observe the evolution of deformation morphology of nanocrystalline Cu during tensile tests, a high-speed camera was used to image the deformed sample for various engineering strains at 10^{-4} s^{-1}. Strain localization in the form of shear band is developed during tensile loading (Fig. 8.15). The onset of plastic instability takes place at around 3% of strain, whereas signs of necking appear at later stages (Pictures 7–9). Although shear localization develops in cryomilled Cu samples, Cheng *et al.* attributed the ability to sustain a flat stress–strain curve over a few percent of plastic strains to larger strain rate sensitivity but smaller activation volume [105]. More research is required to clarify unclear aspects of strain loacalization on the tensile ductility of nanocrystalline metals.

As shear bands tend to form in nanograins that eventually concentrate very large strains, ductilization of nanocrystalline metals can be realized by introducing coarser grains into the nanocrystalline matrix. In this case, a bimodal grain size distribution is developed. It is expected that the larger grains in the nanocrystalline matrix would induce dislocation-mediated plasticity to the material. By controlling the volume fraction, size, shape, and distribution of the coarse grains, it is feasible to tailor for desired properties in nanocrystalline materials [91]. On the basis of simple numerical simulations, Gil Sevillano and Aldazabal demonstrated that significant enhancement of the ductility of nanocrystalline materials can be reached by a dispersion of a moderate volume fraction of coarse grains within nanocrystalline matrix [106]. In practice, bimodal grain size distribution can be tailored by annealing the nanocrystalline metals at an optimum temperature. For example, Wang *et al.* reported that a bimodal grain structure with average grain sizes of ~55 and 275 nm, respectively can be developed in electrodeposited Ni (29 nm) after annealing at 200°C. However, moderate growth of nanograins is expected to occur during annealing at 200°C. Above this temperature, the grains grow abnormally to the range of 1–2 μm. Therefore, a combination of decent strength and ductility can be reached for carefully selected annealing parameters [107,108].

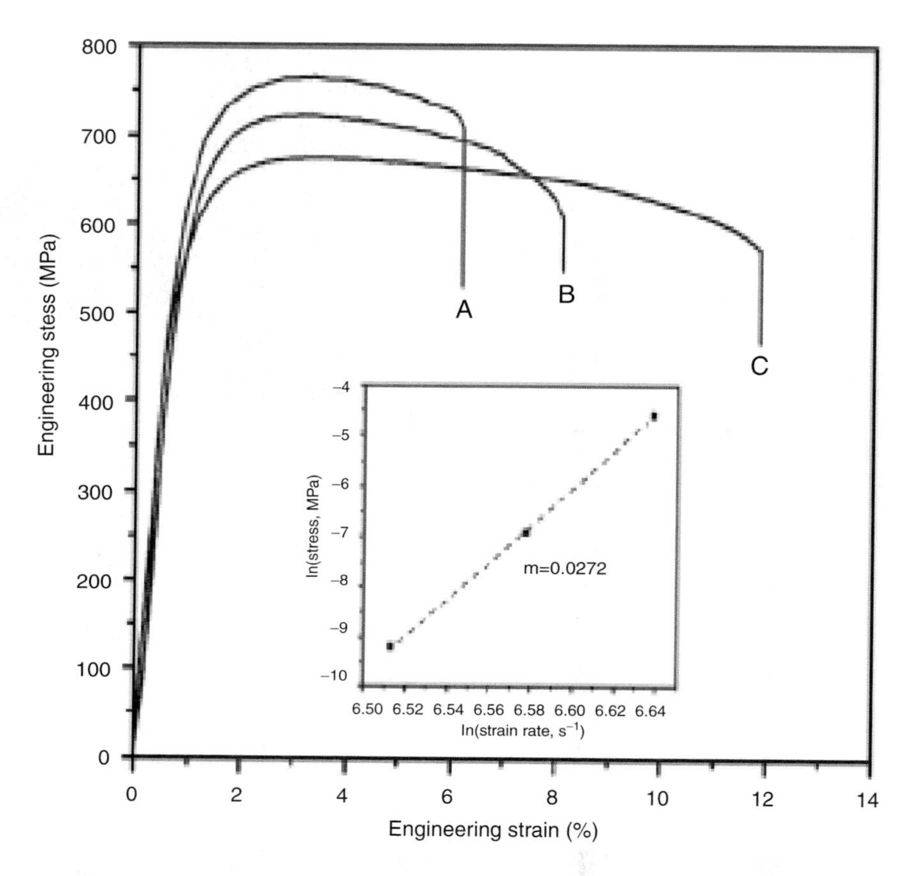

Fig. 8.14. Tensile test of nanocrystalline Cu at room temperature with different strain rates: (A) 10^{-2} s^{-1} (B) 10^{-3} s^{-1}, and (C) 10^{-4} s^{-1}. The inset is used to estimate the strain rate sensitivity, *m*. The samples were ball-milled at liquid nitrogen for 3 h and room temperature for 6 h (reprinted from [105] with permission from Elsevier).

8.7 Conclusions

This chapter presented a review of the reports that have been dedicated to the study of mechanical properties of nanocrystalline metals. The rapid development of nanocrystalline metals with grain sizes in the nanometer regime presents new challenges for understanding the fundamentals of the structure–property relationship. Basic understanding of the deformation mechanisms of nanocrystalline metals is of academic and technological importance. Nanocrystalline metals generally exhibit significantly high yield strength and hardness but very limited tensile plasticity. Low-tensile ductility arises from poor strain-hardening capacity, leading to earlier plastic instabilities in the form of localized deformation, such as necking and shear banding. Such a lack of tensile

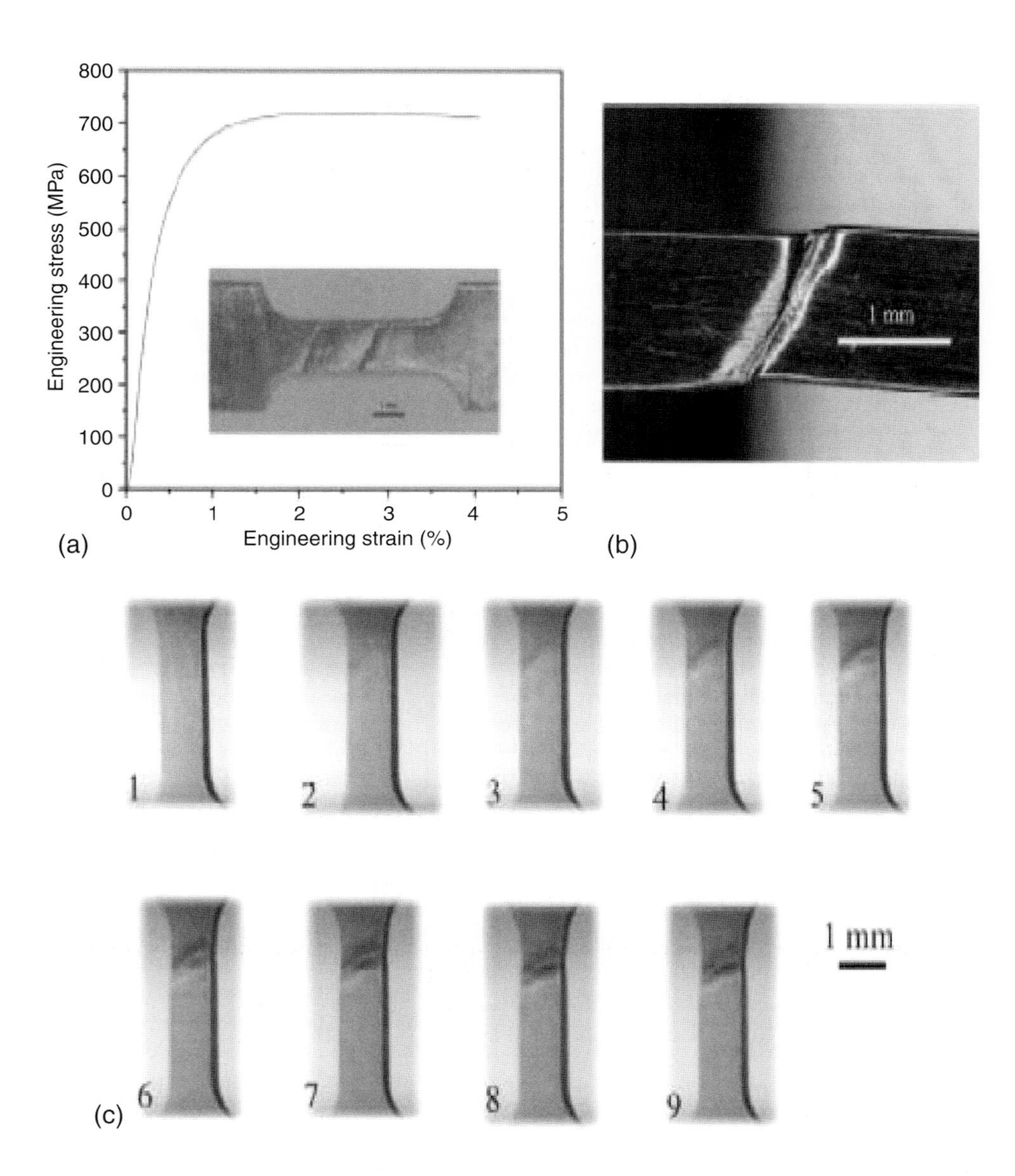

Fig. 8.15. (a) Engineering stress–strain curve of nanocrystalline Cu ball-milled at liquid nitrogen for 3 h and room temperature for 5 h at a strain rate of 10^{-4} s^{-1} showing the formation of shear localization. (b) An optical micrograph showing the shear zone and shear steps before cracking fails the sample. (c) The evolution of shear localization as captured by high speed camera. The corresponding engineering strain as follows: (1) 0%, (2) ~3.2%, (3) ~3.9%, (4) ~4.3%, and (5)–(9) were taken in the strain range of 4.6–7.0% (reprinted from [105] with permission from Elsevier).

ductility seriously hinders the use of nanocrystalline metals for structural applications. Basic understanding of the deformation and failure mechanisms is needed for designing and developing nanocrystalline metals with improved mechanical properties and performances.

The grain-size dependence of hardness or strength of nanocrystaline metals deviates significantly from the classical H–P relation. Below a critical grain size of ~20 nm, a negative H–P slope is observed. Dislocation pile-ups cannot be supported at such a very small grain size. Many attempts have been made to interpret the origin of the observed deviations from the H–P relation. However, the explanations for inverse H–P effect remain controversial. These include the presence of nanopores, grain-boundary sliding, and Coble creep diffusion. The mechanism by which the softening takes place is poorly understood for at least two reasons. First, it has been difficult to fabricate bulk nanocrystalline metals free from processing flaws and impurities for the mechanical tests. Second, the availability of appropriate experimental routes to characterize the intrinsic tensile properties of nanocrystalline materials below the critical grain size is very scarce. At present, the nanocrystalline metals for tensile tests are mostly prepared from consolidation of IGC nanopowders or electrodeposition routes. Consolidated IGC samples are known to possess residual porosity and insufficient interparticle bonding, leading to a substantial decrease in Young's modulus. On the other hand, electrodeposited samples contain impurities derived from both the plating solutions and organic additives. Sulfur impuritiy and its segregation to the grain boundaries of Ni results in low-tensile ductility. However, sulfur segregation to the grain boundaries of Ni at temperatures ≥330°C is beneficial for the occurrence of superplasticity. This is because the grain growth of nanocrystalline Ni at higher temperatures is suppressed by sulfur segregation. In general, superplastic deformation of nanocrystallline metals occurs at temperatures much lower than those of their microcrystalline metals.

As bulk nanocrystalline metals with grain sizes smaller than 20 nm, free from porosities and grain-boundary impurities are difficult to prepare and characterize, MD simulations are particularly useful to elucidate the mechanical deformation mechanisms of nanocrystalline metals accordingly. MD samples are fully dense and free from internal pores and defects with their nanograins separated by high angle boundaries. A critical grain size (d_c) has been predicted by MD simulations for some nanocrystalline metals. The simulations predict that an inverse H–P relation prevails in nanocrystalline metals having grain sizes below d_c in which no dislocation activity exists. In this case, plastic deformation takes place entirely by means of the grain-boundary sliding. The sliding is triggered by atomic shuffling and to some extent by stress-assisted free volume migration. Above d_c, Schockley partial dislocations are reported to be emitted from the grain boundaries. Such partials move through the grains and absorb into the opposite grain boundaries, leaving trails of stacking faults. At even larger grain size, full dislocation activity is expected to occur. It should be noted that MD simulations also have certain drawbacks. MD simulations of the deformation of nanocrystalline materials are performed on idealized model microstructures under high load and extremely high strain rate conditions to meet the capacity of computation. These impose restrictions on the number or the size of the grains to be simulated. The simulated properties of samples with a small number of grains may not represent the actual characteristics of real nanocrystalline materials. The sample must be deformed rapidly because of the short time scale accessible with MD simulations. Further, the deformation can only be initiated at early stages, thus subsequent microstructural evolution and fracture damage resulting from prolonged deformation are excluded from the simulations. Nevertheless, MD simulations provide some insights into the deformation mechanisms of nanocrystalline metals at atomic levels. Much effort is still needed to find out the suitability of MD simulations for prediction of the tensile and creep behavior of nanocrystalline metals at room temperature.

References

[1] S.C. Tjong and H. Chen, Mater. Sci. Eng. R 45 (2004) 1.

[2] R.W. Siegel and G.E. Fougere, Nanostruct. Mater. 6 (1995) 205.

[3] A.H. Chokshi, A. Rosen, J. Karch and H. Gleiter, Scripta Metall. 23 (1989) 1679.

[4] G. Palumbo, S.J. Thorpe and K.T. Aust, Scripta Metall. Mater. 24 (1990) 1347.

[5] C. Suryanarayana, D. Mukhopadhyay, S.N. Patankar and F.H. Froes, J. Mater. Res. 7 (1992) 2114.

[6] P.G. Sanders, C.J. Youngdahl and J.R. Weertman, Mater. Sci. Eng. A 234–236 (1997) 77.

[7] (a) S. Takeuchi, Scripta Mater. 44 (2001) 1483.
(b) R.J. Asaro and S. Suresh, Acta Mater. 53 (2005) 3369.

[8] G.W. Nieman, J.R. Weertman and R.W. Siegel, J. Mater. Res. 6 (1991) 1012.

[9] G.E. Fougere, J.R. Weertman, R.W. Siegel and S. Kim, Scripta Metall. Mater. 26 (1992) 1879.

[10] P.G. Sanders, J.A. Eastman and J.R. Weertman, Acta Mater. 45 (1997) 4019.

[11] H. Gleiter, Deformation of Polycrystals : Mechanisms and Microstructure, Eds. N. Hansen, A. Horsewell, T. Lefferes and H. Lilholt, Riso National laboratory, Roskilde, Denmark, 1981, pp. 15–21.

[12] R. C. Flagan and M.M. Lundes, Mater. Sci. Eng. A 204 (1995) 113.

[13] H. Gleiter, Prog. Mater. Sci. 33 (1989) 223.

[14] Z. Livne, A. Munitz, J.C. Rawers and R.J. Fields, Nanostruct. Mater. 10 (1998) 503.

[15] V.L. Tellkamp and E.J. Lavernia, Nanostruct. Mater. 12 (1999) 249.

[16] T.M. Lillo and G.E. Korth, Nanostruct. Mater. 10 (1998) 35.

[17] J. Rawers, Nanostruct. Mater. 11 (1999) 513.

[18] U. Erb, Nanostruct. Mater. 6 (1995) 533.

[19] A.M. El-Sherik and U. Erb, J. Mater. Sci. 30 (1995) 5743.

[20] U. Erb, G. Palumbo, B. Szpunar and K.T. Aust, Nanostruct. Mater. 9 (1997) 261.

[21] C.C. Kock, Scripta Mater. 49(2003) 657.

[22] X.K. Sun, J. Xu, W.X. Chen and W.D. Wei, Nanostruct. Mater. 4 (1994) 337.

[23] F. Dalla Torre, H. Van, Swygnhoven and M. Victoria, Acta Mater. 50 (2002) 3957.

[24] W.M. Yin, S.H. Whang and R.A. Mirshams, Acta Mater. 53 (2005) 383.

[25] J.K. Heuer, P.R. Okamoto, N.Q. Lam and J.F. Stubbins, Appl. Phys. Lett. 76 (2000) 3403.

[26] M.P. Seah and C. Leach, Phil. Mag. 31(1975) 627.

[27] A.M. El-Sherik, U. Erb, G. Palumbo and K.T. Aust, Scripta Metall. Mater. 27 (1992) 1185.

[28] R.L. Coble, J. Appl. Phys. 34 (1963) 1697.

[29] R.C. Gifkins, J. Am. Ceram. Soc. 51 (1968) 69.

[30] H. Luthy, R.A. White and O.D. Sherby, Mater. Sci. Eng. 39 (1979) 211.

[31] B. Cai, Q.P. Kong, L. Lu and K. Lu, Mater. Sci. Eng. A 286 (2000) 188.

[32] W.M. Yin, S.H. Whang, R. Mirshams and C.H. Xiao, Mater. Sci. Eng. A 301 (2001) 18.

[33] A.K. Mukherjee, Mater. Sci. Eng. A 322 (2002) 1.

[34] F.A. Mohamed and Y. Li, Mater. Sci. Eng. A 298 (2001) 1.

[35] S.X. Mcfadden, R.S. Mishra, R.Z. Valiev, A.P. Zhilyaev and A.K. Mukherjee, Nature 398 (1999) 684.

[36] S.X. Mcfadden, A.P. Zhilyaev, R.S. Mishra, and A.K. Mukherjee, Mater. Lett. 45 (2000) 345.

[37] S.X. Mcfadden and A.K. Mukherjee, Mater. Sci. Eng. A 395 (2005) 265.

[38] F. Dalla Torre, H.V. Swygenhoven, R. Schaublin, P. Spatig and M. Victoria, Scripta Metall. 53 (2005) 23.

[39] M. Ke, S.A. Hackney, W.W. Milligan and E.C. Aifantis, Nanostruct. Mater. 5 (1995) 689.

[40] K.S. Kumar, S. Suresh, M.F. Chisholm, J.A. Horton and P. Wang, Acta Mater. 51 (2003) 387.

[41] C.J. Youngdahl, J.R. Weertman, R.C. Hugo and H.H. Kung, Scripta Mater. 44 (2001) 1475.

[42] M. Legros, B.R. Elliot, M.N. Rittner, J.R. Weertman and K.J. Hemker, Philos. Mag. A 80 (2000) 1017.

[43] H. Van Swygenhoven, D. Farkas and A. Caro, Phys. Rev. B 62 (2000) 831.

[44] H. Van Swygenhoven, M. Spaczer and A. Caro, Acta Mater. 47 (1999) 3117.

[45] H. Van Swygenhoven and A. Caro, Phys. Rev. B 58 (1998) 11246.

[46] H. Van Swygenhoven, M. Spaczer, A. Caro and D. Farkas, Phys. Rev. B 60 (1999) 22

[47] H. Van Swygenhoven and A. Caro, Nanostruct. Mater. 9 (1997) 669.

[48] H. Van Swygenhoven and P.M. Derlet, Phys. Rev. B 64 (2001) 224105.

[49] H. Van Swygenhoven, P.M. Derlet and A. Hasnaoui, Phys. Rev. B 66 (2002) 024101.

[50] J. Schiotz, F.D. Di Tolla and K.W. Jacobsen, Nature 391 (1998) 561.

[51] J. Schiotz, T. Verge, F.D. Di Tolla and K.W. Jacobsen, Phys. Rev. B 60 (1999) 11971.

[52] J. Schiotz and K.W. Jacobsen, Science 301(2003) 1357.

[53] J. Schiotz, Scripta Mater. 51 (2004) 837.

[54] P. Keblinski, S.R. Phillpot, D. Wolf and H. Gleiter, Acta Mater. 45 (1997) 987.

[55] V. Yamakov, D. Wolf, S.R. Phillpot, A.L. Mukherjee and H. Gleiter, Phil. Mag. Lett. 83 (2003) 385.

[56] P. Keblinski, D. Wolf and H. Gleiter, Interf. Sci. 6 (1998) 205.

[57] V. Yamakov, D. Wolf, S.R. Phillpot and H. Gleiter, Acta Mater. 50 (2002) 5005.

[58] V. Yamakov, D. Wolf, S.R. Phillpot and H. Gleiter, Acta Mater. 50 (2002) 61.

[59] M. Chen, E. Ma, K. Hemker, H. Sheng, Y. Wang and X. Cheng, Science 300 (2003) 1275.

[60] (a) S.L. Frederiksen, K.W. Jacobsen and J. Schiotz, Acta Mater. 52 (2004) 5019.
(b) G.P. Zheng, Y.M. Wang and M. Li, Acta Mater. 53 (2005) 3893.

[61] H.S. Kim, Y. Estrin and M.B. Bush, Acta Mater. 48 (2000) 493.

[62] H.S. Kim and Y. Estrin, Acta Mater. 53 (2005) 765.

[63] H.H. Fu, D.J. Benson and M.A. Meyers, Acta Mater. 49 (2001) 2567.

[64] D.J. Benson, H.H. Fu and M.A. Meyers, Mater. Sci. Eng. A 319–321 (2001) 854.

[65] P. Sharma and S. Ganti, J. Mater. Res. 18 (2003) 1823.

[66] R.A. Masumura, P.M. Hazzledine and C.S. Pande, Acta Mater. 46 (1998) 4527.

[67] C.S. Pande, R.A. Masumura and P.M. Hazzledine, Mater. Phys. Mech. 5 (2002) 16.

[68] C.J. Youngdahl, P.G. Sanders, J.A. Eastman and J.R. Weertman, Scr. Mater. 37 (1997) 809.

[69] R. Suryanarayana, C.A. Frey, S.M.L. Sastry, B.E. Waller, S.E. Bates and W.E.. Buhro, J. Mater. Res. 11 (1996) 439.

[70] P.G. Sanders, J.A. Eastman and J.R. Weertman, Acta Mater. 45 (1997) 4019.

[71] P.G. Sanders, J.A. Eastman and J.R. Weertman, Processing and Properties of Nanocrystalline Materials, Eds. C. Suryanarayana, J. Singh and F.H. Froes. TMS, Warrendale, PA 1996, pp. 397–405.

[72] H.H. Fu, D.J. Benson and M.A. Meyers, Acta Mater. 49 (2001) 2567.

[73] D.J. Benson, H.H. Fu and M.A. Meyers, Mater. Sci. Eng. A 319–321 (2001) 854.

[74] T.R. Mallow and C.C. Koch, Acta Mater. 45 (1997) 2177.

[75] J.R. Weertman, D. Farkas, K. Hemker, H. Kung, M. Mayo, R. Mitra and H. Van Swygenhoven, MRS Bulletin, 24 (1999) 44.

[76] P. Feltham and J.D. Meakin, Phil. Mag. 2 (1957)105.

[77] M.A. Meyers, U.R. de Andrade and A.H. Chokshi, Metall. Mater. Trans. A 26 (1995) 2881.

[78] E.P. Abrahamson II, Surfaces and Interfaces, Syracuse Univ. Press: New York, 1968, p 262.

[79] R. Schwaiger, B. Moser, M. Dao, N. Chollacoop and S. Suresh, Acta Mater. 51 (2003) 5159.

[80] S.N.G. Chu and J.C.M. Li, J. Met. Sci. 12 (1977) 2200.

[81] F. Yang and J.C. Li, Scripta Mater. 32 (1995)139.

[82] D.S. Stone and K.B. Yoder, J. Mater. Res. 9 (1994) 2524.

[83] A.A. Elmustafa, J.A. Eastman, M.N. Ritter, J.R. Weertman and D.S. Stone, Scripta Mater. 43 (2000) 951.

[84] A.A. Elmustafa and D.S. Stone, Acta Metall. 50 (2002) 3641.

[85] A.A. Elmustafa and D.S. Stone, J. Mech. Phys. Solids 51 (2003) 357.

[86] F.A. McClintock and A.S. Argon, Mechanical Behavior of Materials, Addison-Wesley, N.Y., 1966, pp. 176–177.

[87] G. Taylor, Prog. Mater. Sci. 36 (1992) 29.

[88] H. Conrad, J. Metals 16(1964)582.

[89] Q. Wei, S. Cheng, K.T. Ramesh and E. Ma, Mater. Sci. Eng. A 381 (2004) 71.

[90] Y.M. Wang and E. Ma, Appl. Phys. Lett. 83 (2003) 3165.

[91] Y.M. Wang and E. Ma, Acta Mater. 52 (2004) 1699.

[92] H. Conrad and J. Narayan, Scripta Mater. 42 (2000) 1025.

[93] P.S. Follansbee and U.F. Kocks, Acta Metall. 36 (1988) 81.

[94] F. Dalla Torre, P. Spatig, R. Schaublin and M.Victoria, Acta Mater. 53 (2005) 2337.

[95] E. Ma, Science 305 (2004) 623.

[96] L. Lu, R. Schwaiger, Z.W. Shan, M. Dao, K. Lu and S. Suresh, Acta Mater. 53 (2005) 2169.

[97] L. Lu, Y. Shen, X. Chen and K. Lu, Science 304 (2004) 422.

[98] G.E. Dieter, Mechanical Metallurgy, McGraw Hill, New York, 1988, pp. 289–290.

[99] D. Jia, K.T. Ramesh and E. Ma, Scripta Mater. 42 (2000) 73.

[100] Q.M. Wei, D. Jia, K.T. Ramesh and E. Ma, Appl. Phys. Lett. 81 (2002) 1240.

[101] D. Jia, K.T. Ramesh and E. Ma, Acta Mater. 51 (2003) 3495.

[102] Y.M. Wang and E. Ma, Mater. Sci. Eng. A 375–377 (2004) 46.

[103] Y.M. Wang, S. Cheng, Q.M. Fei, E. Ma, T.G. Nieh and A. Hamza, Scripta Mater. 51 (2004) 1023.

[104] K.M. Youssef, R.O. Scattergood, K.L. Murty and C.C. Koch, Appl. Phys. Lett. 85 (2004) 929.

[105] S. Cheng, E. Ma, Y.M. Wang, L.J. Kecskes, K.M. Youssef, C.C. Koch, U.P. Trociewitz and K. Han, Acta Mater. 53 (2005) 1521.

[106] J. Gil Sevillano and J. Aldazabal, Scripta Mater. 51 (2004) 795.

[107] Y.M. Wang, M.W. Chen, F. Chou and E. Ma, Nature 419 (2002) 912.

[108] Y.M. Wang, S. Cheng, Q.M. Wei, E. Ma, T.G. Nieh and A. Hamza, Scripta Mater. 51 (2004) 1023.

CHAPTER 9

Semiconductor Nanoparticle-Polymer Composites

Mikrajuddin Abdullah[1], **Khairurrijal**[1], **Ferry Iskandar**[2], **and Kikuo Okuyama**[2]
[1]*Department of Physics*
Institute of Technology Bandung
Jalan Ganeca 10 Bandung 40132, Indonesia

[2]*Department of Chemical Engineering*
Graduate School of Engineering
Hiroshima University, 1-4-1 Kagamiyama
Higashi-Hiroshima 738-8527, Japan

9.1 Polymer Nanocomposites

Composites consisting of an insulating polymer matrix filled with nanosized particles are of interest because their long-term stability and because they offer new strategies for influencing interactions that may take place between the matrix and the nanoparticles. By integrating two or more materials with complementary properties, composite materials offer the potential to perform at a level far beyond that of the constituent materials. For example, ferroelectric ceramics possess a very high dielectric constant but are brittle and have a low dielectric strength. On the other hand, polymers are flexible, easy to process at low temperatures and possess a high dielectric breakdown field. By combining these two properties, the possibility of developing new material with a high dielectric constant and a high dielectric breakdown field might be feasible. An epoxy resin composite filled with SiO_2 particles is an important material system used in electrical devices, such as insulators and LSI package materials [1,2]. The large SiO_2 primary particle size (smaller specific surface area) decreases the viscosity of the composite system, because the resin supplied for the particle surface per unit area apparently increases [3,4]. A broadened particle size also decreases the viscosity [4–6]. Device fabrication with composites of a conjugated polymer and fullerene C_{60} as the active layer with efficient photo-induced

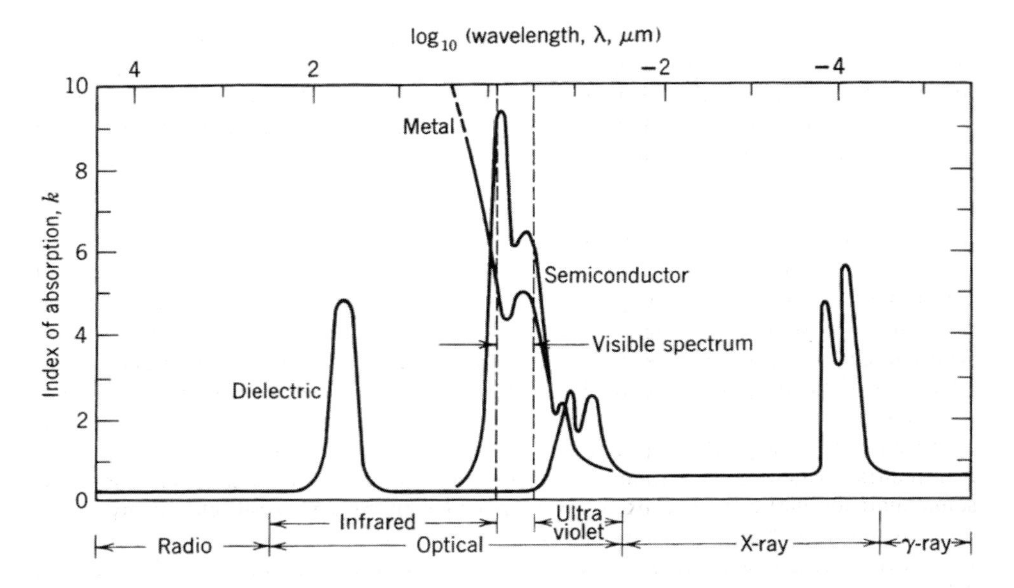

Fig. 9.1. Typical index of absorption of a metal, semiconductor, and dielectric. Reprinted with permission from [9].

charge transfer for preventing the initial electron-hole recombination, was a significant advancement in the exploration of polymeric photodiodes and photovoltaic cells [7,8].

The optical properties of dielectric materials are also of general interest because of their good transmission in the optical region of the spectrum as compared with other classes of materials. At short wavelengths, this desirable transmission is terminated at the ultraviolet absorption edge where the absorption of energy arises from electronic transitions between levels in the valence band to unfilled states in the conduction band. At long wavelengths, the relatively good transmission of dielectrics is terminated by the elastic vibration of ions in resonance with the imposed radiation. Figure 9.1 shows typical index of absorption, k (complex dielectric constant is expressed as $n^* = n - ik$) at various wavelengths [9].

A number of methods are available for producing nanocomposites of semiconductor nanoparticles in a polymer matrix. A very simple one is achived by first preparing nanoparticles and then dispersing them in a solution of a polymer, followed by drying. This method in relatively easy, and commercially available nanoparticles can be used. However, the method requires extensive mixing to disperse the nanoparticle homogeneously in the polymer matrix. Flocculation or agglomeration sometimes occurs, resulting in inhomogenity of these composites. A better dispersability might be achieved by first treating the surface of nanoparticles with a material that is compatible with the polymer matrix. The treating material acts as a lubricant to induce easy movement of the nanoparticle in the matrix.

If the homogeneity of a particle dispersion is not a critical factor, this method of synthesis is very promising. On the other hand, if the homogeneity of a material is a critical factor, synthesis of nanocomposites by the simple dispersion of the prepared nanoparticles in polymer matrix is not so promising. A critical problem might be encountered when the material is used in optical devices, such as fiber optics and other transparent materials. Inhomogenity of particle distribution leads to disadvantageous scattering, which reduces the performance of the material.

It would be expected that growing nanoparticles in a polymer matrix might lead to the production of nanocomposites consisting of well-dispersed nanoparticles. In this method, extensive mixing to disperse the nanoparticles is not required. This method can be performed in several ways. One potential way is by dispersing polymer in one precursor solution until homogenous solution is obtained and then mixing this solution with another precursor. The reaction of the first precursor that had been mixed homogeneously with the polymer with the second precursor might lead to the production of nanoparticles that are dispersed homogeneously in the matrix [10–12]. The second approach is to first produce a composite of metal nanoparticles dispersed in a polymer matrix, which is then followed by the transformation of metallic particles into semiconductor particles. For example, composites of AgS nanoparticles in nylon 11 were synthesized by first producing composites of nanoparticles of Ag in nylon 11. [13]. Silver reacts with H_2S in the presence of oxygen according to the reaction $4Ag + 2H_2S + O_2 \rightarrow 2Ag_2S + 2H_2O$ [14]. The composite of Ag–nylon 11 is placed in a glass cell and exposed to a mixture of H_2S and O_2.

The matrix-mediated control of growth and morphology has drawn considerable attention among various groups of researchers since it offers a new route to material synthesis [15–17]. Different types of materials such as $CaCO_3$, $CuCl_2$, K_2CO_3, CdS, $CaSO_4$, etc. have been prepared in situ within a polymer matrix such as polyethylene oxides with a modified or controlled morphology, crystalline phase, orientation, and growth habit of these compounds [18–20].

The polymer matrix can also control the size of nanoparticles. For example, mixing calcium chloride with polyethylene oxide followed by blending with amorphous polyvinyl acetate (PVAc) results in different nanoparticle sizes; this is achieved by changing the weight ratio of polymer entities. Table 9.1 shows an example of the dependence of polymer weight ratio on the size of nanoparticles dispersed in a polymer matrix [21].

Table 9.1. Different sizes of nanoparticles obtained when polymers are used in different molar ratios [21].

Composition	Crystalline size determined with Scherrer formula (nm)
PEO (0%)	83
PEO:PVAc(100:0)	12
PEO:PVAc(80:20)	9
PEO:PVAc(60:40)	8
PEO:PVAc(40:60)	8
PEO:PVAc(20:80)	7

9.1.1 Measuring particle sizes in composites

Size is an important parameter for describing nanoparticle properties. Understanding the size and size distribution help us understand the enormous properties of nanoparticles; since the physical and chemical properties of nanoparticles are significantly dependent on size, especially when the sizes are smaller than 10 nm. In this size range, the bulk properties of the materials gradually disappear and approach molecular behavior when the sizes approach about 1 nm.

A number of methods have been introduced to determine nanoparticle size, and the method of choice depends on the nature of nanoparticles being examined. At present, the well known and possibly the best method to determine the size of nanoparticles in a composite is transmission electron microscopy (TEM). This equipment provides very high-resolution pictures. Today, high-resolution TEMs can be used to identify even the position of individual atoms in a particle so that the crystalline structure of particles can be identified. The sizes of particles are determined by measuring the size of the TEM image of the composite (Fig 9.2). This method also permits the size distribution of nanoparticles to be determined. By measuring hundreds to thousands of nanoparticles, the size distribution of nanoparticles can be determined by fitting the size distribution data to an appropriate distribution function.

Fig. 9.2. Image of nanoparticles observed using a HRTEM. K. Okuyama *et al.* (unpublished).

Another method for determining nanoparticle size is based on X-ray diffraction (XRD). Although this method is only an approximate one, it becomes a useful alternative when TEM equipment is not available. This method permits the size of crystals to be determined, but not the size of the particle itself. Crystalline size is usually smaller than particle size. For single crystalline nanoparticles, the crystalline size approximates the particle size. Fortunately, if a nanoparticle is crystalline, it usually appears as single crystal so that the crystalline size is the same as the particle size.

The size of a nanocrystal can be predicted using XRD data by combining the Scherrer and Warren formulas. The broadening of the XRD peaks reflects either crystallinity or the size of nanocrystal. Assuming that the crystallinity of nanoparticles are not too different, the broadening of the XRD peaks reflects the size of nanocrystals only: smaller nanocrystals have a wider reflection peak. The crystalline sizes in a sample from XRD patterns is predicted using the Scherrer formula

$$d = \frac{0.9\lambda}{\mathrm{B} \cos \theta_{\mathrm{B}}} \tag{9.1}$$

where d is the crystalline diameter, λ the wavelength of X-ray, and θ_{B} the Bragg angle. B, the line broadening, obeys the Warren formula

$$B^2 = B_{\mathrm{M}}^2 - B_{\mathrm{S}}^2 \tag{9.2}$$

where B_{M} is the measured peak width at half the peak height of the sample and B_{S} the corresponding width of a standard material having a large crystalline size mixed with the sample with diffraction peak near the relevant peak of the sample. Figure 9.3 illustrates these peaks. For large crystalline materials, the XRD peaks are very narrow, while for nanocrystalline materials, the XRD peaks are broad. Thus, an approximation of $B_{\mathrm{M}} \gg B_{\mathrm{S}}$ or $B \cong B_{\mathrm{M}}$ is usually used [22].

The size distribution of a colloid of CdS nanoparticles prepared by flowing a mixture of HeS and He gases into a solution of $Cd(ClO_4)_2$ can be readily fitted using a normal distribution [23]

$$f(d) = \frac{1}{\sqrt{2\pi}\sigma} \exp\left[-(d - d_{\mathrm{av}})^2/(2\sigma^2)\right] \tag{9.3}$$

where d, d_{av}, are the particle diameter, the average particle diameter, and the variance, respectively. Sampling of particle size can be performed on the TEM images. Yao and Kitamura fitted the size distribution of CdS nanoparticles in a chelate polymer using a log-normal size distribution [24]

$$f(d) = \frac{1}{(2\pi)^{1/2}\sigma} \exp\left[-\frac{\ln^2(d/d_0)}{2\sigma^2}\right] \tag{9.4}$$

where d_0 is the center of the diameter and σ is the geometrical standard deviation. Sampling of particle size was also performed on the TEM images. The size distribution

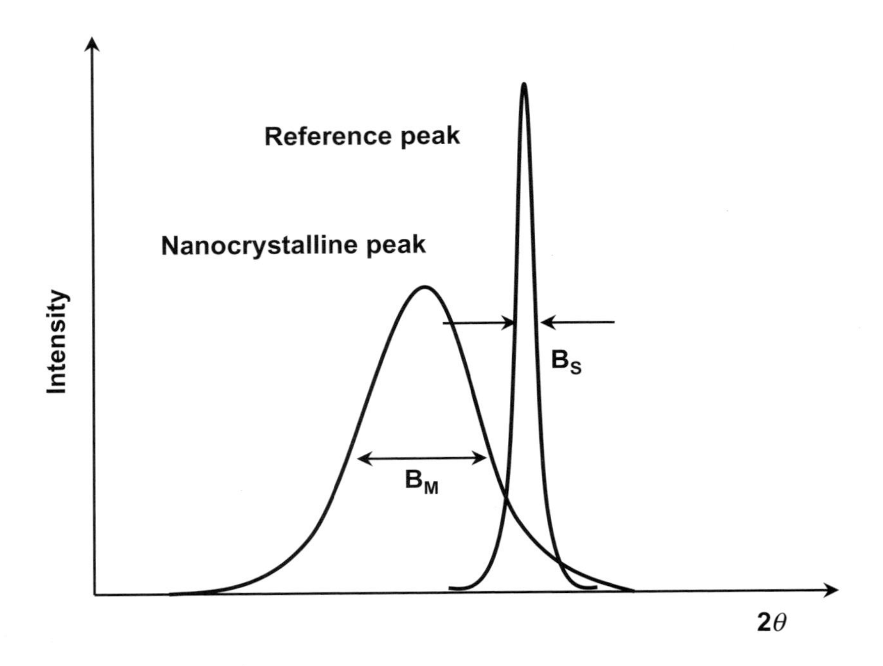

Fig. 9.3. Illustration of XRD peaks for nanoparticles and a reference sample.

of Eq. 9.4 gives an average particle diameter of

$$\langle d \rangle = d_0 \exp\lfloor \ln^2 \sigma/2 \rfloor \tag{9.5}$$

Figure 9.4 shows the size distribution of nanoparticles obtained from TEM images and the corresponding fitting curve.

The properties of nanoparticle composites are strongly influenced by the size of nanoparticles, the size distribution of nanoparticles, as well as interactions between nanoparticles. In the following section, such parameters that affect the properties of composites are discussed briefly.

9.1.2 Effect of size

Energy band gap. The absorption of a composite depends on the optical properties of the filler particles as well as matrix. Typically, in a composite of nanoparticle dispersed in a polymer matrix, a strong absorption of polymer occurs in the infrared region due to atomic vibration. In the visible region, however, optical absorption is dominated by the absorption of nanoparticles caused by the transition of electrons between valence and conduction bands or some exciton states located near the edge of the conduction band. In recording the absorption spectra at various light energies at around band gap energy of

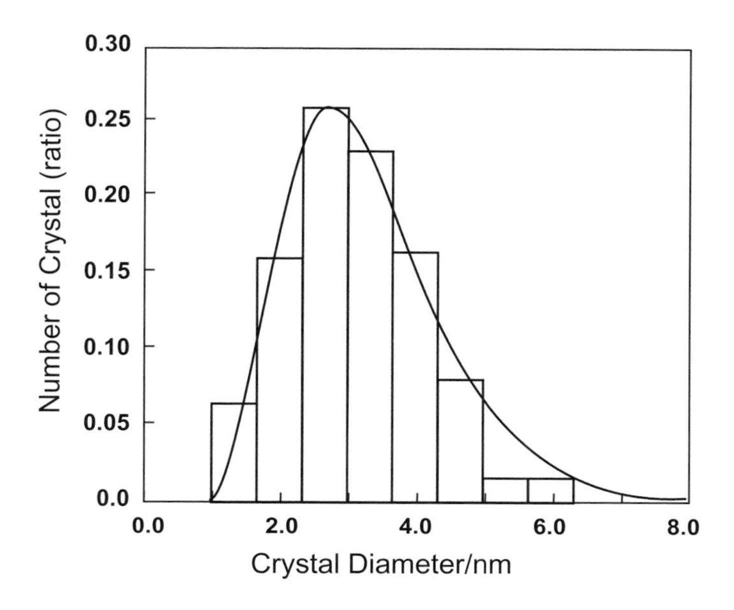

Fig. 9.4. Size distribution of nanoparticles in composites of CdS nanoparticles in a chelate polymer. Reprinted with permission from [24].

nanocrystals, it has been observed that an almost zero absorption is observed when the photon energy is less than the band gap energy of the nanoparticles. A sudden increase in absorption is observed when the photon energy surpasses the band gap energy of the nanoparticles. The shape of the absorption cross section with respect to photon energy when photon energy is above the band gap energy of nanoparticles differs for different materials. Generally however, the shape fits into two general classes depending on the type of interband transition of electrons, i.e. whether it is a direct band gap transition or an indirect band gap transition. The direction of the two transitions is illustrated in Fig. 9.5. The adsorption coefficient near the absorption edge for a direct interband transition is given by [25]

$$\alpha = \frac{A(\hbar\omega - E_g)^{1/2}}{\hbar\omega} \tag{9.6}$$

and for an indirect band gap transition, it can be expressed as

$$\alpha = \frac{A(\hbar\omega - E_g)^2}{\hbar\omega} \tag{9.7}$$

Examples of direct and indirect band-gap semiconductors and the corresponding energy gap of bulk materials are shown in Table 9.2 [26].

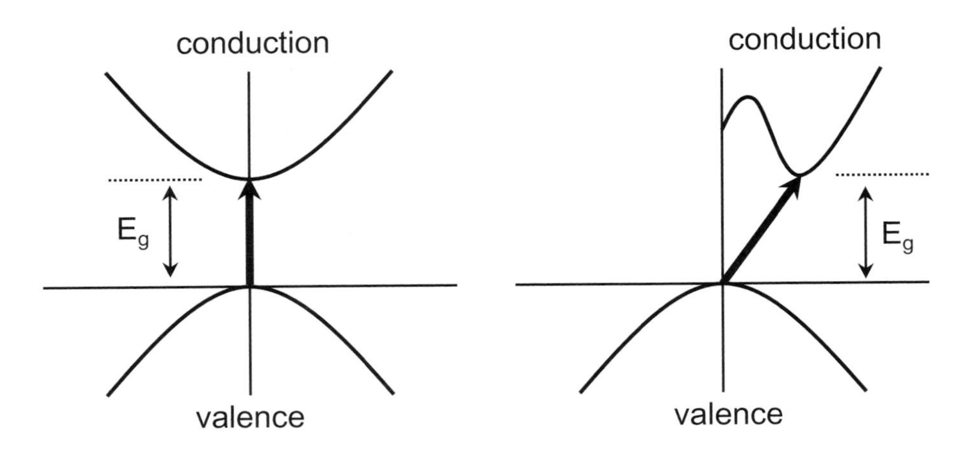

Fig. 9.5. Direct and indirect band gap transition.

Table 9.2. Band gap of materials: direct and indirect [26].

Materials	Direct (d) or indirect (i)	$E_g(\infty)$ (eV) at 0 K	$E_g(\infty)$ (eV) at 300 K
Si	i	1.17	1.11
Ge	i	0.744	0.66
GaP	i	2.32	2.25
AlSb	i	1.65	1.6
SiC (hex)	i	3.0	0.27
PbSe	i	1.65	–
PbTe	i	0.19	0.29
InSb	d	0.27	0.17
InAs	d	0.43	0.36
InP	d	1.42	1.27
GaAs	d	1.52	1.43
GaSb	d	0.81	0.68
PbS	d	0.286	0.34–0.37
CdS	d	2.582	2.42
CdSe	d	1.840	1.74
CdTe	d	1.607	1.44
SnTe	d	0.3	0.18
Cu_2O	d	2.172	–

The band gap of the nanoparticles are those given by Eqs. 9.6 and 9.7. For nanometer-sized particles, the band gap opening depends on particle size. The band gap opening increases with decreasing particle size. For example, HgSe particles with a diameter of 50 nm have a band gap of 0.3 eV. When the diameter is reduced to about 3 nm, the band gap opening increases to around 3.2 eV. An equation for approximating the size-dependent band gap of nanoparticles on particles size was first derived by Brus using a

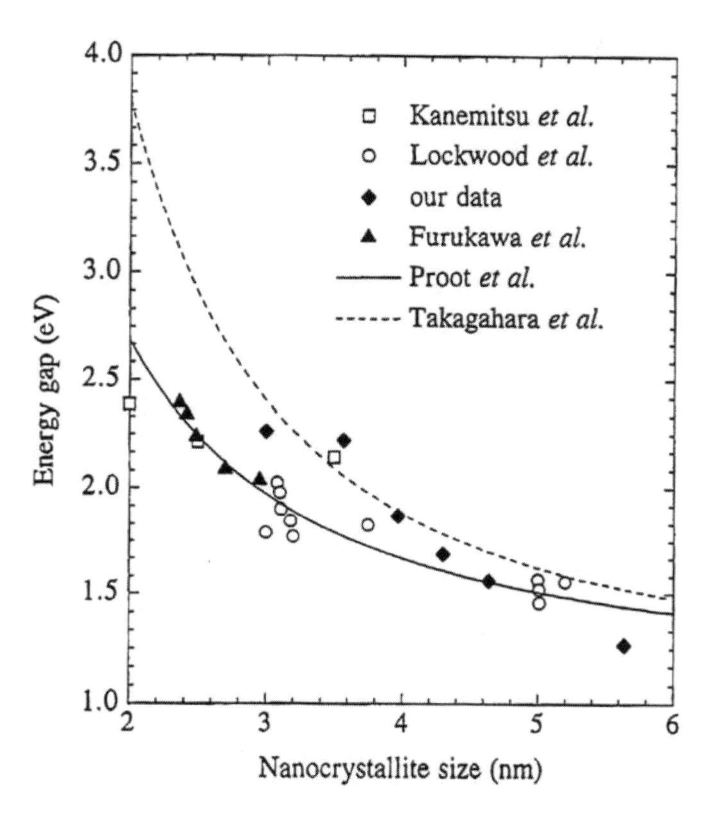

Fig. 9.6. Effect of particle size on the band gap of nanoparticles. Reprinted with permission from [29].

mass effective approximation [27,28].

$$E_g(R) = E_g(\infty) + \frac{\pi^2 h^2}{2R^2}\left(\frac{1}{m_e^*} + \frac{1}{m_h^*}\right) - \frac{1.8\,e^2}{\kappa R} \tag{9.8}$$

with E_g(R) is the band gap of nanoparticles having radius R, $E_g(\infty)$ is the band gap of the bulky material, m_e^* the effective mass of an electron, m_h^* the effective mass of a hole, e the charge on an the electron , and κ the dielectric constant of the material. Some measured and calculated data on the effect of particle size on the band gap of nanoparticles is displayed in Fig. 9.6 [29].

Band gap energy can be determined by plotting $(\alpha.\hbar\omega)^2$ against $\hbar\omega$. Figure 9.7 shows such a plot for PbS nanoparticles of different sizes [30]. Connecting the measured data with a straight line, one obtains the band gap of nanoparticles at the intersection of the curve with the horizontal axis. Carefully measuring the absorption data near the band edge, a tail at the band edge position can be observed, which tends to deflect from

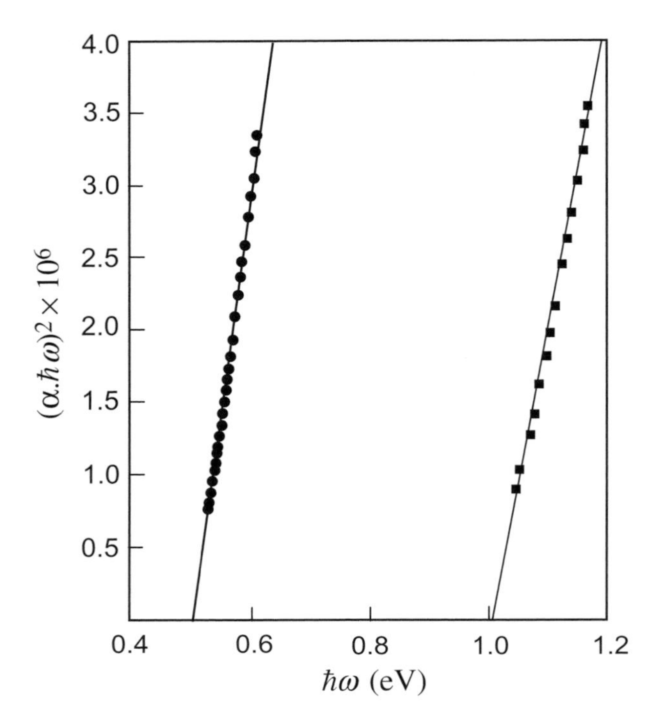

Fig. 9.7. Plot of $(\alpha \cdot \hbar\omega)$ with respect to $(\hbar\omega)$ for PbS composites consisting of two different particle sizes. Data points were extracted from [30].

crossing the horizontal axis, known as an Urbach tail, the value of which depends on the temperature. However, the band gap of a material can be determined by using only linear data as displayed in Fig. 9.7. This figure also shows the dependence of band gap on particle size.

Narrow peaks are occasionally apparent just below the band gap edge. This is due to exciton transition. Exciton energy in nanocrystals measured from the top of the valence band is given by [31,32]

$$E_{\mathrm{ex}}(d) = E_{\mathrm{g}}(d) - \frac{3.572}{d\varepsilon_{\mathrm{nc}}(d)} \tag{9.9}$$

Clearly, exciton energy also depends on particle size as well as dielectric constant. In addition, the dielectric constant also depends on particle size.

If the size of particles dispersed in the matrix is in the order of nanometer, the dielectric constant of the particles is dependent on size. Many expressions for obtaining the size-dependent dielectric constant have been proposed. For example, the size-dependent dielectric constant on a nanocrystal is approximated by [33]

$$\frac{1}{\varepsilon_{\mathrm{nc}}(a)} = \frac{1}{\varepsilon_{\mathrm{nc}}^{\infty}(a)} - \beta(a)\left[\frac{1}{\varepsilon_{\mathrm{nc}}^{\infty}(a)} - \frac{1}{\varepsilon_{\mathrm{nc}}^{\infty}(a) + 3.5}\right] \tag{9.10}$$

and the electronic contribution to the total polarizability is

$$\varepsilon_{\mathrm{nc}}^{\infty}(a) = 1 + \frac{\varepsilon_{\mathrm{bulk}}^{\infty} - 1}{1 + (3.75/a)^{1.2}} \tag{9.11}$$

with a is in Angstrom units, and $\beta(a)$ indicates how much the ions participate in the screening.

Another form of size-dependent dielectric constant than has been proposed is the following [34]

$$\varepsilon(R) = 1 + \frac{\varepsilon_{\mathrm{bulk}}^{\infty} - 1}{1 + (\alpha/a)^{\ell}} \tag{9.12}$$

where $\varepsilon_{\mathrm{bulk}}^{\infty}$ is the dielectric constant of the bulky material, α and ℓ are constants. For example, for silicon $\varepsilon_{\mathrm{bulk}}^{\infty} = 11.4$, $\alpha = 1.098$ nm, and $\ell = 1$. Figure 9.8 shows a plot of the dielectric constant of silicon nanoparticles with respect to particle size.

The dependence of energy band gap on particle size induces the dependence of electrical conductivity on the particle size. For a bulky material, the effect of temperature on conductivity is given by

$$\sigma(T) = f(T)\exp\left[-\frac{E_{\mathrm{g}}}{kT}\right] \tag{9.13}$$

where k is the Boltzmann constant, T the temperature and $f(T)$ a factor that depends slowly on the temperature. Since the energy band gap of nanoparticles depends on particle size, it is clear that the electrical conductivity of a nanoparticle also depends on particle size. For percolating composites consisting of a semiconductor filler, since the conductivity of a single nanoparticle depends on size, the conductivity of a composite depends on the size of distributed nanoparticles as long as the electrical conductivity of the nanoparticles is larger than that of the matrix. Consequently, the electrical conductivity of composites depends on the size of nanoparticles dispersed in that composite.

Scattering. Light scattering occurs as a consequence of a fluctuation in optical properties of a medium, a completely homogeneous material can scatter light only in the forward direction [35]. This can be demonstrated with the aid of Fig. 9.9, which shows a completely homogeneous medium being illuminated by a plane wave. The volume element dV_1 scatters light in the θ direction. However, for any direction, except the exact forward direction ($\theta = 0$) there must be a nearby volume element dV_2, the scattered field of which interferes destructively with that of dV_1. Since the same argument can

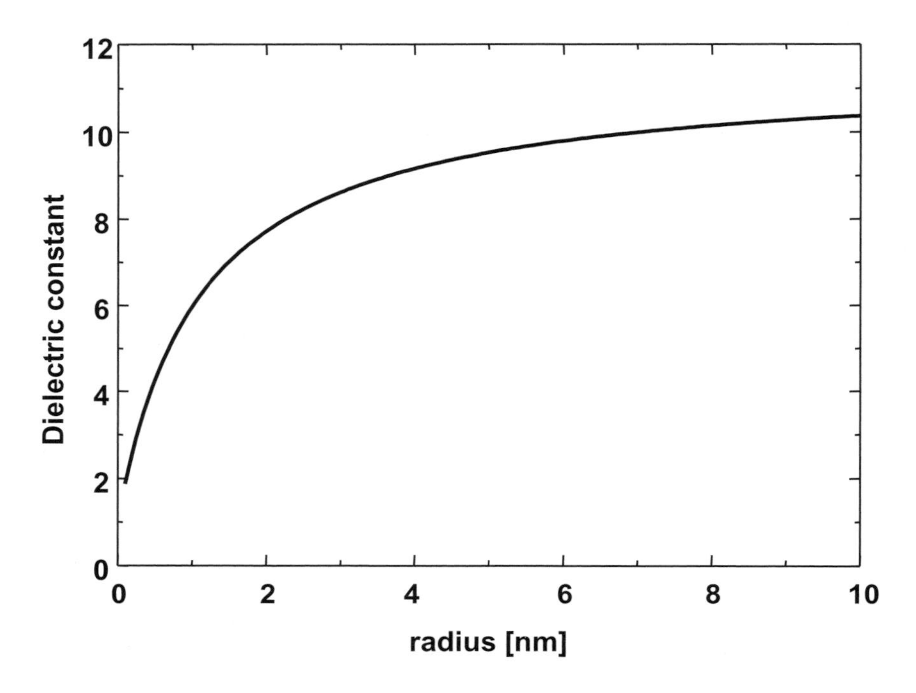

Fig. 9.8. Dielectric constant of silicon nanoparticles as a function of particle size calculated using Eq. 9.12.

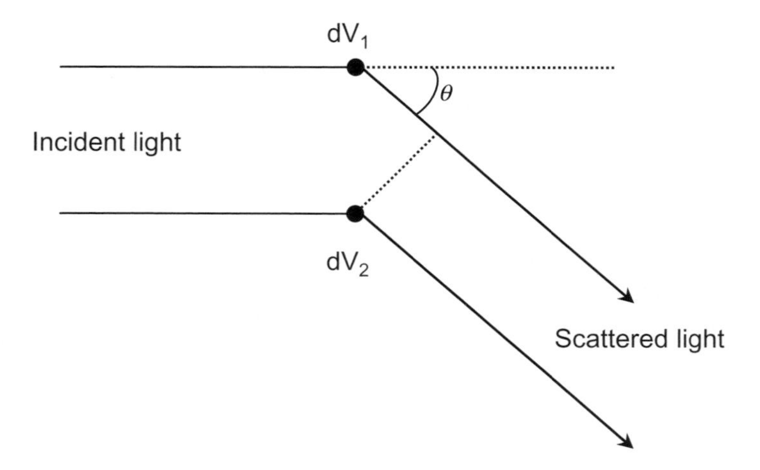

Fig. 9.9. The direction of scattered light in which $\theta = 0$ interferes destructively in homogeneous materials.

be applied to any volume element in the medium, we conclude that there can be no scattering in any direction except $\theta = 0$.

Scattering can occur as the result of fluctuations in any of the optical properties of the medium. For example, if the density of the medium is not uniform due to the dispersion of nanoparticles in polymer matrix, the total number of particles in the volume element dV_1 may not be equal to the number of particles in dV_2, and consequently, the destructive interference between the fields scattered by these two elements will not be exact.

Consider the scattering of polarized light by a composite consisting of a dilute particle concentration. We assume that the size of the particle is far smaller than the wavelength of light. This assumption is usually fulfilled by composite of nanoparticles since the size of particles (less than 100 nm) is far less than the wavelength of optical light (several nanometers). The differential cross section of scattering is given by [36]

$$\frac{d\sigma}{d\Omega} = \frac{\omega^4}{c^4} |\alpha(\omega)|^2 \sin^2\phi \tag{9.14}$$

where ϕ is the angle between the induced dipole moment of particles and the direction to the point of observation, ω the frequency of light, c the speed of light in a vacuum and $\alpha(\omega)$ the electric polarizability of a particle at frequency ω. The angular dependence of $d\sigma/d\Omega$ is contained entirely in the $\sin^2\phi$ term. Therefore we can immediately obtain an expression for the total scattering cross section by integrating $d\sigma/d\Omega$ over all solid angles, yielding

$$\sigma = \int \frac{d\sigma}{d\Omega} d\Omega = \frac{8\pi}{3} \frac{\omega^4}{c^4} |\alpha(\omega)|^2 \tag{9.15}$$

The scattering is strongly dependent on particle size, and the maximum scattering occurs when the particle size is of the same magnitude as the wavelength of the radiation. For particle sizes much smaller than the wavelength of the incident radiation, the scattering constant increases with particle size. The scattering constant reaches a constant value for particle sizes that are substantially larger than the wavelength of the incident radiation.

Assume a composite consisting of spherical-shaped dielectric spheres. We take ε_1 to be the dielectric constant of particles and to ε be that of the matrix (surrounding material). The polarizability of a sphere can be calculated using the electrostatic method. For a dilute concentration of nanoparticles, the polarizability of each sphere is given by [37,38]

$$\alpha = \frac{\varepsilon_1 - \varepsilon}{\varepsilon_1 + 2\varepsilon} a^3 \tag{9.16}$$

where a is the radius of the particles. The dependence of α on frequency holds for any possible frequency dependence of ε or ε_1. Substituting Eq. 9.16 into Eq. 9.15, the total cross section of a composite consisting of a dilute concentration of nanoparticles is given by

$$\sigma = \frac{8\pi}{3} \frac{\omega^4}{c^4} a^6 \left(\frac{\varepsilon_1 - \varepsilon}{\varepsilon_1 + 2\varepsilon} \right)^2 \tag{9.17}$$

Differences in the effect of the scattering of composites consisting different sizes of particles is shown in Fig. 9.10. The material is an epoxy resin consisting of nanoparticles of silica. A different cross section results in a difference in transmittance. A composite consisting of smaller nanoparticles transmits more light (small cross section) than composites consisting large particle sizes.

9.1.3 Effect of size distribution

The quantum-size effect depends on particle size. Therefore, unique size-dependent properties can be obtained via the use of monodisperse nanoparticles. However, truly monodisperse nanoparticles are almost impossible to produce. No preparation method

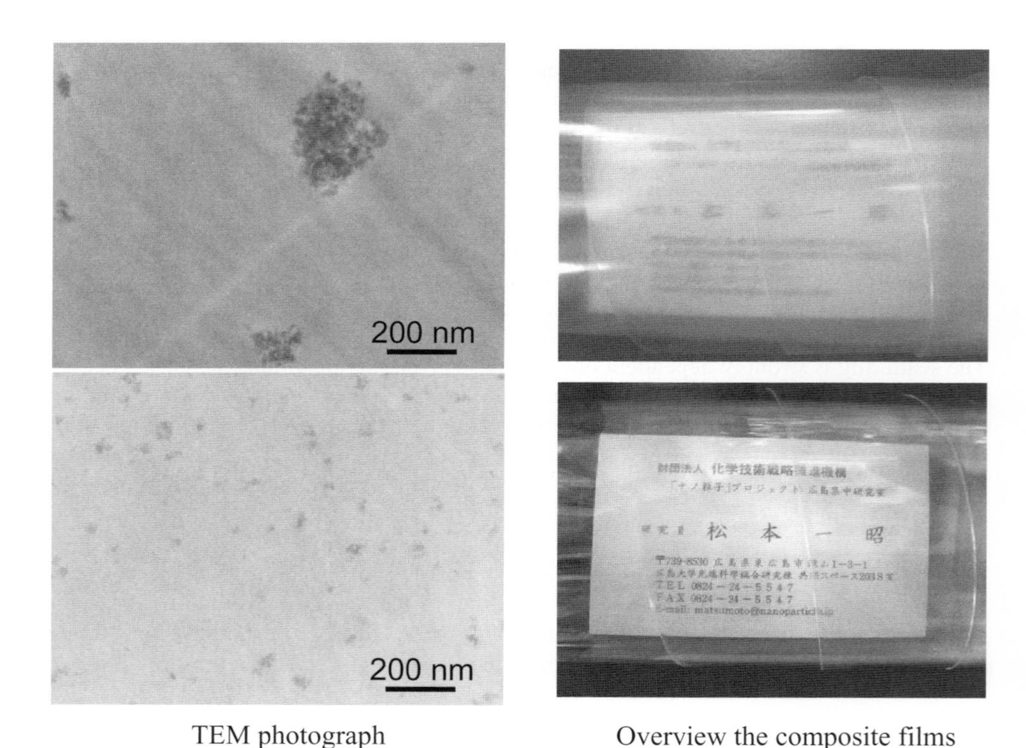

Fig. 9.10. Left: TEM pictures of particles used to prepare composites: large particles (top) and small particles (bottom). Right: transparencies of the produced composites in the visible region; more transparent for a composite made from smaller particles (bottom). K. Okuyama *et al.* unpublished.

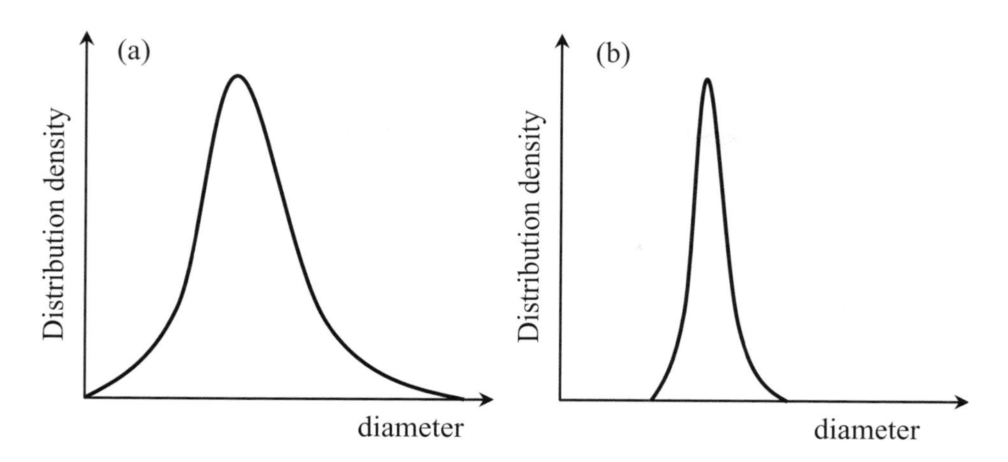

Fig. 9.11. Size distribution of particles: (a) broad size distribution and (b) narrow size distribution. A very narrow size distribution are usually referred to as monodisperse nanoparticles.

consistently produces distributed particle sizes. Figure 9.11 shows a typical size distribution for some such produced nanoparticles. Monodispersed nanoparticles from many reports are usually referred to as nanoparticles having a narrow size distribution, not a single size nanoparticles (Fig. 9.11b). Therefore, the entire properties of composites should be contributed by the properties of the individual particles.

In the case where there is no interaction between particles, the entire properties of the composite can be expressed as a simple superposition of the properties of individual particles. If $\varphi(a)$ denotes the properties of a single particle, the entire properties of the composite, ψ, can be written as,

$$\psi = \int_0^\infty f(a)\varphi(a)da \tag{9.18}$$

where $f(a)$ is the frequency distribution of particle with radius a.

Yao and Kitamura [24] attempted to predict the absorption spectra of nanocomposite of CdS in polymer using the relation

$$\alpha(\omega) = k \int \alpha_d(\omega) f(d) d\ln(d) \tag{9.19}$$

where $\alpha_d(\omega)$ is the absorption spectra of particles of diameter d, k a constant, and $f(d)$ a log-normal size distribution as expressed in Eq. 9.4. Figure 9.12a shows absorption spectra of various composites of nanoparticles and the corresponding fitted curves. For calculating the fitted curves, a log-normal size distribution of nanoparticles as expressed in Eq. (9.4) was used. The size distributions are shown in Fig. 9.12b.

(a)

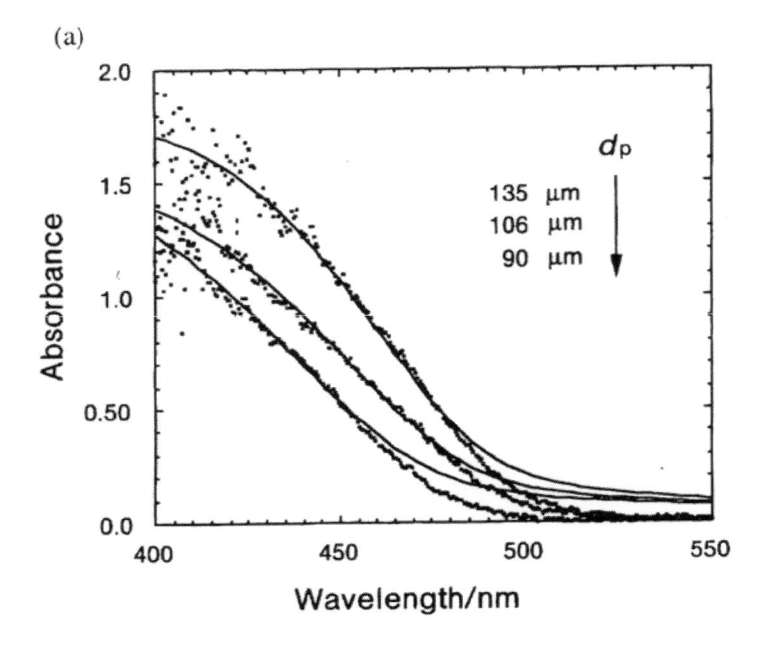

(b)

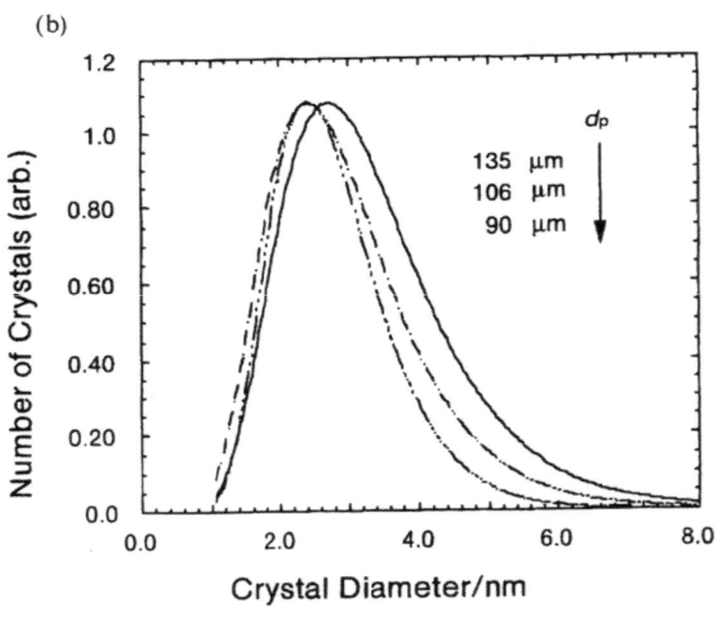

Fig. 9.12. (a) Absorption spectra of composites of CdS nanoparticles in a chelate polymer and the corresponding fitted curve calculated using Eq. 9.19. (b) Size distribution of nanoparticles to obtain the fitted curves. Reprinted with permission from [24].

The parameters for the composites used for size distribution are A: $(2R_0 = 2.7;\ \sigma = 0.37)$; B: $(2R_0 = 2.4;\ \sigma = 0.37)$; C: $(2R_0 = 2.4;\ \sigma = 0.31)$.

9.2 Effect of Interactions between Particles

To show that the interaction of particles significantly affects the properties of composites, consider a simple case, the effect of particle interaction on the coefficient of absorption of a composite. Suppose particles with dielectric constant ε are dispersed in a background matrix with dielectric constant ε_b. For an isolated particle with radius a, the application of an electric field E leads to the occurrence of an induce dipole moment

$$p_0 = \frac{2\tau}{3-\tau} E a^2 \tag{9.20}$$

where we have defined a scaled dielectric contrast parameter

$$\tau = \frac{\varepsilon - \varepsilon_b}{\varepsilon + \varepsilon_b} \tag{9.21}$$

To demonstrate the effect of interactions of dielectric properties, consider a simple case of a pair of spherical particles of the same size separated by a certain distance that might vary from zero (touching) to infinity (isolated). This condition is illustrated in Fig. 9.13. The dipole moment of the pairs are calculated using the method of images [39–41]. When a uniform longitudinal field is used, the resulting induced dipole moment in each particle is given by [42]

$$p_{\mathrm{L}} = p_0 \sum_{n=0}^{\infty} \tau^n \left[\frac{\sinh^3 \vartheta}{\sinh^3 (n+1)\vartheta} + \frac{\sinh\vartheta \sinh^2 n\vartheta}{\sinh^3 (n+1)\vartheta} \right] - p_0 \left[\sum_{n=0}^{\infty} \tau^2 \frac{\sinh\vartheta \sinh(n\vartheta)}{\sinh^2 (n+1)\vartheta} \right]^2 \Big/ \left[\sum_{n=0}^{\infty} \tau^n \frac{\sinh\vartheta}{\sinh(n+1)\vartheta} \right] \tag{9.22}$$

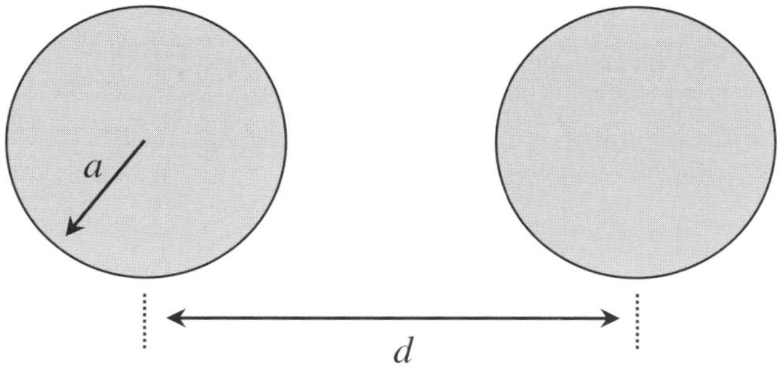

Fig. 9.13. Two identical spheres of radius a separated by d.

with ϑ is the separation parameter given by the relation

$$\cos\vartheta = \frac{d}{2a} \tag{9.23}$$

The separation parameter approaches zero as the gap between the pair approaches zero ($d = 2a$). When the field is applied in the transverse direction, the induced dipole moment satisfies

$$p_T = p_0 \sum_{n=0}^{\infty} (-\tau)^n \left[\frac{\sinh\vartheta}{\sinh(n+1)\vartheta} \right]^3 \tag{9.24}$$

If the pair is not touching, a useful approximation can be obtained by replacing the summation with an integral. The integration is performed on the first term of the MacLaurin expansion for the series and assuming that it correctly gives the asymptotic behavior of the series, the integral result for a longitudinal field is

$$p_L \approx p_0 \left[2\zeta(3) + \frac{\pi^4}{36\left[\ln(2\vartheta) + \psi\left(s + \frac{1}{2}\right)\right]} \right] \tag{9.25}$$

with

$$s = -\frac{\ln\tau}{2\vartheta} \tag{9.26}$$

where $\psi(z)$ digamma function and $\zeta(n)$ the Riemann zeta function [43]

At long wavelengths, the complex contrast τ can be approximated by the series [40]

$$\tau = 1 + i\frac{\lambda_0}{\lambda} + A\frac{\lambda_0^2}{\lambda^2} + \ldots \tag{9.27}$$

The coefficient λ_0 can be obtained by fitting the expression to the experimental results. If particles are isolated (no interaction), the total dipole moment of the particles is Np_0 and the absorption is given by

$$\alpha_{\text{eff}} = \frac{\pi}{\lambda}\text{Im}\left[\frac{Np_0}{E}\right] = \frac{9f\lambda_0}{8\lambda^2} + O\left[\frac{1}{\lambda^3}\right] \tag{9.28}$$

For a pair of particles, due to the presence of interactions, the equation for optical absorption changes. This coefficient depends on the separation of particles. By defining a critical wavelength, $\lambda_c = \lambda_0/\vartheta$, the coefficient of absorption for an array of spherical particles can be expressed as

$$\alpha_{\text{eff}} \cong \begin{cases} f\dfrac{\lambda_0}{\lambda^2}\dfrac{\pi^6}{144\vartheta\,[\gamma - \ln(\vartheta/2)]^2}, & \lambda \gg \lambda_c \\ f\dfrac{1}{\lambda}\dfrac{\pi^5}{72\left[\pi^2/4 + \ln(\lambda_0/\lambda)^2\right]}, & \lambda \ll \lambda_c \end{cases} \tag{9.29}$$

In the case of dense composites, the coefficient of absorption has a different form. A specified particle can interact with a large number of particles since the distance between the particles is small. However, by assuming that the polarization of a specified particle is predominantly determined by the adjacent particles, the coefficient of absorption can be approximated as

$$\alpha_{\text{eff}} \cong \begin{cases} \dfrac{\pi^2}{2\lambda}\left[\dfrac{M}{\ln(\lambda/\lambda_0)}\right]^{1/2}, & \lambda \gg \lambda_c \\ \dfrac{\pi^3}{2}\ln\left[\dfrac{3M}{8\ln(\lambda_c/\lambda)}\right]\dfrac{\lambda_c}{\lambda^2}, & \lambda \ll \lambda_c \end{cases} \tag{9.30}$$

9.3 Polymer Electrolyte Nanocomposites

Worldwide research is currently being focused on the development of high power and high energy density polymer electrolytes with major emphasis on safety, performance, and reliability. A battery contains two electrodes: positive and negative (both sources of chemical reactions), separated by an electrolyte that contains dissociated salts through which ion carriers flow. Once these electrodes are connected to external circuits, chemical reactions occur at both electrodes resulting in the delivery of electrons to the external circuits. The properties of a battery strongly depend on the electrolyte, anode, and cathode. With the use of polymer electrolytes in lithium batteries, a high specific energy and specific power, safe operation, flexibility in packaging, and low cost in fabrication can be expected [44]. In addition, a low internal voltage drop at relatively large current withdrawal is expected.

Another potential application of polymer electrolyte nanocomposites is for manufacturing solar cells. Dye-sensitized solar cells have attracted great scientific and technological interest as potential alternatives to classical photovoltaic devices. The mechanism of cell operation involves the absorption of visible light by a chemisorbed dye, following by electron injection from the excited synthesizer into the semiconductor conduction band. The selection of liquid electrolyte, usually containing an organic solvent such as acetonitrile and propylene carbonate, assures the perfect regeneration of the dye by the direct interaction of the dye in the oxidized state and an I^-/I_3^- redox couple thus leading to impressively high solar-to-electrical conversion efficiencies (7–11%) [45,46]. However, the stability and long-term operation of such cells are affected by solvent evaporation or leakage. Thus the commercial exploitation of these devices requires that the liquid electrolyte be replaced by a solid charge transport medium, which not only offers hermetic sealing and stability but also reduces design restrictions and endows the cell with choices in shape and flexibility. A solid state dye-sensitized solar cell uses composite polymer electrolytes of PEO and TiO_2 in the presence of an I^-/I_3^- redox couple [47].

In the case of polymer electrolytes, the cations are coiled by the polymer segment leaving the anions to occupy separate positions [48]. Battery performance is limited by the speed of diffusion of the cations. The transport of cations takes place if there is a relaxation in the polymer segments so that cations are released from a segment and then occupy another segment. Segmental relaxation requires the presence of a free volume in the polymer matrix, a condition that can be attained only if the polymer is in an amorphous state. Unfortunately, most HMWPs crystallize at ambient temperatures. Ions are transported with difficulty in a crystalline matrix, since no chain relaxation occurs and, the conductivity of polymer electrolytes in this phase (at ambient temperature) is depressed. The transport of ions in this state is dominated by the jumping of cations to the nearest location, which depends on the blocking potential (activation energy). This results in conductivity of the order of 10^{-8} S/cm, a value that is far below the desired value of about 10^{-4} S/cm [49]. When it enters the amorphous state, i.e. at temperatures above the melting point, it results in high conductivity. For a commonly used polymer, i.e. polyethylene oxide, the melting temperature is 65°C. This is, of course, impractical since the operating temperature for most electronic devices is at room temperature. In addition, at temperatures above the melting point, the polymer becomes soft, causing the solid state properties to degrade. Initiated by the work of Wright and Armand [50,51], several types of polymer electrolytes have been extensively investigated around the world. Table 9.3 shows examples of polymer electrolytes and their measured conductivities at around room temperature [44].

9.3.1 Improvement of ambient temperature conductivity

Improvements in the electrical conductivity of polymer electrolytes at ambient temperature is therefore of critical importance for technological applications. Several approaches have been introduced to improve the conductivity of polymer electrolytes. One such strategy for achieving this is enhancing the amorphous state at low temperatures. The first approach involves initially preparing a polymer with a low degree of crystallinity. To date, some of these are two cross-linked polymers [52,53], the synthesis of new polymers, high molecular weight polymer cross-linked by γ-irradiation [54], the addition of plasticizers to polymer electrolytes, the addition of fillers, the blending of two polymers [55]. Another strategy is to prepare an amorphous polymer so as to obtain a polymer that is comprised of four to five monomeric units. For this system, the chains must be sufficiently long to effectively complex cations but sufficiently short to crystallize at low temperatures. Thus, the matrix would still be in the amorphous state even at low temperatures. The polymer host serves as a solvent and no organic liquids are included.

An alternative way to decrease the crystallinity of a polymer matrix is to introduce side-chains to the polymer main chain. Chain ends and branches can be thought of as

Table 9.3. Examples of polymer electrolytes.

Polymer Host	Repeat Unit	Example Polymer Electrolyte	Conductivity (S/cm) at 20°C
Poly(ethylene oxide), PEO	$-[CH_2CH_2O]_n-$	$(PEO)_8$:$LiClO_4$	10^{-8}
Poly(oxymethylene), POM	$-[CH_2O]_n-$	POM:$LiClO_4$	10^{-8}
Poly(propylene oxide), PPO	$-[(CH_3)CH_2CH_2O]_n-$	$(PPO)_8$:$LiClO_4$	10^{-8}
Poly(oxymethylene-oligo-ethylene), POO	$-[(CH_2O)(CH_2CH_2O)]_n-$	$(POO)_{25}$:$LiCF_3SO_3$	3×10^{-5}
Poly(dimethyl siloxane), DMS	$-[(CH_3)_2SiO]_n-$	DMS:$LiClO_4$	10^{-4}
Unsaturated ethylene Oxide segmented, UP	$-[HC{=}CH(CH_2)_4O(CH_2CH_2O)_n(CH_2)_4]_x-$	UP:$LiClO_4$ (EO:Li^+ = 32:1)	10^{-5}
Poly[(2-methoxy)ethyl glycidyl ether], PMEGE	$-[CH_2CHO]_n-$ with side chain $CH_2(OCH_2CH_2)_2OCH_3$	$(PMEGE)_8$:$LiClO_4$	10^{-5}
Poly[(methoxy) poly(ethylene glycol)] methacrylate, PMG_n	$-[CH_2C(CH_3)]_n-$ with side chain $C(=O)O-(CH_2CH_2O)_xCH_3$	PMG_{22}:$LiCF_3SO_3$ (EO:Li^+=18:1)	3×10^{-5}
(PEO–PPO–PEO)–SC SC=siloxane crosslinked	PEO–$(CH_2)_3$–Si(CH_3)–O–Si(CH_3)–$(CH_2)_3$–PEO, each Si linked through O to PEO–$(CH_2)_3$–Si(CH_3)–O–Si(CH_3)–$(CH_2)_3$–PEO	PEO–PPO–PEO)–SC: $LiClO_4$ (4:1 molar)	$1\text{–}3 \times 10^{-5}$
PEO grafted polysiloxane, PGPS	$-[Si(CH_3)O]_n-$ with side chain CH_2CH_2PEO	PGPS:$LiClO_4$	10^{-4}
Poly[bis-2-(2-methoxyethoxy) Ethoxy))phosphazene, MEEP	$-[P{=}N]_n-$ with two side chains $OCH_2CH_2OCH_2CH_2OCH_3$	$(MEEP)_4$:$LiBF_4$	2×10^{-5}
		$(MEEP)_4$: $LiN(CF_3SO_2)_4$	5×10^{-5}
		$(MEEP)_4$: $LiC(CF_3SO_2)_4$	10^{-4}

impurities, which depress the melting point of the polymer. Ikeda *et al.* observed that the presence of side-chains also promotes the solvation of a salt [56,57]. A side-chain has a shorter relaxation time compared to the main chain. The coupling of the side-chain with the ion carrier, therefore, results in an increase in conductivity. Watanabe *et al.* designed a comb-shaped polyether host with a short polyether side chain [58]. However, the mechanical properties decreased and even the conductivity increased. A high conductivity with good mechanical properties was obtained for a high molecular weight polymer with trioxyethylene side-chains, as reported by Ikeda *et al.* [59].

A composite of a polymer with room temperature molten salt is also an interesting approach to improve the conductivity of polymer electrolytes. Watanabe *et al.* reported on a composite consisting of chloroaluminate molten salt that had a conductivity of 2×10^{-3} S/cm at room 303 K [60,61]. However, the disadvantage of chloroaluminate is its hygroscopic properties, which are impractical for many applications. The use of non-chloroaluminate molten salt, therefore avoids the hygroscopic problem. Tsuda *et al.* reported a conductivity of 2.3×10^{-2} S/cm for a composite of a polymer and room temperature molten fluorohydrogenates [62].

The third approach appears to be the simplest since a pre-produced polymer can be used to prepare the polymer electrolytes. Previously, low molecular weight polymers were usually used to reduce the operational temperature of polymer electrolytes. Low molecular weight polymers, which were added to the matrix of a high molecular weight polymer to reduce crystallinity at low temperatures, are frequently known as liquid plasticizers. Feullade and Perche demonstrated the concept of plasticizing the polymer with an aprotic solution containing an alkali metal salt in which the organic solution of the alkali metal salt remained trapped within the matrix of the solid polymer matrix. Such mixing results in the formation of gels with ionic conductivity close to that of liquid electrolytes [63]. Solvents that evaporate more slowly, such as ethylene carbonate (EC), propylene carbonate (PC), dimethyl formamide (DMF), diethyl phthalate (DEP), diethyl carbonate (DEC), methyl ethyl carbonate (MEC), dimethyl carbonate (DMC), γ-butyrolactone (BL), glycol sulfide (GS), and alkyl phthalates have been investigated as plasticizers for gel electrolytes.

However, an improvement in conductivity is adversely accompanied by degradation in solid state configuration and a loss of compatibility with the lithium electrode [49], particularly when the fraction of plasticizer is too high. For example, the modulus of elasticity and elastic strength significantly decreases on the addition of a plasticizer. This is because the plasticizers are usually low molecular weight polymers with low mechanical strength. Therefore, the addition of plasticizers decreases the mechanical strength of the host polymer. The use of moderate or large quantities of plasticizer results in the production of a gel electrolyte. The presence of some plasticizers may also give rise to problems caused by their reaction with the lithium anode. The poor mechanical stability appears to be mainly due to the solubility of the polymer matrix in the plasticizer [64]. Cross-linking of polymers with ultraviolet radiation [65], thermally [66], photo-polymerization [67] or electron beam radiation polymerization [68] was found to reduce the solubility of the polymer in the solvent and also helped liquid electrolytes to be trapped within the polymer matrix.

Currently, one popular approach to improve conductivity involves dispersing ceramic fillers (solid plasticizers) in a polymer matrix producing what is currently known as composite polymer electrolytes. This approach was first introduced by Weston and Steele [69]. A ceramic filler was used to reduce the glass transition temperature and the crystallinity of the polymer thus allowing the amorphous polymer to maintain liquid-like characteristics at the microscopic level. Ceramic fillers that are frequently used have particle sizes in the range of about several nanometers up to one micrometer. Fortunately, such filler materials are commercially available in various sizes at low prices.

The inorganic filler acts as a support matrix for the polymer, so that even at high temperature, the composite remains solid. However, at the microscopic level, it maintains a liquid-like structure, which is important for sufficient conductivity. The filler particles, due to their high surface area, prevent re-crystallization of polymer when annealed above the melting point. Acid–base interactions between filler surface groups and the oxygen of the PEO lead to a Lewis acid characteristic of the inorganic filler and favor the formation of complexes with PEO. The filler then acts as cross-linking center for the PEO reducing the tension of the polymer for self-organization and promoting stiffness. On the other hand, acid–base interactions between the polar surface group of the filler and electrolyte ions probably favor the dissolution of the salt.

9.3.2 Measuring the electrical conductivity

One important parameter of polymer electrolyte nanocomposites is electrical conductivity. The difficulty in making a DC measurement is in finding an electrode material that is compatible with the electrolyte composites. For example, if stainless steel electrodes are attached to an electrolyte composite, as shown in Fig. 9.14a, and a small voltage is applied across the electrodes, Li^+ ions migrate preferentially toward the cathode, but pile up without being discharged at the stainless–electrolyte interface. A Li^+ ion deficient layer forms at the electrolyte–stainless steel interface. The cell therefore behaves like a capacitor. There is an accumulation of ions at the interface region of the electrode and composite. A large instantaneous current I_0 occurs when the cell is switched on, the magnitude of which is related to the applied voltage and the resistance of the electrolytes, but then falls exponentially with time, as illustrated in Fig. 9.14b. The characteristic time for the current to decrease is relatively fast so that it is difficult to make an accurate measurement.

Therefore, an AC method is commonly used to make a measurement over a wide range of frequencies. The DC value of the conductivity can be extracted from the corresponding AC data. Many AC measurements are performed with blocking electrode such that no discharge or reaction occurs at the electrode–electrolyte interface. Because the current will flow back and forth, no ion pile up is found on the electrode surface, especially when a high AC frequency is used. Commonly used electrodes are platinum, stainless steel, gold, and indium tin oxide (ITO) glass.

The complex impedance method is widely used to determine the resistance of a sample.

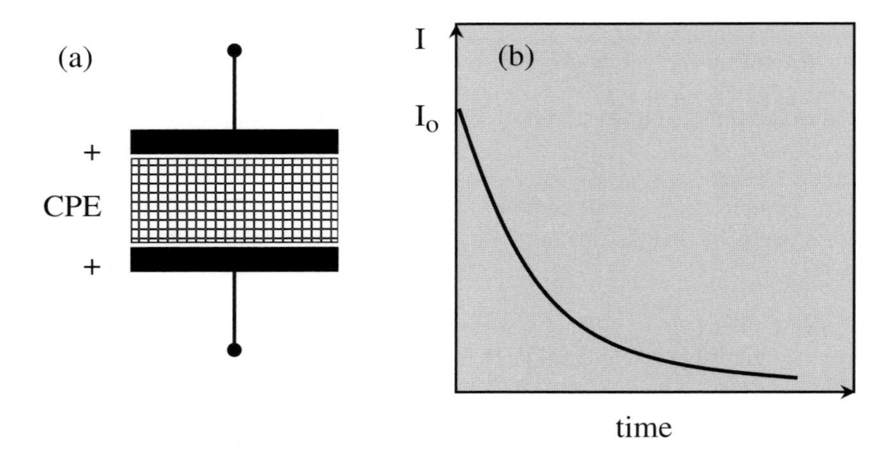

Fig. 9.14. Polymer electrolyte nanocomposite sandwiched between two blocking electrodes. (b) Current decay when a constant dc voltage is applied between two electrodes.

The principle of the method is based on the measurement of cell impedance, which is taken over a wide frequency range and then analyzed in the complex impedance plane, which is useful for determining the appropriate equivalent circuits for a system and for estimating the values of the circuit parameters.

Impedance is simply the AC resistance of the cell. The value in general contains a real and an imaginary part. An electrochemical cell, in general, exhibits resistive, capacitive, as well as inductive properties. The resistive property contributes to the real part of the impedance, while the capacitive and the inductive properties contribute to the imaginary part of the impedance. Therefore, an electrochemical cell can be considered to be a network comprised of a resistor, a capacitor, as well as a conductor. The arrangement for such a cell is usually determined after performing a measurement and analyzing the form of the impedance curve. The capacitor present as an open circuit in a DC network and an inductor that appears as a straight conductor wire in a DC circuit, both appear as imaginary resistors in an AC circuit. Until now, the inductive properties of electrochemical cells are ignored so that the polymer electrolyte composite is considered only to be a network of resistors and capacitors.

The complex impedance can be written in a general form as

$$Z(\omega) = Z'(\omega) - iZ''(\omega) \tag{9.31}$$

ω, the frequency, $Z'(\omega)$, the real part of impedance, contributed by the resistive part, $Z''(\omega)$, the imaginary part of the impedance, contributed by the capacitive part, $i = \sqrt{-1}$, the imaginary number.

As an illustration, Fig. 9.15 shows examples of simple RC circuits and corresponding plots of impedance (Nyquist plot). For a serial arrangement of a resistor and a capacitor,

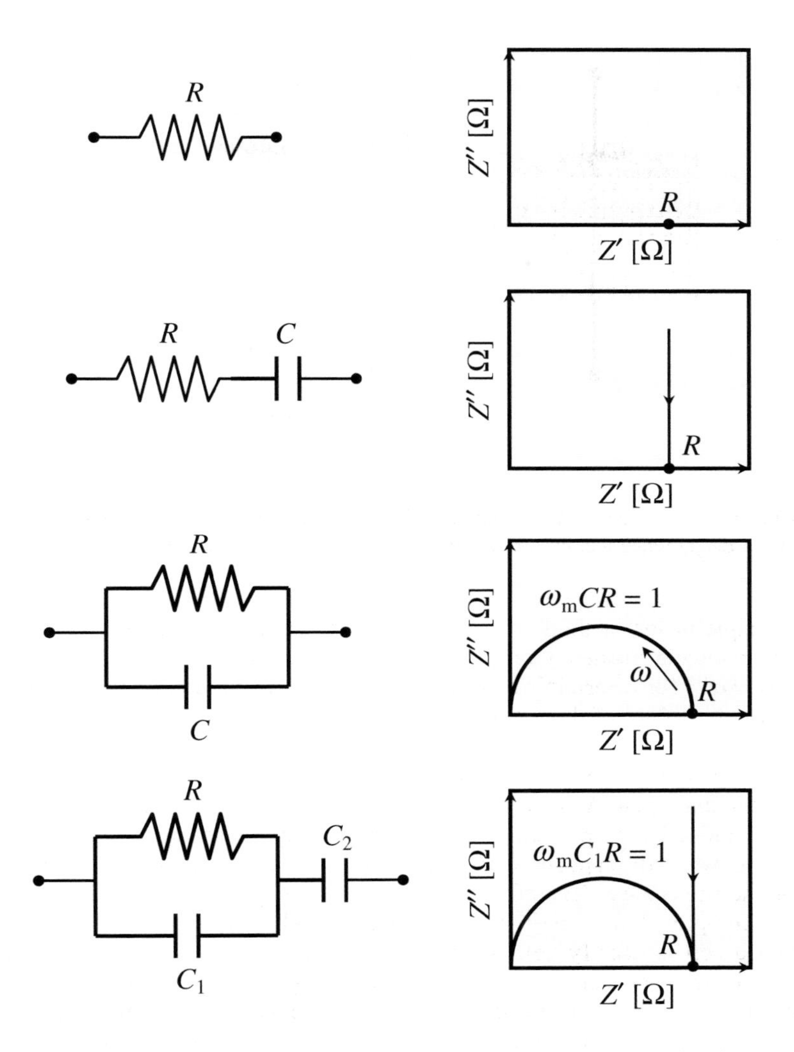

Fig. 9.15. Example of simple *RC* circuits and corresponding Nyquist plots.

as displayed in Fig. 9.15b (left), the impedance can be written as

$$Z = R - \frac{i}{\omega C} \tag{9.32}$$

or

$$Z' = R \tag{9.33a}$$

$$Z'' = \frac{1}{\omega C} \tag{9.33b}$$

It is clear that the real part of impedance is constant and independent of the frequency, while the imaginary part is dependent on the frequency. For a very small frequency, the imaginary part is very large and this value decreases inversely with frequency. When the frequency approaches infinity, the imaginary part of impedance approaches zero and the impedance value at this very high frequency is equal to the resistance. Thus the Nyquist plot for this arrangement appears as a vertical straight line, starting from a lower frequency value at the upper part downwards when the frequency increases, as shown in Fig. 9.15b (right). The intersection of this line with the horizontal axis (the real value of impedance) corresponds to the resistance.

For a parallel arrangement of a resistor R and capacitance C, as shown in Fig. 9.15c (left), the real and imaginary parts of the impedance are given by

$$Z' = \frac{R}{1 + (\omega RC)^2} \tag{9.34a}$$

and

$$Z'' = R\frac{\omega RC}{1 + (\omega RC)^2} \tag{9.34b}$$

and the corresponding Nyquist plot appears in Fig 9.15c (right). The Nyquits plot appears as an arc. The intersection of this arc with the vertical axis at a low frequency (right arc) corresponds to the resistance. The frequency at the peak of the arc, ω_{m}, satisfies the relation

$$\omega_{\mathrm{m}} RC = 1 \tag{9.35}$$

From the intersection point in the low frequency region and the position of the arc peak, the resistance and the capacitance of the system can be determined.

For a more complex arrangement, a more complex expression for impedance is needed. For example, a combination of a serial and parallel circuit, as shown in Fig. 9.15d (left) has the impedance as

$$Z = \left(\frac{1}{R} + i\,\omega C_1\right)^{-1} + \frac{1}{i\omega C_2} \tag{9.36}$$

while the corresponding Nyquist plot appears in Fig. 9.15d (right). It contains a vertical line that intersects the horizontal axis at $Z' = R$, and an arc with the peak satisfies $\omega_{\mathrm{m}} RC_1 = 1$. Again, from these two values, one can determine R and C_1. The value of C_2 is determined by measuring the vertical component of the impedance at a certain frequency, say ω^*. If the vertical component of the impedance at this point is Z''^*, the value of C_2 satisfies,

$$Z''^* = \frac{1}{\omega^* C_2} \tag{9.37}$$

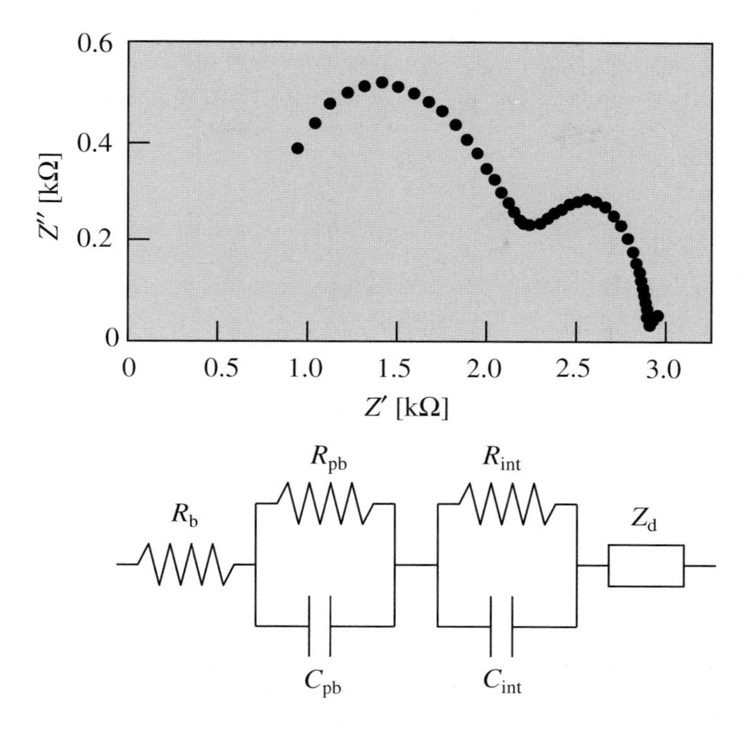

Fig. 9.16. (a) Impedance plot of PEO:$LiCF_3SO_3$ containing $Li_{1.4}Al_{0.4}Ge_{1.6}(PO_4)_3$ fillers obtained from experiment. Data points were extracted from [68]. (b) The suggested RC circuit for data in (a).

Sometimes, the form of the curve is not as simple as that described here. However, in principle, it is possible to find some circuit arrangement such that the theoretical Nyquist plot is in agreement with the measured data. Some computer software programs are commercially available for extracting the equivalent circuit for the measured data. As an example, an impedance measurement for a system of PEO:$LiCF_3SO_3$ containing $Li_{1.4}Al_{0.4}Ge_{1.6}(PO_4)_3$ fillers is displayed in Fig. 9.16a [70]. The corresponding AC circuit that produced this impedance data appears in Fig. 9.16b.

where R_b is the bulk resistance, R_{pb} the phase boundary resistance, R_{int} the interfacial resistance, C_{pb} the phase boundary capacity, C_{int} the interfacial capacity, and Z_d the diffusive impedance. The corresponding parameter values that can properly fit the measured data are $R_b = 593\,\Omega$, $R_{pb} = 1637\,\Omega$, $C_{pb} = 31$ nF, and $C_{int} = 1.9\,\mu$F [70]. From the measured resistance of the polymer electrolytes, the electrical conductivity can be calculated using a simple equation

$$\sigma = \frac{1}{R_e}\frac{\ell}{A} \tag{9.38}$$

with R_e the resistance of polymer electrolyte, ℓ material thickness (electrode spacing), and A material cross section.

9.3.3 *Effect of filler volume fraction on conductivity*

The electrical conductivity of polymer electrolyte composites strongly depends on the filler content. Initially, the conductivity increases with filler content and then decreases with further increases in filler content, reaching a maximum value at a specific filler content. This behavior can be explained by a simple effective medium approximation method.

We used a modified effective medium approximation to predict the conductivity of composites [11,71]. We restricted our attention to composites consisting of monosized-filler particles to simplify the formulation of the model. First, we consider the case of low temperatures. The presence of particles in the polymer electrolyte induces the formation of an amorphous region around the particles, which leads to the occurrence of a high conductivity layer at the interface between the particles and electrolyte. We call this region a high conductive region. At regions far removed from the particle, the polymers are in a crystalline state such that the conductivity is lower than that of the interface region. We call this region a medium conductive region. When two particles make contact, the electrolyte medium between those particles is removed and the transport of ions does not occur. The conductivity in this region is reduced to close to zero (approximately equal to that of the insulator particles), and we call this a low conductive region. The overall system can then be modeled as a three-conductivity system. If the probability for the existence of low, medium, and high conductive regions are P_ℓ, P_m, and P_h, respectively, an effective medium approximation can be written as [11,71]

$$P_\ell \frac{\sigma_\ell - \sigma_e}{\sigma_\ell + (z/2 - 1)\sigma_e} + P_m \frac{\sigma_m - \sigma_e}{\sigma_m + (z/2 - 1)\sigma_e} + P_h \frac{\sigma_h - \sigma_e}{\sigma_h + (z/2 - 1)\sigma_e} = 0 \tag{9.39}$$

where σ_ℓ, σ_m, and σ_h are the conductivities of the low, medium, and high conductivity mediums, respectively, σ_e the effective conductivity of composite and z the coordination number. We predicted these three probabilities using a simple combination approach as following.

The composite was divided into a number of similar cells. Each cell will be occupied either by a particle or be empty (fully occupied by matrix). We select the shape of the cell so as to be similar to a Wigner–Seitz cell. Suppose the total number of cells is N and the total number of particles is N_1. The total number of matrix-filled cells is $N_2 = N - N_1$. The probability for particles being in contact is proportional to the number of permutations of $N_1 - 2$ particle-filled cells and N_2 matrix-filled cells, i.e.

$$P_\ell = A \frac{(N - 2)!}{(N_1 - 2)! N_2!} \tag{9.40}$$

where A is a normalization factor. The probability for matrix-filled cells being in contact is proportional to the number of permutation of N_1 particle-filled cells and N_2-2 matrix-filled cells, i.e.

$$P_{\mathrm{m}} = A\frac{(N-2)!}{N_1!(N_2-2)!} \tag{9.41}$$

The probability for particles and matrix-filled cells to be in contact is proportional to double the number of permutations of N_1-1 particle-filled cells and N_2-1 matrix-filled cells, i.e.

$$P_{\mathrm{h}} = 2A\frac{(N-2)!}{(N_1-1)!(N_2-1)!} \tag{9.42}$$

The factor 2 was introduced because of the possibility of interchanging the position of a particle-filled cell and a matrix-filled cell being in contact. Since the total probability must be unity, the expression for A is

$$\frac{1}{A} = \frac{(N-2)!}{(N_1-2)!N_2!} + \frac{(N-2)!}{N_1!(N_2-2)!} + 2\frac{(N-2)!}{(N_1-1)!(N_2-1)!}$$

For common composites, $N_1 \gg 1$, $N_2 \gg 1$, $N \gg 1$ and $N^2 \gg N$, such that the approximation for A is given by

$$A = \frac{1}{N^2}\frac{N_1!N_2!}{(N-2)!} \tag{9.43}$$

Substituting Eq. (9.7) into Eqs. (9.3), (9.4), and (9.5) one obtains

$$P_\ell = \frac{1}{N^2}\frac{N_1!N_2!}{(N-2)!}\frac{(N-2)!}{(N_1-2)!N_2!} = \frac{(N_1-1)N_1}{N^2} \cong \frac{N_1^2}{N^2} \tag{9.44}$$

$$P_{\mathrm{m}} = \frac{1}{N^2}\frac{N_1!N_2!}{(N-2)!}\frac{(N-2)!}{N_1!(N_2-2)!} = \frac{(N_2-1)N_2}{N^2} \cong \frac{N_2^2}{N^2} \tag{9.45}$$

$$P_{\mathrm{h}} = 2\frac{1}{N^2}\frac{N_1!N_2!}{(N-2)!}\frac{(N-2)!}{(N_1-1)!(N_2-1)!} = 2\frac{N_1N_2}{N^2} \tag{9.46}$$

Suppose u_0 is the volume of one cell, and V is the total volume of the composite. Apparently a cell cannot be fully occupied by a spherical particle. Rather, only a fraction of (packing fraction) of particle-filled cells is occupied by particle material and the remaining fraction $(1-f)$ are occupied by insulator material. Therefore, the total volume of particles in the composite is only $V_1 = N_1 f u_0$. Since the total volume of composite is given by $V = Nu_0$, then $N_1/N = V_1/fV = v/f$, $N_2/N = 1 - N_1/N = 1 - v/f$ where v is the volume fraction of filler particles. Finally, one obtains the probability for obtaining three kinds of contact as

$$P_\ell = \frac{v^2}{f^2}, \quad P_m = \left(1 - \frac{v}{f}\right)^2, \quad P_h = 2\left(\frac{v}{f}\right)\left(1 - \frac{v}{f}\right) \tag{9.47}$$

We simulated the variation conductivity for various particle volume fractions. For simplicity we used a simple cubic packing of particles such that $z = 6$. The three conductivities were set to satisfy $\sigma_\ell/\sigma_h = 0.02$, $\sigma_m/\sigma_h = 0.1$ and for a simple cubic packing we used $z = 6$ and $f = \pi/6$. Figure 9.17, curve (a) shows the variation in conductivity at low temperatures. A bell-shaped conductivity curve is obtained as the volume fraction of particles increases. At zero volume fraction of the particles, the conductivity of the composites is equal to that of polymer in the crystalline phase. There is a fraction of particles such that the conductivity reaches a maximum value and there is threshold volume fraction of particles such that the conductivity tends toward zero. A similar shape has been observed for various composites of insulating particles and solid-state ionics matrix [72–75].

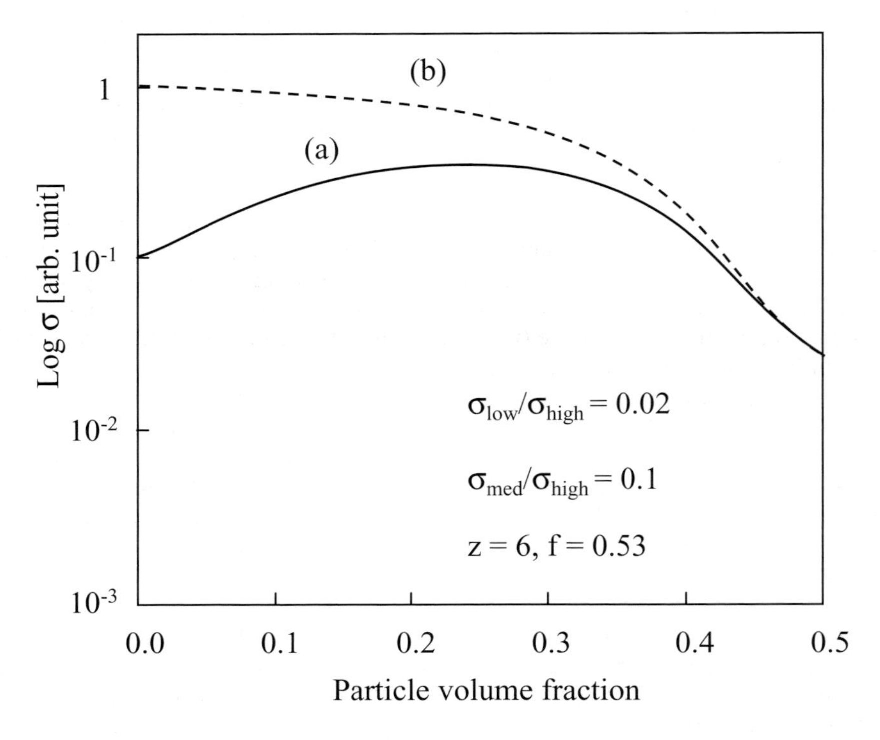

Fig. 9.17. Calculated variation in electrical conductivity with respect to the volume fraction of nanoparticles at (a) low temperatures and (b) high temperatures [11].

At high temperatures, on the other hand, not only are the regions around the particle surface amorphous, but nearly, all parts of the polymer are changed to an amorphous state. Therefore, the conductivity of the high and medium conductivity regions becomes similar, or $\sigma_{\rm m} = \sigma_{\rm h}$. Using this value and maintaining $\sigma_{\ell}, /\sigma_{\rm h} = 0.02$, we have curve (b) in Fig. 9.17. The presence of nanoparticles is simply to block the transfer of ion carriers. The conductivity always decreases with an increase in the volume fraction of particles.

9.4 Luminescent Polymer Electrolyte Nanocomposites

It has been shown that the addition of inorganic fillers (particles in the micrometer range down to nanometer sizes) improves the conductivity of polymer electrolytes, even at ambient temperature, due to the induction of an amorphous phase around the interface between the polymers and the filler surface. In order to produce luminescent polymer electrolytes with a high mobility, the luminescence source should be attributed to filler and lithium ions are used as ion carriers.

Since ZnO nanoparticles emit a blue to green luminescence (depends on the crystalline size) under ultraviolet excitation, we were able to produce a luminescence polymer electrolyte with filler as luminescence centers. Figure 9.18 shows the luminescence spectra of composites of ZnO nanoparticles dispersed in polyethylene glycol containing Li ions as carriers [12]. Pictures of the samples under the ultraviolet region are shown in Fig. 9.19. The difference in color is due to the difference in ZnO particle size dispersed in the matrix. A smaller size particle emits a blue color while large-sized particle emits a green-yellow color.

Polymer electrolytes emitting luminescence using rare earth ions as luminescence centers have been investigated by many authors [76–80]. Color constituents from rare earth elements are characterized by an incomplete f shell. Rare earth ions usually have dual functions: as ion carriers and as luminescence centers. However, since the atomic masses of the rare earth ions are very large (for example the atomic mass of europium is 151.96) the mobility of the ions becomes lower, resulting in a decreased electrical conductivity. For example, Bermudez *et al.* observed conductivities of less than 10^{-4} S/cm in samples of Eu^{3+}-doped hybrid organic/inorganic materials [77] and Silva and Smith reported conductivities of less than 10^{-5} S/cm in polymer electrolytes based on europium picrate [78]. Highly conductive electrolytes can be obtained using lithium as an ion carrier (the atomic mass of lithium is only 6.94). Therefore, in order to produce high conductivity, the rare earth ions should serve as luminescence centers only and not as ion carriers.

Rare-earth-ion-doped oxides have been used as phosphors in displays because they produce a very sharp and intense emission [81]. Europium-doped yttrium oxide (Y_2O_3:Eu), for example, produces sharp emission centered at 612 nm when excited using a wavelength of around 254 nm. It would be interesting to investigate the possibility of using Y_2O_3:Eu nanoparticles as fillers in polymer electrolyte composites. Combined with the use of lithium ions as carriers, it might be possible to produce luminescent polymer

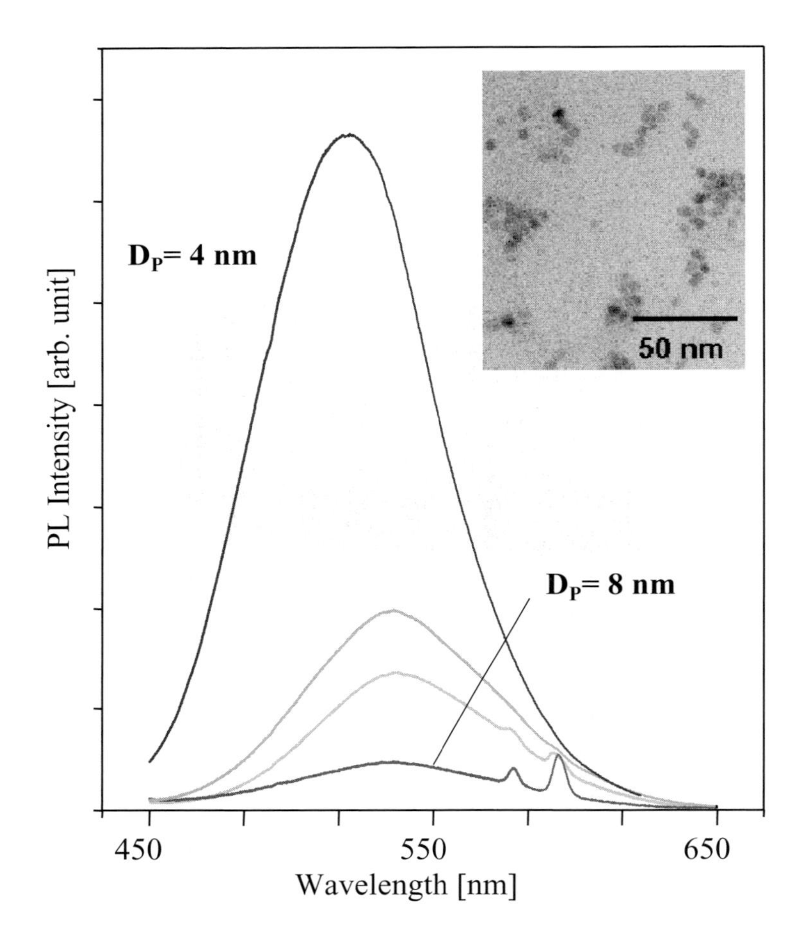

Fig. 9.18. Luminescence spectra of ZnO nanoparticles dispersed in polyethylene glycol with various sizes of ZnO nanoparticles. Inset is a SEM picture of one sample [12].

electrolyte composites exhibiting both a high luminescence intensity and high electrical conductivity.

It has been proposed that visible luminescence is caused by the disintegration of excitons and an electron then jumps to a state located near the center of the gap [12]. The increase in band gap on the reduction in particle size shifts the position of the luminescence spectra to shorter wavelengths. If the time evolution of the luminescence spectra of a ZnO colloid prepared by Spanhel and Anderson's method is observed, it can be seen that the fresh colloid emits a nearly blue color, which then changes to green, and finally a yellow-green color. This definitive change is very interesting. If the size of particles in a ZnO colloid could be stopped at various aging times, ZnO particles producing a variety of colors ranging from nearly blue to nearly yellow could be produced.

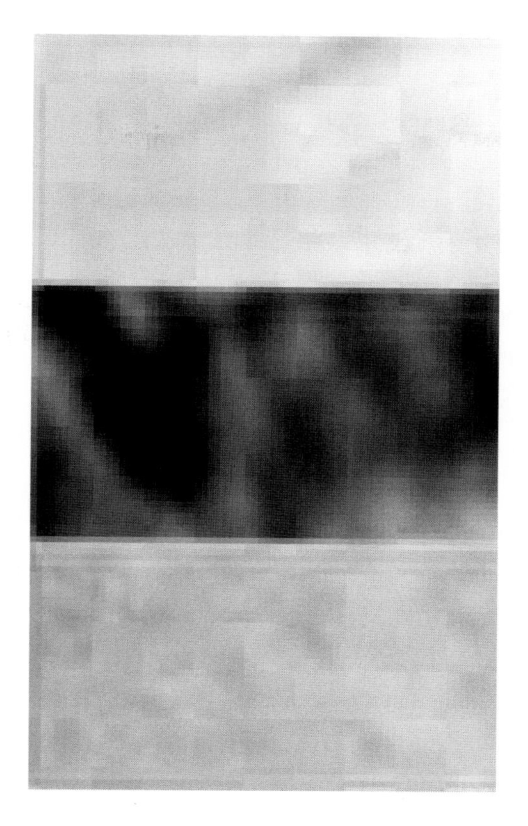

Fig. 9.19. Pictures of ZnO nanoparticles/polyethylene glycol containing different size ZnO particles under ultraviolet illumination [12].

References

[1] S.W. Shang, J.W. Williams, K.-J. Soderholm, J. Mater. Sci. 30 (1995) 4323.

[2] S. Nakajima, Filler 3 (1998) 160.

[3] K. Otsuka, Bull. Ceram. Soc. Jpn. 28 (1993) 1124.

[4] T. Kitano, Filler 3 (1998) 96.

[5] S. Hagiwara, Plastics 49 (1998) 58.

[6] T. Yoshida, Plastics 49 (1998) 63.

[7] N.S. Sariciftci, L. Smilowitz, A.J. Heeger and F. Wudl, Science 258 (1992) 1474.

[8] G. Yu, J. Gao, J.C. Hummelen, F. Wudl and A.J. Heeger, Science 270 (1995) 1789.

[9] W.D. Kingery, H.K. Bowen and D. R. Uhlmann, Introduction to Ceramics, John Wiley: NY, 1976.

[10] Mikrajuddin, I.W. Lenggoro, K. Okuyama and F.G. Shi, J. Electrochem. Soc. 149 (2002) 107.

[11] M. Abdullah, I.W. Lenggoro, K. Okuyama and F.G. Shi, J. Phys. Chem. B, 107 (2003) 1957.

[12] M. Abdullah, T. Morimoto and K. Okuyama, Adv. Funct. Mater. 13 (2003) 800.

[13] K. Akamatsu, S. Takei, M. Mizuhata, A. Kajinami, S. Deki, S. Takeoka, M. Fujii, S. Hayashi and K. Yamamoto, Thin Solid Films 359 (2000) 55.

[14] H.J. Wasserman and J.S. Vermaak, Surf. Sci. 22 (1970) 164.

[15] L. M. Sherman, Plast. Technol. 45 (1999) 53.

[16] J.R. Heath, Acc. Chem. Res. 32 (1999) 388.

[17] R. Dagani, C&E News, 77 (1999) 25.

[18] S.J. Radhakrisnan, J. Cryst. Growth, 141 (1994) 437.

[19] S.J. Radhakrisnan, J. Cryst. Growth, 129 (1993) 191.

[20] S. Radhakrisnan, J.M. Schultz, J. Cryst. Growth. 116 (1992) 378.

[21] C. Saunjaya, Ashamol, S. Padalkar and S. Radhakrishnan, Polymer 42 (2001) 2255.

[22] Y. Ito, M. Abdullah and K. Okuyama, J. Mater. Res. 19 (2004) 1077.

[23] T. Hayashi, H. Mizuma, H. Yao and S. Takahara, Jpn. J. Appl. Phys. 37 (1998) 2660.

[24] H. Yao and N. Kitamura, Bull. Chem. Soc. Jpn. 69 (1996) 1227.

[25] N. Serpone, D. Lawlwss and R. Khairutdinov, J. Phys. Chem. 99 (1995) 16646.

[26] Kittel, Introduction to Solid State Physics: John Wiley: NY, 1996.

[27] L.E. Brus, J. Chem. Phys. 80 (1984) 4403.

[28] L.E. Brus, J. Chem. Phys. 79 (1983) 5566.

[29] W.H. Lee, C. Lee, and J. Jong, J. Non-Crystalline Solids 198–200 (1996) 911.

[30] Y. Wang, A. Suna, W. Mahler, and R. Kawoski, J. Chem. Phys. 87 (1987) 7315.

[31] L.E. Brus, Acc. Chem. Res. 23 (1990) 183.

[32] M.G. Bawendi, M.L. Stiegerwald and L.E. Brus, Annu. Rev. Phys. Chem. 41 (1990) 477.

[33] E. Rabani, B. Hetenyi, B.J. Berne and L.E. Brus, J. Chem. Phys. 110 (1999) 5355.

[34] S.L. Lai, J.Y. Guo, V. Petrova, G. Ramanah and L.H. Allen, Phys. Rev. Lett. 77 (1996) 99.

[35] I.L. Fabelinskii, Molecular Scattering of Light, Plenum Press: New York, 1968.

[36] R. W. Boyd, Nonlinear Optics, Academic Press: Boston, 1992.

[37] J.D. Jackson, Classical Electrodynamics, Wiley: New York, 1982.

[38] J.A. Stratton, Electromagnetic Theory, McGraw Hill: New York, 1941.

[39] L. Poladian, Q.J. Mech. Appl. Math. 41 (1988) 395.

[40] R.C. McPhedran, L. Poladian and G.W. Milton, Proc. R. Soc. London, Ser. A 415 (1988) 185.

[41] L. Poladian, Proc. R. Soc. London, Ser. A 426 (1989) 343.

[42] L. Poladian, Phys. Rev. B 44 (1991) 2092.

[43] M. Abramowitz and I. A. Stegun, Handbook of Mathematical Functions, Dover: NY, 1965.

[44] F.B. Dias, L. Plomp and J.B.J. Veldhuis, J. Power Sources 88 (2000) 169.

[45] B. O'Regan and M. Gratzel, Nature 353 (1991) 737.

[46] M.K. Nazeeruddin, A. Kay, I. Rodicio, R. Humphry-Baker, E. Muller, P. Liska, N. Vlachopoulos, M. Gratzel, J. Am. Chem. Soc. 115 (1993) 6382.

[47] G. Katsaros, T. Stergiopolous, I.M. Arabatzis, K.G. Papagodokostaki and P. Falaras, J. Phochem. Photobiol. A, 149 (2002) 191.

[48] P. Lighfoot, M.A. Metha and P.G. Bruce, Science 262 (1993) 883.

[49] F. Croce, G.B. Appetecchi, L. Persi and B. Scrosati, Nature 394 (1998) 456.

[50] D.E. Fenton, J.E. Parker and P.V. Wright, Polymer, 14, 589, 1973; P.V. Wright, Br. Polym. J. 7 (1975) 319.

[51] M.B. Armand, J.M. Chabagno and M.J. Duclot, Eds. Fast Ion Transport in Solids, M.J. Duclot, P. Vashishta, J.M. Mundy and G.K. Shenoi, Elsevier, New York, 1979.

[52] J.R. MacCalum, M.J. Smith and C.A. Vincent, Solid State Ionics 11 (1981) 307.

[53] M. Watanabe, S. Nagano, K. Sanui and N. Ogata, J. Power Sources 35 (1987) 327.

[54] Y. Song, X.Peng, Y. Lin, B. Wang and D. Chen, Solid State Ionics 76 (1995) 35.

[55] O. Inganas, Br. Polym. J. 20, 233 (1988); D.W. Kim, J.K. Park and H.W. Rhee, Solid State Ionics 83 (1996) 49.

[56] A. Nishimoto, M. Watanabe, Y. Ikeda and S. Kohjiya, Electrochim. Acta 43 (1998) 1177.

[57] Y. Ikeda, Y. Wada, Y. Mataba, S. Murakami and S. Kohjiya, Electrochim. Acta 45 (2000) 1167.

[58] K. Motogami, M. Kono, S. Mori, M. Watanabe and N. Ogata, Electrochim. Acta 37 (1992) 1725.

[59] Y. Ikeda, H. Masuo, S. Syoji, T. Sakashita, Y. Matoba and S. Kohjiya, Polym. Int. 43 (1997) 269.

[60] M. Watanabe, S. Yamada, K. Sanui and N. Ogata, J. Chem. Soc. Chem. Commun. (1993) 929.

[61] M. Watanabe, S. Yamada and N. Ogata, Electrochem. Acta 40 (1995) 2285.

[62] Y. Tsuda, T. Nohira, Y. Nakamori, K. Matsumoto, R. Hagiwara and Y. Ito, Solid State Ionics 149 (2002) 295.

[63] G. Feullade and P. Perche, J. Appl. Electrochem. 5 (1975) 63.

[64] S. Chintapalli and R. French, Solid State Ionics 86–88 (1996) 341.

[65] H.W. Rhee, W.I. Jung, M.K. Song, S.Y. Oh and J.W. Choi, Mol. Cryst. Liq. Cryst. Sci. Technol. A294 (1997) 225.

[66] M. Morita, T. Fukumasa, M. Motoda, H. Tsutsumi, Y. Matsuda, T. Takahashi and H. Ashitaka, J. Electrochem. Soc. 137 (1990) 3401.

[67] D.W. Xia, D. Soltz and J. Smidt, Solid State Ionics 14 (1984) 221.

[68] M.S. Michael, M.M.E. Jacob, S.R.S. Prabaharan and S. Radhakrishna, Solid State Ionics 98 (1997) 167.

[69] J.E. Weston and B.C.H. Steele, Solid State Ionics 7 (1982) 75.

[70] C.J. Leo and G.V.S. Rao, B.V.R. Chowdary, Solid State Ionics 148 (2002) 159.

[71] Mikrajuddin, F.G. Shi and K. Okuyama, J. Electrochem. Soc. 147 (2000) 3157.

[72] J. Jow and J.B. Wagner, J. Electrochem. Soc. 126 (1979) 1963.

[73] B. Kumar and L.G. Scanlon, J. Power Source 52 (1994) 261.

[74] P. Przylusky, M. Siekiersk and W. Wieczorek, Electrochim. Acta 40 (1995) 2102.

[75] S. Furusawa, S. Miyaoka and Y. Ishibashi, J. Phys. Soc. Jpn. 60 (1991) 1666.

[76] L.C. Carlos, A.L.L. Videira, M. Assuncao and L. Alcacer, Electrochim. Acta 40 (1995) 2143.

[77] V. de, Z. Bermudez, L.D. Carlos, M.C. Duarte, M.M. Silva, C.J.R. Silva, M.J. Smith, M. Assuncao and L. Alcacer, J. Alloys Comp. 275–277 (1998) 21.

[78] M.M. Silva and M.J. Smith, Electrochim. Acta 45 (2000) 1463.

[79] L.D. Carlos, M. Assuncao, P.M. Mourao and L. Alcacer, Electrochim. Acta 43 (1998) 1365.

[80] M. Furlani, A. Ferry, A. Franke, P. Jacobson and B.-E. Mellander, Solid State Ionics 113–115 (1998) 129.

[81] G. Blasse and B.C. Grabmaier, Luminescent Materials, Springer–Verlag: Berlin, 1994.

CHAPTER 10

Synthesis and Structure–Property Characteristics of Clay–Polymer Nanocomposites

S.C. Tjong

Department of Physics and Materials Science
City University of Hong Kong
Tat Chee Avenue, Kowloon, Hong Kong

10.1 Introduction

Many low-cost clay minerals such as kaolinite, talc, mica, etc. have been used as inorganic fillers for conventional composites to enhance specified desired properties such as modulus, strength, dimensional stability, wear resistance, etc. Such properties can be tailored by changing the volume fraction, shape, and size of the filler particles as well as the bulk properties of the polymer matrix. A fine and uniform dispersion of the clay particles within the polymeric matrix, and adequate interfacial bonding are needed for achieving better mechanical strength [1–3]. However, the typical filler content needed for significant enhancement of these properties can be as high as 20% by volume, leading to poor processability of composite materials. The reinforcing effect of clay fillers with sizes within micrometer regime is relatively low due to a lack of intense interaction between the filler and polymer and to low aspect ratio of the fillers. Moreover, the constituents are immiscible, resulting in a coarsely blended microstructure with chemically distinct phases. By decreasing the dimensions of inorganic fillers to nanometer regime, stronger interaction between nanofiller and polymer is expected. In this aspect, most of the polymer is associated with nanofiller interface. As a consequence, new nanocomposite materials with unusual chemical, physical, and mechanical properties are produced by dispersing low volume fraction of nanofillers with large surface area in polymer matrices. However, the basic reason for the improvement in properties is still far from being understood.

Clay minerals are hydrous aluminum silicates and are generally classified as phyllosilicates, or layer silicates. They are constructed by a combination of tetrahedral and octahedral sheets. Silica is the main component of a tetrahedral sheet whilst octahedral sheet comprises diverse elements such as Al, Mg, and Fe. A natural stacking of tetrahedral and octahedral sheet occurs in the specific ratios and modes, leading to the formation of the 1:1 and 2:1 layer silicates. Typical example of 1:1 layer silicate is kaolinite clay having the formula $Al_2(Si_2O_5)(OH)_4$. The silica tetrahedral layer, represented by $(Si_2O_5)^{2-}$, is made electrically neutral by connecting to an adjacent $Al_2(OH)_4^{2+}$ layer to form a composite sheet. Thus, the crystal plates of kaolinite are made of a series of parallel sheets (up to about 50) bonded together by weak secondary bonds [4]. On the other hand, the phyllosilicate 2:1 layer clays include mica, smectite, vermiculite, and chlorite. Smectite group can be further divided into montmorillonite (MMT), nontronite, saponite, and hectorite species [5,6]. A tetrahedral sheet of smectites is composed of corner-linked tetrahedral, whose central ions are Si^{4+} or Al^{3+} and sometimes Fe^{3+}. The basal oxygens of a tetrahedron are shared by the neighboring tetrahedral, forming hexagonal pattern. Thus, the crystal lattice of 2:1 phyllosilicate consists of two-dimensional layers where an edge-shared octahedral sheet of aluminum or magnesium hydroxide is fused to two silica tetrahedra by the tip such that the oxygen ions of the octahedral sheet also belong to the tetrahedral sheets. Figure 10.1 shows the crystalline structure of MMT. The layer thickness of the sheets is ~1 nm corresponding to the *c*-axis dimensions of its unit cell. The lateral dimensions of these layers vary from 250–400 nm. The sheets layers are stacked along the *c*-axis either regularly with 0° rotation between the successive layers, semi-random stacking with random degree of rotations, or turbostratic stacking with completely random rotations and translation between the layers [7]. These layer stacks are commonly referred to as tactoids. Isomorphic substitution within the layers, e.g. Al^{3+} replaced by Mg^{2+} or Fe^{2+}, or Mg^{2+} replaced by Li^{+} generates negative charges on the surface, defined through the cationic exchange capacity (CEC), depending on the mineral origin. Neutrality on the surface and in the interlayer (gallery) is balanced by alkali cations or hydrated cations. The electrostatic forces holding the layers together are relatively weak, and the interlayer distance varies depending on the nature of the cation present and its degree of hydration. For example, the stacks of Na^{+}-MMT swell easily in water and the layers can be delaminated by shearing. When the hydrated cations are ion-exchanged with organic cations such as bulky alkylammonium, it generally results in a larger interlayer spacing. The cation exchange capacity (CEC), expressed as mequiv/100gm-clay is used to characterize the degree of isomorphous substitution. Typical CEC values of 2:1 phyllosilicates are listed in Table 10.1. MMT has been employed in different applications varying from thixotropic agents in paints and cosmetics to fillers in polymer nanocomposites [8,9]. It is environmentally friendly, readily available in large quantities with low cost, and its intercalation chemistry has been well investigated [8]. Vermiculite (VMT) is a clay mineral which has a crystalline structure similar to MMT. However, the electrostatic charge of the VMT layer is larger than that of MMT due to a higher substitution of Si atoms with Al and Mg atoms.

Intercalation of molecular species into layered inorganic clays is a method of forming ordered inorganic–organic assemblies with unique microstructures controlled by host–guest and guest–guest interactions [10]. Smectite clays possess unique intercalation, swelling, ion exchange, and large surface area properties are attractive for organizing organic guest species. Na^{+}MMT clay expands the interlayer spaces readily when immersed in a hydrophilic solvent such as water. Alkylammonium ions of different

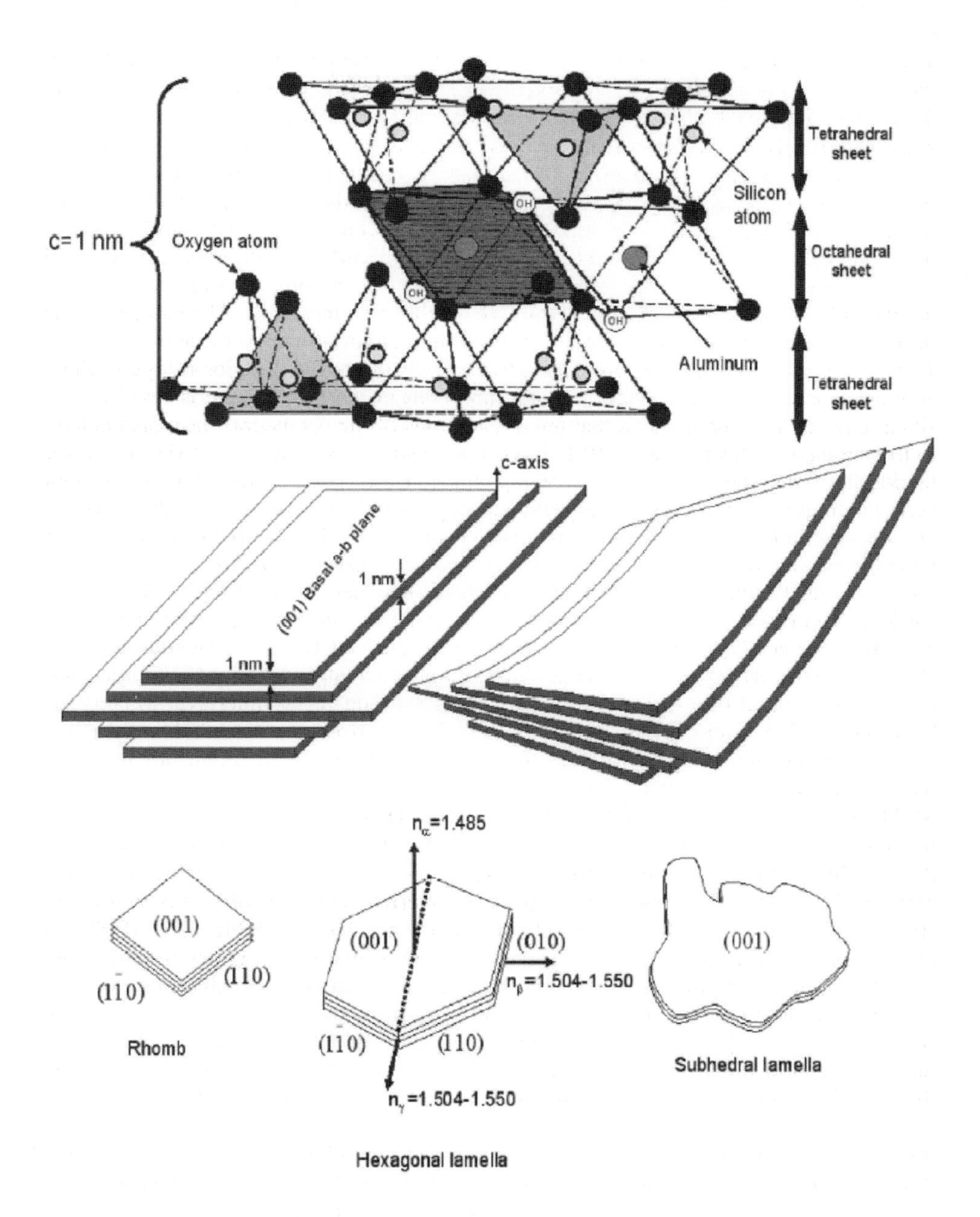

Fig. 10.1. Structure of montmorillonite basic crystal unit and its morphological variations from a perfect hexagonal habit (reprinted from [7] with permission from Elsevier).

Table 10.1. Chemical formula and cation exchange capacity of 2:1 phyllosilicates.

Silicate	Formula	CEC (mequiv/100g)
Montmorillonite	$M_x(Al_{4-x}Mg_x)Si_8O_{20}(OH)_4$	110
Hectorite	$M_x(Mg_{6-x}Li_x)Si_8O_{20}(OH)_4$	120
Saponite	$M_xMg_6(Si_{8-x}Al_x)Si_8O_{20}(OH)_4$	86.6
Vermiculite	$(Mg,Fe,Al)_3[(Al,Si)_4O_{10}](OH)_2M_x.nH_2O$	100–150

M represents exchangeable cation and x is the layer charge.

chain lengths can displace hydrated cations on the clay layers. In this case, the clay surfaces become hydrophobic, and the basal spacing of clays increases dramatically. The interlayer spacing can be controlled by the size of the organic cations used. The obtained product is known as 'organoclay'. The conversion of hydrophilic inorganic clay to a hydrophobic organoclay also improves the interfacial adhesion properties between the organic and inorganic phases when a hydrophobic polymer matrix is involved. On the other hand, the charge-compensating cations of the clay layers can also be exchanged with larger inorganic hydroxyl cations, formed by hydrolysis of metal salts. The intercalation of inorganic clusters into the clay layers and subsequent thermal activation leads to the formation of the so-called pillared clay. Upon calcination, these metal hydroxyl cations undergo dehydration and dehydroxylation, creating a material with stable metal oxide clusters, and acting as pillars to separate the clay layers. Pillared clays find widespread application as catalysts and catalyst supports due to their uniform pore structures [11,12]. A greater understanding of the manipulation of layered silicates at the nanoscale has led to the development of novel materials with better mechanical and physical properties.

Recently, there has been a growing interest in the development of novel inorganic–organic hybrids using layered ceramics, nanoparticles, and nanotubes as reinforcement materials [13–35]. These layered materials include clay silicates [13–20], manganese oxides [21–23], titanates [24], layered phosphates, [25] and layered double hydroxides [26]. Among layered materials, clay–polymer nanocomposites reinforced with minor amounts of silicate loading, e.g. 3–6 wt% show great promise for industrial applications. Such lower loading facilitates processing and reduces the component weight. The dispersion of low loading level of layered silicates as individual flakes in a polymeric matrix on the nanometer scale and with a large aspect ratio leads to large increases in tensile strength, thermal stability, and flame retardancy. Producing flame-retardant polymers for household goods could eventually save human lives. Currently, flame-retardant plastics contain bromine compounds, which produce poisonous gases when burned. Using clay as filler is a green alternative to current practices and it reduces flammability in various plastics. The nanocomposites also serve as a new system to study fundamental scientific studies concerning the confinement of polymer chains in galleries and the specific polymer–surface interaction not normally observed in the bulk.

The clay-polymer nanocomposites are also characterized by significant reductions in liquid and gas permeabilities. The improvement can be attributed to tortuous path experienced by a gas or liquid which penetrates deep into nanocomposites. The tortuousity factor can be as high as several-hundred-fold for impermeable platelets with aspect ratios of 100–500 at low loading levels [36]. Polymers are currently used in packaging applications, particularly food and drug packaging, where enhanced gas barrier, good mechanical properties, and transparency are important criteria. Clay–polyethylene terephthalate (PET) nanocomposite with enhanced gas barrier resistance shows potential application in food-processing industry as a material for beer or wine container. The nanocomposites protect the beverage from the effects of oxygen penetration [37].

Successful pioneering work on the clay-polymer nanocomposites was conducted by Toyota research group on in-situ polymerization of the MMT–nylon 6 system [38–40]. They demonstrated that exfoliated organoclay (organoMMT) platelets dispersed in a nylon-6 matrix greatly improved the thermal, mechanical, barrier, and even the flame-retardant properties of the polymer. Now, these materials are used in under-the-hood applications in the automobile industry. Since then, numerous studies have been carried out on the synthesis and the properties of various polymers reinforced with 2:1 phyllosilicates, particularly those reinforced with MMT [13–20, 41–51]. On the other hand, the intercalation of the polymer chains is greatly impeded in kaolinite (1:1 phyllosilicate) as a consequence of the high layer-to-layer interactions between $(Si_2O_5)^{2-}$ and $Al_2(OH)_4^{2+}$ and low cation exchange capacity. In spite of its great abundance, high crystallinity, and high purity when compared with other mineral clays, very few studies have been conducted on the intercalation of kaolinite with organic polymers [52–54].

Depending on the structure of dispersed clay platelets in the polymer matrix, the composites can be classified as intercalated or exfoliated nanocomposites. The intercalated clay-polymer nanocomposites have clay layers dispersed in a polymer matrix with large gallery spacing associated with the insertion of polymer chains into the gallery. In the exfoliated structure, the silicate layers are completely delaminated and dispersed individually as nanoscale platelets in a polymer matrix. Each platelet of large surface area interacts with the matrix, thereby improving the mechanical and physical properties of the nanocomposites more effectively. The nanoscale structure and morphology in polymer nanocomposites are commonly characterized by transmission electron microscopy (TEM). However, the sample preparation procedures for TEM observation are often tedious and difficult. In this case, X-ray diffraction (XRD) is a complementary structural technique to electron microscopy. For an intercalated structure, the (001) peak of the clay tends to shift to lower angles due to the expansion of the basal spacing. Although the layer spacing increases, there still exists an attractive force between the silicate layers to stack them in an ordered structure. In contrast, basal peak of the clay disappears in the exfoliated nanocomposites due to loss of the structural registry of the layers. It is noted that the disappearance of a basal peak in the XRD pattern alone should not be intended as a clear sign of exfoliation. Only TEM observation can confirm that exfoliation has occurred. Correlation between morphology evaluated by TEM and structure by XRD and final properties enables us to design functional clay-polymer nanocomposites with desirable characteristics.

10.2 Organoclay

Pristine layered silicate surfaces are hydrophilic, thus they are not compatible with most polymers. Cation exchange offers an effective way of modifying the galleries of the clays, thereby rendering the silicate surfaces more organophilic and therefore, more compatible with polymers. Smectic clays can undergo cation exchange with organic molecules, leading to the intercalation of organic molecules into their galleries. Long chain alkylamines are excellent cationic surfactants which are easily intercalated by numerous layered oxides such as silicates, phosphates, arsenates, etc [55]. The alkyl chain can intercalate the interlayer under proper processing conditions, leading to expansion of the basal spacing of MMT in the range 19–40Å. The organic cationic surfactants used include the primary, secondary, tertiary, and quaternary alkylammonium cations [56]. Because of the nonpolar nature of the alkyl chain, they reduce the electrostatic interaction between the silicate layers, thus facilitating diffusion of the polymer into the galleries. The space between the silicate layers depends greatly on the length of the alkyl chain and the ratio of cross-sectional area to available area per cation [57]. In the former case, the longer the alkyl chain, the wider the gap between the silicate layers.

Nowadays, several organoclays are available commercially at a relatively low cost. These include Cloisite 15A, 20A, and 30B produced from Southern Clay Products (USA) [58]. Cloisite 15A and 20A, are apolar Na^{+}-MMT clays modified with dimethyl, dehydrogenated tallow, quaternary ammonium (2M2HT). HT stands for a tallow-based compound (~65%C18, ~30%C16, and ~5%C14) in which a majority of the double bonds have been hydrogenated. The modifier concentration of Cloisite 15A and 20A is 125 and 95 meq/100g clay, respectively. The respective basal spacing (d_{001}) of the Cloisite 15A, 20A, and 30B is 3.23, 2.42, and 1.85 nm. This is reported in the product literature [58]. The cation molecular structure of Cloisite 15A and 20A is:

$$
\begin{array}{c}
CH_3 \\
| \\
CH_3\text{—}N^{+}\text{—}HT \\
| \\
HT
\end{array}
$$

Cloisite 30B is polar organoclay with the organic modifier methyl tallow bis-2-hydroxyethyl ammonium (MT2EtOH) having the formula:

$$
\begin{array}{c}
CH_2\text{—}CH_2\text{—}OH \\
| \\
H_3C\text{—}N^{+}\text{—}T \\
| \\
CH_2\text{—}CH_2\text{—}OH
\end{array}
$$

where T stands for tallow.

Selection of organoclays depends mainly on the type of polymer matrix used. Paul and coworkers investigated the effect of the structure of alkylammonium compounds on the dispersion of MMT in polyamide-6 during melt compounding [59]. They reported that the alkylammonium compound consisting of one alkyl tail is more effective than the quaternary cation having two alkyl tails in forming exfoliated nanocomposites. They explained this in terms of the competition between the effects of platelet–platelet interactions and the interaction of the polymer with the organoclay platelet. Polyamide-6, because of its polarity or strong hydrogen-bonding characteristic, has some affinity for the pristine surface of the clay, even more than for the largely aliphatic organic modifier. In this case, the organic modifier consisting of two alkyl tails (e.g. Closite 20A) shields more silicate surface than one alkyl tail, thereby precluding desirable interactions between the polyamide and the clay surface. On the other hand, nanocomposites made from a non-polar polymer like linear low density polyethylene (LLDPE) showed complete opposite trends. In that case, the two-tailed organoclay formed nanocomposites exhibit better exfoliation and mechanical properties than a one-tailed organoclay [60].

Accordingly, the interaction between the clay surfaces and the intercalants plays a key role in the formation of exfoliated structure. Therefore, understanding the structure of organoclays and the interaction of surfactant-clay is of crucial importance in design, fabrication, and characterization of exfoliated nanocomposites. Various structural models have been proposed for the molecular conformation of the surfactant, including monolayer, bilayer, paraffin-type monolayer and bilayer (Fig. 10.2) [55]. Alkylamine molecules in the interlayer space commonly aggregate to simple mono- or bimolecular structures. In a paraffin-type structure, the alkyl chains are considered to radiate away from the clay surface. Such structural models are considered to be too simple and idealistic. Vaia et al. employed transmission Fourier transform infrared spectroscopy (FTIR) to determine the molecular conformation of intercalated alkylammonium silicates by monitoring frequency shifts of the CH_2 stretching and scissoring vibrations as a function of the interlayer packing density, chain length, and temperature [61]. Under most

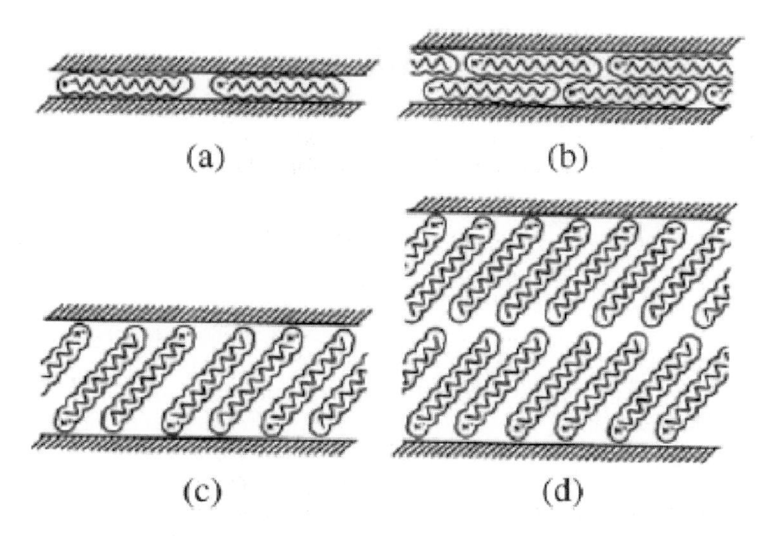

Fig. 10.2. Arrangement of alkylammonium ions in 2:1 layered silicates (reprinted from [61] with permission from The American Chemical Society).

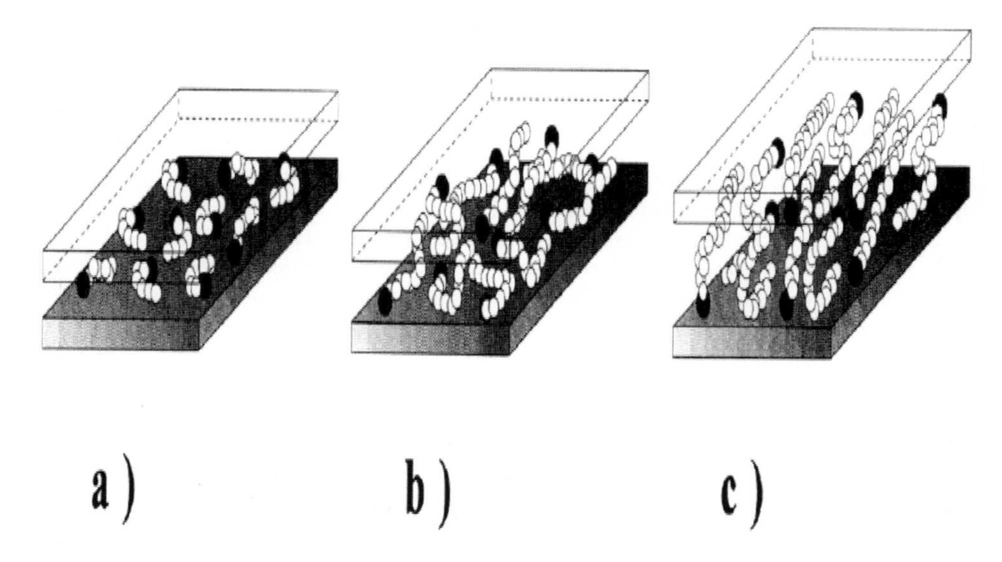

Fig. 10.3. Alkyl chain aggregation models : (a) The molecules are effectively isolated from each other at the shortest lengths, (b) quasi-discrete layers form with various degree of in plane disorder and interdigitation between the layers at medium lengths, and (c) interlayer order increases leading to a liquid crystalline polymer environment at long lengths. Open circles represent CH_2 segment while cationic head groups are represented by filled circles. (reprinted from [61] with permission from The American Chemical Society).

conditions, a wide range of molecular arrangements varying from solid-like to liquid-like or even to intermediate liquid crystalline case, depending on the packing density and chain length (Fig. 10.3). Increasing the packing density or chain length, improves the ordering of the chains. However, high temperature favors the disordered, liquid-like conformation. Wang et al. [62] used the nuclear magnetic spectroscopy (NMR) to probe the conformation of alkylammonium surfactant molecules in MMT. They demonstrated the coexistence of ordered and disordered conformations. Two main resonance peaks are resolved and associated with the backbone of alkyl chains. The resonance at 33 ppm corresponds to the ordered conformation (all-trans), and the resonance at 30 ppm corresponds to the disordered conformation (mixture of trans and gauche). Osman et al. [56] used IR, NMR, XRD, and differential scanning calorimetry (DSC) to study the structure and chain dynamics of self-assembled monolayers of mono-, di-, tri-, and tetraalkylammonium cations of varying length (C, C_8, and C_{18}) on MMT platelets. At ambient temperatures, alkylammonium monolayers assembled on MMT adopt a two-dimensional ordered or a disordered state, depending on the cross-sectional area of the molecules, the area/cation available on the substrate and the alkyl length. At low temperatures, the alkyl chains preferentially assume an all-trans conformation. Conformation transformation of the chain takes place with increasing temperature, leading to a disordered phase (liquid-like) in which the chains assume a random conformation. From the XRD measurements, the dependence of the (001) basal spacing of MMT on the number of alkyl chains of different length is shown in Fig. 10.4. At a chain length of four carbon atoms (C4), there

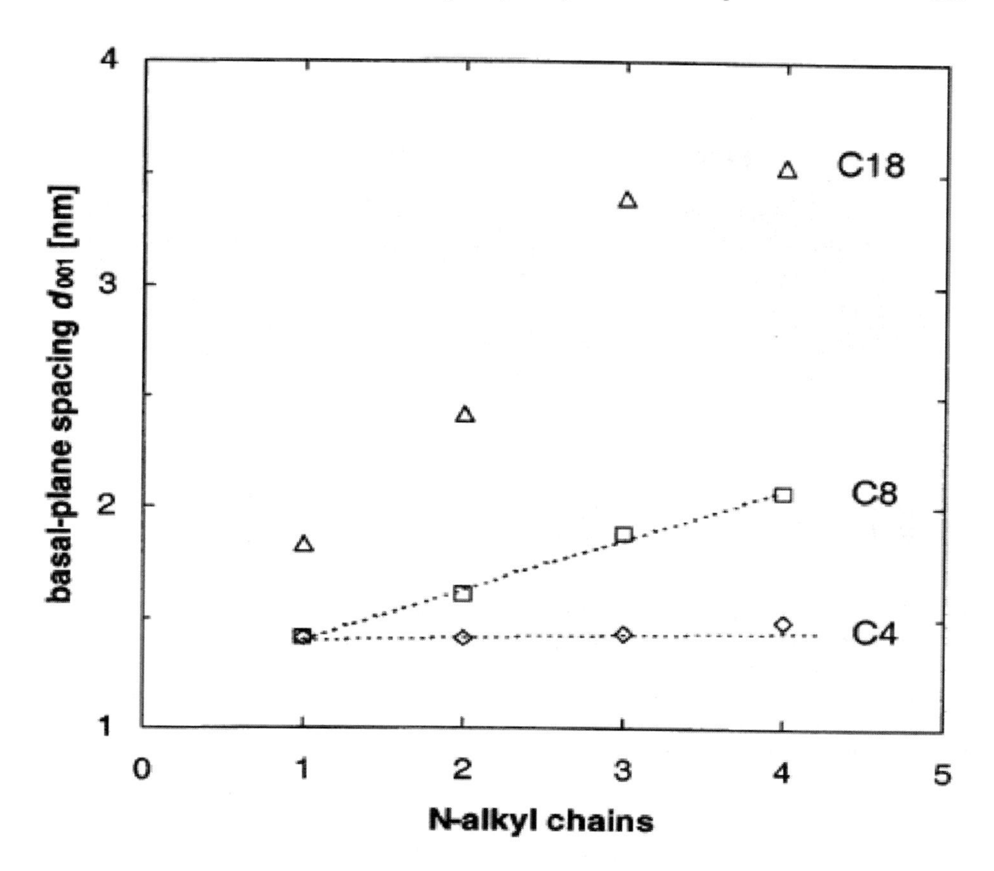

Fig. 10.4. Dependence of the basal-plane spacing on the number of alkyl chains, of different length, in the organic cation at room temperature (reprinted from [56] with permission from The American Chemical Society).

is nearly no change in the basal-plane spacing of MMT. In the octyl series (C8–4C8), the d-spacing increases linearly with growing number of chains. In the octadecyl series (C18–4C18), the d-spacing increases markedly and non-linearly. Such a large increase in the d-spacing of MMT modified with octadecyl series is beneficial in the synthesis of exfoliated polymer nanocomposites [56].

Despite these efforts, the exact molecular conformation of surfactant molecules in silicates is still far from fully understood. In this case, molecular dynamics (MD) simulations is a powerful tool to model the molecular configuration of surfactant at atomic scale and to study the mechanisms of interaction between the clay, surfactant, and polymer. Giannelis and coworkers [63] used MD simulations to study the static and dynamic properties of 2:1 silicates modified with alkylammonium surfactants. Their simulations revealed that the organically modified layers self-assemble parallel to each other to form alternating, well-ordered organic/inorganic multilayers [63]. Recently, Zeng et. al. used MD

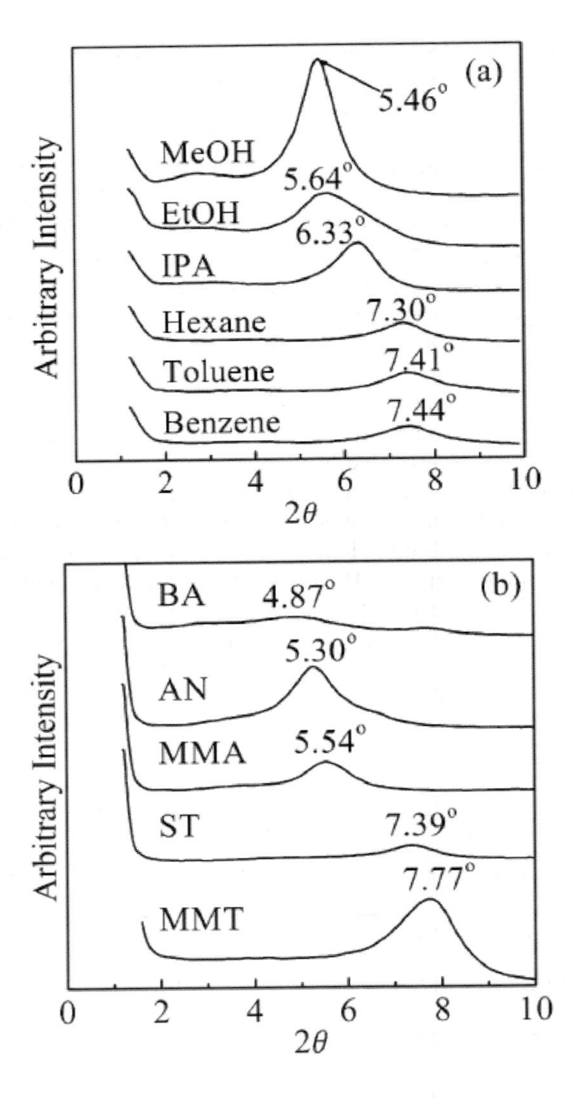

Fig. 10.5. XRD patterns of liquid MMT with a weight ratio of 30:3 for (a) organic solvents and (b) vinyl monomers (reprinted from [67] with permission from The American Chemical Society).

simulations to investigate the effects of alkyl chain length of quaternary alkylammonium compound on the basal spacings of MMT [64]. In their simulation, the Dreiding force field incorporating the parameters for the stretching, angle, torsion of the bonds, and nonbonded interaction parameters consisting of the van der Waals and the electrostatic components was used [65]. They demonstrated that the basal spacings of MMT increase with the increase of alkyl chain length and layer charge. The simulated basal spacings of organoclays agree satisfactorily with some experimental results of other researchers published in the literature [64].

Apart from the structural arrangement of tethered surfactant molecules, the interaction between the surfactant molecules and matrix polymer chains also affects primarily the tactoid formation or exfoliated structure of the clay platelets. Organic molecules and monomers can intercalate the spaces between the layers of the clay by interacting with the clay surface via ion–dipole, dipole–dipole, and hydrogen bonding [66]. Recently, Hansen solubility parameters, $\delta_t^2 = \delta_d^2 + \delta_p^2 + \delta_h^2$ have been adopted by the researchers to characterize the dispersion of organoclays in organic solvents [66,67]. The parameters take the dispersive (δ_d), polar (δ_p) and hydrogen bonding (δ_H) forces acting together to disperse MMT in various solvents into account. These three components can explain the mixing behavior of polar solvents or polymers. The Hansen solubility parameters are somewhat similar to the solubility parameter $\delta = (E/V)^{1/2}$ (E is the molar cohesive energy and *V* the molar volume) originally proposed by Hildebrand and Scott [68,69]. More recently, Choi et al. determined the dispersion and basal spacings of Na^+-MMT in various solvents and monomers [67]. They reported that some liquids with strong hydrogen-bonding groups (δ_h) showed little suspended MMT, but liquids with moderate or poor hydrogen-bonding groups precipitated MMT completely. This indicated that δ_h is an important factor for the dispersion state of MMT in a liquid (Table 10.2). The basal spacing expansion also depends on polar components (δ_p) and hydrogen-bonding components (δ_h) of organic liquids. Figures 10.5a and b show the XRD patterns of liquid- MMT with a weight ratio of 30:3 for organic solvents and vinyl monomers, respectively. The (001) basal spacing of MMT located at 7.77° tends to shift to lower angle regime after dispersion of MMT in a solvent or monomer. This corresponds to the expansion of basal spacing. The d_{001} spacing MMT dispersed in solvent appears in the following order:

Table 10.2. Dispersion states of MMT in various liquids and the Hansen solubility parameters (reprinted from [67] with permission from The American Chemical Society).

Organic liquids	δ_d, MPa$^{1/2}$	δ_p, MPa$^{1/2}$	δ_h, MPa$^{1/2}$	δ_t, MPa$^{1/2}$	H-bonding group	Dispersion state
Water	15.5	16.0	42.4	47.9	s	Viscous paste
Methyl alcohol (MeOH)	15.1	12.3	22.3	29.7	s	Slightly suspended
Ethyl alcohol (EtOH)	15.8	8.8	19.4	26.6	s	Slightly suspended
2-propyl alcohol (IPA)	15.8	6.1	16.4	23.5	s	Slightly suspended
n-butyl acrylate (BA)	14.0	8 .3	6.8	17.7	m	Precipitated
Methyl methacrylate (MMA)	13.7	9.8	6.1	17.9	m	Precipitated
Acrylonitrile (AN)	16.5	17.4	6.8	24.8	p	Precipitated
Styrene	18.6	1.0	4.1	19.0	p	Precipitated
Benzene	18.4	0	2.0	18.6	p	Precipitated
Toluene	18.0	1.4	2.0	18.2	p	Precipitated
n-hexane	14.9	0	0	14.9	p	Precipitated

The notations s, m, and p in the H–bonding group column stand for strong, moderate, and poor.

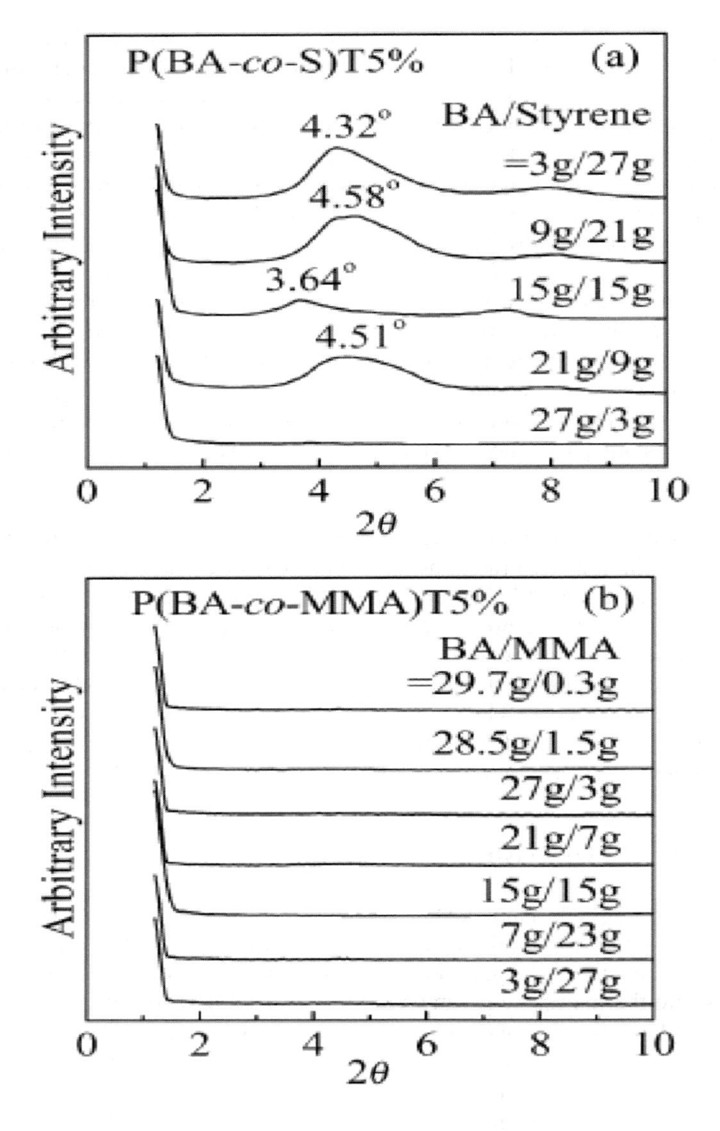

Fig. 10.6. XRD patterns of (a) P(BA–co–S) and (b)P(BA–co–MMA)-MMT nanocomposites with various monomer ratios at 5 wt% of MMT loading (reprinted from [67] with permission from The American Chemical Society).

MeOH (1.62 nm) > EtOH (1.57 nm) > IPA (1.40 nm) > n-hexane (1.21 nm) > toluene (1.19 nm) – benzene (1.19 nm). For the monomers having almost the same δ_h values, Fig. 10.5b reveals the basal spacing of MMT in the following order: BA (1.81 nm) > AN (1.66 nm) > MMA (1.60 nm) > styrene (1.20 nm). These results demonstrate that the δ_p and δ_h of solvents are the key parameters that affect the basal spacing expansion of MMT.

Choi et al. synthesized P(BA–co–S)-MMT and P(BA–co–MMA)-MMT nanocomposites reinforced with 5 wt% MMT via emulsion polymerization. The dipole moment values of BA, MMA, and styrene are 2.31, 1.67, and 0.12, respectively. The XRD patterns of most P(BA–co–S)-MMT nanocomposites indicate that the d_{001} spacing of MMT shifts to lower angles of 3.64–4.58°, indicating formation of the intercalated structure. The only exception is the nanocomposite synthesized with BA and styrene at a weight ratio of 27:3, showing no XRD peaks (Fig. 10.6a). On the other hand, XRD peaks are not observed in the P(BA-co-MMA)-MMT nanocomposites owing to the formation of exfoliated structure (Fig. 10.6b). This is because MMA has a higher dipole moment value than styrene. They concluded that monomers with high dipole moments showed large basal spacings before polymerization and produced exfoliated nanocomposites, whereas those with low dipole moments showed smaller basal spacings and produced intercalated nanocomposites. Their results are summarized in a schematic diagram as shown in Fig. 10.7 [67]. The Hansen solubility parameter and its associated components are useful indicators for making the clay-polymer nanocomposites via the solution intercalation method [70,71].

Organoclays having appropriate functional molecules in between the silicate layers are considered to exhibit a wide range of novel characteristics such as ion conductivity and unusual nonlinear optical (NLO) properties [10,72]. Smectic clays with superior swelling behavior and ion exchange properties are ideal materials for organizing organic guest species. In this regard, functional nanocomposite materials can be developed by intercalation of NLO chromophores into the interlayer space of MMT and then dispersing it in a compatibilized PP matrix. Thus, guest–host nanocomposites with a good distribution of chromopheres are obtained. The chromophore-MMT polymer nanocomposites exhibit interesting color properties on the basis of UV/Vis measurement [72].

10.3 Synthesis of Clay-Polymer Nanocomposites

In general, clay–polymer nanocomposites can be prepared via in-situ intercalative polymerization of monomers, solution, and melt intercalation. In-situ polymerization involves inserting a monomer into the galleries of silicates and then expanding and dispersing the clay layers into the matrix by polymerization. For commercial production, in-situ polymerization and solution intercalation routes are limited because a suitable monomer and solvent are not always readily available. The cost of monomer or solvent is relatively high. Furthermore, they are not compatible with current polymer processing techniques. Melt intercalation is broadly applicable to many commodity and engineering polymers, from non-polar polyolefin, weakly polar PET to strong polar polyamide. Melt compounding is a well-known process to fabricate various thermoplastics into useful shapes with low cost and high productivity. Moreover, the high shear environment of the melt extruder can assist delamination or exfoliation of the clay platelets.

10.3.1 Solution intercalation

This route involves the dispersion of the organoclay and the polymer in water or polar organic solvent. Water soluble polymer such as polyethylene oxide (PEO) has been

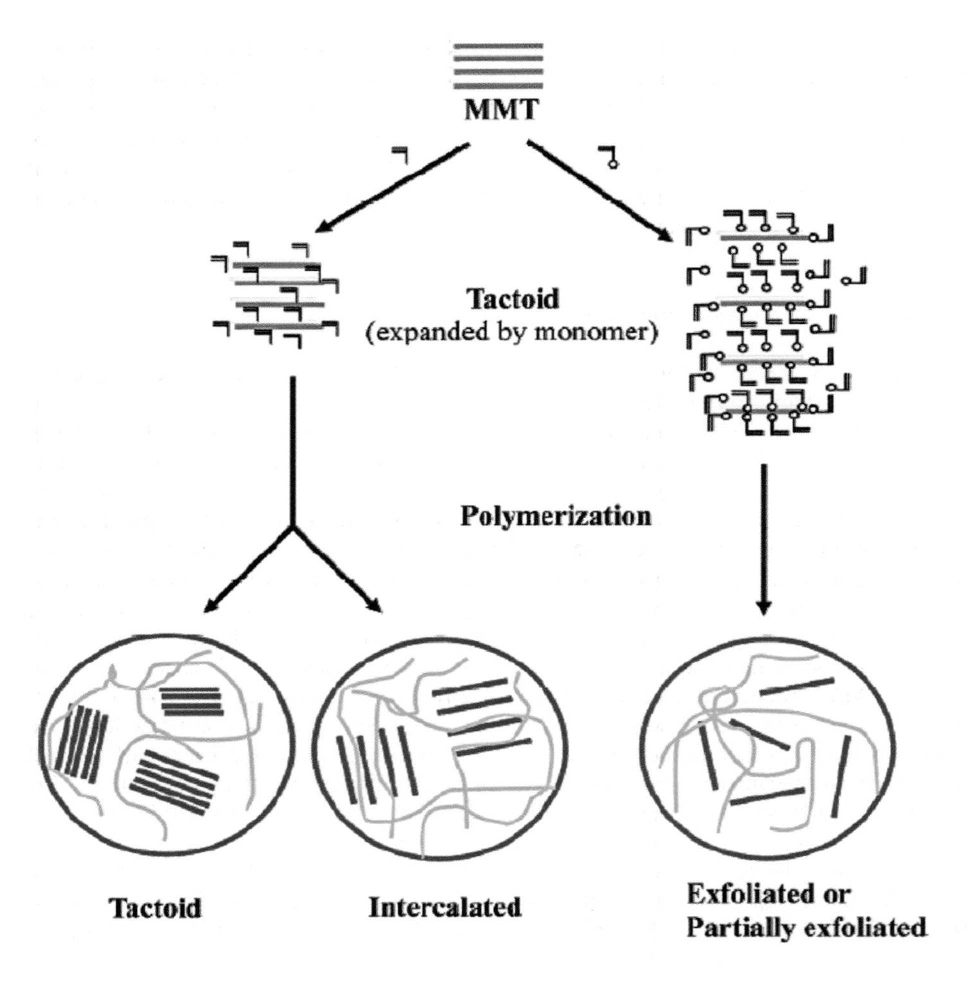

Fig. 10.7. Schematic diagram showing correlation between the basal expansion by monomers and the structure of polymer–MMT nanocomposites (reprinted from [67] with permission from The American Chemical Society).

intercalated into the clay gallery spacing using this technique. The high polarity of water results in swelling of Na^{+}MMT. The layered silicates owing to their unique feature can be dispersed easily in an adequate solvent [70]. The layered silicates owing to their unique feature can be dispersed easily in an adequate solvent. The polymer dissolves in the solvent, then adsorbs onto the expanded silicate sheets. When the solvent is evaporated, the sheets reassemble, sandwiching the polymer to form the intercalated structure. This behavior is referred to as the exfoliation-adsorption solution intercalation [70]. From the thermodynamic viewpoint, the driving force for polymer intercalation into the gallery spacing is the entropy gained by desorption of solvent molecules is compensated for the entropy loss of the confined polymer chains in the gallery. In general, intercalation only occurs for certain polymer/solvent or monomer/solvent pairs. Selection of a proper solvent is the primary criterion to achieve the desired level of exfoliation of organoclays for dispersion into the polymers. Recently, Morgan and Harris reported the synthesis of the MMT–polystyrene (PS) nanocomposites via solution intercalation route aided by sonication [73]. Prior to sonication, all the samples showed large clay tactoids, indicating poor dispersion of the clays in chlorobenzene solvent. Sonication during solvent blending with an imidazolium modified MMT was found to produce an exfoliated PS nanocomposite.

10.3.2 Intercalative polymerization

Polyamides are one of the few polymer types that readily form well-exfoliated nanocomposites due to their polar molecular chains. Pioneering work in this area was carried out by the Toyota Research group to produce clay/polyamide-6 nanocomposites [38,39]. In the process, MMT was cation exchanged with the ammonium cations of various ω-amino acids ($H_3N^+(CH_2)_{n-1}COOH$, $n = 2,3,4,5,6,8,11,12$, and 18). The modified MMTs were intercalated by ε-caprolactam at 100°C in which ring opening polymerization of ε-caprolactam occurred. The number of carbon atoms in ω-amino acids has a strong effect on the swelling behavior of MMT at 100°C. The (001) basal spacing of MMT tends to increase dramatically when $n \geq 11$. This implies that a large amount of monomer can be intercalated into the gallery of MMT when the number of carbon atoms in the ω-amino acids is high [38]. From this, they synthesized the 12-aminolauric acid modified MMT (denoted as 12-MMT) and ε-caprolactam in a mortar. The content of 12-MMT ranged from 2 to 70 wt%. The carboxyl end group (–COOH) of 12-aminolauric acid initiated ring-opening polymerization of ε-caprolactam in the presence of a small amount of aminocaproic acid. Reicherd et al. used a similar process to prepare MMT-polyamide12 nanocomposites [45]. This approach is capable of producing well-exfoliated nanocomposites and has been applied to a wide range of polymer systems. Other than MMT, saponite, hectorite, and mica were used by Usuki and coworkers to synthesize the clay-PA6 nanocomposites. They reported that the tensile strengths of PA6 nanocomposites reinforced with 5 wt% phyllosilicates (2:1) at 23 and 120°C decrease in the order: MMT > mica > saponite > hectorite [40].

Polystyrene is one of the most mass-productive and commercialized polymers. A common method to prepare the clay-PS nanocomposites consists of inserting an organoclay with styrene monomer, and subsequently carry out a free radical polymerization. When the catalyst or the reactive site for polymerization resides in the gallery, in-situ polymerization takes place in the gallery space and the system is driven to disperse the silicate platelets by the driving force of polymerization reaction. Fu and Qutubuddin [74] modified Na-MMT and Ca-MMT with vinylbenzyldimethyldodecylammonium chloride (VDAC) by cationic exchange between inorganic ions of MMT and ammonium ions of VDAC in an aqueous medium. Dispersion of Org-MMT in styrene monomer followed by polymerization in the presence of free radical initiator (2,2-azobis-isobutyronitrile; AIBN) results in exfoliated nanocomposites (Fig.10.8). The exfoliated nanocomposites exhibit higher dynamic modulus (Fig. 10.9) and higher thermal degradation temperature than PS as expected. Similarly, Weimer et al. prepared the delaminated PS-silicate nanocomposites by anchoring a living free radical polymerization initiator into silicate layers followed by bulk polymerization [75]. Alternatively, it is reported that a one-step emulsion polymerization

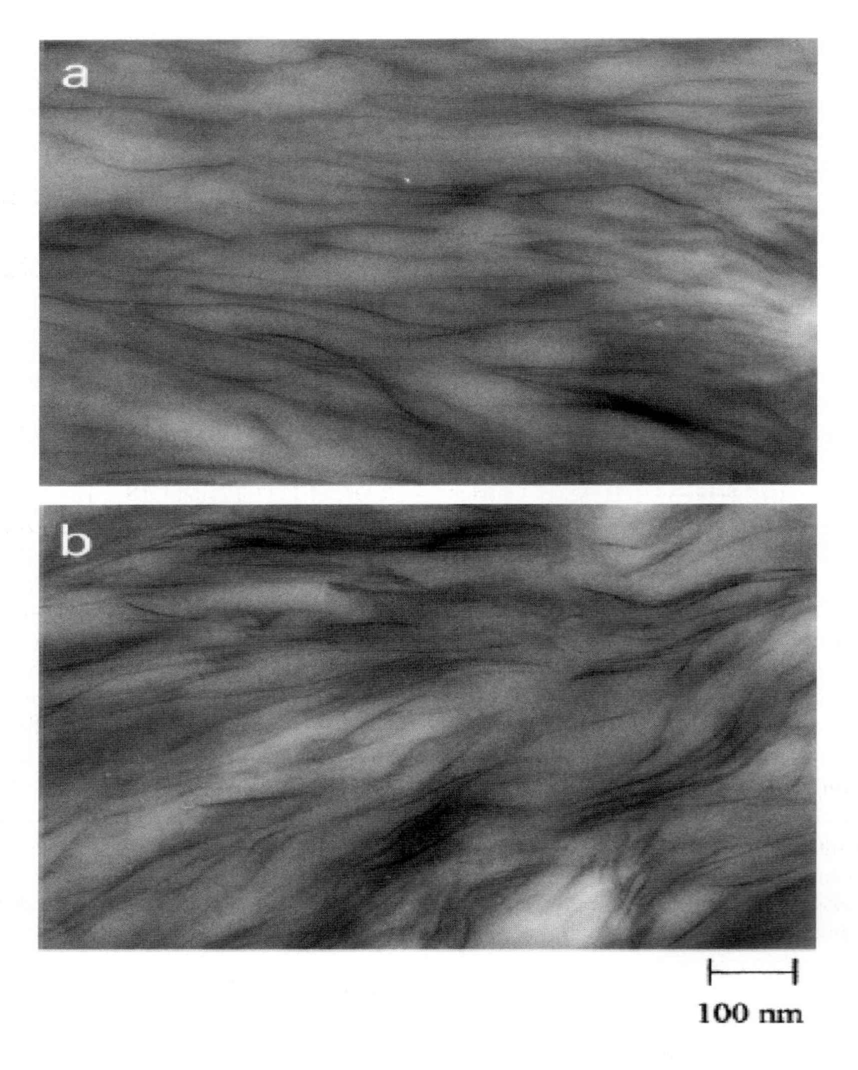

Fig. 10.8. TEM micrographs of polystyrene-clay nanocomposites: (a) 5.5 wt% VDAC–NaMMT, and (b) 5.6 wt% VDAC–CaMMT (reprinted from [74] with permission from Elsevier).

could offer a simple approach to the synthesis of clay–PS nanocomposites [76–79]. The important advantage of emulsion polymerization is the employment of water as a dispersion medium. As mentioned above, the basal spacing expansion of MMT depends on polar components (δ_p) and hydrogen-bonding components (δ_h) of solvents [70]. Thus water with large polar components (δ_p) and hydrogen-bonding components (δ_h) is an ideal solvent to expand the basal spacing of MMT. Some workers have used water to widen the basal space of pristine clays without any chemical treatment and performed the polymer intercalation into the silicate layers by emulsion polymerization [76,77]. In the process, styrene monomer was dispersed in water under agitation and polymerized with

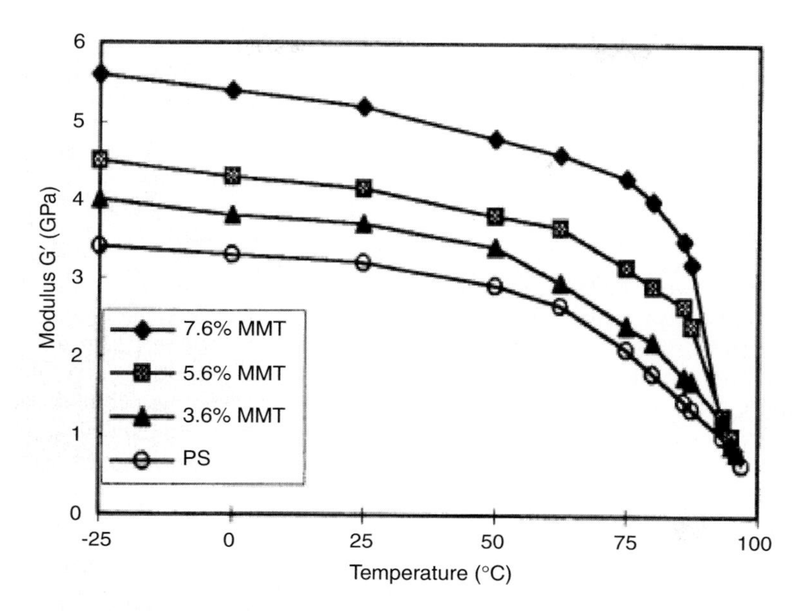

Fig. 10.9. Dynamic storage modulus vs. temperature for polystyrene-clay nanocomposites containing various Na-MMT loadings (reprinted from [74] with permission from Elsevier).

a water soluble radical initiator (e.g. potassium persulfate) in the presence of pristine MMT. However, strong hydrophobicity of styrene monomer makes it difficult to penetrate into the clay galleries, thereby resulting in the formation of intercalated MMT–PS nanocomposites. On the other hand, Chung and coworkers prepared exfoliated PS–MMT nanocomposites by emulsion polymerization using a reactive surfactant, 2-acrylamido-2-methyl-1-propanesulfonic acid (AMPS) [78]. AMPS molecules can be tethered on silicate layers due to strong interaction of amido group with pristine silicates. Sulfonic acid acts as a surface-active material. They concluded that water, styrene monomer, and reactive surfactant made the basal space of silicate wide before polymerization, and then the silicate host lost regularity and was exfoliated during the polymerization. The extent of exfoliation depended on the amount of AMPS. As the amount of AMPS increased, it was easy to enlarge and exfoliate the clay layers. Great improvements in storage modulus and thermal stability can be achieved in the exfoliated nanocomposites. In a later study, polar MMA monomer has been introduced to copolymerize with styrene to increase the polarity of the polymer matrix for the synthesis of exfoliated MMT–P(S-co-MMA) nanocomposites [79].

Despite the prime importance of polypropylene and polyethylene for a wide variety of engineering applications, the synthesis of polyolefin-silicate nanocomposites remains a scientific challenge. Exfoliation of the silicate platelets by non-polar polyolefins is rather difficult to achieve. Thus addition of an appropriate catalyst is required during polymerization. O'Hare and coworkers have reported the intercalation of synthetic layered silicates with a cationic Ziegler-Natta catalyst after protecting the internal surfaces with methylaluminoxane. However, a dispersed nanocomposite was not obtained. [80]. Bergman et al. intercalated a cationic palladium-based Brookhart catalyst into galleries of an organically modified fluorohectorite. An exfoliated polyethylene-silicate nanocomposite was achieved when exposed to olefinic monomer [81].

10.3.3 Melt intercalation

From an earlier work in 1993, Giannelis et al. reported that it is possible to melt-mix polymers [e.g. polystyrene (PS), poly(ethylene oxide)] with organoclays to form the nanocomposites [82,83]. Polystyrene intercalation was achieved by annealing the PS pellets with organosilicate at temperatures greater than the glass transition temperature of PS. Since then, the high incentive for industrial applications has motivated vigorous research on the nanodispersion of silicate layers in various polymers by melt intercalation. Figure 10.10 shows typical XRD patterns for PS-fluorohectorite composites

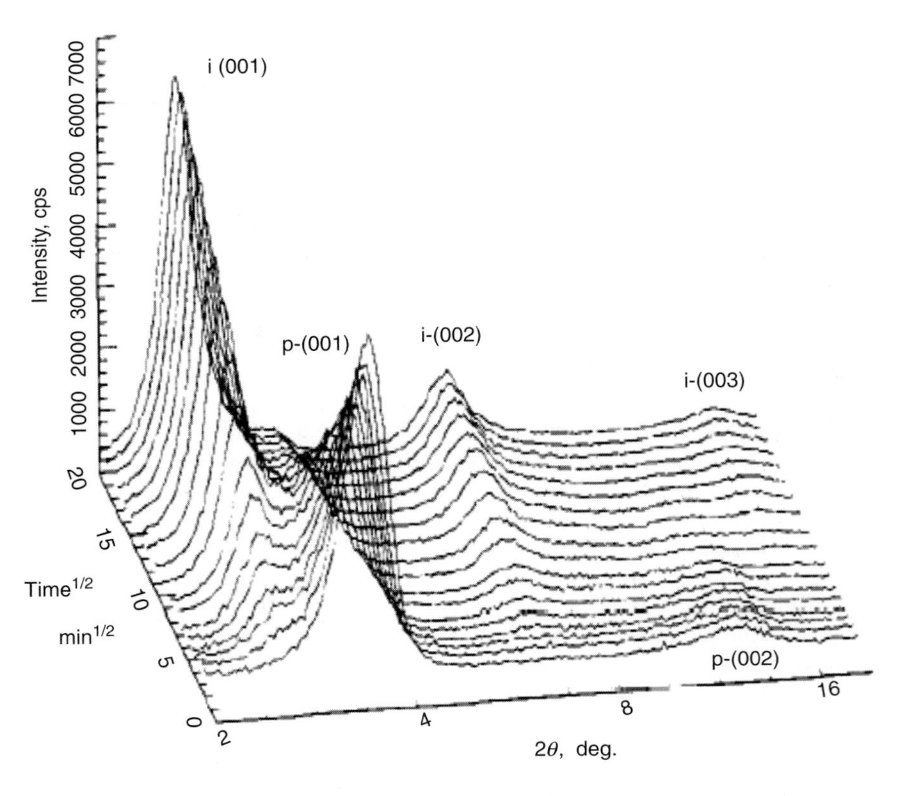

Fig. 10.10. Temproral series of XRD patterns for the polystyrene/octadecylammonium-exchanged fluorohectorite pellets annealed in-situ at 160°C for various periods of time (reprinted from [83] with permission from The American Chemical Society).

annealed in vacuum at 160°C for various periods of time. Fluorohectorite was modified with octadecylammonium compound. Initially, basal reflections from the pristine silicate, p-(001) and p-(002) are observed at $2\theta = 4.15°$ (d = 2.13 nm) and $2\theta = 48.03°$, respectively. During annealing, the intensity of the diffraction peaks corresponding to the pristine silicate is progressively reduced while a set of new peaks from the intercalated silicate appears at $2\theta = 2.82$, 5.66, and 8.07°, respectively. The peak at low angle ($2\theta = 2.82°$) with a basal spacing of 3.13 nm clearly demonstrates that the molecular chains of PS have intercalated into the galleries space of silicate during annealing [83].

With the development of clay-PA6 nanocomposites by Toyota Research Center via intercalative polymerization, there has been increasing attention to this system in the past decade. An effective approach for making such a composite system is by melt intercalation. At present, one of the main limitations to large-scale commercialization of melt-compounded PA6-clay nanocomposites arises from the difficulty in controlling the morphology of dispersed clay, degree of exfoliation and clay orientation, and crystallinity of polymer matrix [84]. Nevertheless, nanoclay platelets offer exceptional reinforcing effect for polyamides at very low loading level. Figure 10.11 shows the effectiveness

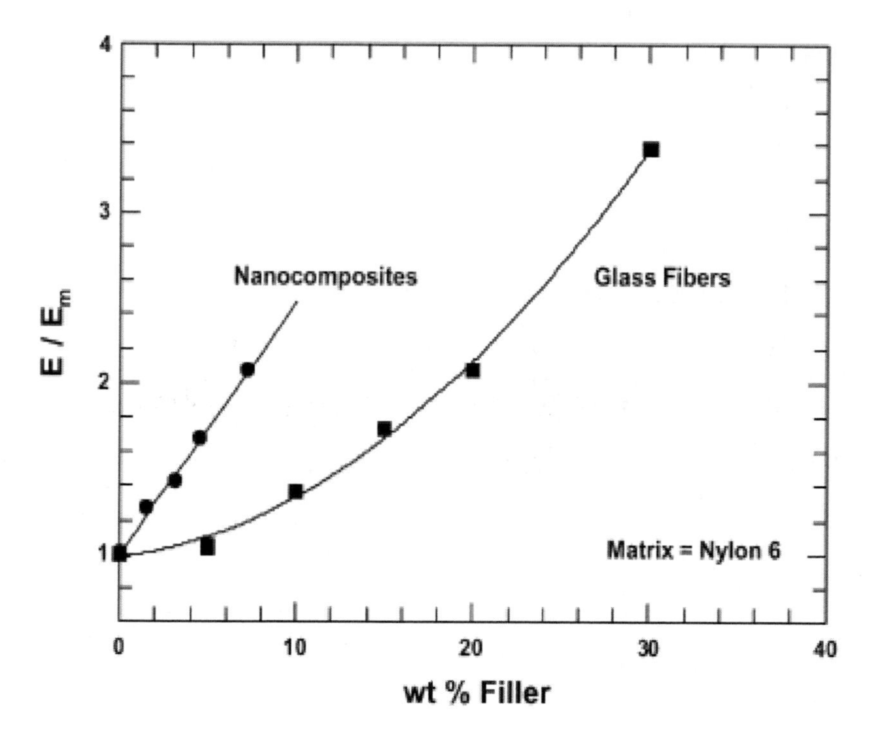

Fig. 10.11. Comparison of the reinforcement of PA6 by organoclays and glass fibers. The elastic modulus of composite (E) is normalized to the elastic modulus of matrix (E_m) (reprinted from [85] with permission from Elsevier).

of the reinforcement of PA6 provided by the organoclay compared to glass fibers [85]. Apparently, the modulus of PA6 increases much more rapidly by addition of organoclay than by glass fibers. A twofold increase in the stiffness of PA6 requires about 6.5 wt% organoclay whilst it needs approximately 20 wt% glass fiber to achieve the same increase. This implies that structural components with higher stiffness and lower weight can be fabricated with PA6 having nanoclay low loading.

According to the literature, well-exfoliated PA6-MMT nanocomposites can be obtained from melt compounding and they exhibit comparable mechanical properties to those prepared by the intercalative polymerization. However, the extent of exfoliation of the MMT layers during melt compounding depends greatly on the type and structure of organoclays used and the molecular weight of polyamides. Paul and coworkers investigated systematically the structure–property relationships for nanocomposites formed by melt compounding from a series of MMT modified with various amine compounds and PA6 of high and low molecular weights [59,60,85–89]. The degree of exfoliation of silicate platelets blended with PA6 of high molecular weight (HMW) is related to three issues. These include the use of one long alkyl tail on the ammonium ion rather than two, use of methyl rather than 2-hydroxy-ethyl groups, and use of an equivalent amount of amine surfactant on the clay as opposed to an excess amount. As the PA6 molecules would react with pristine MMT via polar interaction, the organic modifier consisting of two alkyl tails shields more silicate surfaces than one alkyl tail, thereby precluding polar interaction between the PA6 and the clay surface [86]. The XRD pattern and morphology of nanocomposites based on HMW PA6 and the quaternary amine containing one or two long alkyl tails are shown in Figs. 10.12a–c. Apparently, the nanocomposite derived from one-tailed hydrogenated tallow organoclay, $M_3(HT)_1$, exhibits no characteristic X-ray peaks (Fig. 10.12a). This implies the formation of an exfoliated nanostructure. The nanocomposite based on the two-tailed amine, $M_2(HT)_2$, has a peak at 2.15°, which corresponds to an interlayer spacing of 4.1 nm. This spacing is larger than the basal spacing of the pristine organoclay. TEM micrographs confirm the XRD results. The nanocomposite from $M_3(HT)_1$ consists predominantly of exfoliated MMT platelets (Fig. 10.12b), and the composite from the $M_2(HT)_2$ contains mainly intercalated multi-layer stacks (Fig. 10.12c). The effects of number of alkyl tails attached to quaternary amines on the tensile modulus and elongation at break of nanocomposites are summarized in Figs. 10.13a–a′. The exfoliated nanocomposite derived from the one-tailed $M_3(HT)_1$ exhibits higher stiffness than that from the two-tailed $M_2(HT)_2$. But the tensile ductility of the former nanocomposite is lower than that from the two-tailed $M_2(HT)_2$ as expected. Similar trends in stiffness and tensile ductility are observed for the composites derived from tertiary amines (Figs. 10.13b–b′). The presence of 2-hydroxy-ethyl group in surfactant obviously leads to lower stiffness but higher elongation at break than those of the nanocomposite based on organoclay modified with methyl group (Figs. 10.13c–c′). This is due to the hydroxy-ethyl group of larger size that occupies more space than the methyl counterpart, thereby shielding off the desirable polyamide–silicate interaction. Fig. 10.13d shows the organic surfactant loading level (MER) has little effect on modulus at low MMT content. At higher MMT content, the organoclay with excess surfactant leads to a slightly lower modulus. There is no advantage in using excess surfactant to disperse the MMT platelets. Similar trends, but lower extent of exfoliation, were observed for PA6 having low molecular weight. From this, it appears that with proper selection of PA6, organoclay modifier

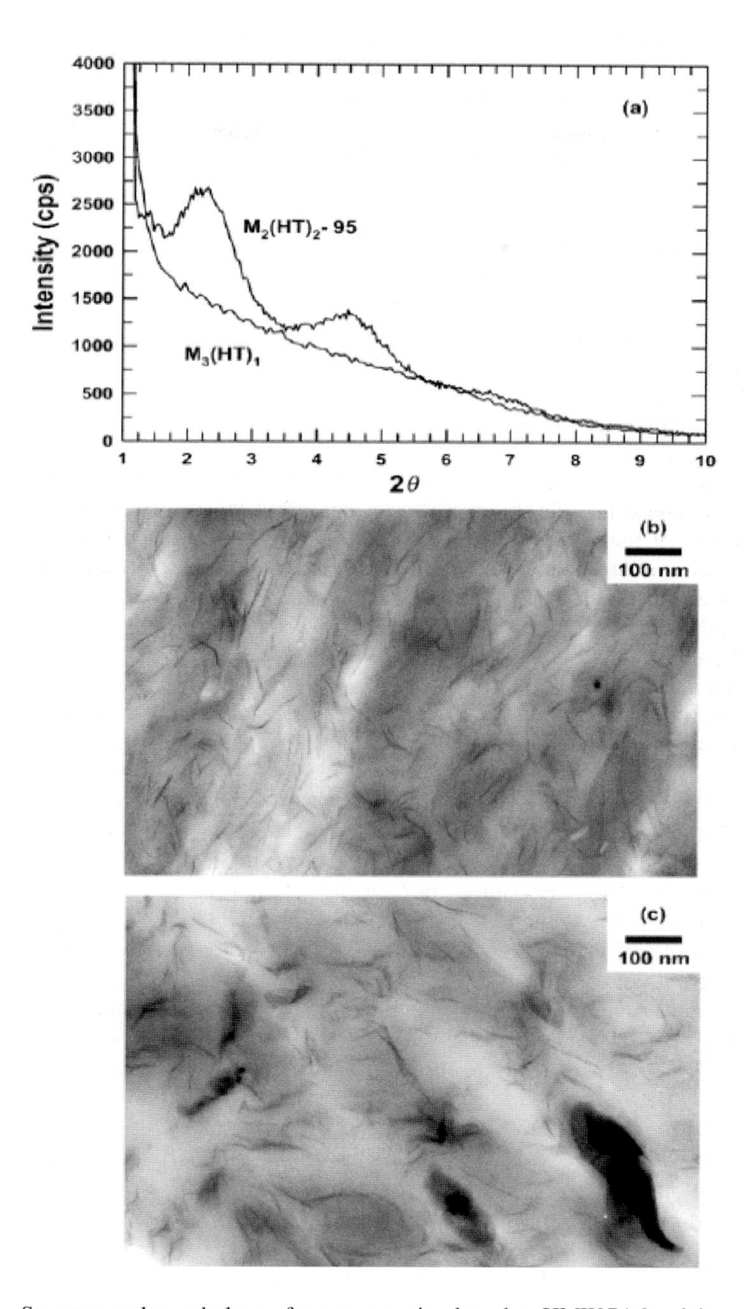

Fig. 10.12. Structure and morphology of nanocomposites based on HMW PA6 and the organoclays $M_3(HT)_1$ and $M_2(HT)_2$ (Cloisite 20A). (a) XRD pattern and TEM micrographs of (b) $M_3(HT)_1$ and $M_2(HT)_2$ based composites. The concentration of MMT in the $M_3(HT)_1$ $M_2(HT)_2$ are 2.9 and 3.0 wt%, respectively (reprinted from [86] with permission from Elsevier).

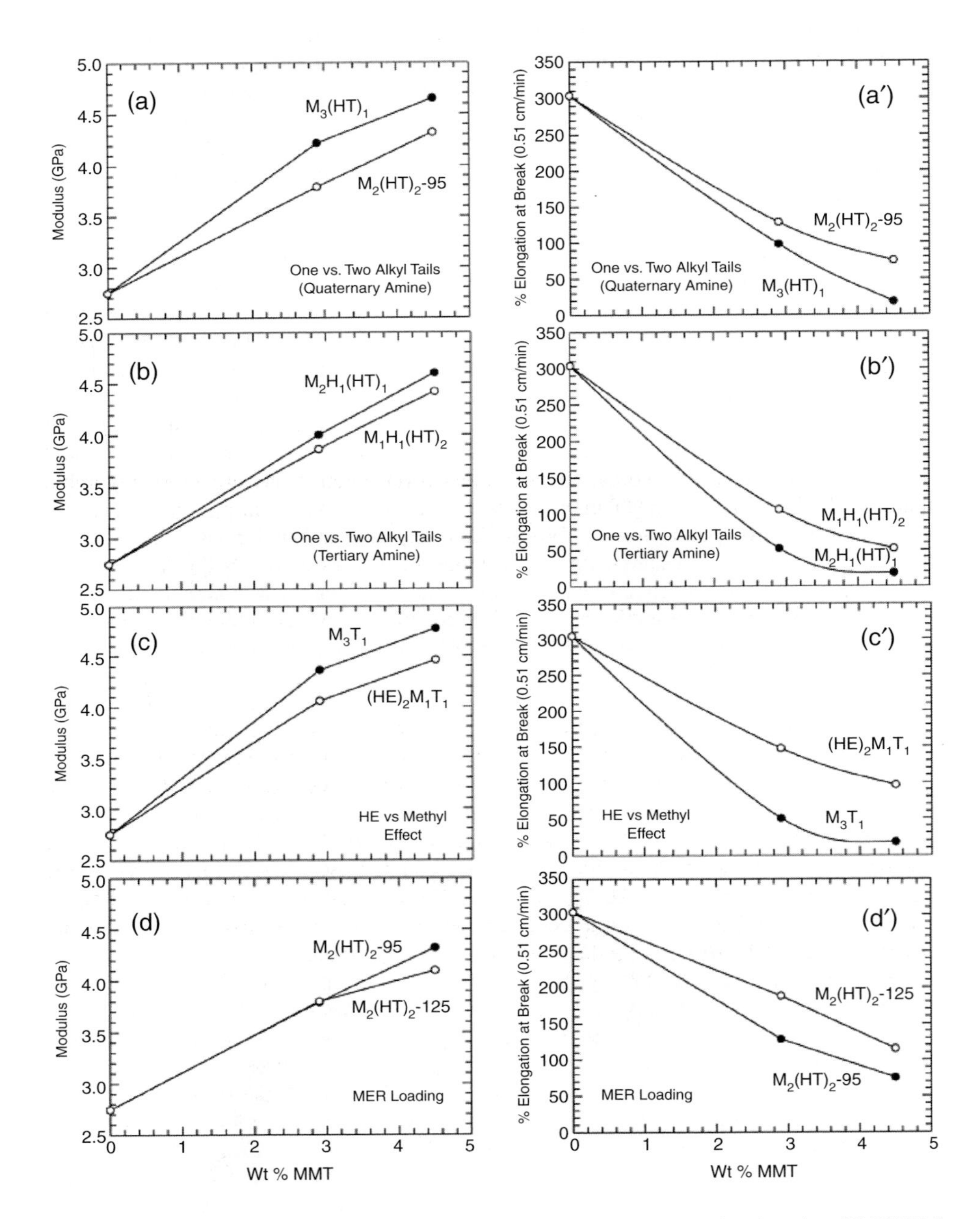

Fig. 10.13. Modulus (left) and elongation at break (right) of nanocomposites based on HMW PA6 and organoclays showing the effects of (a,a′) quaternary amines having one vs. two alkyl tails, (b,b′) tertiary amines having one vs. two alkyl tails, (c,c′) the effect of 2-hydroxy-ethyl vs. methyl groups, and (d,d′) the effect of MER loading (reprinted from [86] with permission from Elsevier).

and processing conditions can lead to the formation of exfoliated nanocomposites with enhanced mechanical properties. From Figs. 10.13 a′–d′, it is apparent that the tensile ductility or elongation at breaks of the nanocomposites decreases dramatically with the addition of low level of MMT, i.e. 3 wt%. Thus, the enhanced modulus of the nanocomposites associated with the incorporation of low loading of MMT is achieved at the expense of tensile ductility. This is the general characteristics of polymer–clay nanocomposites.

From the thermodynamic analysis, both the entropic and enthalpic factors play important roles in controlling the formation of intercalated or exfoliated nanocomposites [90–92]. In principle, a sufficiently favorable enthalpic contribution from the interaction between the polymer and the clay is required to overcome the entropy loss of the polymer matrix in forming the nanocomposites. Favorable enthalpy of mixing for the organoMMT/clay is achieved when the polymer–MMT interactions are more favorable when compared to the surfactant–MMT interactions [90–92]. For polar polymers, an alkylammonium surfactant is adequate to offer sufficient excess enthalpy and promote the formation of exfoliated nanocomposites. In other words, the polymer–MMT interactions are more favorable than the alkylammonium–MMT interactions. In the case of nonpolar polypropylene, the alkylammonium compound also exhibits the same aliphatic and apolar nature as PP. Thus, there is no favorable excess enthalpy to promote the dispersion of MMT clay platelets. In this case, it is necessary to improve the interactions between the polymer and the MMT so as to become more favorable than the alkylammonium–MMT interactions. This can be achieved by functionalization of PP via introducing polar groups in the polymer such as maleic anhydride (MA) [93–99].

Usuki and coworkers from Toyota Research laboratories also carried out extensive study on the effect of MA functional group compatibilizer on the structure and mechanical properties of the PP–MMT nanocomposites via melt intercalation [93–97]. In the process, MA-grafted PP (PP–MA) oligomer was first compounded with octadecylammonium-exchanged MMT (C_{18}-MMT), forming a master batch. Pure PP pellets were then blended with master batch at 200–230°C using a twin-screw extruder to yield the nanocomposites [94]. Figure 10.14 shows the XRD patterns of master batch samples containing 2.1 and 5.3 wt% organoclay (traces a and b), respectively. There are no peaks in the XRD patterns, indicating there are no stacked clay platelets in these samples. In contrast, the XRD pattern of the composite prepared from neat PP and C_{18}-MMT (trace c) is exactly identical to that of C_{18}-MMT (trace d). This indicates that PP was not intercalated into the gallery space of MMT without the aid of MA functional group. Further study by the same group demonstrated that stacked silicate layers are formed in the composite containing C_{18}-MMT content $\geq$4 wt% on the basis of TEM observation. Such stacked silicate layers become more apparent when the C_{18}-MMT content reaches 7.5 wt% [96]. The tensile modulus and strength of composites tend to increase with increasing the clay content (Fig. 10.16). The mechanism of dispersing MMT platelets depends on the maleic anhydride content, clay content, and processing conditions. These results are summarized schematically as shown in Fig. 10.15.

It is worth noting that MA can also be used both as a modifying additive for the PP polymeric matrix and as a swelling agent for the vermiculite silicate [13]. Tjong and

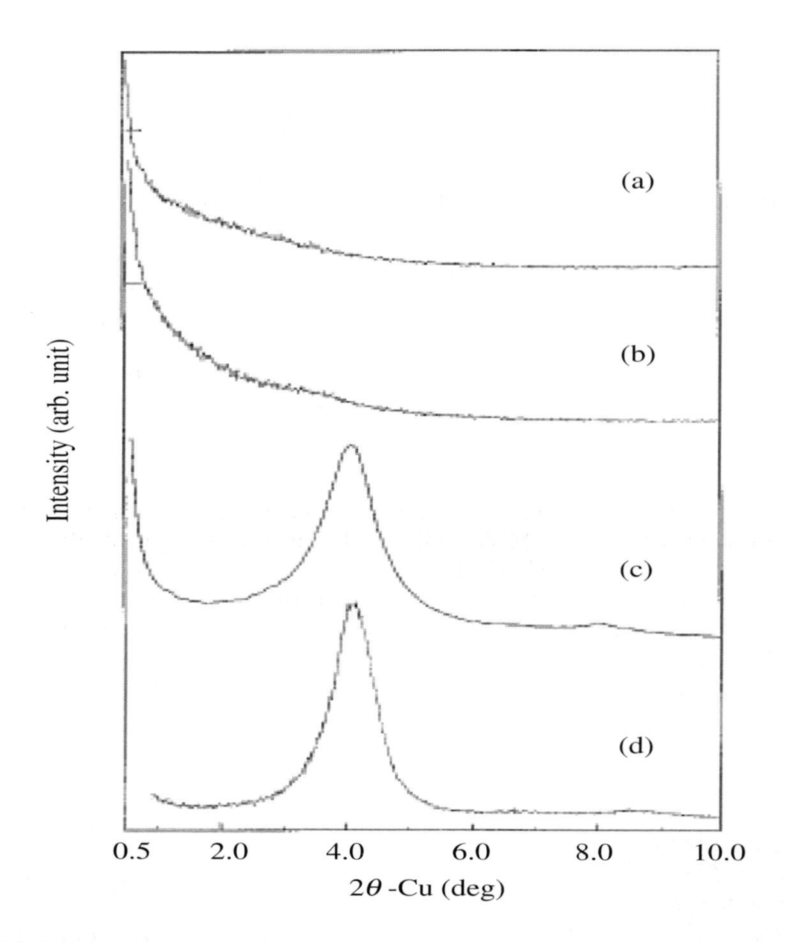

Fig. 10.14. XRD patterns for PP nanocomposites functionalized with MA group and reinforced with (a) 2 wt% and (b) 5 wt% C_{18}-MMT. (c) XRD trace for PP composite prepared from melt-mixing pure PP and C_{18}-MMT. (d) XRD pattern for C_{18}-MMT (reprinted from [95] with permission from Wiley).

coworkers pretreated vermiculite initially with hydrochloric acid solution. The acid-delaminated vermiculite was further treated with MA in the presence of acetic acid to form an organoclay. In this case, MA can easily enter the galleries of acid-treated vermiculite because solvent (acetic acid) can act as a carrier to transport MA into hydrophilic vermiculite. The absence of basal (001) peak in the XRD pattern of MA-treated vermiculite provides a substantial evidence for this. Vermiculite-PP nanocomposites can be prepared by direct melt mixing of MA-modified vermiculite (MAV) with PP. In such compatibilized nanocomposites, grafted MA acts a center or bridge to bond the vermiculite and PP together. A model for this structure is depicted as follows,

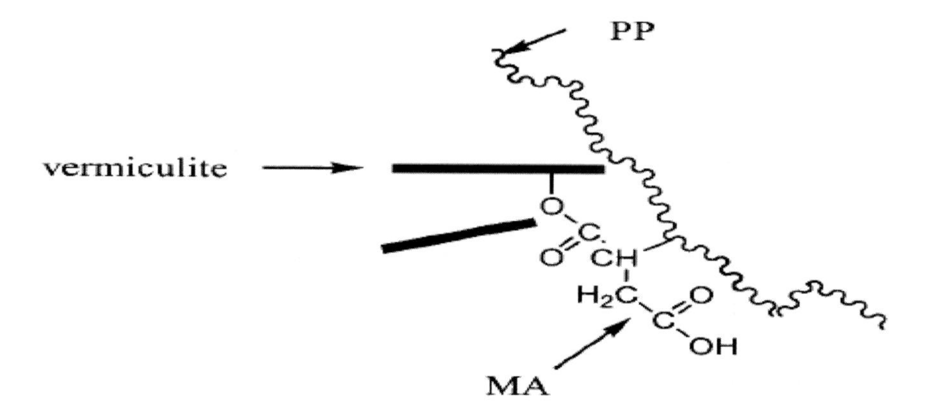

Tensile tests show that the tensile strength and modulus of the nanocomposites tend to increase dramatically with MAV addition. The tensile strength of PP is increased by 18.3% with the addition of only 2% MAV and by up to 29.5% when 5% MAV is introduced. Such enhancement is mechanical properties results from the formation of mixed intercalated-exfoliated reinforcement in the nanocomposites. However, the tensile ductility decreases sharply with MAV additions. The tensile properties of MAV-PP nanocomposites are summarized in Table 10.3. To overcome the problem of low tensile

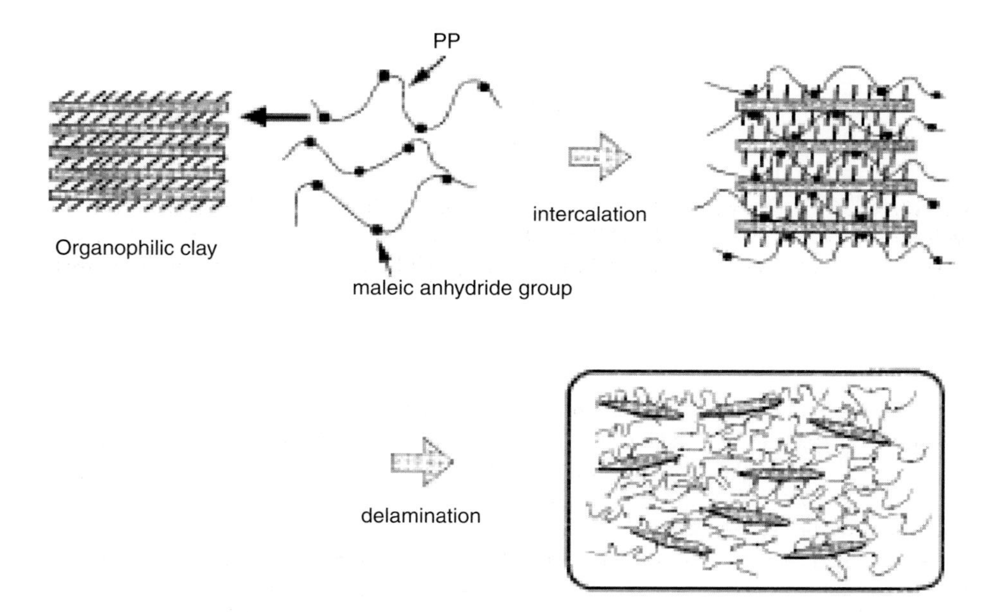

Fig. 10.15. A schematic diagram showing the clay dispersing process during melt intercalation (reprinted from [95] with permission from Wiley).

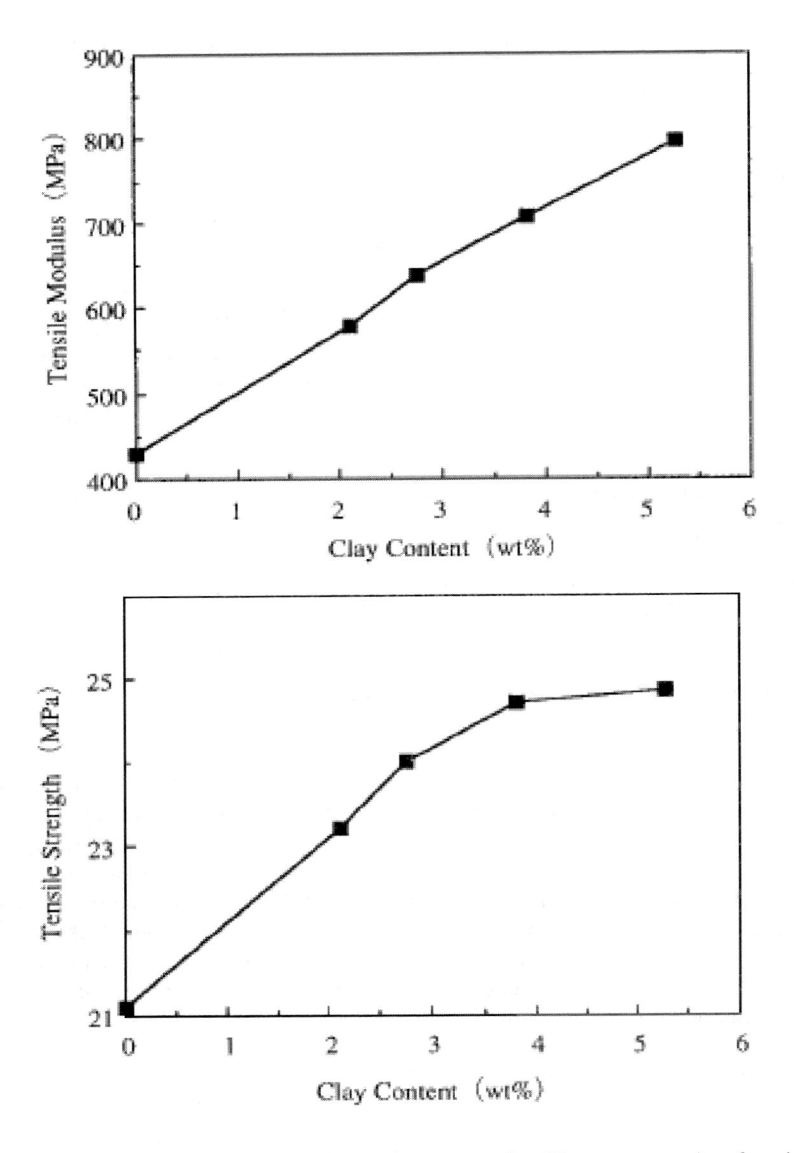

Fig. 10.16. Tensile modulus and strength vs. clay content for PP nanocomposites functionalized with MA group (reprinted from [95] with permission from Wiley).

ductility, elastomer particles such as styrene-ethylene-butylene-styrene (SEBS) are incorporated into the matrices of clay-polymer nanocomposites [17,20]. With the incorporation of SEBS particles, the tensile ductility and impact strength of clay-polymer nanocomposites can be restored at the expenses of tensile strength and stiffness (Table 10.4). A balance between the tensile strength and ductility can be maintained by proper control of elastomer content.

Table 10.3. Mechanical properties of MAV-PP nanocomposites [13].

Sample	Tensile strength, (MPa)	Tensile modulus, (GPa)	Elongation at break, (%)	Storage modulus*, (GPa)
PP	28.8	0.84	670	0.78
2% MAV-PP	34.1	1.01	18.89	2.37
5% MAV-PP	37.3	1.30	10.23	3.32

*Determined from the dynamic mechanical analysis at 25°C.

Table 10.4. Mechanical properties of MAV/PP-(SEBS-g-MA) nanocomposites [17].

Sample	Tensile strength, (MPa)	Tensile modulus, (GPa)	Elongation at break, (%)	Storage modulus*, (GPa)
4% MAV-PP	36.3	1.20	12.8	3.02
4% MAV-(95% PP-5% SEBS-g-MA)	32.5	0.79	88.4	1.62
4% MAV-(90% PP-10% SEBS-g-MA)	29.9	0.72	127	1.53
4% MAV-(85% PP-15% SEBS-g-MA)	28.1	0.68	168	1.42

*Determined from the dynamic mechanical analysis at 25°C.

10.4 Thermal Stability and Other Materials Properties

Dynamic mechanical analysis is a useful tool for investigating relaxation in polymers. Analysis of the storage modulus, loss modulus, and loss tangent provides the information on the changes in the solid structure of a polymer under stress and temperature. The variation of storage modulus with temperature for the MAV–PE nanocomposites is shown in Fig. 10.17. It is apparent that the storage modulus of the nanocomposites increases with increasing MAV content, particularly at lower temperature regime. The storage modulus of the nanocomposite with only 4% MAV is 1.53 GPa at 25°C, which is 52% higher than that of pure PE. Thus, the direct melt blending of PE with vermiculite appears to be an effective way to improve the mechanical properties of PE. Figure 10.18 shows the typical tan delta vs. temperature curves for pure PE, PE/2% MAV and PE/4% MAV nanocomposites. Two glass transition temperatures (Tgs) are observed, corresponding to the glass transition temperatures of amorphous and crystalline PE phases, respectively. For MAV–PE nanocomposites, very small amount of MAV addition results in an increase of the Tg of PE amorphous phase from -119 to -112°C. However, there is no effect of MAV additions on the Tg of PE crystalline phase (Table 10.5). Cho and Paul reported that the presence of nanoscale particle does not affect the Tg of the polyamide 6 matrix, but slightly reduces its crystalline melting point [44]. This is because the polymer material wholly confined within the silicate layers generally does not contribute to the heat flow related to the glass transition of DMA signals [100].

Compared to organic materials, the inorganic fillers exhibit higher thermal stability and thermal-resistant properties. Therefore, the introduction of layered silicates with unique physical properties would improve the thermal stability of organic polymer materials. For example, the heat distortion temperature (HDT) of PA6 (65°C) can be increased

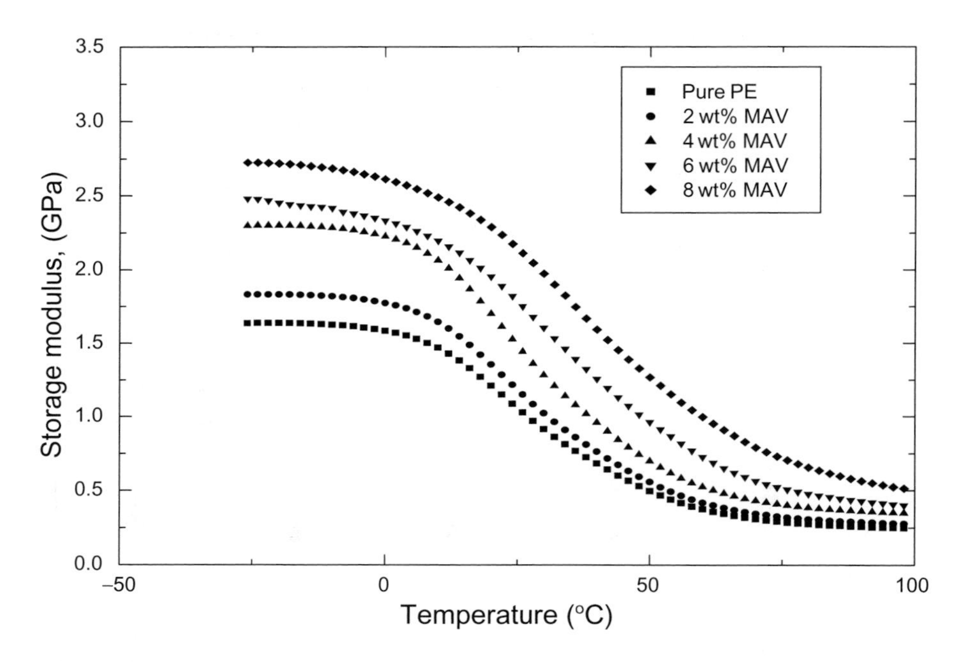

Fig. 10.17. Storage modulus vs. temperature for PE and MAV–PE nanocomposites [16].

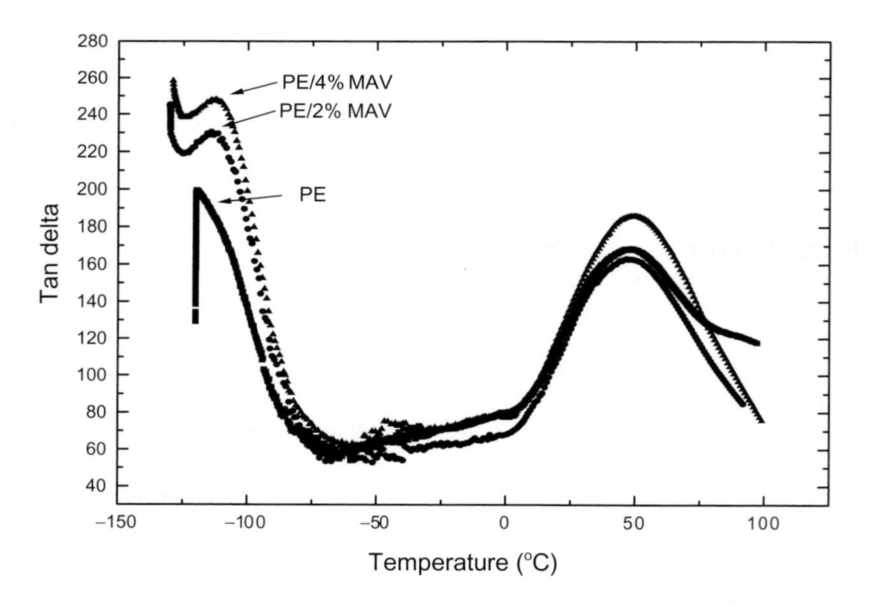

Fig. 10.18. Loss modulus vs. temperature for pure PE, 2% MAV–PE and 4% MAV–PE nanocomposites [16].

Table 10.5. T_g values of PE/MAV nanocomposites determined from DMA measurements [16].

Specimen	T_g, °C	T_g, °C
Neat PE	−118.9	48.24
PE/0.5% MAV	- - - -	- - - -
PE/2% MAV	−113.46	46.71
PE/4% MAV	−112.46	45.92
PE/6% MAV	−112.41	48.16
PE/8% MAV	−112.66	47.34

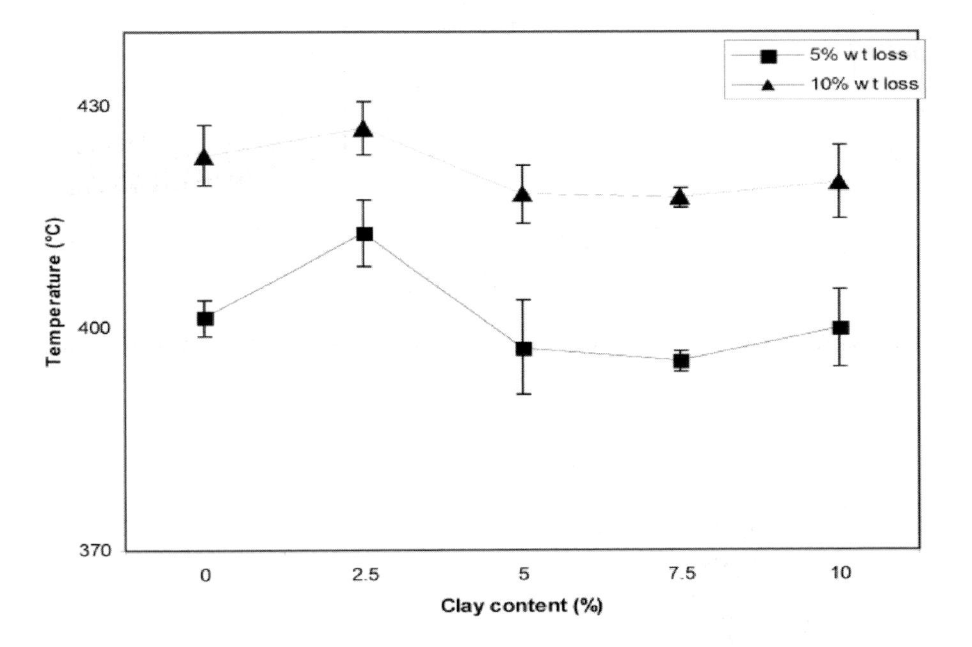

Fig. 10.19. The 5 and 10% weight loss temperature vs. clay content (wt%) (reprinted from [101] with permission from Elsevier).

markedly to 152°C by adding 5 wt% MMT. The effectiveness of different types of clays in enhancing the HDT of PA6 is in the order: MMT > mica > saponite > hectorite [40]. HDT of the polymers is considered as an index of heat resistance under applied load. Recently, Pramoda et al. investigated the thermal degradation behavior of clay–PA6 nanocomposites [101]. They reported that only exfoliated clays dispersed in polymer matrices exhibit improved thermal stability. Figure 10.19 shows the variations of 5 and 10% weight loss temperature with the clay content [101]. Apparently, both weight loss temperatures of the nanocomposite with 2.5% clay are higher than those for neat PA6 resulting from the formation of exfoliated structure in the nanocomposite. The onset temperatures remained nearly unchanged or somewhat lower for samples with high clay loading because of distinct clay agglomeration. Zhou et al. also indicated that addition of 5 wt% MMT to PP can increase the degradation temperature considerably [102]. The enhanced thermal stability of the clay–polymer nanocomposites is attributed to lower permeability of oxygen and the diffusibility of the degradation products from the bulk of the polymer caused by the exfoliated clay in the composites. With the dispersion of nanoclays throughout the polymer matrix, the barrier properties of the nanocomposites are expected to be strongly enhanced compared to the respective neat polymer. The dispersion of exfoliated clay platelets in the polymeric matrices can reduce the gas and water vapor permeability by more than 60% due to the geometrical detour effect of the nanoclays [40,103].

The burning behavior of polymers is expressed in terms of their ability to generate flammable volatile products subject to thermal combustion. Thus, the combustion of organic polymers is a complicated process consisting of two stages: thermal oxidative degradation and normal burning [104]. The burning process involves a series of steps, such as heat transfer, thermal-oxidative decomposition to generate flammable volatile products, diffusion of gaseous products in solid state and gaseous state, and combustion reaction of mixture involving volatiles and oxygen [105]. Clay–polymer nanocomposites are known to display excellent flame–retardant characteristics. Flame–retardant properties of nanocomposites can be evaluated using the cone calorimetry. This method determines the rate of heat release of a sample during its combustion, and the time to ignition under a heat flux of 35 W/m^2 according to ASTM E1354. Such heat flux is equivalent to that typical of a small fire [106]. An important feature of flame-retardant properties is the maximum amount of energy that a material may release during combustion. This is characterized by the peak of heat release rate (PHRR) in cone calorimetric curves. Figure 10.20 shows the typical combustion curves for PA6

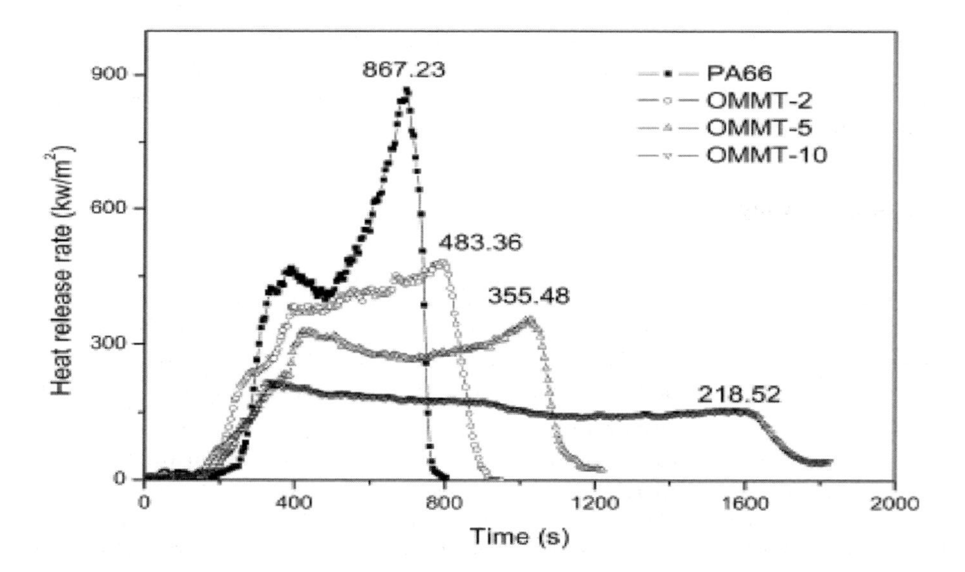

Fig. 10.20. Heat release rate plots for PA6 and its nanocomposites reinforced with 2, 5, and 10% organoMMT at a heat flux of 35 kW/m^2 (reprinted from [107] with permission from Elsevier).

and its nanocomposites [107]. The times to ignition of the nanocomposites are delayed by several seconds. Moreover, the reduction in the PHRR increases as the clay content increases. The PHRR of PA6-10% clay is reduced by about 60% compared to neat PA6. The reduction in the PHRR is attributed to the formation of a high performance carbonaceous-silicate char barrier that retards mass transfer of degrading polymer to the vapor phase. The barrier also shields the underlying polymer from the external thermal radiation, acting as a thermal insulation layer [108,109].

The thermal decomposition of alkylammonium salts is well known to take place with the Hoffman mechanism leading to volatilization of ammonia and the corresponding olefin [110]:

$$\begin{aligned} LS^{-+}[NH_3 - CH_2 - (CH)_n - CH_3] &\rightarrow LS^{-+}H + NH_3 + CH_2 \\ &= CH - (CH_2)_{n-1} - CH_3 \end{aligned}$$

An acidic site is thus created on the clays during heating. Zanetti et al. investigated the thermal behavior of clay–PP nanocomposite in which the clay was modified with protonated octadecyl amine (C18) [110]. They reported that the thermal oxidation process of the polymer is strongly slowed down in the nanocomposite with high char yield both by a physical barrier effect, enhanced by ablative reassembling of the silicate, and by a chemical catalytic action due to the silicate and to the strongly acid sites created by thermal decomposition of protonated amine silicate modifier. More recently, Qin et al. investigated the influence of compatibilizer (MA-functional group), alkylammonium, organoclay on flame retardant mechanism of the MMT–PP nanocomposites [105]. They reported that a decrease in PHRR of the nanocomposites is mainly due to the delay of thermo-oxidative decomposition of the composites. The active acidic sites created from the thermal degradation of organoclay can catalyze the formation of a protective coating char on the nanocomposites, and can also catalyze the dehydrogenation as well as cross-linking of polymer chains. Accordingly, the thermal-oxidative stability is increased and PHRR is decreased.

10.5 Anionic Clay

Layered double hydroxides (LDHs) are minerals and synthetic materials often regarded as ‘anionic clays’ in contrast to smectic ‘cationic clays’. They have found widespread applications as flame retardants, anion scavengers, and in catalysis. LDHs are layered materials constituted by a stacking of positive layers represented by the formula $[M^{2+}_{1-x}M^{3+}_{x}(OH)_2]^{x+}$ separated by anionic species and water molecules $(A^{n-}_{x/n}) \cdot mH_2O$ where M^{2+} is a divalent cation (e.g. Mg^{2+}, Zn^{2+}, Ni^{2++}), M^{3+} is a trivalent cation (e.g. Al^{3+}, Fe^{3}, Ga^{3+}), x is the ratio $M^{3+}/(M^{2+} + M^{3+})$. Partial substitution of M^{2+} by M^{3+} requires the incorporation of anions, such as CO_3^{2-}, Cl^-, SO_4^{2-}, NO_3^-, between the interlayers to balance the resulting positive charge. The high charge density of LDH layers and the high content of anionic species result in stronger interlayers electrostatic interactions between the sheets. This leads to exfoliation of LDHs, which is more difficult to achieve compared to the layered silicates. However, conversion of inorganic LDHs to nanosheets by exfoliation is a necessary step to prepare polymer nanocomposites.

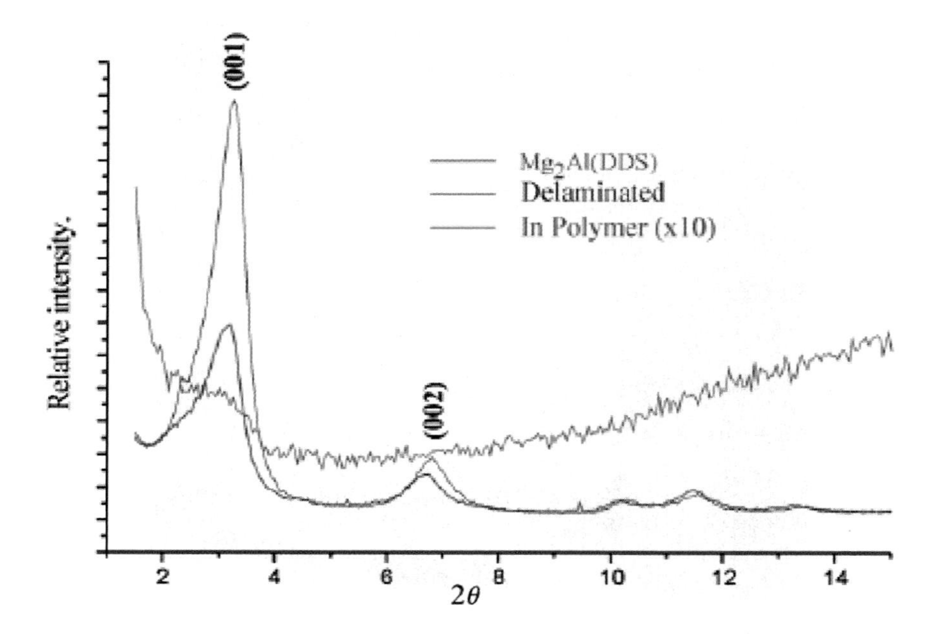

Fig. 10.21. XRD patterns of the as synthesized $Mg_2Al(OH)_6(C_{12}H_{25}SO_4)$ [Mg_2Al(DDS)], Mg_2Al(DDS) after delamination in 2-hydroxyethyl methacrylate (HEMA), and Mg_2Al(DDS) (5%) in polymerized HEMA (x10) (reprinted from [118] with permission from The Royal Society of Chemistry).

The inorganic anions can be exchanged with limited organic substances such as carboxylic acids and anionic surfactants [111,112]. The introduction of such species weakens the interlayer interactions and stacking of the layers. The rate of ionic exchange reaction in LDHS is relatively slow as compared to the MMT intercalation. There are only a few reports on the exfoliation and the preparation of LDH-polymer nanocomposites in the literature.

LDH materials can be synthesized with several preparation routes. These involve coprecipitation [113–115] and hydrothermal methods [116,117]. Recently, O'Leary et al. reported the synthesis, delamination of $Mg_2Al(OH)_6(C_{12}H_{25}SO_4)$ in polar acrylate monomers and subsequent polymerization of monomers to form the nanocomposites [118]. The loadings of MgAl LDH were between 1 to 10 wt%. Figure 10.21 shows the XRD patterns of synthesized MgAl LDH and its nanocomposite. It can be seen that the intensity of the (001) reflection of the nanocomposite reduces dramatically as compared to that of synthesized LDH, indicating that there is little long-range order in the *c*-direction. TEM micrograph shows that the LDH presents as individual sheets or in stacks of 2–5 layers (Fig. 10.22). More recently, Chen et al. synthesized ZnAl LDH and then exfoliated it in a xylene solution of linear low-density polyethylene (LLDPE) solution [119]. As the development of LDH-nanocomposites is still at an early stage, large efforts are still needed to prepare these nanocomposites via melt-compounding and to study their associated mechanical as well as physical properties.

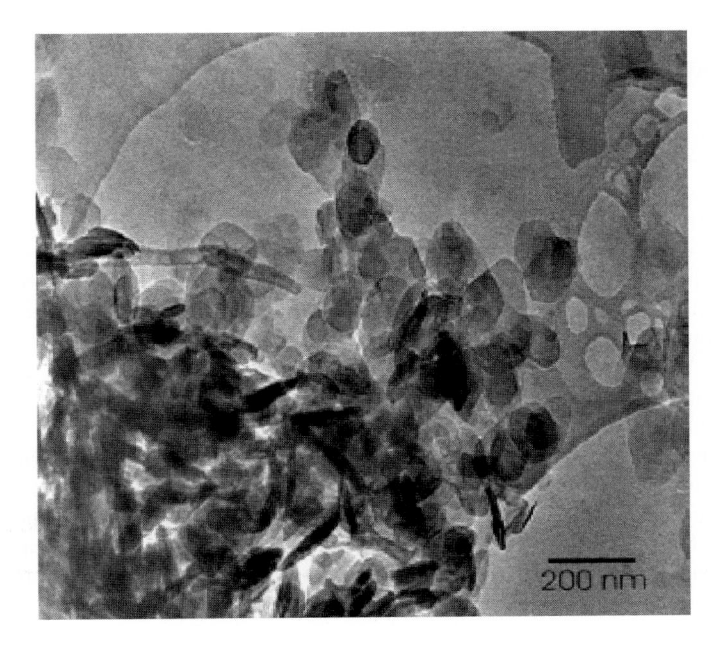

Fig. 10.22. TEM micrograph of a sample of poly-HEMA containing nanodispersed $Mg_2Al(DDS)$ (reprinted from [118] with permission from The Royal Society of Chemistry).

10.6 Conclusions

Low-cost 2:1 phyllosilicates are attractive inorganic fillers for polymers because they exhibit attractive features such as large surface area, superior swelling behavior, and ion exchange properties for organizing organic guest species. Accordingly, they can be broken down easily into nanoscale building blocks under proper processing conditions. Their structure, composed of either an octahedral sheet of aluminum or magnesium hydroxide is fused to two silica tetrahedral sheets. Isomorphous substitution within the silicate layer generates a net negative charge on that layer, which is counterbalanced by hydrated alkali metal cations in the gallery. Pristine layered silicates are hydrophilic and incompatible with most polymers. The clays are commonly modified with intercalation agents to render their surfaces more organophilic. Several kinds of molecular compounds with varied polarity have been selected to swell the clay layers. These include alkylammonium, water, ω-amino acids, etc. Formation of nanostructured composites depends mainly on the homogeneous dispersion of the clay platelets into extremely large aspect ratio in the polymer matrix. In this case, most of the polymer is associated with nanoclay interfaces. Converting bulk polymer into interfacial polymer represents an interesting route to diversified polymers with novel characteristics. Therefore, delamination of clay platelets is of fundamental importance in designing materials with desired properties. Exfoliated nanostructures can be designed with a high degree of complexity by combining several processes from clay chemistry to surfactant chemistry to polymer chemistry. Structural control at the nanometer level and tailoring of the desired properties are two important issues facing the development of polymer–clay nanocomposites. Comprehensive understanding of the structure–property relationships enables one to design novel polymer–clay nanocomposites with superior chemical, physical, and mechanical properties.

Synthesis of clay–polymer nanocomposites is of interest from the viewpoints of both basic research and practical applications. Polymer-layered silicate nanocomposites are found to present a unique combination of reduced flammability, enhanced thermal stability, and improved mechanical properties. They show potential applications as high performance engineering materials in automotive, healthcare, and food-processing industries. Good improvement in the properties of polymer-silicate nanocomposites can be achieved with the addition of the clays at low loading levels. Prospects of the polymer–clay nanocomposites depend mainly on the success in preparation procedures. The preparation of clay–polymer nanocomposites generally involves either intercalation of a suitable monomer and its subsequent polymerization, solution intercalation, as well as direct melt intercalation. Melt-compounding is more attractive because it can produce large volume scale of the polymer–clay nanocomposites of various shapes by using conventional extrusion technique. In the process, polymer chains are transported from the bulk melt, enter into expanded clay galleries and push the end of the platelets apart. The driving force is either physical or chemical affinity of the polymer molecules for organoclay surface. The enthalpy of mixing could be rendered favorable by maximizing the driving force. A typical example is the polar groups of polyamides that can interact with polar hydroxyl groups on the silicate clay surfaces. The PA6–organoMMT nanocomposites prepared via melt-mixing have been shown to exhibit exfoliated structure. Therefore, PA6–organoMMT nanocomposites exhibit enhanced Young's modulus and tensile strength but lower tensile ductility compared to neat PA6. For polyolefin polymer, it is necessary to improve the interactions between the polymer and MMT to facilitate melt intercalation of polyolefin molecules. This can be achieved by introducing polar group such as maleic anhydride (MA) into the polymer chains.

Several examples of polymer intercalation in layered silicates and layered double hydroxides have been reviewed and discussed in this chapter. It is concluded that novel inorganic–organic nanocomposites with attractive functional properties can be designed by tailoring the intercalation reactions in both the layered silicates and layered double hydroxides. Understanding the correlation between structures, physical and mechanical properties is crucial for the synthetic design and control of environmentally friendly clay–polymer nanocomposites.

References

[1] S.C. Tjong and R.K.Y. Li, J. Vinyl Add. Technol. 3 (1997) 89.

[2] F. Hindryckx, Ph. Dubois, M. Patin, R. Jerome, Ph. Teyssie and M.G. Marti, J. Appl. Polym. Sci. 56 (1995) 1093.

[3] D. Gan, S. Lu, C. Sung and Z. Wang, Macromol. Mater. Eng. 286 (2001) 296.

[4] William D. Callister, Jr, Materials Science and Engineering, 2nd Ed. John Wiley & Sons: New York, 1991; pp. 408–409.

[5] S.W. Bailey, Structures of layer silicates, in Eds. G.W. Brindley and G. Brown, Crystal Structures of Clay Minerals and their X-ray Identifications, Monograph No. 5, Mineralogical Society: London, 1980, pp. 1–123.

[6] D.M. Moore and R.C. Reynolds, Jr. X-ray Diffraction and the Identification and Analysis of Clay Minerals, Oxford University Press: Oxford, 1997; Ch. 4, pp. 104–120.

[7] B. Yalcin and M. Cakmak, Polymer 45 (2004) 6623.

[8] B.K.G. Theng, The Chemistry of Clay-Organic Reactions, Adam Hilger, London, 1974.

[9] L.A. Perez-Maqueada, M.C. Jimenez de Haro, J. Poyato and J.L. Perez-Rodriguez, J. Mater. Sci. 39 (2004) 5347–5351.

[10] M. Ogawa and A. Ishikawa, J. Mater. Chem. 8 (1998) 463.

[11] N.D. Hutson, D.J. Gualdoni and R. T. Yang, Chem. Mater. 10 (1998) 3707.

[12] G.R. Rao and B.G. Mishra, Mater. Chem. Phys. 89 (2005) 110.

[13] S.C. Tjong, Y.Z. Meng and A.S. Hay, Chem. Mater. 14 (2002) 44.

[14] S.C. Tjong, Y.Z. Meng and Y. Xu, J. Appl. Polym. Sci. 86 (2002) 2330.

[15] S.C. Tjong, Y.Z. Meng and Y. Xu, J. Polym. Sci. Part B: Polym. Phys. 40 (2002) 2860.

[16] S.C. Tjong and Y.Z. Meng, J. Polym. Sci. Part B: Polym. Phys. 41 (2003) 1476.

[17] S.C. Tjong and Y.Z. Meng, J. Polym. Sci. Part B: Polym. Phys. 41 (2003) 2332.

[18] S.C. Tjong and S.P. Bao, J. Polym. Sci. Part B: Polym. Phys. 42 (2004) 2878.

[19] S.C. Tjong and S.P. Bao, J. Polym. Sci. Part B: Polym. Phys. 43 (2005) 253.

[20] S.C. Tjong and S. P. Bao, J. Polym. Sci. Part B: Polym. Phys. 43 (2005) 585.

[21] Z.H. Liu, X.J. Yang, Y. Makita and K. Ooi, Chem. Lett. 7 (2002) 680.

[22] Y. Omomo, T. Sasaki, L. Wang and M. Watanabe, J. Am. Chem. Soc. 125 (2003) 3568.

[23] X. Yang, Y. Makita, Z. Liu, K. Sakane and K. Ooi, Chem. Mater. 16 (2004) 5581.

[24] N. Sukpirom and M.M. Lerner, Chem. Mater. 13 (2001) 2179.

[25] T. Nakato, Y. Furumi, N. Terao and T. Okuhara, J. Mater. Chem. 10 (2000) 737.

[26] F. Leroux, M. Adachi-Pagano, M. Intissar, S. Chauviere, C. Forano and J.P. Bess, J. Mater. Chem. 11 (2001) 105.

[27] X.H. Li, S.C. Tjong, Y.Z. Meng and Q. Zhu, J. Polym. Sci. Part B: Polym. Phys. 41 (2003) 1806.

[28] Z.M. Dang, C.W. Nan, D. Xie, Y.H. Zhang and S.C. Tjong, Appl. Phys. Lett. 85 (2004) 97.

[29] Z.M. Dang, Y.H. Zhang and S.C. Tjong, Synthetic Metals 146 (2004) 79.

[30] Q.X. Zhang, Z.Z. Yu, X.L. Xie and Y.W. Mai, Polymer 45 (2004) 5985.

[31] C.G. Ma, M.Z. Rong, M.Q. Zhang and K. Friedrich, Polym. Eng. Sci. (2005) 529.

[32] P. Musto, G. Ragosta, G. Scarinzi and L. Mascia, Polymer 45 (2004) 1697.

[33] D. Ma, Y. A. Akpalu, Y. Li, R. W.Siegel and L.S. Schadler, J. Polym. Sci. Part B: Polym. Phys. 43 (2005) 488.

[34] G.L. Hwang, Y.T. Shieh and K.C. Hwang, Adv. Funct. Mater. 14 (2004) 487.

[35] K.W. Putz, C.A. Mitchell, R. Krishnamoorti and P.F. Green, J. Polym. Sci. Part B: Polym. Phys. 42 (2004) 2286.

[36] E.L. Cussler, S.E. Hughes, W.J. Ward and R.J. Aris, J. Membr. Sci. 38 (1988) 161.

[37] J.C. Matayabas and S.R. Turner, Polymer-Clay Nanocomposites, Eds. T.J. Pinnavaia and G.W. Beall, John Wiley & Sons: New York, 2001, pp. 207–226.

[38] A. Usuki, M. Kawasumi, Y. Kojima, A. Okada, T. Kurauchi and O. Kamigaito, J. Mater. Res. 8 (1993) 1174.

[39] Usuki, Y. Kojima, M. Kawasumi, A. Okada, Y. Fukushima, T. Kurauchi and O. Kamigaito, J. Mater. Res. 8 (1993) 1179.

[40] Usuki, N. Hasegawa and M. Kato, Adv. Polym. Sci. 179 (2005) 135.

[41] S.D. Burnside, and E.P. Giannelis, Chem. Mater. 7 (1995) 1597.

[42] H. Shi, T. Lan and T.J. Pinnavaia, Chem. Mater. 8 (1996) 1584.

[43] X. Kornmann, L.A. Berglund, J. Skerte and E.P. Giannelis, Polym. Eng. Sci. 38 (1998) 1351.

[44] H.R. Fischer, L.H. Gielgens and T.P.M. Koster, Acta Polymer 50 (1999) 122.

[45] P. Reichert, J. Kressler, R. Thomann, R. Mulhaupi and G. Stoppelmann, Acta Polymer 49 (1998) 116.

[46] D.J. Suh, Y.T. Lim and O. Park, Polymer 41 (2000) 8557.

[47] J.W. Cho and D.R. Paul, Polymer 42 (2001) 1083.

[48] Yalcin, D. Valladares and M. Cakmak, Polymer 44 (2003) 6913.

[49] S. Hotta and D.R. Paul, Polymer 45 (2004) 7639.

[50] M. Kurian, A. Dasgupta, F.L. Beyer and M.E. Galvin, J. Polym. Sci. Part B: Polym. Phys. 42 (2004) 4075.

[51] R.W. Truss and A.C. Lee, Polym. Int. 52 (2003) 1790.

[52] J.T. Tunney and C. Detellier, Chem. Mater. 8 (1996) 827.

[53] J.E. Gardolinski, L.C.M.Carrera, M.P. Cantao and F. Wypych, J. Mater. Sci. 35 (2000) 3113.

[54] L. Cabedo, E. Gimenez, J.M. Lagaron, R. Gavara and J.J. Saura, Polymer 45 (2004) 5233.

[55] G. Lagaly, Solid State Ionics 22(1986) 43.

[56] M.A. Osman, M. Ploetze and P. Skrabal, J. Phys. Chem. B 108 (2004) 2580.

[57] T.D. Fornes, P.J. Yoon, D.L. Hunter, H. Keskula and D.R. Paul, Polymer 43 (2002) 5915.

[58] http://www.nanoclay.com/.

[59] T.D. Fornes, D.L. Hunter and D.R. Paul, Macromolecules 37 (2004) 1793.

[60] S. Hotta and D.R. Paul, Polymer 45 (2004) 7639.

[61] R.A. Vaia, R.K. Teukolsky and E.P. Giannelis, Chem. Mater. 6 (1994) 1017.

[62] L.Q. Wang, J. Liu, G.J. Exarhos, K.Y. Flanigan and R. Bordia, J. Phys. Chem. B 104 (2000) 2810.

[63] E. Hackett, E. Manias and E.P. Giannelis, J. Chem. Phys. 108 (1998) 7410.

[64] Q.H. Zeng, A.B. Yu, G.Q. Lu and R.K. Standish, Chem. Mater. 15 (2003) 4732.

[65] S.L. Mayo, B.D. Olafson and W.A. Goddard, J. Phys. Chem. 94 (1990) 8897.

[66] D.L. Ho and C.J. Glinka, Chem. Mater. 15 (2003) 1309.

[67] Y.S. Choi, H.T. Ham and I.J. Chung, Chem. Mater. 16 (2004) 2522.

[68] (a) C.M. Hansen, J. Paint Technol. 39 (1967) 104; (b) C.M. Hansen, J. Paint Technol. 39 (1967) 505.

[69] (a) J.H. Hildebrand, R.L. Scott, Solubility of Non-Electrolytes, 3rd Ed., Reinhold: New York, 1950; (b) J.H. Hildebrand, R.L. Scott, Regular Solutions, Prentice-Hill: Englewood Cliffs, N.J., 1962.

[70] Y. Li and H. Ishida, Polymer 44 (2003) 6571.

[71] A. Ghosh and E.M. Woo, Polymer 45 (2004) 4749.

[72] M.M. Dudkina, A.V. Tenkovtsev, D. Pospiech, D. Jehnichen, L. Haubler and A. Leuteritz, J. Polym. Sci. Part B: Polym. Phys. 43 (2005) 2493.

[73] A.B. Morgan and J.D. Harris, Polymer 45 (2004) 8695.

[74] X. Fu and S. Qutubuddin, Polymer 42 (2001) 807.

[75] M.W. Weimer, H. Chen, E.P. Giannelis and D.Y. Sogah, J. Am. Chem. Soc. 121 (1999) 1615.

[76] M.W. Noh and D.C. Lee, Polym. Bull. 42 (1999) 619.

[77] T.H. Kim, L.W. Jang, D.C. Lee, H.J. Choi and M.S. Jhon, Macromol. Rapid Commun. 23 (2002) 191.

[78] Y.K. Kim, Y.S. Choi, K.H. Wang and I.J. Chung, Chem. Mater. 14 (2002) 4990.

[79] M. Xu, Y.S. Choi, Y.K. Kim, K.H. Wang and I.J. Chung, Polymer 44 (2003) 6387.

[80] J. Tudor, L. Willington, D. O'Hare and B. Royan, Chem. Commun. (1996) 2031.

[81] J.S. Bergman, H. Chen, E.P. Giannelis, M.G. Thomas and G.W. Coates, Chem. Commun. (1999) 2179.

[82] R.A. Vaia, H. Ishii and E.P. Giannelis, Chem. Mater. 5 (1993) 1694.

[83] R.A. Vaia, K.D. Jandt, E.J. Kramer and E.P. Giannelis, Macromolecules 28 (1995) 8080.

[84] L. Incarnato, P. Scarfato, L. Scatteia and D. Acierno, Polymer 45 (2004) 3487.

[85] T.D. Fornes and D.R. Paul, Polymer 44 (2003) 4993.

[86] D. Fornes, P.J. Yoon, D.L. Hunter, H. Keskkula and D.R. Paul, Polymer 43 (2002) 5915.

[87] T.D. Fornes, P.J. Yoon and D.R. Paul, Polymer 44 (2003) 7545.

[88] T.D. Fornes, P.J. Yoon and D.R. Paul, Polymer 45 (2004) 2321.

[89] H.R. Dennis, D.L. Hunter, D.Chang, S. Kim, J.L. White, J.W. Cho and D.R. Paul, Polymer 42 (2001) 9513.

[90] R.A. Vaia and E.P. Giannellis, Macromolecules 30 (1997) 7990.

[91] R.A. Vaia and E.P. Giannellis, Macromolecules 30 (1997) 8000.

[92] A.C. Balazs, C. Singh and E. Zhulina, Macromolecules 31 (1998) 8370.

[93] M. Kato, A. Usuki and A. Okada, J. Appl. Polym. Sci. 66 (1997) 1781.

[94] N. Hasegawa, M. Kawasumi, M. Kato, A. Usuki and A. Okada, J. Appl. Polym. Sci. 67 (1998) 87.

[95] N. Hasegawa, H. Okamoto, M. Kato and A. Usuki, J. Appl. Polym. Sci. 78 (2000) 1918.

[96] P.H. Nam, P. Maiti, M. Okamoto, T. Kotaka, N. Hasegawa and A. Usuki, Polymer 42 (2001) 9633.

[97] P. Maiti, P.H. Nam, M. Okamoto, N. Hasegawa and A. Usuki, Macromolecules 35 (2002) 2042.

[98] Manias, A. Touny, L. Wu, K. Strawhecker, B. Lu and T.C. Chung, Chem. Mater. 13 (2003) 3516.

[99] F.C. Chiu, S.M. Lai, J.W. Chen and P.H. Chu, J. Polym. Sci. Part B: Polym. Phys. 42 (2004) 4139.

[100] R.A. Vaia, B.B. Sauer, O.K. Tse and E.P. Giannelis, J. Polym. Sci. Part B: Polym. Phys. 35 (1997) 59.

[101] K.P. Pramoda, T. Liu, Z. Liu, C. He and H.J. Sue, Polym. Degr. Stab. 81 (2003) 47.

[102] Y. Zhou, V. Rangari, H. Mahfuz, S. Jeelani and P.K. Mallick, Mater. Sci. Eng. A 402 (2005) 109.

[103] K.E. Strawhecker and E. Manias, Chem. Mater. 12 (2000) 2943.

[104] J. Troitsch, International Plastic Flammability Handbook, 2nd Ed., Hanser, Munich, 1990.

[105] H. Qin, S. Zhang, C. Zhao, G. Hu and M. Yang, Polymer 46 (2005) 8386.

[106] V. Babrauskas, Fire Mater. 19 (1995) 243.

[107] H. Qin, Q. Su, S. Zhang, B. Zhao and M. Yang, Polymer 44 (2003) 7533.

[108] J.W. Gilman, C.L. Jackson, A.B. Morgan, R. Harris Jr, E. Manias, E.P. Giannelis, M. Wuthenow, D. Hilton and S. H. Phillips, Chem. Mater. 12 (2000) 1866.

[109] T. Kashiwagi, R.H. Harris Jr, X. Zhang, R.M. Briber, B.H. Cipriano, S.R. Raghavan, W.H. Awad and J.R. Shields, Polymer 45 (2004) 881.

[110] M. Zanetti, G. Camino, P. Reichert and R. Mulhaupt, Macromol. Rapid Commum. 22 (2001) 176.

[111] S. Aisawa, S. Takahashi, W. Ogasawara, Y. Umetsu and E. Narita, J. Solid State Chem. 162 (2001) 52.

[112] N. Iyi, K. Kurashima and T. Fujita, Chem. Mater. 14 (2002) 583.

[113] V. Rives, Layered Double Hydroxides: Present and Future, Nova Science Publishers: New York, 2001.

[114] Canvani, F. Trifiro and A. Vaccari, Catal. Today 11 (1991) 173.

[115] J.J. Bravo-Suarez, E.A. Paez-Mozo and S.T. Oyama, Chem. Mater. 16 (2004) 1214.

[116] S.P. Newman, W. Jones, P. O'Connor and D.N. Stamires, J. Mater. Chem. 12 (2002) 153.

[117] Z.P. Xu and G. Q. Lu, Chem. Mater. 17 (2005) 1055.

[118] S.O'Leary, D. O'hare and G. Seeley, Chem. Commun. (2002) 1506.

[119] W. Chen, L. Feng and B. Qu, Chem. Mater. 16 (2004) 368.

Index